Handbook of Experimental Pharmacology

Volume 148

Editorial Board

G.V.R. Born, London
D. Ganten, Berlin
H. Herken, Berlin
K. Starke, Freiburg i. Br.
P. Taylor, La Jolla, CA

Springer-Verlag Berlin Heidelberg GmbH

The Pharmacology of Functional, Biochemical, and Recombinant Receptor Systems

Contributors

J.A. Angus, J.W. Black, A.J. Barr, R.A. Bond, M. Bouvier,
M. Corsi, T. Costa, S. Cotecchia, D.M. Ignar, C.K. Jayawickreme,
S.S. Karnik, A.J. Kaumann, T. Kenakin, S.M. Lanier, M.J. Lew,
M.R. Lerner, J.J. Linderman, M. Lutz, D.R. Manning,
G. Milligan, P. Morgan, H.O. Onaran, D.M. Perez, S. Rees,
A. Scheer, D.G. Trist, J. Weiss, R.T. Windh

Editors:
T. Kenakin and J.A. Angus

 Springer

Terry Kenakin, Ph.D.
Principal Research Scientist
Department of Receptor Biochemistry
Glaxo Wellcome Research and Development
5 Moore Drive
Research Triangle Park, NC 27709
USA
e-mail: tpk1348@glaxowellcome.com

Professor James A. Angus, Ph.D. FAA
Head, Department of Pharmacology
The University of Melbourne
Parkville, VIC 3010
AUSTRALIA
e-mail: j.angus@pharmacology.unimelb.edu.au

With 132 Figures and 20 Tables

ISBN 978-3-642-63028-6

Library of Congress Cataloging-in-Publication Data

The pharmacology of functional, biochemical, and recombinant receptor systems / contributors, J.A. Angus ... [et al.]; editors, T. Kenakin and J.A. Angus.
  p. cm. – (Handbook of experimental pharmacology; v. 148)
  Includes bibliographical references and index.
  ISBN 978-3-642-63028-6   ISBN 978-3-642-57081-0 (eBook)
  DOI 10.1007/978-3-642-57081-0
  1. Drug receptors – Research – Methodology   I. Kenakin, Terrence P.   II. Angus, James A.   III. Series.
  [DNLM: 1. Receptors, Drug – physiology.   2. Dose-Response Relationship, Drug.   3. Pharmaceutical Preparations – metabolism.   4. Receptors, Cell Surface – drug effects.   5. Technology, Pharmaceutical – methods. QV 38 P5367 2000]
  QP905.H3 vol. 148
  [RM301.41]
  615'.7 – dc21
                                                                                                    00-036585

This work is subject to copyright. All rights are reserved, whether the whole or part of the material is concerned, specifically the rights of translation, reprinting re-use of illustrations, recitation, broadcasting, reproduction on microfilms or in any other way, and storage in data banks. Duplication of this publication or parts thereof is permitted only under the provisions of the German Copyright Law of September 9, 1965, in its current version, and permission for use must always be obtained from Springer-Verlag. Violations are liable for Prosecution under the German Copyright Law.

© Springer-Verlag Berlin Heidelberg 2000
Originally published by Springer-Verlag Berlin Heidelberg in 2000
Softcover reprint of the hardcover 1st edition 2000

The use of general descriptive names, registered names, etc. in this publication does not imply, even in the absence of a specific statement, that such names are exempt from the relevant protective laws and regulations and free for general use.

Product liability: The publishers cannot guarantee the accuracy of any information about dosage and application contained in this book. In every individual case the user must check such information by consulting the relevant literature.

Coverdesign: design & production GmbH, Heidelberg

Typesetting: Best-set Typesetter Ltd., Hong Kong

SPIN: 10650831   27/3020–5 4 3 2 1 0 – printed on acid-free paper

Preface

This, the 148th volume of the Handbook of Experimental Pharmacology series, focuses on the very core of pharmacology, namely receptor theory. It is fitting that the originator of receptor pharmacology, A. J. CLARK, authored the fourth volume of this series 63 years ago. In that volume CLARK further developed his version of receptor theory first described four years earlier in his classic book *The Mode of Action of Drugs*. An examination of the topics covered in volume 4 reveals a striking similarity to the topics covered in this present volume; pharmacologists today are still as interested in unlocking the secrets of dose-response relationships to reveal the biological and chemical basis of drug action as they were over half a century ago. Sections in that 1937 volume such as "Curves relating exposure to drugs with biological effects" and "Implications of monomolecular theory" show Clark's keen insight into the essential questions that required answers to move pharmacology forward.

With the advent of molecular biological cloning of human receptors has come a transformation of receptor pharmacology. Thus the expression of human receptors into surrogate host cells helped unlock secrets of receptor mechanisms and stimulus-transduction pathways. To a large extent, this eliminates the leap of faith required to apply receptor activity of drugs tested on animal receptor systems to the human therapeutic arena. However, a new leap of faith concerning the veracity of the effects found in recombinant systems with respect to natural ones is now required.

The use of recombinant systems allows manipulation of the composition and stoichiometry of receptor systems; this, in turn, can tell us much about the inner workings of receptors. It also allows this knowledge to be applied to natural systems and thus far has shown how intricate natural systems can be with respect to numerous controls and feedback systems needed for delicate influenced response to chemical signals. This book discusses three aspects of receptor pharmacology related to this new recombinant age. The volume's first section takes up synoptic physiological systems and how receptors and receptor stimulus-response systems are integrated and interact with each other. Its second section describes the theoretical models of receptor function that have resulted from recent knowledge of receptor systems. Its third section depicts the new technological approaches to the study of receptors and their function.

Sixty three years after Clark's volume, technology has revolutionized pharmacology. However, while the means to answer the questions have completely changed, the questions remain the same.

Research Triangle Park, NC, USA TERRY KENAKIN
Parkville, Vic, Australia JAMES A. ANGUS

List of Contributors

ANGUS, J.A., Department of Pharmacology, University of Melbourne,
Parkville 3010, Victoria, Australia
e-mail: j.angus@pharmacology.unimelb.edu.au

BLACK, J.W., James Black Foundation, 68 Half Moon Lane,
Dulwich SE24 9JE, United Kingdom
e-mail: james.black@kcl.ac.uk

BARR, A.J., Department of Pharmacology, University of Pennsylvania School
of Medicine, 3620 Hamilton Walk, Philadelphia, PA 19104-6084, USA

BOND. R.A., Department of Pharmacological and Pharmaceutical Sciences,
University of Houston, 4800 Calhoun, Houston, TX 77204-5515, USA
e-mail: RABond@UH.EDU

BOUVIER, M., Department of Biochemistry and Groupe de Reserche sur le
Système Nerveux Autonome, Université de Montreal, Montreal,
CANADA H3C 3J7

CORSI, M., Pharmacology Department, Glaxo Wellcome S.P.A. Medicine
Research Centre, I-37135 Verona, Italy

COSTA, T., Istituto Superiore di Sanità, Dept. of Pharmacology, Viale Regina
Elena 299, I-6, Rome, Italy
e-mail: tomcosta@iss.it

COTECCHIA, S., Institut de Pharmacologie et de Toxicologie, University of
Lausanne, Lausanne, Switzerland

IGNAR, D.M., Department of Receptor Biochemistry, Glaxo Wellcome
Research and Development, 5 Moore Drive, Research Triangle Park, NC
27709 USA
e-mail: DIT12222@glaxowellcome.com

JAYAWICKREME, C.K., Glaxo Wellcome Inc., Receptor Biochemistry 3.2134B,
Division of Biochemistry, 5 Moore Drive, Durham, NC 27709, USA
e-mail: ckj10747@glaxowellcome.com

KARNIK, S.S., Department of Molecular Cardiology, The Lerner Research Institute, The Cleveland Clinic Foundation, 9500 Euclid Ave., Cleveland, OH 44195, USA
e-mail: karnik@ccf.org

KAUMANN, A.J., Babraham Institute, Cambridge, CB2 4AT, United Kingdom and Physiological Laboratory, University of Cambridge, Downing Street, Cambridge, CB2 3EG, United Kingdom
e-mail: ajk41@hermes.cam.ac.uk,

KENAKIN, T., Department of Receptor Biochemistry, Glaxo Wellcome Research and Development, 5 Moore Drive, Research Triangle Park, NC 27709 USA
e-mail: TPK1348@glaxowellcome.com

LANIER, S.M., Department of Pharmacology, Medical University of South Carolina, 171 Ashley Avenue, Charleston, SC 29425, USA
e-mail: laniersm@musc.edu

LERNER, M.R., University of Texas South Western Medical School, F4.100 Department of Dermatology, 5323 Harry Hines Blvd., Dallas, Texas 75235, USA

LEW, M.J., Department of Pharmacology, University of Melbourne, Parkville 3052, Victoria, Australia
e-mail: M.Lew@Pharmacology.unimelb.edu.au

LINDERMAN, J.J., Department of Chemical Enginering, University of Michigan, Ann Arbor, MI 48109-2136, USA
e-mail: linderma@umich.edu

LUTZ, M., Department of Research Computing, Glaxo Wellcome Research and Development, 5 Moore Drive, Research Triangle Park, NC 27709 USA

MANNING, D.R., Department of Pharmacology, University of Pennsylvania School of Medicine, 3620 Hamilton Walk, Philadelphia, PA 19104-6084, USA
e-mail: manning@pharm.med.upenn.edu

MILLIGAN, G., Molecular Pharmacology Group, Division of Biochemistry and Molecular Biology, Institute of Biomedical and Life Sciences, University of Glasgow, Glasgow G12 8QQ, Scotland, United Kingdom
e-mail: G.Milligan@bio.gla.ac.uk

List of Contributors

MORGAN, P., Department of Research Computing, Glaxo Wellcome Research and Development, 5 Moore Drive, Research Triangle Park, NC 27709 USA

ONARAN, H.O., Faculty of Medicine, Department of Pharmacology and Clinical Pharmacoloy. Ankara University, Ankara, Turkey

PEREZ, D.M., Department of Molecular Cardiology, The Lerner Research Institute, The Cleveland Clinic Foundation, 9500 Euclid Ave., Cleveland, OH 44195, USA
e-mail: Perezd@ccf.org

REES, S., Receptor Systems Unit, Glaxo Wellcome Medicines Research Centre, Gunnels Wood Road, Stevenage, Herts SG1 2NY,
United Kingdom
e-mail: esr1353@glaxowellcome.co.uk

SCHEER, A, Institut de Pharmacologie et de Toxicologie, University of Lausanne, Lausanne, Switzerland

TRIST, D.G., Pharmacology Department, Glaxo Wellcome S.p.A., Medicine Research Centre, I-37135 Verona, Italy
e-mail: DGT0669@glaxowellcome.co.uk

WEISS, J., Biomathematics Program, Department of Statistics, North Carolina State University, Raleigh, NC, USA

WINDH, R.T., Department of Pharmacology, University of Pennsylvania School of Medicine, 3620 Hamilton Walk, Philadelphia, PA 19104-6084, USA

Contents

Introduction: Bioassays – Past Uses and Future Potential
J.W. BLACK 1

Section I: Classical Pharmacology and Isolated Tissue Systems

CHAPTER 1

Human Vascular Receptors in Disease: Pharmacodynamic Analyses in Isolated Tissue
J.A. ANGUS. With 25 Figures 15

A. Introduction 15
B. Receptors Mediating Coronary Artery Contraction: Role in
   Variant Angina 16
   I. Large Coronary Arteries 16
     1. Ergometrine 16
     2. 5-HT and Endothelium-Derived Relaxing Factor 16
   II. Small Resistance Arteries 21
   III. Summary and Future Work 27
C. Vascular Reactivity in Human Primary Hypertension and
   Congestive Heart Failure 28
   I. The Technique, Function and Structure 28
   II. Remodelling 29
   III. Endothelial Dysfunction 31
   IV. Forearm Veins in Primary Hypertension 33
   V. Chronic Heart Failure 34
D. Pharmacology of Vascular Conduits for Coronary-Bypass
   Graft Surgery 37
   I. Introduction 37
   II. Internal Mammary Artery 38
   III. Saphenous Vein 41
E. Human Vascular-to-Cardiac Tissue Selectivity of L- and T-Type
   VOCC Antagonists 41
References 47

CHAPTER 2

Problems in Assigning Mechanisms: Reconciling the Molecular and Functional Pathways in α-Adrenoceptor-Mediated Vasoconstriction
M.J. LEW. With 6 Figures 51

A. Introduction 51
B. Coupling Mechanisms at the Molecular and Cellular Levels 52
  I. α_1-Adrenoceptors 52
  II. α_2-Adrenoceptors 54
C. Coupling Mechanisms in the Intact Animal 56
D. Reconciliation 59
E. Myogenic Activation 61
F. Autocrine and Paracrine Activation 62
G. Functional Antagonism 64
H. Adenylate Cyclase and α_2-Adrenoceptors 66
I. Conclusions 68
References 69

CHAPTER 3

G_s Protein-Coupled Receptors in Human Heart
A.J. KAUMANN. With 10 Figures 73

A. Introduction 73
B. Receptor Subtypes 73
C. β-Adrenoceptor Subtypes 74
  I. Comparison of β_1- and β_2-Adrenoceptors 74
    1. Localisation 74
    2. Function of β_1- and β_2-Adrenoceptors 74
    3. Selective Coupling of β_2-Adrenoceptors 75
D. Is There a Functional role for Cardiac β_3-Adrenoceptors? 80
  I. Evidence Against Cardiostimulation 80
  II. Evidence for Cardiostimulation 81
  III. Evidence for Cardiodepression 81
  IV. Evidence Against Cardiodepression 82
E. Cardiostimulant Effects Through the Putative β_4-Adrenoceptor... 82
  I. Non-Conventional Partial Agonists 82
  II. The Putative β_4-Adrenoceptor Resembles – But Is Distinct from – the β_3-Adrenoceptor 84
  III. Which Endogenous Agonist for the Putative β_4-Adrenoceptor? 87
  IV. The Putative β_4-Adrenoceptor is a Special State of the β_1-Adrenoceptor 88
F. 5-HT$_4$ Receptors 90
  I. Coupling to a cAMP Pathway 90
  II. 5-HT$_4$-like Receptors 91

Contents XIII

G. Cross-Talk Between Cardiac G_s-Coupled Receptors, as
   Revealed by Chronic Blockade of β_1-Adrenoceptors 95
H. Physiological, Pathophysiological and Therapeutic Relevance 101
   I. β_1- and β_2-Adrenoceptors 101
   II. Putative β_4-Adrenoceptors 105
   III. 5-HT$_4$ Receptors 105
I. Epilogue 107
References 108

Section II: New Theoretical Concepts and Molecular Mechanisms of Receptor Function

CHAPTER 4

Kinetic Modeling Approaches to Understanding Ligand Efficacy
J.J. LINDERMAN. With 19 Figures 119

A. Introduction 119
B. Background 120
   I. Efficacy 120
   II. Modeling 121
      1. Equilibrium, Steady State, and Kinetic Models 122
      2. Diffusion-Versus Reaction-Controlled Events 123
      3. Model Structures and Dose–Response Curves 124
C. Parameters Contributing to Ligand Efficacy 126
   I. The Lifetime of the Individual Receptor–Ligand Complex
      ($1/k_r$) 126
   II. The Receptor Desensitization Rate Constant k_x 131
   III. The Ligand Binding and Dissociation Rate Constants at
        Endosomal pH 134
   IV. Rate Constants in a Ternary Complex Model 139
D. Concluding Remarks 143
References 144

CHAPTER 5

The Evolution of Drug-Receptor Models: The Cubic Ternary-Complex Model for G Protein-Coupled Receptors
T. KENAKIN, P. MORGAN, M. LUTZ, and J. WEISS. With 7 Figures........ 147

A. Receptor Models 147
B. The Ternary Complex Model of Receptor Function 148
C. Two-State Theory 149
D. The Extended Ternary Complex Model 151
E. The CTC Model 152

F. General Application of the Cubic Model 155
   I. Evidence for the AriG Complex 158
G. Conclusion 163
References 163

CHAPTER 6

Inverse Agonism
R. BOND, G. MILLIGAN, and M. BOUVIER 167

A. Background 167
B. Overexpression, Spontaneous Activity and Inverse Agonism 168
C. Mutations and Diseases of Spontaneous Receptor Activity 173
D. Modulation of Receptor Function by Agonists and
   Inverse Agonists 174
E. Models 177
F. Summary and Conclusions 178
References 178

CHAPTER 7

Efficacy: Molecular Mechanisms and Operational Methods of Measurement. A New Algorithm for the Prediction of Side Effects
T. KENAKIN. With 17 Figures 183

A. Introduction 183
B. The Molecular Nature of Efficacy 183
C. Positive and Negative Efficacy 186
D. The Operational Measurement of Relative Efficacy 188
   I. Binding Studies 188
      1. Guanosine Triphosphate γS Shift 188
      2. High-Affinity Selection Binding 191
   II. Function 192
      1. The Method of Furchgott 195
      2. The Method of Stephenson 196
      3. Comparison of Relative Maximal Responses 197
E. Limitations of Agonist Potency Ratios 201
F. Why Measure the Relative Efficacies of Agonists? 204
G. A Simple Algorithm for the Prediction of Agonist Side Effects
   Using Efficacy and Affinity Estimates 206
   I. Therapeutic Versus Secondary Agonism: Side Effect Versus
      Coupling Constant Profiles 208
   II. Algorithm for Calculation of Relative β_{50} 210
   III. Application of the Algorithm to β_3-Adrenoceptor
      Agonists 212
   IV. Limitations of the Algorithm 214

H. Conclusions 215
References 215

CHAPTER 8

A Look at Receptor Efficacy. From the Signalling Network of the Cell to the Intramolecular Motion of the Receptor
H.O. ONARAN, A. SCHEER, S. COTECCHIA, and T. COSTA.
With 12 Figures 217

A. Introduction 217
B. Biological Receptors and the Dualism of Affinity and Efficacy ... 217
   I. Signal Transfer and Conformational Change in Membrane Receptor 217
   II. The Distinction Between Affinity and Efficacy 218
   III. Generality of the Concept of Efficacy in Functional Proteins 219
   IV. Asking the Questions 219
   V. Two Flavours in the Definition of Efficacy: Biological and Molecular 220
C. Biological Definitions of Efficacy 221
   I. The Nature of Signal Strength 221
   II. Stimulus–Response Relationships 222
   III. The Scale of Agonism and Antagonism 223
   IV. Steps of Signal Transduction and the Indeterminacy of Stimulus–Response Relationships 225
D. Molecular Definitions of Efficacy 227
   I. A Molecular Link Between Affinity and Efficacy 227
   II. Allosteric Equilibrium, Free-Energy Coupling, and Thermodynamic Definitions of Efficacy 227
   III. Linkage Between Macroscopic Perturbations in the Receptor 231
   IV. Functional and Physical States in Proteins 233
   V. Microscopic Interpretation of Allosteric Equilibrium 235
E. A Stochastic Model of Molecular Efficacy 238
   I. Protein Motion and Fluctuations in Its Conformational Space 238
   II. Probability Distribution of Microscopic States and Derivation of Macroscopic Constants 238
   III. Probabilistic Interpretation of Ligand Efficacy 241
   IV. Relationship Between Physical States and Biological Function 244
   V. Relationship Between Efficacy and Fluorescence Changes in β_2-Adrenoceptors 244
   VI. Correlated Macroscopic Changes in Constitutively Active Adrenoceptors 248
References 254

CHAPTER 9

Mechanisms of Non-Competitive Antagonism and Co-Agonism
D.G. TRIST and M. CORSI. With 9 Figures 261

A. Non-Competitive Antagonism 261
   I. Definition 261
   II. Analysis of the Effect of Non-Competitive Antagonists 261
B. Co-Agonism 267
   I. Theory as Applied to the Glutamate NMDA Receptor 267
     1. Evidence for Co-Agonism 267
     2. The Theory of Co-Agonism 268
     3. The Effect of Competitive Antagonism on Co-Agonism ... 271
     4. The Effect of a Non-Competitive Antagonist on
       the Response of Co-Agonists 273
   II. Experimental Data Supporting the Concept of Co-Agonism
     and Its Antagonism 274
     1. Recombinant Experiments 274
     2. Tissue Experiments 276
   III. Implications of Co-Agonism 278
C. General Observations 279
References 280

CHAPTER 10

Mechanisms of Receptor Activation and the Relationship to Receptor Structure
D.M. PEREZ and S.S. KARNIK. With 2 Figures 283

A. Introduction 283
B. Common GPCR Structure/Function 283
C. Rhodopsin and bR Activation: Light as the Ligand 286
   I. Direct Structural Information 287
   II. The Activation Mechanism 288
     1. How Retinal Binds 288
     2. Salt-Bridge Constraining Factor: Movement of TM3 and
       TM6 289
D. AR Activation: Small Organics as Ligands 291
   I. Important Binding Contacts of the Endogenous Ligands 291
   II. Insights on How Epinephrine Activates 293
     1. Release of Constraining Factors 293
     2. Evidence for a Salt-Bridge as a Constraint: Movement of
       TM3 and TM7 294
     3. Evidence for Multiple Activation States or Mechanisms ... 296
     4. Evidence for Additional Constraining Factors: Movement
       of TM5 and TM6 296
E. Angiotensin-Receptor Activation: Peptides as Ligands 298

	I. Peptide-Hormone GPCRs: How the Peptide Binds	298
	II. The Ang-II Receptor	298
	III. How AngII Peptides and Non-Peptides Bind	299
	IV. Insights into AngII Activation	300
	1. Role of His256 in TM6	300
	2. Release of Constraining Factors: Role of TM3	300
F.	The Ties that Bind: Concluding Remarks	302
	I. Conservation of Critical Binding Contacts and Resulting Helical Movements	302
	II. Conservation of Switches that Control Attainment of the Active State(s)	302
References		304

Section III: New Technologies for the Study of Drug Receptor Interaction

CHAPTER 11

The Assembly of Recombinant Signalling Systems and Their Use in Investigating Signaling Dynamics
S.M. LANIER. With 4 Figures... 313

A.	Introduction	313
B.	Assembly of Recombinant Signaling Systems	314
	I. Stable Transfection	314
	II. Transient Expression Systems	316
C.	Drug–Receptor Interactions in Recombinant Signaling Systems	318
	I. Cell-Type-Specific Signaling Events	318
	II. Influence of Accessory Proteins	323
D.	Perspective	328
References		328

CHAPTER 12

Insect Cell Systems to Study the Communication of Mammalian Receptors and G Proteins
R.T. WINDH, A.J. BARR, and D.R. MANNING. With 4 Figures 335

A.	Introduction	335
	I. Selectivity of Receptor–G Protein Interactions	335
	II. Effector Modulation as a Measure of G Protein Activity	336
	III. Direct Measures of G Protein Activity	337
B.	Insect Cell Expression Systems	338
	I. Expression of Receptors in Insect Cell Lines	338

II. Interaction of Receptors with Endogenous G Proteins and
      Effectors 342
  III. Quantitation of Coupling Using Radioligand Binding 343
C. Reconstitution of Mammalian Receptors and G Proteins:
   Reconstituted Properties of Ligand Binding 345
   I. Influence of Heterotrimeric G Proteins on Binding of
      Agonists 346
   II. Influence of Individual G Protein Subunits on Binding of
       Agonists 348
   III. Characterization of Inverse Agonism 349
D. Reconstitution of Mammalian Receptors and G Proteins:
   G Protein Activation 350
   I. Activation Following Co-Expression of Receptor and
      G Protein 350
   II. Activation Following Addition of Purified G Protein to
       Membranes 354
   III. Limitations and Technical Considerations 355
E. Conclusions 356
F. References 356

CHAPTER 13

Altering the Relative Stoichiometry of Receptors, G Proteins and Effectors: Effects on Agonist Function
G. MILLIGAN 363

A. Introduction 363
   I. Background 363
   II. Systems to Modulate GPCR–G Protein–Effector
       Stoichiometries 364
   III. Cellular Distribution of Elements of G Protein-Coupled
        Signaling Cascades 365
B. GPCR–G Protein Fusion Proteins. A Novel Means to Restrict
   and Define the Stoichiometry of Expression of a GPCR and a
   G Protein α Subunit 367
C. G Protein-Coupled Receptors 370
D. $G_s\alpha$ 376
E. Effector Enzymes: Adenylyl Cyclase 380
F. Conclusions 383
References 383

CHAPTER 14

The Study of Drug–Receptor Interaction Using Reporter Gene Systems in Mammalian Cells
D.M. IGNAR and S. REES. With 10 Figures 391

- A. Reporter Systems 391
  - I. What is a Reporter Gene System? 391
  - II. Detection Methods 391
    1. Enzymatic 391
    2. Non-Enzymatic 393
  - III. Measurement of Intracellular Signaling 393
    1. Inducible Reporter Genes 393
       a) Responsive Promoters 393
       b) Chimeric Transcription Factors 395
    2. Reporter Proteins 397
  - IV. Transient Versus Stable Expression of Reporter Genes ... 400
    1. Transient Expression 400
    2. Stable Expression 400
- B. Use of Reporter Gene Systems in Pharmacology 401
  - I. Drug Discovery 401
  - II. Receptor Pharmacology 401
    1. Use of Reporter Gene Assays to Assess Receptor Agonism 402
       a) Example 1. 6CRE-Luciferase as a Reporter for $G\alpha_s$-Coupled Receptor Signaling 402
       b) Example 2. Aequorin as a Reporter for $G\alpha a_{q/11}$-Coupled Receptor Signaling 402
       c) Example 3. The Use of a Gal4/Elk-1 Chimera to Report Opioid-Receptor-Like Receptor-1 Activation of MAPK 405
    2. Use of Reporter Gene Assays to Evaluate Receptor Antagonism 405
    3. Simultaneous Detection of Multiple Signals 405
       a) Example 1. Dual Reporter Assays 405
       b) Example 2. Combination of Reporter Assays with Other Assay Types 407
    4. Measurement of Constitutive Activity 409
    5. Measurement of Efficacy 411
    6. Assessment of New Signaling Pathways 411
- References 412

CHAPTER 15

Melanophore Recombinant Receptor Systems
C.K. JAYAWICKREME and M.R. LERNER. With 7 Figures 415

A. Introduction 415
B. Cellular Signaling in Melanophores 417
   I. Signaling Pathways 417
   II. Endogenous Receptor Signaling 419
      1. Melatonin 1c Receptor 421
      2. α-Melanocyte Stimulating Hormone (MSH) Receptor 421
      3. Endothelin-C Receptor 422
      4. Serotonin Receptor 422
      5. β-Adrenoreceptor 422
      6. VIP Receptor 423
C. Melanophore Assay Technology 423
   I. Cell Culture and Related Techniques 423
      1. Preparation of Cultures of Melanophore 423
      2. Continuous Culturing of *Xenopus laevis* Melanophores . . . 424
      3. Receptor Expression 424
      4. Preparation of Stable Melanophore Lines Expressing
         Exogenous Receptors 425
   II. Signal Detection 425
      1. Transmittance Reading 425
      2. Digital Imaging 426
   III. Screening Formats 427
      1. Microtiter Well Format 427
      2. Lawn Format 428
   IV. Receptor Cloning and Mutagenesis Studies 428
D. Receptor Studies and Applications 429
   I. Characterization of Novel GPCRs 431
   II. Lawn Format Screen System 432
   III. Single Transmembrane Receptors 433
   IV. Receptor-Ligand Interaction Studies 434
E. Summary 435
References 435

Subject Index 441

Introduction:
Bioassays – Past Uses and Future Potential

J.W. BLACK

Pharmacology as an experimental science began in the nineteenth century in the wake of advances in physiology. Pharmacologists described the effects of drugs on physiological systems such as the respiratory, cardiovascular and locomotor systems. Animals were used as pharmacological detectors. The use of animals as pharmacological measuring instruments was not developed seriously until the 1920s. Quantitative pharmacology became vitally important with the discovery of insulin in 1921. Although crystalline insulin became available in 1926, the chemical structure of insulin only became known in 1955. For many years after its introduction into clinical practice, even crystallised insulin contained many impurities. As a consequence, the activity of these extracts had to be assayed and standardised using experimental animals.

Insulin produces hypoglycaemia. Hypoglycaemia induces convulsions. So the first attempts to assay the amount of insulin in pancreatic extracts tried to find out how much of the extract was needed to produce convulsions in rabbits. The method was later refined to measure 1 unit as the amount of the extract needed to reduce the blood sugar level to a given concentration. However, these methods, which try to estimate the potency of a substance in terms of its effects, ran into problems. Over and above within-experiment variances in the susceptibility of individual animals to the test substance, pharmacologists had to learn that a response parameter, measured repeatedly in a population, was rarely stable. Repeated measurements throughout the year disclosed seasonal changes. Between laboratories, there were variations due to diet and breed plus variations due to goodness knows what. The attempt to measure the activity of an extract by an animal's reaction to it, often expressed in "animal units", was eventually found to be too unreliable. Reproducibility and accuracy were only achieved when it was realised that when an animal's reactions are used to compare the activity of a standard and test preparation within a balanced experiment, then these population variations cancel out. Experiments, in which a comparison is made between a standard and a test preparation, where the test contains an unknown amount of the standard and where the biological experiment is an exercise in analytical chemistry, are known as "analytical dilution assays". In these assays, provided that the active substance is the same in both standard and test preparations, the results will be independent of both the species and the effect used in the comparison. The experimental design and related statistical analysis of analytical dilution assays were progressively refined between 1920 and 1950 to turn them into very successful chemical measuring instruments. Analytical bioassays have had an important pharmacopoeial role.

Today, various types of chromatography and radioimmunoassay have largely taken over the pharmaceutical standardisation of biological products such as erythropoetin and the interferons. Nevertheless, the sophisticated experimental designs and corresponding statistical methods that were developed during the era of analytical dilution bioassays were invaluable in the development of quantitative bioassays for other purposes.

The sound logical and mathematical basis for analytical bioassays does not apply to assays in which standard and test are chemically different. To quote Schild (1950), "The assay then ceases to be an analytical method and becomes a comparison of biological activity in which species, end-point and experimental conditions become all-important". GADDUM (1953) wrote that "comparative assays have no satisfactory logical basis and their only justification lies in the fact that the greatest contributions of pharmacology to medicine have been based upon them". He was, of course, referring to drug screening tests. Nevertheless, comparative assays to establish rank orders of agonists on different tissues have been much used by pharmacologists to expose receptor heterogeneity. While this method cannot claim to be robust, it may reasonably claim to have had some pragmatic utility. Still, the confusing literature on agonist potency ratios should warn us that these methods may have similar logical flaws to the use of "animal units".

In the early days of the development of bioassays, when pharmacologists were struggling to define "animal units", the emphasis was on the "dose" of drug that was given to the animal. The assays were being used because no chemical analytical methods were available. Therefore, when comparative assays were introduced, the reference standard was usually a specified weight of dried extract. These assays were being used to specify drug dosages, so those units by weight were appropriate. However, analytical dilution assays were also being developed for another, non-pharmacopoeial purpose. Small transmitter molecules, such as acetylcholine, histamine, noradrenaline and 5-hydroxytryptamine, were originally discovered in tissues by their biological effects. However, they were found to be present in tissues at concentrations much lower than could be detected by the standard chemical analytical methods then available. Consequently, bioassays had to be developed that would estimate the concentrations of these substances in tissues. As the standard substances were available as pure chemicals, accurate concentrations of them could be produced for calibrating the assays. Only in organ baths could tissues be exposed to the standard solutions. So these assays spearheaded the development of intact tissues in vitro as sensitive and accurate chemical-measuring instruments.

The use of sensitive tissues or intact animals as a litmus test in a dose-titration assay involved no theory. The biosystem was just a black box that had the property of responding reliably and sensitively to an applied chemical stimulus. However, about a hundred years ago, J.N. LANGLEY tried to illuminate the black box with an idea. The idea was that certain substances produce their effects by combining with specific chemical sites in tissues. He referred to these sites as receptive substances. This mere idea eventually revolutionised the science of pharmacology. LANGLEY (1878) proposed that drugs, like pilocarpine, act by forming compounds with these receptive substances in tissues, compounds that are formed "according to some law of which their relative mass and chemical affinity for the substance are factors". His argument was based on analogy with inorganic chemistry. A.V. HILL, a pupil of LANGLEY, turned this idea of receptors into algebraic form. "Receptors" became a concentration parameter, a mathematical operator, invented to relate agonist concentration to effect. Mathematical operators are theoretical concepts that do not have an independent physical existence. Thus, in mechanics, "force" is the concept that relates mass and acceleration and, in electricity, "resistance" relates voltage and current flow. HILL (1906), in fact, measured the contractile responses of the rectus abdominis muscle of the frog. In the classical "organ bath" experiment, he recorded both the time course and size of the contractions produced by different concentrations of nicotine. He then showed, mathematically, that either of the concentration-dependent measurements could be "explained" by assuming that the interaction between nicotine and the hypothetical "receptors" was governed by the laws governing chemical interactions (otherwise known as the Law of Mass Action). LANGLEY's idea, that the effects of alkaloids such as muscarine and nicotine were due to a chemical interaction rather than a physical process, such as diffusion, was thus substantiated.

Hill imported into pharmacology the utility of measuring complete dose–response curves, the value of recording the time-course of responses and the significance of using algebraic curve-fitting procedures to describe these relationships. Most important of all, he showed how an "as if" model could be used to interpret the data. The attempts to interpret drug actions, pharmacological hermeneutics if you like, in terms of underlying, hypothetical mechanisms I have referred to elsewhere as Analytical Pharmacology. This branch of pharmacology was pioneered by A.J. CLARK in the 1920s and by J.D. GADDUM in the 1930s. First, Clark showed that dose–response data could be adequately described by at least four quite different saturation functions and that the quality of the data was not good enough to distinguish between them. The arbitrary nature of the choice of curve-fitting procedures means that the analyst is free to choose one that refers to the simplest generative model. In practice, the hyperbolic function (or, its more-generalised form, the logistic function) is usually chosen because each also describes Mass Action-determined, reversible, chemical interactions. Clark also studied the interaction between agonists and antagonists. Although the mutual antagonism of agonist–antagonist pairs, such as atropine and pilocarpine, had been well-recognised for over 50 years, Clark was able to show that the dose–response curves of acetylcholine on the frog heart were displaced in parallel by increasing concentrations of atropine, that is the antagonistic effects of atropine were wholly surmountable. However, he was unable to conclude that the effects of atropine were due to "an antagonism of effects rather than of combination".

GADDUM (1936), however, solved the problem of the quantitative expectations of mutual, competitive antagonism. He assumed that the effects of an agonist, A, are due to its ability to induce a tissue response in proportion to the fraction of the (hypothetical) receptors, R, which are occupied when the response reaches a steady state; that the receptor population is homogeneous; and that the fraction of receptors occupied is determined by the Law of Mass Action. In other words, he assumed a model of agonism in which A is able to activate R by occupation. In the simplest case, where the agonist concentration–effect relationship is hyperbolic, the relation can be characterised by a single parameter, the K_A, the concentration of the agonist needed to occupy half of the receptors at equilibrium. If, now, a molecule, B, can occupy the same receptors as A but where, unlike A, receptors occupied by B, BR, are not activated, then the effects of A are antagonised when B is allowed to compete with A for receptor occupation. Gaddum showed that B has the effect of "diluting" the effects of A such that the concentration–occupancy curve is displaced in parallel along the concentration axis. The new K_A, $K_{A(B)}$, is equal to K_A multiplied by a factor that contains only antagonist-related elements. He showed that the concentration ratio, $K_{A(B)}/K_A$, is equal to the normalised concentration of B plus 1 (the Gaddum equation), where the normalised concentration is the ratio of the concentration of B to its dissociation ("affinity") constant, K_B. By using the ratio of agonist concentrations needed to produce equal effects in the presence and absence of antagonist, that is a null method, the agonist "affinity" parameter is cancelled out. Therefore, the characteristics of the measured dose–response curves are irrelevant, only their displacement by B is important. Finally, the advances made by Clark and Gaddum were codified by H.O. SCHILD (1954). Schild rearranged Gaddum's equation and then expressed it in logarithmic form. He showed that the relationship between antagonist concentration and the measured concentration ratio minus 1 would, for a simple competitive antagonist, be a straight line with a slope of 1 and an intercept on the concentration axis equal to the antagonist pK_B ($\log K_B$), the dissociation constant. Antagonist data plotted in this way is universally known as a Schild plot.

The combined discoveries of Hill, Clark, Gaddum and Schild revolutionised the usefulness of quantitative bioassays using intact tissues. Intact-tissue bioassays moved from instruments that could be used for estimating drug concentrations to instruments that could be used for analysing mechanisms of drug actions. To do this, create a family of agonist concentration–effect curves measured in the presence of increasing concentrations of an antagonist. Plot them in semi-logarithmic space. If the curves are displaced in parallel, this is sufficient evidence that the antagonism is surmountable. Measure the concentration ratios, r, that is, the shift in location of the concentration–response curves produced by each concentration of antagonist. Plot log (r-1) versus each antagonist concentration [B]. If the resulting Schild plot is linear over at least 2 log units and has a slope not different from unity, then the K_B of the antagonist for a specified receptor can be calculated. The numbers that are estimated by these assays were eventually validated by the development of radio-ligand binding technology. This is quite remarkable. How can a tissue, such as the classical "guinea pig ileum preparation", a tissue of great cellular and chemical complexity, behave so simply?

Part of the answer lies in the nature of the agonist–receptor interaction. Physiological agonists are variously classified as hormones, neurotransmitters, autacoids, growth factors, chemokines and so on. These classes of chemicals have in common the concept of "messenger molecules". The concept of "messenger molecule" was invented by BAYLISS and STARLING in 1902 to describe the physiological properties of secretin. They discovered that acid in the duodenum released "secretin" into the circulation, which then stimulated the pancreas to secrete water and bicarbonate into the duodenum, so that the lumenal acid was neutralised, the first example of single loop negative feedback control. Subsequently, they invented the word "hormone", derived from the Greek word "hormao", meaning "I excite", to define this new physiological concept. CLARK included the neurotransmitters acetylcholine and adrenaline into the generic class of hormones. Unfortunately, no one else seems to have followed his example. So the overarching beauty of the messenger (hormone)–receptor concept and the entailed (implicit) quantitative relationships underlying all of them have become obfuscated. Investigators who work on chemokines usually have a different background and training from people who work on neurotransmitters. Specialists who work on hormones, endocrinologists, differ in training and experience from the immunologists who study lymphokines. There seems to be no deep appreciation of the quantitative expectations (and models) that underlie all of their work. Failure to share basic chemical concepts, analytical techniques and derived models, a failure that deprives us of potentially fruitful intellectual interactions, is the consequence of taxonomic slackness. Having failed over the years in persuading biologists to see the merit of using "hormone-receptor" as an umbrella classification, perhaps I will be more successful, in the age of Information Technology, in getting acceptance for "messenger–receptor" systems.

All messenger–receptor conjugates express both selectivity and specificity. Selectivity is the property that allows a messenger molecule to be recognised in the first place, that is, when a substance extracted from a tissue is found to have a tissue-selective ability to stimulate or inhibit the beat of pacemakers, the generation of nerve impulses, the contractility of muscles, the secretion of glands or the movement of cells. Thus, selectivity is an empirical observation. Specificity, on the other hand, is the analytical attempt to explain the basis of selectivity, that is, to try to specify the combination of receptors (and their subtypes), second messengers and metabolic sinks that achieve physiologically relevant selective activity. The important point here is that messenger molecules can have the property of selectively "lighting up" a single class of proteins in a tissue. So how are homogeneous classes of protein receptors specified, operationally in the first instance, in intact-tissue bioassays? The answer – by using "specific" antagonists, that is, chemicals that meet the criteria of simple competitive antagonism so that a K_B can be estimated. An antagonist K_B is a receptor-specific parameter. Therefore, estimated K_B's can be used to explore the homogeneity of receptors across tissues and between species. Antagonists classify receptors; agonists expose them. A corollary to this, and a beautiful test, is that an antagonist K_B is independent of the specific agonist used in the measurement. The whole exercise flirts with circularity!

Based on the old "lock and key" idea, the basic antagonist–receptor interaction is conceptually simple. Although binding interactions obviously involve molecular interactions, the concept of "binding" does not entail mechanism. How the antagonist manages to exclude the agonist from its receptor is irrelevant. Competitive antagonism is an operational definition.

Although the pharmacological concept of competitive antagonism was developed using bioassays, they have no monopoly in exploiting that concept. "Binding" is a chemical concept and is best measured by chemical methods. Usually, the antagonist binding parameter is measured by arranging for the test ligand to compete with a radiolabelled ligand for receptors expressed in homogenised tissues. Progressive refinement towards an unambiguous chemical measurement can be achieved by using solubilised receptors or even pure proteins; non-specific binding appears to be reduced without improved precision in parameter estimation. However, there is a trade-off. Increase in the explicitness of chemical measurements is associated with the loss of the informativeness of bioassays. Some years ago, Colquhoun, commenting on pharmacological models, wrote "The Schild method is beautiful, and it works . . ." I agree, but, in my experience of assaying several thousand different ligands, it doesn't work very often. Indeed, the Schild method is most powerful when the model doesn't fit! Thus, the informativeness of intact-tissue bioassay has allowed us to expose indirect competitive antagonism, various receptor subtypes and plural drug actions that combine as pharmacological resultant activity. None of these would have been discovered using explicit chemical methods.

Bioassays have another advantage over chemical assays. Over 50 years ago, it became clear that while the elementary "binding" model was a satisfactory description of antagonist–receptor interactions, this model was an inadequate descriptor of hormone–receptor interactions. Within a given isolated-tissue bioassay, Ariens and Stephenson showed, independently, that agonists not only varied in their potency (as judged by the half-maximal location of their dose–response curves, the A_{50}) but also in their varying capacity to generate a maximal response (as judged by the value of α in a logistic function fitted to the corresponding dose–response curve). Agonists whose maximal response was submaximal in a tissue became known as "partial agonists". Agonists were postulated to have the property of "efficacy" as well as affinity. In a systematic series of "hormone" analogues, efficacy was often found to vary independently from affinity. Au fond, agonism was conceived as a sequential two-stage process. The big surprise was that efficacy was differentially expressed both between tissues and between species. Thus, the adrenaline analogue, dichloroisoprenaline, is a "full" agonist in the pacemaker of the guinea pig heart but a simple competitive antagonist to adrenaline in ventricular muscle. So a model with a new parameter was invented by STEPHENSON to describe agonist–receptor interactions. In this model, the efficacy, "e", is a global parameter subsuming chemical elements intrinsic to the ligand, to receptor density and to unspecified tissue factors that combine to convert occupied receptors into effect. This is still the basic "standard model" used today to define the expectations for agonist–receptor interactions. Note that efficacy is a systems phenomenon and is not seen at the level of binding interactions. Bioassays at the level of intact tissues or cell suspensions are needed.

In addition to exploring different tissues and species for efficacy detection, biotechnology has now made the efficacy of receptor systems a manipulable property. Receptor densities can now be over- or under-expressed in transgenic mice. Producing transgenic animals is now a huge industry. Ironically, the new technology has increased the need for the traditional technology of isolated-tissue bioassays. In addition to being useful for exposing low levels of partial agonist activity, overexpression of receptors has disclosed the new pharmacological property of inverse agonism, that is, the discovery that ligands previously classified as competitive antagonists switch off spontaneously active receptors. In my lab, we are now studying transgenic mice that have cardiac-specific overexpression of human adrenergic β_2 receptors. If these mice are allowed to grow old they develop heart failure. Physiologically, isoprenaline, a powerful cardiac stimulant, activates both β_1 and β_2 receptors. In hearts isolated form these old transgenic mice, isoprenaline now inhibits cardiac contractions. Treatment with a selective β_2-receptor antagonist converts the effects of isoprenaline to cardiac stimulation. This stimulant action is mediated by β_1 receptors. β_1 receptors are known to couple only to the excitatory G_s proteins while β_2 receptors can couple to both excitatory G_s and inhibitory G_i proteins. Our explanation is that activation of overexpressed β_2 receptors steals G_s proteins from the β_1 receptors. The point is that this is a system phenomenon that can only be disclosed in the intact tissue.

This property of bioassays, the ability to expose the efficacy of messenger molecules, has traditionally been invaluable for the invention of new drugs. Drugs act at the micro (molecular) but their desired effects have to be manifest at the macro (physiological) level. The macro level of cells, tissues and organs (ordered into physiological systems) is characterised by organisational complexity. This complexity, a compound of histological structures and chemical messengers, is organised into linear components, local feedback loops and management hierarchies. To illustrate this, I have a slide that I use a lot in lectures – it is simply a list. The list has intact animals at the top; followed by isolated, perfused organs; pieces of tissue, with their cellular architecture intact, suspended in organ baths; cells in tissue culture; homogenised cells; and, finally, purified proteins. I used to say that this was my "reductionism" slide, if viewed from the top down, . . . or my "hierarchies" slide, if viewed from the bottom up. However, I now say that this is my "fractal" slide because I have come to realise that reductionism in biology does not achieve the aim of simplification. Reductionism in biology merely replaces one type of complexity by a different kind of complexity. No one level is more reliably informative then any other. So I strongly believe that pharmacology needs to be studied at all levels, the choice of level being dictated by the nature of the question being asked.

In applied pharmacology, the question most commonly asked today is "How can we find a drug to interact with a specific gene product?" It so happens that we now have the technology to study the interaction of new compounds with gene products at the level of pure chemistry. This is the entry level of pharmacological screening in industrial pharmaceutical research at the present time. To do this, huge libraries of molecules are being generated by the ingenious techniques of combinatorial chemistry (combichem). Then these libraries are rapidly scanned for selective interactions with the target molecules. The process is known as HTS, high throughput screening. The idea is that a successful binding interaction, known as a hit or a lead, then becomes the basis for the conventional, systematic, iterative, optimisation process that medicinal chemists excel at. As a lead generator, combichem plus HTS undoubtedly works. Perhaps this is not surprising. Complexity specialists, such as Kaufman, argue that life began as an exercise in combinatorial chemistry. Even so, single-cell primitive tube-like forms and bacteria seem to have been around for 2.5 billion years before the Cambrian explosion started structural evolution 500 million years ago. Maybe natural selection had first to operate at the level of combinatorial chemistry, chemical evolution, to develop populations of molecules that were comfortable with each other before structural evolution could take off. My reading of the history of drug inventions suggests that the most selective drugs, with the widest therapeutic ratio, have come when the initial lead was a native, physiological molecule. Perhaps drugs that are crafted round a natural template retain some of the parental selectivity. So I wonder if the leads discovered by the combichem plus HTS strategy will all have the same quality. Will the epitope on a receptor that selects a lead from a combichem library necessarily be the same as the site that recognises the natural messenger?

However, my main concern about the current thrust of drug research is that it is rooted in targeting components rather than systems. The importance of working with systems was discovered by Furchgott nearly 20 years ago, work that led to the extraordinary discovery of nitric oxide as a universal chemical messenger. Here is the story. When acetylcholine is infused intraarterially, the blood vessels dilate. Furchgott studied the effects of drugs on arterial muscle isolated in organ baths. At one time, the only way to measure the effects of drugs on arterial muscle was by measuring changes in its length. To get measurable shortening, the blood vessel was cut into a long spiral strip. In this preparation, Furchgott found that acetylcholine had no relaxant effects, indeed the muscle usually contracted. When instruments for measuring tension became available, isometric measurements could be made on rings of arterial muscle. Imagine his surprise when Furchgott found out that acetylcholine now had the expected relaxant effects on arterial muscle. Furchgott showed that the endothelium was intact in the muscle ring but destroyed in the muscle spiral. He showed that acetylcholine relaxes arterial muscle indirectly by stimulating the endothelium to secrete a relaxing factor which diffused into the adjacent muscle layer. This seminal discovery was only possible when the muscle and endothelium were combined in a system. We now know that the relaxing factor is a gas, nitric oxide, which escapes when endothelial cells are cultured in vitro. So this astonishing discovery could not have been made by the powerful, component-directed, techniques of molecular and cellular biology.

From the point of view of pharmaceutical research, I want to make a more general and speculative point about the contrast between systems and components. As far as I can see, the new drugs that the pharmaceutical industry is seeking, to treat disorders such as asthma, dementia and cancer, are expected to be similar to the ones that they have already invented to reduce high blood pressure, heal stomach ulcers and relieve pain. I am concerned that this expectation may not be fulfilled. Physiology is about how cells use chemicals to talk to each other. Sometimes, the message has the shape of a command, such as "contract" or "secrete"! Thus, adrenaline is the final messenger to the pacemaker of the heart in emergency situations. So a drug that blocks the effects of adrenaline on the heart effectively controls cardiac stress responses. Note that the heart must react to stress reliably, on cue, but if it beats faster inappropriately nothing very bad happens. The process of heart rate changes is inherently reversible. Some of our most useful drugs act by interfering with chemical commands.

However, there is another kind of physiological system that must also be activated reliably, on cue, but which, if activated inappropriately can have damaging, even lethal, effects. Examples of these systems are commitment of bone marrow stem cells, activation of killer lymphocytes, cell division and the growing of new blood capillaries. Once initiated, these are inherently irreversible processes. So, how are these physiological processes controlled such that they can be activated on cue but never inappropriately? The striking feature of these irreversible processes is that many chemical messengers are involved, each having a different cellular origin. They are often described as cascades but they are unlikely to operate in sequence as implied by that word. Another feature of these messengers is that they can often be shown to potentiate each other. So I imagine a process that I call "convergent control". I imagine that an effective stimulus might involve the co-operative interaction of more than one agent, involving addition or amplification of, individually, subliminal stimuli. For example, I imagine a growth factor giving a stem cell, not a command, but a piece of advice, such as "Other things being equal, you should start dividing". The other equal things are other chemical messengers, which have to impinge on the cell at the same time to achieve its activation. Unlike the command-control action of adrenaline-like molecules, such an advise–consent arrangement leads to information-rich management. Physiological control by chemical convergence entails the possibility of redundancy. Therefore, annulling the action of a single component may be disappointing. Biotechnology has been hugely successful at blocking various molecules that are overproduced by the immune system in septic shock. In every case, laboratory success has ended up in clinical failure. At some point we must ask whether the model or our way of thinking is wrong. I believe that the way forward will lie, not in the discovery of more and more components, but in improving our understanding of how components are organised into systems.

The principles involved in this approach can be illustrated with reference to asthma. Experimental models of asthma have, to a great extent, focused on allergic aspects of asthma, although allergic (atopic) subjects comprise only one subset of the asthmatic population. In allergic asthma, a reductionist approach to this disease has prevailed, resulting in many groups placing heavy emphasis on individual components of the allergic response, such as mast cell degranulation, specific mediators (e.g. leukotrienes) or specific cell types (e.g. eosinophils). Whilst this has undoubtedly led to a detailed understanding of the individual components of the allergic response, most of the drugs that have been developed as highly specific antagonists or down-regulators of these components have proved ineffective in clinical trials in subjects with asthma. In fact, apart from the recent introduction of leukotriene receptor antagonists, no new drug class has been introduced for the treatment of asthma in the last 20 years; indeed, even the leukotriene antagonists only give relief to a small subset of patients. All of these therapeutic approaches have been based on the assumption that asthma is a "component" problem. The time seems right to try to tackle asthma as a "systems" problem.

A "systems" approach might go something like this: The large number of molecular components recognised by reductionism are produced locally by several different cell types. Others reach the tissues from the capillary circulation. With asthma in mind, the architectural relations of these cell types, such as afferent and efferent nerve fibres, mast cells and eosinophils and their relation to bronchial muscle cells, can be maintained in intact-tissue bioassays from human or animal lungs. Various messenger molecules can be selectively released locally from these cell types and the resultant effects on the targeted muscle cells can be studied. Messenger molecules and drugs added to the organ bath might be considered dynamically equivalent to their delivery to the tissues via the circulation. This intact biosystem can now be manipulated in various ways and its behaviour exposed as families of frequency- or concentration–response relationships. An attempt to interpret these data can now be made in terms of hypothesised interactions among locally released messenger molecules, "circulation-delivered" agents and the target cells. As with the simple models, when a system is interpreted by an "as if" scenario, the idea can always be given algebraic expression and manipulated as a potential model of the data. The importance of the algebraic model is that it not only takes all the vagueness out of the concept but it also makes the concept testable. When the model turns out to be a poor fit to the data, a new or modified concept is needed.

In the 1996 volume of Annual Reviews of Pharmacology, I wrote:

"Mathematically, this model of convergent threshold amplification works. Bioassay is the appropriate experimental tool. Bioassays can be designed to mimic and analyse such convergent control systems. They could be used to explore new pharmaceutical strategies. Hence, my belief is that bioassay and analytical pharmacology have an exciting and important future."

Bioassays have been at the heart of pharmacology for most of this century. The uses of bioassays have kept changing. The theme of this essay is that some of the traditional uses of bioassays and related models in analytical pharmacology are still enjoying uses that cannot be achieved any other way. More than this, I think that isolated-tissue bioassays are set to lead the way into the study of the pharmacology of complex convergent-control systems.

Section I
Classical Pharmacology and Isolated Tissue Systems

CHAPTER 1
Human Vascular Receptors in Disease: Pharmacodynamic Analyses in Isolated Tissue

J.A. ANGUS

A. Introduction

The pharmacologist's goal is to discover new medicines. It follows that, at an early stage in the discovery process, the potential drug must be tested against the target disease in humans to confirm its selectivity and specificity (BLACK 1996). Because experiments on disintegrated systems have greater analytical power than those on integrated systems, scientists habitually employ experiments at the molecular and cellular levels. In reality, however, tissue-based assays isolated from the target organs of humans with or without disease offer essential pharmacodynamic knowledge of the new medicine's activity. The use of human tissue in pharmacodynamic studies is becoming appreciated as we learn that experimental animal-tissue assays do not always reflect human receptor homology and tissue structure. In addition, so-called animal disease models of human disease are usually found wanting and cannot reflect ageing, genetic differences and so on.

While there are clearly a number of theoretical advantages of testing pharmacodynamics in human isolated tissue, in reality, this approach is full of difficulties. For example, patient care is obviously paramount; tissue can only be removed if this is normally done, and the tissue must then be discarded. Surgeons do not set their timetables to suit a laboratory; furthermore, pre-existing diseases, multiple therapeutic drug treatment and age and sex differences play havoc with experimental design and population sampling. "Normal" human tissue is probably the most difficult to obtain in many instances. With these issues in mind, the human-tissue experiment may not meet "ideal" assay design, and the results are necessarily less robust.

In addition to testing agonist or antagonist receptor selectivity and specificity in human-tissue assays, the pharmacologist must first "calibrate" the tissue in terms of its reactivity to stimuli. Often, comparing pharmacological reactivity in normal versus diseased tissue offers a clue to the causes or consequences of the disease.

This chapter will review approaches and findings of four areas of pharmacodynamics in human isolated tissues of interest to the author. These areas of study are centred on (1) the cause of variant angina and the analysis of reactivity in large and small coronary arteries; (2) the vascular reactivity of human gluteal resistance arteries in patients with essential hypertension and chronic heart failure (CHF); (3) a suitable vasodilator cocktail to prevent spasm of the internal mammary artery (IMA) and saphenous vein grafts during coronary artery bypass graft surgery; and finally (4) the development of a robust measure of the vascular-to-cardiac selectivity ratio of calcium-channel antagonists.

B. Receptors Mediating Coronary Artery Contraction: Role in Variant Angina

I. Large Coronary Arteries

1. Ergometrine

Variant (or Prinzmetal's) angina is characterised by short-lived spasm (zero flow) of a large coronary artery, accompanied by acute chest pain with the patient at rest (PRINZMETAL et al. 1959). This spasm can be reproduced in the coronary-catheter laboratory by intracoronary injection of ergometrine (ergonovine) (CURRY et al. 1977). Often, the angiogram will show a small atherosclerotic lesion at the locus of the spasm in response to ergonovine. In normal patients, ergonovine only causes a diffuse narrowing of the large coronary arteries, without any ischaemia. Candidate receptors for ergometrine are serotonin (5-HT) receptors and α-adrenoceptors, given the earlier pharmacology of ergot derivatives. In dog isolated coronary arteries, it was established that ergometrine had no affinity for α-adrenoceptors but was a partial 5-HT-receptor agonist with very slow onset to equilibrium compared with 5-HT (MÜLLER-SCHWEINITZER 1980; BRAZENOR and ANGUS 1981). This finding could imply that endogenous 5-HT released from aggregating platelets on an unstable atherosclerotic intima or from nerve varicosities or other cells in the vessel wall was triggering vasospasm in patients with variant angina. However, early clinical trials of the 5-HT_2 receptor antagonist ketanserin against ergometrine-induced ischaemia (FREEDMAN et al. 1984) or variant angina (DE CATERINA et al. 1984) were disappointing. Clearly, 5-HT_2 receptors were not involved or were not the only 5-HT receptors being activated by ergometrine.

2. 5-HT and Endothelium-Derived Relaxing Factor

The discovery of endothelium-derived relaxing factor (EDRF), its subsequent identification as nitric oxide (NO) and the discovery of various endothelium-derived hyperpolarising factors (EDHFs) have added a major insight to this field. 5-HT can activate the receptors 5-HT_{2A} and 5-HT_{1D}, which mediate contraction or relax the vessel directly (through 5-HT_4 receptors on the smooth muscle) or indirectly (by the release of EDRF through 5-HT_{2A} or 5-HT_{2B} receptors on endothelium; COCKS and ANGUS 1983; MARTIN 1998). Theoretically, endothelial dysfunction or loss of endothelial cells could tip the balance towards contraction for two reasons (1) loss of endogenous basal NO; and (2) loss of receptor-stimulated release of NO counteracting smooth muscle-cell contraction. This has been confirmed experimentally; ergometrine caused exaggerated constriction of endothelium-denuded coronary arteries in conscious canine coronary arteries after intimal thickening following balloon endothelial cell denudation and high-cholesterol feeding (KAWACHI et al. 1984).

Data from experimental animals confirmed that high-cholesterol diets will amplify the contraction resulting from 5-HT. For example, in rabbit isolated left main coronary artery mounted in a myograph, 5-HT was a very weak agent, causing contractions in less than 5% of vessels. However, in vessels taken 16 weeks after beginning a 1%-cholesterol-supplemented diet, the concentration–contraction curve was markedly amplified (Fig. 1; ANGUS et al. 1989). Similarly, in long-term cholesterol-fed cynomolgus monkeys, the coronary arteries display hyperreactivity to U46619 and 5-HT (QUILLEN et al. 1991).

By exploiting explant heart tissue from the heart-transplant program, we were able to secure left and right main-epicardial coronary artery segments with mild to moderate atheromatous plaque. Three-millimetre-long ring segments were mounted on wires in conventional organ baths and stretched to a normalised wall tension for isometric force recording. This tissue came mostly from hearts transplanted due to cardiac failure caused by cardiomyopathy or ischaemic heart disease. On a few occasions (16%), unused donor hearts were available. In general, the force records of the large arteries tended to break into phasic contractions when activated, making analysis of concentration–response curves difficult. Nifedipine ($0.1\,\mu M$), while attenuating the range of the contraction curves, was a convenient tool to prevent the phasic contraction. As far as 5-HT receptors were concerned, we found that 5-HT caused a maximum contraction to only 40% of the maximum contraction in response to K^+ (K^+_{max}), while the 5-HT$_{1D}$ agonist sumatriptan and the mixed agents methysergide and ergometrine caused contractions less than 10% of those caused by K^+. However if the arteries were exposed to the stable thromboxane mimetic U46619 (1 nM), which caused a small rise in force to $4 \pm 1\%$ of K^+_{max} in arteries that did not develop phasic contractions, then the four agonists caused substantial contractions, e.g. the contraction caused by sumatriptan was 5%–49% of K^+_{max}, and that caused by 5-HT was 40%–92% of K^+_{max} (ANGUS and COCKS 1996; Fig. 2).

This synergy may help explain how a combination of local factors could give rise to variant angina. The synergy with 5-HT agonists is not confined to U46619, since endothelin-1 has also been shown to enhance contractions to noradrenaline (NA) and 5-HT in human isolated coronary and IMAs (YANG et al. 1990). However, these experiments, by necessity, are conducted in arteries probably harbouring a moderate degree of intimal pathology. To test the degree of endothelial dysfunction, we applied the "acetylcholine test". In 15 patients, multiple ($n = 100$) 3-mm-long ring segments of large coronary arteries were

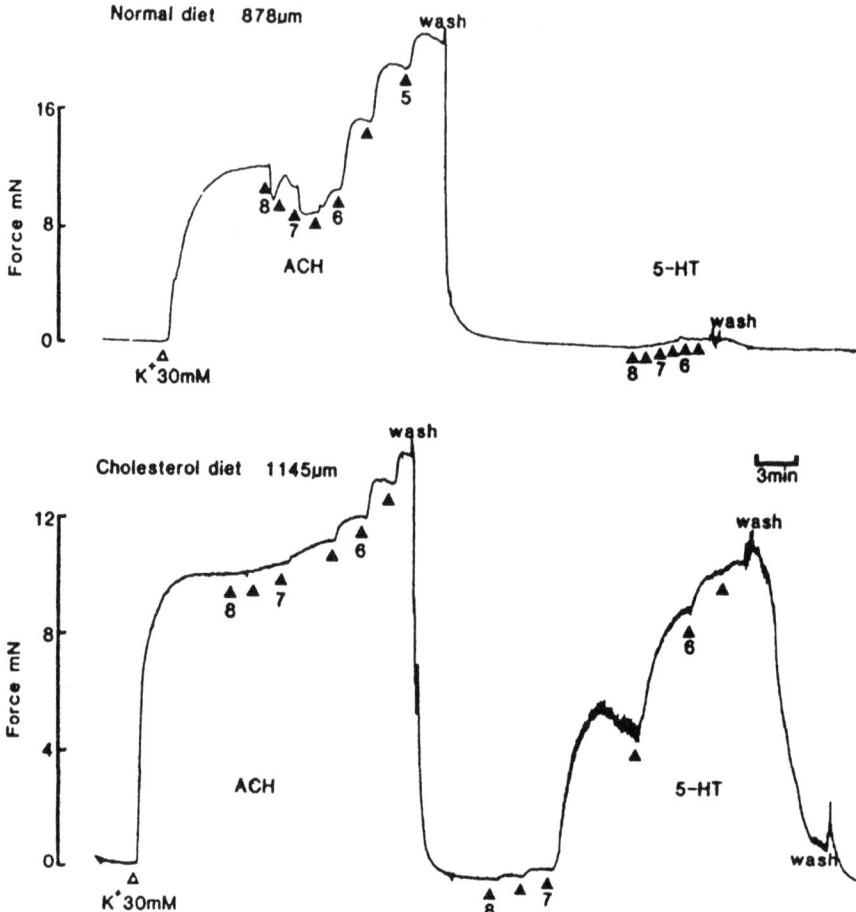

Fig. 1. Chart records of isolated coronary arteries from rabbits fed a normal diet (*top*) or a 1%-cholesterol-supplemented diet for 16 weeks (*bottom*). Acetylcholine (*ACH*) caused relaxation and contraction in normal rings (*top*) but only contraction in the atheromatous artery (*bottom*). Serotonin (*5-HT*) was a powerful constrictor in the atheromatous vessel. Concentrations are −log M, in half-log-unit increments. Artery-lumen internal diameters: normal diet, 878 μm; cholesterol, 1145 μm

pre-contracted to 50%–80% of the maximum contraction with U46619 (3–10 nM), followed by construction of an acetylcholine concentration–response curve. Acetylcholine relaxed only 50% ($n = 31$) of the vessels, while the remainder either had no change ($n = 31$) or contracted further ($n = 38$ rings; Fig. 3). Substance P and bradykinin concentration–relaxation curves were slightly less potent (right shifted), without a major change in range for the arteries that were not relaxed by acetylcholine compared with the change in range for those that were. There may, of course, be a major role for EDHF (in addition to NO) in the human large coronary, since N^G-nitro-L-arginine (0.1 mM) only partially attenuated the relaxation cause by acetylcholine, substance P and bradykinin

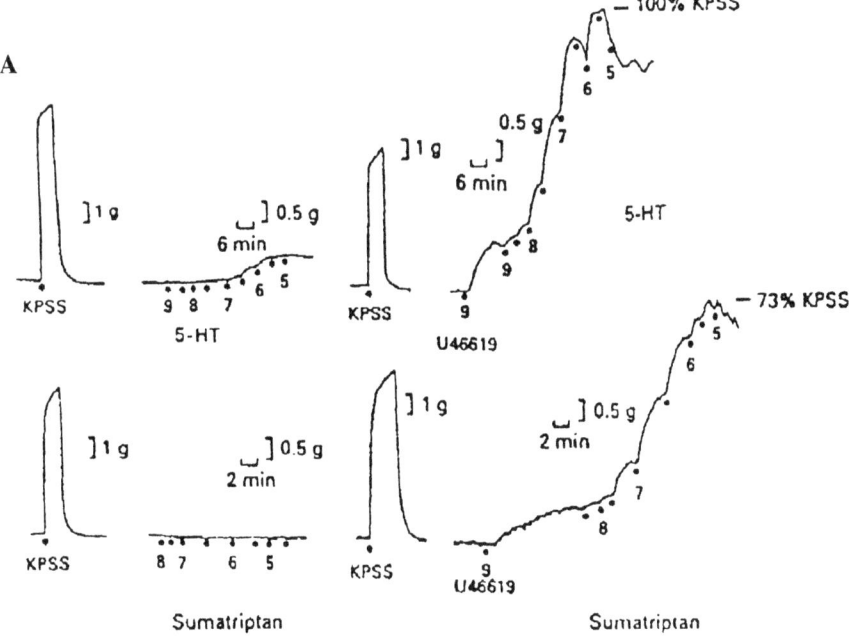
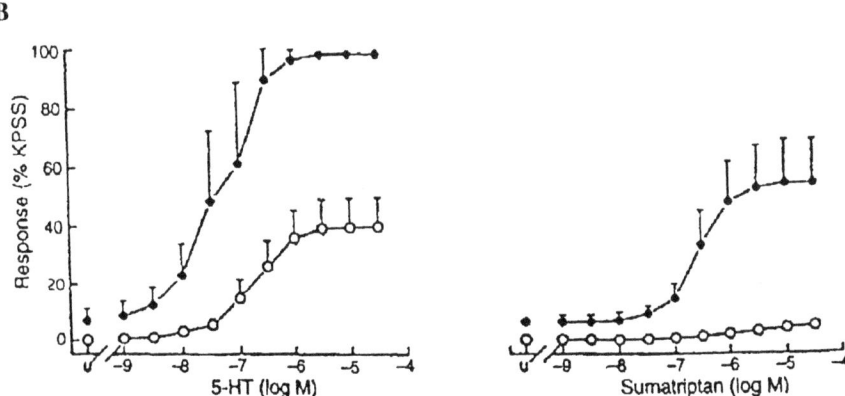

Fig. 2. Original traces (**A**) and group data (**B**) showing the effect of pre-contraction with the thromboxane A_2 mimetic U46619 (1 nM; *solid symbols*) on contractions in response to serotonin and sumatriptan in isolated rings of human epicardial coronary artery. Contractions are expressed as percentages of the maximum contraction in response to 124 mM K^+ Krebs' solution (*KPSS*; COCKS et al. 1993)

in arteries pre-contracted with U46619 (STORK and COCKS 1994; Fig. 4). Given that acetylcholine can activate NO release and contract the artery through muscarinic receptors on smooth muscle, many believe that acetylcholine offers a better test for identifying endothelial dysfunction. However, this test is also subject to uncertainty. For example, the agonist is biologically unstable due to

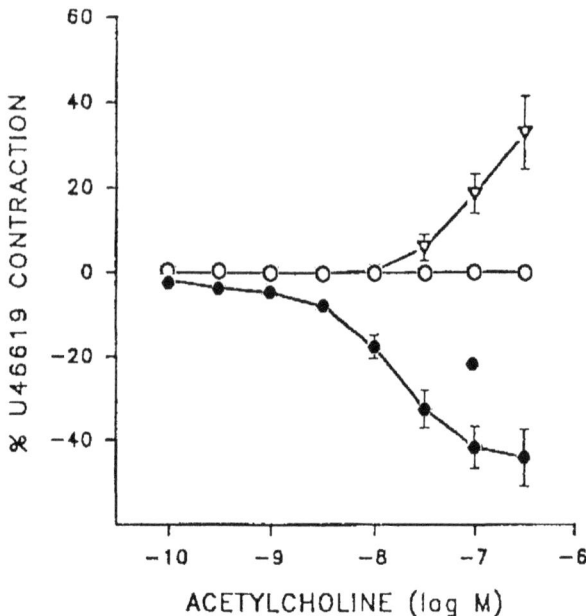

Fig. 3. Acetylcholine responses in rings of human epicardial coronary artery rings precontracted by U46619 isolated from 15 patients. Acetylcholine caused relaxation (○; $n = 31$ rings), no response (●; $n = 31$ rings) or contraction (△; $n = 38$ rings)

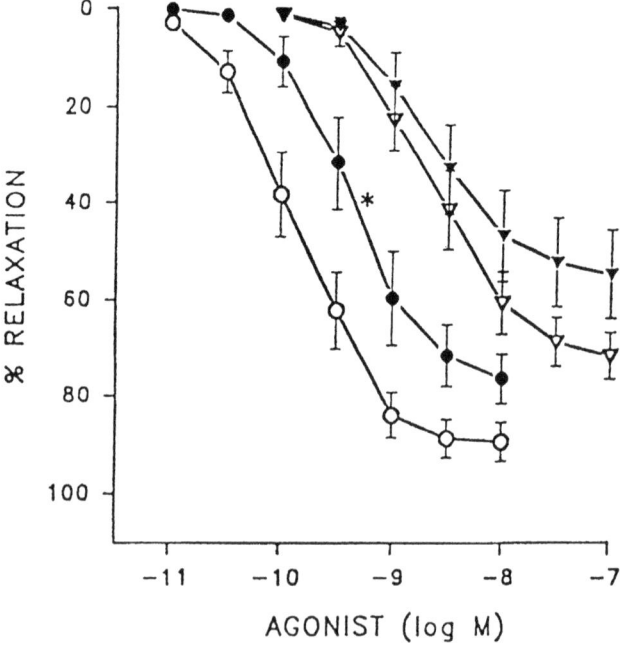

Fig. 4. Substance P (*circles*) and bradykinin (*triangles*) responses in human coronary artery rings sequential to rings which relaxed (*open symbols*) or failed to relax (*closed symbols*) in response to acetylcholine (ACH). Asterisks indicate significantly different ($P < 0.05$) pEC_{50} for vessels which did not relax in response to ACH compared with those that did relax

cholinesterases. If injected, the distal vasculature may not be exposed to the same concentrations as the upstream endothelium. Variation may occur in the muscarinic population in endothelial cells compared with that in smooth muscle cells in different parts of the vasculature (see below). Finally, basal NO arising from endothelial NO synthase (NOS) activity can affect reactivity in a different manner than receptor-stimulated NO release (ANGUS and LEW 1992).

II. Small Resistance Arteries

To investigate the pharmacodynamics of the important small resistance coronary arteries, we developed a cooperative routine with cardiac surgeons in nearby city hospitals. As patients were prepared for routine heart–lung bypass surgery for coronary bypass grafting or valve replacement, the surgeons, with Human Ethics Committee consent, removed the tip of the right atrial appendage (0.5–1.0 cm) and placed it directly into cold Krebs' physiological salt solution instead of discarding the tissue. Generally one or sometimes two suitable arteries [100–300 μm internal diameter (i.d.)] were dissected under microscopes and mounted as 2-mm-long ring segments on parallel, 40-μm-diameter stainless steel wires in a Mulvany-Halpern myograph (MULVANY and HALPERN 1977).

Similarly sized arteries were removed from the apices of dog hearts, left ventricles of rabbit hearts or atria or ventricles of pig (domestic large white) hearts. As far as pathology was concerned, the human small coronary vessels were remarkably free of atheroma, only showing the occasional lipid droplet in a smooth muscle cell, even though these patients were undergoing bypass graft surgery for extensive *large artery* atheroma.

The concentration–response curves to acetylcholine were markedly different. No attempt was made to remove the endothelium, as this destroyed the contractility of these small vessels. As expected, acetylcholine caused concentration-dependent relaxation in dog arteries, similar to the results of subsequent test with substance P. In contrast, human arteries pre-contracted with K^+ (30 mM) contracted further with acetylcholine. This occurred despite the subsequent finding that substance P relaxed the artery, indicative of the presence of endothelium and the capacity to release NO. Rabbit coronary arteries were biphasic in response; they were first relaxed and then contracted by acetylcholine at concentrations greater than 1 μM (ANGUS et al. 1991b) (Fig. 5). This acetylcholine paradox in human microcoronary arteries could be explained by a decreased or absent muscarinic-receptor population on the endothelium, compared with a significant number of receptors mediating contraction on smooth muscle cells (Fig. 6). This is a clear example of how coronary microvessel pharmacology varies among three species and makes extrapolation from animals to man quite hazardous.

We have investigated the pharmacology of human "normal" coronary microvessels by simultaneously studying electrophysiological measurements

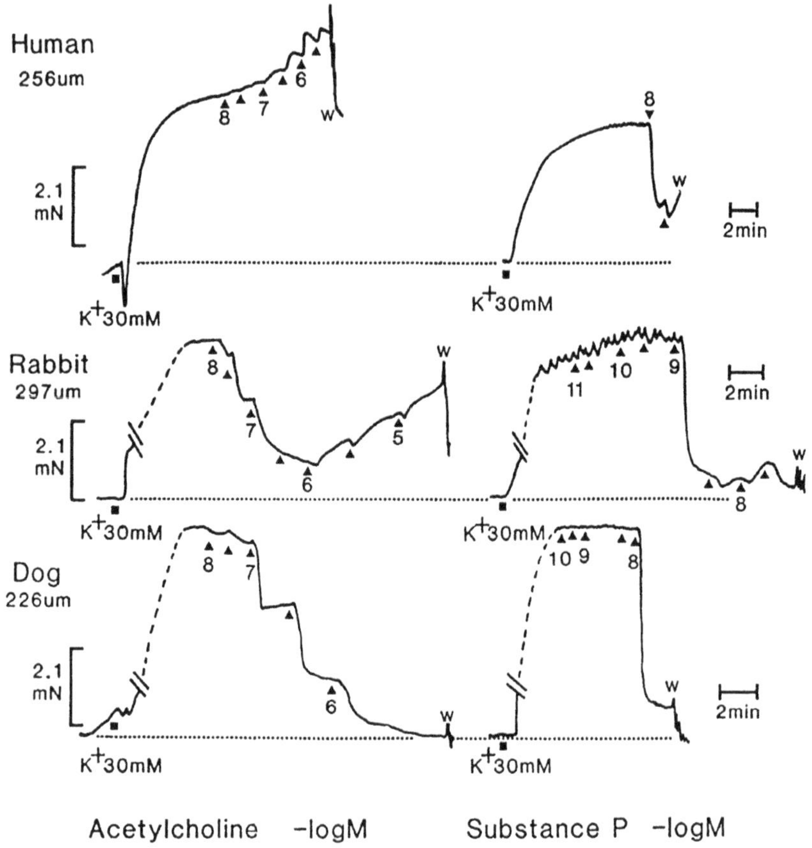

Fig. 5. Chart records of isometric force (in millinewtons [mN]) in similarly sized small-resistance arteries (256, 297 and 226 μm internal diameter) isolated from human atria, rabbit ventricle and dog ventricle. Vessels were pre-contracted with K^+ 30 mM before applying acetylcholine (*left*) or substance P (*right*) in half-log-unit increments

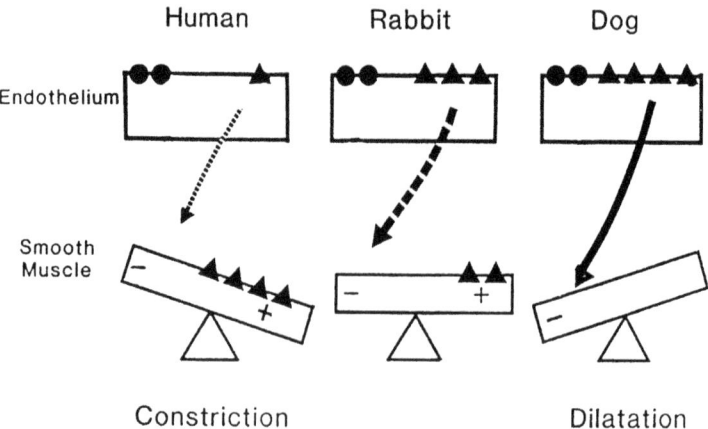

Fig. 6. Possible variations in number and location of acetylcholine (▲) and substance-P (●) receptors on endothelium and smooth muscle cells of coronary-resistance arteries from humans, rabbits and dogs to explain the observed responses in Figs. 1–5

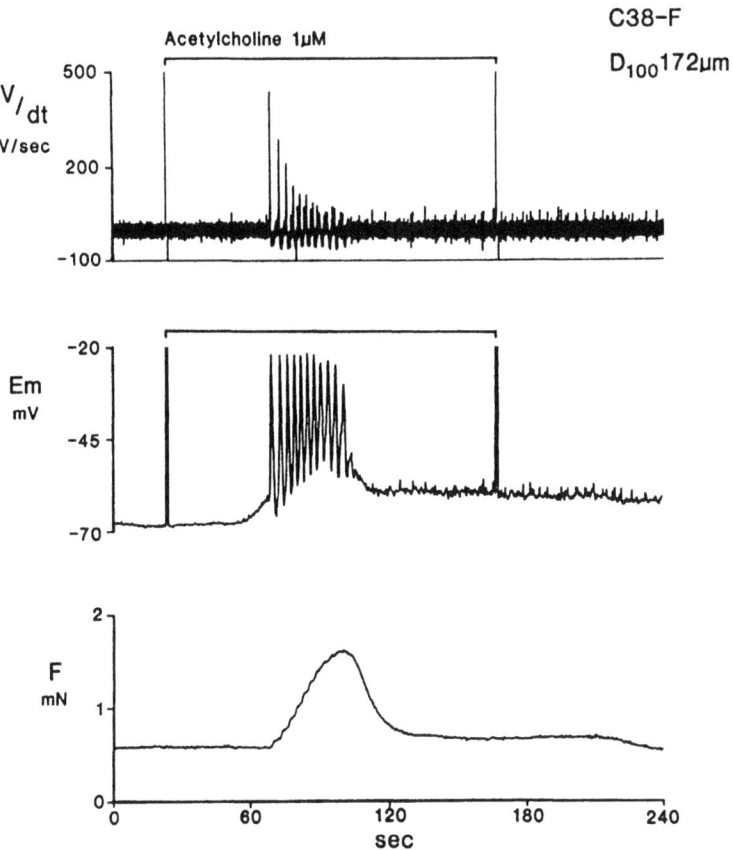

Fig. 7. Computer-regenerated recordings of membrane potential (E_m), its derivative, dV/dt, and isometric force (F) in a human small coronary artery. Acetylcholine infusion (1 μM) was applied between the marks, as indicated, with a 43-s delay before the drug entered the bath. The dV/dt was calculated from six point-slope values of E_m originally digitised at 200 Hz, progressing one point per calculation. The resting value of force is arbitrary

by conventional glass microelectrodes (tip resistance = 100 MΩ) and force measurements in arteries from right atrial appendages mounted in a Mulvany myographs (ANGUS et al. 1991a). Acetylcholine (0.1–1 μM) applied to the inlet solution flowing over the artery caused a distinctive, rapid, short-lived depolarisation from a resting membrane potential (E_m) of –62 ± 3 mV to as high as –20 mV before repolarising to nearly the resting E_m and then depolarising again, at a repetitive frequency of approximately 0.5 Hz. These regular oscillations were accompanied by small-step increases in contractile force, with the maximum rate of change of E_m (dV/dt) occurring with the first depolarisation (Fig. 7). When acetylcholine (1 μM) was infused, the oscillations were accompanied by a steady rise in force to a plateau level. Substance P (30 nM) infused with the acetylcholine caused a rapid repolarisation and relaxation of the vessel, indicative of NO/EDHF release from intact endothelium (Fig. 8). NA

(1–5 μM) also caused this unique oscillatory pattern in E_m in the human small coronary arteries, but K$^+$ (124 mM) infusion only depolarised and steadily contracted these vessels (Fig. 8). Like human large and small coronary arteries, pig (large white) coronary arteries contract in response to acetylcholine. However, we have not been able to reproduce the unique oscillation of E_m (observed in human arteries) in the pig small atrial or ventricle coronary arteries; in the pig arteries, we observed only steady depolarisation with the occasional spike in E_m in ventricular arteries (Fig. 9). This human small-coronary artery (110–250 μm i.d.) oscillatory E_m response to acetylcholine and NA may be related to inositol 1,4,5-trisphosphate activation and calcium-induced calcium release from the sarcoplasmic reticulum, since the E_m oscillations and

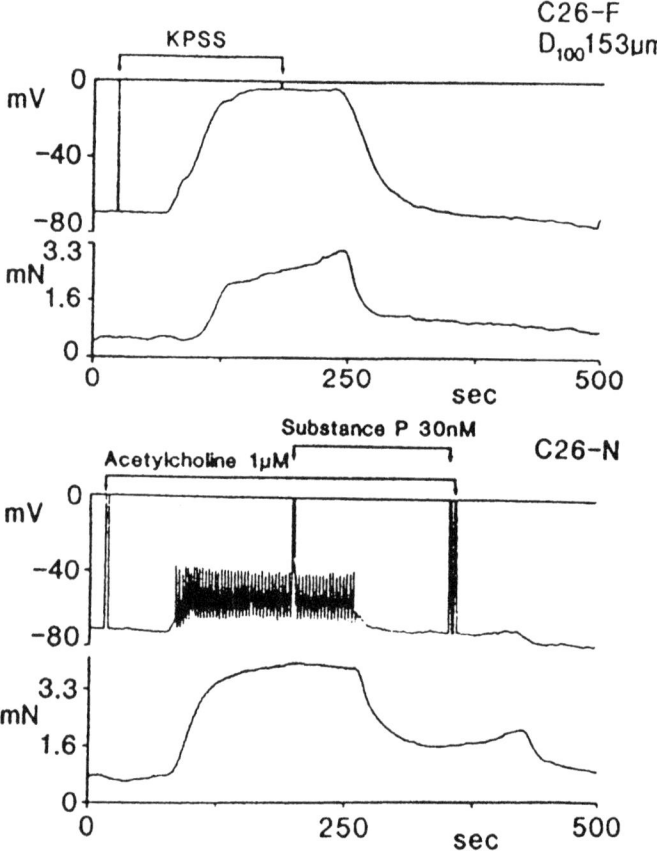

Fig. 8. Simultaneous records of membrane potential (E_m, in millivolts [mV]; *top traces*) and active force (in millinewtons [mN]; *bottom traces*) recorded from the same human small coronary artery (153 μm diameter) in response to K$^+$ Krebs' solution (KPSS, 124 mM; *top*) and acetylcholine with substance P (*bottom*). The *bar* indicates the infusion period, and there was a 60-s delay between the pump switching and the drug change in the myograph chamber

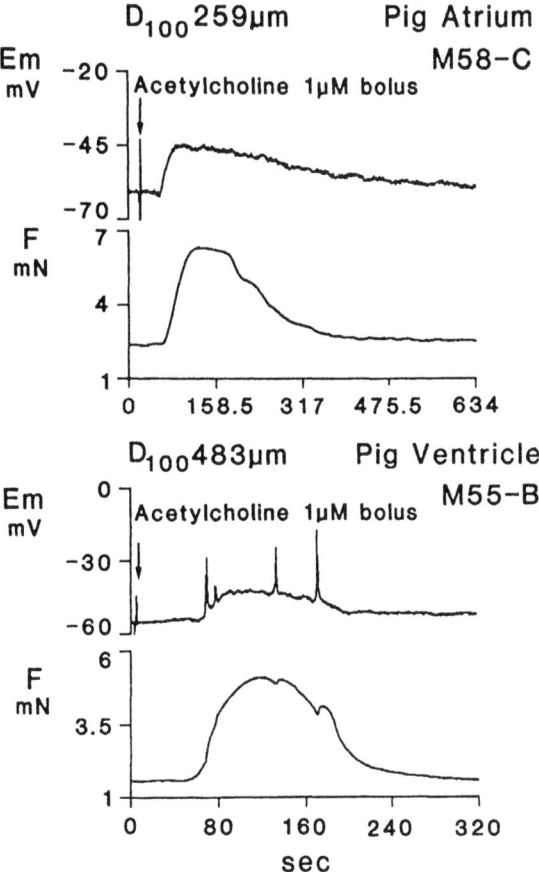

Fig. 9. Membrane-potential (E_m) and isometric-force (F) records from small arteries from pig atrium and ventricle. At the *arrow*, a bolus of acetylcholine was applied to the chamber and assumed to come into concentration equilibrium (1 μM). The chamber (7 ml) was perfused at 5 ml/min

contractions were sensitive to 1,2-bis(*o*-aminophenoxy)ethane-*N*,*N*,*N'*,*N'*-tetraacetic acid acetoxymethyl ester (20 μM) and ryanodine (10 μM; ANGUS and COCKS 1996).

The acetylcholine-induced contractions and E_m oscillations were blocked by atropine (0.1 μM) and are probably mediated by muscarinic M2 receptors, since 4-diphenylacetoxy-*N*-methylpiperidine 0.01 μM gave an estimated pA_2 of 9.2. Methocratamine (1 μM) and pirenzepine (0.03 μM) did not affect the responses. We proposed that acetylcholine activated muscarinic receptors on the smooth muscle to cause a sharp rise in [Ca^{++}]$_i$, which would, in turn, allow Ca^{++}-activated K^+-selective channels to open, causing repolarisation. To date, glibenclamide, charybdotoxin, apamin, tetrodotoxin and ω-conotoxin GVIA have not altered the E_m oscillation, while felodipine (0.1 μM) decreased the force by 60% but did not decrease the E_m response to acetylcholine (ANGUS and COCKS 1996).

To show how different the coronary small artery is from other human resistance arteries in response to acetylcholine, we took, in parallel experiments, a small thumbnail-sized piece of gluteal skin with underlying fat, removed from volunteers under local anaesthetic. These arteries hyperpolarised to acetylcholine and, if pre-contracted with the thromboxane mimetic U46619, relaxed in a concentration-dependent manner (Fig. 10).

We also tested the reactivity to 5-HT, ergometrine and sumatriptan (GR43175) in these human atrial vessels. Only 39% ($n = 7$) of all arteries ($n = 18$) responded to 5-HT (0.01–10 μM), and the maximum in these responders was only $43 \pm 14\%$ of the contraction in response to K^+ (124mM). Neither sumatriptan (0.01–10 μM) nor ergometrine (0.001–1 μM) contracted these vessels (Fig. 11). Some synergy was observed. When acetylcholine (0.3 μM) was used to pre-contract the vessels ($n = 5$) to 6%–17% (mean 10%) of K^+_{max}, sumatriptan (3 μM) caused the force to rise to $16 \pm 4\%$.

Human small coronary arteries studied in the isometric myograph are generally less reactive to 5-HT than are conduit arteries. There is no reason to doubt that the small arteries are viable given their reactivity to both acetylcholine and substance P. Intriguing, however, is the variable response to acetylcholine in human blood vessels. Only relaxation was observed in buttock skin small arteries; however, only contraction (not relaxation) of the large coronary arteries was comparable to that in the atrial small coronary arteries. How this observation relates to resistance changes in vivo or the role of vagal effects on small-coronary reactivity is unknown.

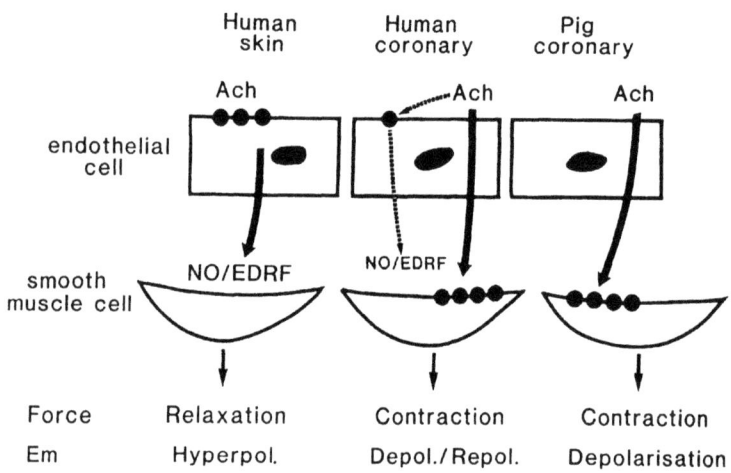

Fig. 10. Possible location of acetylcholine receptors in small coronary arteries from human right atrium and pig atrium compared with human gluteal skin arteries. The force and membrane responses are summarised for each vessel type

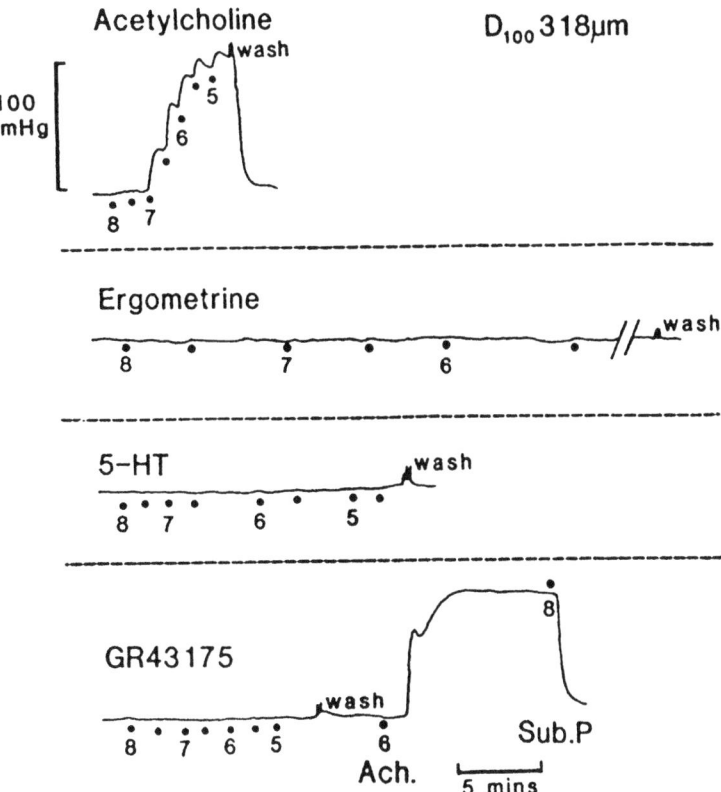

Fig. 11. Chart record of force changes in a human small coronary artery (318 μm diameter) in response to acetylcholine, ergometrine, serotonin and sumatriptan (*GR43175*). Endothelium was intact, given the relaxation in response to substance P. Concentrations are –logM in 0.5-unit steps. Force change is in active pressure units (ΔP = Δtension/radius)

III. Summary and Future Work

The pharmacodynamic results from human large and small coronary arteries mounted under isometric conditions in organ chambers have left many questions unanswered. Most of our small coronary arteries were from the right atrial appendage and the large arteries of explant hearts. We need to learn, if we can, the reactivity of normal small and large arteries taken from various sites in the human heart and measured under both isobaric and isometric conditions (LEW and ANGUS 1992). If each vascular segment has unique pharmacology and the acetylcholine responses in right atrial vessels point to such a trend, then there is much to do.

Fundamental research is required to further unravel the receptor profiles of the endothelium and smooth muscle cells and the interaction of age, endothelial dysfunction and the effect of long-standing influences of hypercholesterolaemia, hypertension and stress. Of singular importance is the need for rigorous attention to previous drug treatment, tissue procurement, dissection and mounting of human vessels to reduce experimental artefacts; these are more easily controlled when using experimental animal tissue prepared at the experimenter's convenience.

Finally, returning to variant angina in patients with coronary spastic angina, receptor specificity is critical. KUGIYAMA and colleagues (1999) have shown that acetylcholine (50–100 μg), as a bolus intracoronary injection, induced coronary large-artery spasm and myocardial ischaemia in all patients with spastic angina. In contrast, α-adrenoceptor stimulation by intracoronary phenylephrine (PE) caused low-level constriction of a similar magnitude in normal patients and in patients with spastic angina. The loss of endothelial factors normally buffering direct contraction by acetylcholine and/or increase in smooth muscle cell receptors or transduction of receptor activation could explain this intriguing phenomenon.

C. Vascular Reactivity in Human Primary Hypertension and Congestive Heart Failure

I. The Technique, Function and Structure

Buttock skin (gluteal biopsy), removed under local anaesthetic, has become a useful tissue with which to explore the pharmacology of human resistance arteries. The biopsies can provide one or two small arteries (100–300 μm i.d.) of 2-mm length for mounting on parallel 40-μm stainless steel wires in a myograph.

AALKJAER and colleagues (1987) have perfected a method of stretching the vessel to a level of passive force equivalent to 100 mmHg transmural pressure. In addition to measuring isometric force changes in response to constrictor or dilator drugs or transmural, perivascular nerve stimulation, the morphology of the wire-mounted vessel can be determined by light microscopy. Under normal illumination, the border between the medial smooth muscle and the adventitia is clearly visible at a total magnification of 400×, as is the outer wire edge touching the intima. Multiple measurements by means of an ocular micrometer allow calculations of medial thickness (w) and lumen radius (r_i) to give the ratio w/r_i normalised to an equivalent transmural pressure of 60 mmHg.

This w/r_i ratio is used as an index of vascular hypertrophy when the artery is actively relaxed. A number of studies have addressed the issue of remodelling in the human vasculature by antihypertensive-drug therapy. These in vitro experiments offer a focus for controlled measurement of the efficacy of antihypertensive therapy at the site of the small resistance artery conveniently dissected from a small biopsy from buttock skin. Thus, the angiotensin-I converting enzyme inhibitor (ACEI) drug cilazapril, given over 2 years to patients with essential hypertension, restored the w/r_i from $7.5 \pm 0.3\%$ before treatment to normal values of $5.8 \pm 0.2\%$ in buttock skin resistance arteries, which was not different from the w/r_i value of $5.2 \pm 0.2\%$ in normotensive subjects (SCHIFFRIN et al. 1995).

Confirming the results of studies in rats, not all antihypertensive drugs of equivalent blood-pressure-lowering efficacy have the same effect on remodelling the vasculature. In the same study, patients treated with atenolol did not show structural remodelling in their skin resistance arteries despite the similar falls in arterial pressure. Values were: 146.4 ± 2.5 mmHg/99.8 ± 1.7 mmHg before cilazapril treatment; 130.2 ± 3.0 mmHg/85 ± 3 mmHg after cilazapril treatment; 151.3 ± 5.5 mmHg/99.4 ± 3 mmHg before atenolol treatment; and 131.5 ± 4.7 mmHg/83 ± 2.8 mmHg after atenolol treatment (SCHIFFRIN et al. 1995). Similarly, THYBO et al. (1994) treated patients with essential hypertension with the ACEI perindopril (4 mg/day) or atenolol (50 mg/day) for 1 year. Both treatments effectively lowered systolic and diastolic blood pressure (BP), but only perindopril significantly lowered the w/r_i ratio and increased the lumen diameter. These studies raise several issues.

II. Remodelling

It was established from model analysis (KORNER and ANGUS 1992) that small decreases in internal radius r_i accompanied by medial hypertrophy to give an increased w/r_i ratio are a much more "economical" way of normalising wall stress in cases of high BP than production of more wall material (increased wall thickness) by the artery while keeping the same r_i. During drug treatment, artery remodelling would presumably occur from both a decrease in wall material and an enlarged radius, both lowering w/r_i. What is intriguing is that, presumably, all the effective antihypertensive drugs decrease wall stress as transmural pressure falls, but long-lasting pressure attenuation and the opportunity for a drug-free lifestyle (that is, no rebound hypertension on ceasing therapy) will occur with drugs that also remodel the resistance arteries. The fact that ACEIs appear to structurally remodel the skin vessels suggests that ACEIs may have additional properties of significance to the vessel wall. Obvious candidates are the prevention of breakdown of bradykinin by the inhibition of converting enzyme and the release of NO and EDHF. These signals would not only cause local vasodilatation but could also slow smooth muscle cell proliferation, actions not shared with β-adrenoceptor antagonists.

Adaptation of vascular smooth muscle-changing haemodynamics, such as flow and pressure, is now a widely accepted phenomenon (ANGUS 1994). An increase in flow and sheer stress would be detected by the endothelium, causing *outward* growth, increased internal radius and, perhaps, hypertrophy (Fig. 12E, lower right; LANGILLE 1993). Normal age-dependent growth of the vasculature to keep pace with body size would cause a rise in cardiac output and adaptation of the distributing arteries, with increased lumen (r_i) and wall

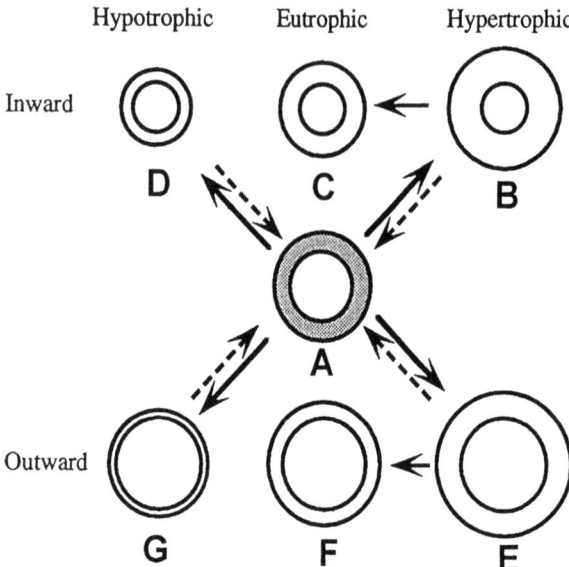

Fig. 12. This diagram is modeled after Mulvany and colleagues (1996). As a starting point, the cross-section of a blood vessel (*A*) is at the centre (*shaded*). If the cross-sectional area were doubled, as in hypertrophic remodelling, the vessels would be as depicted in the right column (vessels *B*, *E*); if the area were halved, as in hypotrophic remodelling, the vessels would be as depicted in the left column (vessels *D*, *G*). If there is a 30% reduction in lumen diameter, inward remodelling occurs (*top row*); if there is a 30% increase in lumen diameter, outward remodelling occurs (*bottom row*). If further remodelling towards the starting vessel occurs by growth or apoptosis, the resultant vessel may have the same cross-sectional area as at the start, but with a smaller (*top*) or larger (*bottom*) lumen diameter. This is eutrophic remodelling. The *arrows* indicate likely changes in the appearance (KORNER and ANGUS 1997)

thickness (w) as BP (and wall stress) rises. Normally, there would be increased vascular length (KORNER and ANGUS 1992). If the hypertrophic, outwardly remodelled artery lost wall volume, returning to the starting volume, the vessel would have undergone eutrophic remodelling around an increased r_i (Fig. 12F, lower middle panel). Finally, in arteries subjected to pressure load from hypertension (increased wall stress), the lumen would narrow, and wall thickness and volume would increase to restore normal wall stress (Fig. 12B, upper right). Under drug treatment (ACEI), the artery could remodel to normal (Fig. 12A, centre) or remain with reduced r_i (Fig. 12C, upper middle). Debate has arisen over whether the normal artery could remodel directly to the eutrophic vessel (Fig. 12A–C; MULVANY et al. 1996). This could occur by de novo rearrangement of the same wall volume by inward growth and outward apoptosis, an event inconsistent with early development. We favour a sequence in which, in hypertension, arteries remodel to the hypertrophic-inward model (Fig. 12A, B) before hypertrophy could regress to Fig. 12C or back to Fig. 12A under drug treatment (KORNER and ANGUS 1997). Thus, measuring r_i and medial thickness in relatively few 2-mm segments of buttock skin resistance arteries, if done with utmost care at normalised transmural pressure in patients with essential hypertension, is fraught with experimental error from sampling, technical difficulties of accurately measuring the medial–adventitial border and extrapolation from skin vessels to the rest of the circulation.

III. Endothelial Dysfunction

The discovery of EDRF prompted scientists to examine whether endothelial dysfunction could explain the cause of hypertension. Many studies in laboratory animals with experimental hypertension induced by coarctation of the aorta (LOCKETTE et al. 1986), desoxycorticosterone acetate salt models, or one-kidney one-clip (VAN DE VOORDE and LEUSEN 1986) or genetic hypertension in rats (LÜSCHER and VANHOUTTE 1986) have shown that isolated thoracic aortae in organ baths relaxed poorly in response to acetylcholine compared with the relaxation exhibited by normotensive control aortae. These findings from relatively robust experimental models concur with human forearm plethysmography studies in which PANZA and colleagues (1990) reported that the increase in forearm blood flow following intra-arterial acetylcholine infusion was attenuated in patients with hypertension, while responses to the non-endothelium-dependent dilator nitroprusside were unaltered. Similarly, in the coronary circulation, acetylcholine responses were blunted in patients with essential hypertension (TREASURE et al. 1992).

However, other studies show that one has to be cautious in describing this attenuated response as evidence for endothelial dysfunction shown indirectly and presumed to be due to a loss of NO and EDHF. For example, prostacyclin, contracting factors (such as superoxide anion), thromboxane A_2/prostaglandin H_2 and endothelin-1 are also released from the endothelium (IMAOKA et al. 1999). Furthermore, some authors have reported that the length of the forearm (and presumably the destruction of intra-arterial acetylcholine as a result of cholinesterase) has a strong negative correlation with vasodilatation in response to acetylcholine (CHOWIENCZYK et al. 1994). Importantly, COCKCROFT et al (1994) found that, in an extensive study of essential hypertensives, they could find no evidence for endothelial dysfunction. Ageing, period and level of hypertension and underlying hypercholesterolaemia could alter the endothelial response and/or the reactivity of the underlying smooth muscle to NO, EDHF or other local factors.

We attempted to assess the endothelial functional response to acetylcholine in buttock skin biopsies from volunteer patients with untreated primary hypertension. We divided the arteries from the biopsies into "large" and "small" at the arbitrary cut-off diameter of 500 μm i.d., the diameter measured when normalised at the equivalent transmural pressure of 100mmHg (ANGUS et al. 1992). Eight patients (mean age 48 years, mean supine BP 116.5 ± 2.5mmHg) and five normotensive volunteers (mean age 50 years, mean

BP 95.2 ± 1.5mmHg) had small vessels of 231.1 ± 23.9µm i.d. ($n = 6$) in the normotensive and 315.8 ± 40.7µm i.d. ($n = 12$) in the hypertensive patient groups. The arteries were contracted to a submaximal steady force by the thromboxane mimetic U46619 (10–30nM) and were stimulated with acetylcholine. We found that all the small arteries (<500µm i.d.) relaxed to acetylcholine with the same range and sensitivity in the two patient groups (Fig. 13).

In vessels where logistic curves could be fitted, the E_{max} relaxation and EC_{50} values for acetylcholine were 78.8 ± 13.3% and 7.29 ± 0.19 (*p*log M) for normotensive (N) and 80.5 ± 10.9% and 7.05 ± 0.23 *p*log M for hypertensive (H) arteries. In the larger arteries (>500µm i.d.), no relaxation or contraction to acetylcholine was observed despite the relaxation in response to substance P (10nM; Figs. 13, 14). The morphometry of these arteries showed a slightly greater medial thickness in H arteries (26.9 ± 2.6µm) compared with the N arteries (21.3 ± 2.9µm), presumably because of the larger diameters of the H arteries. This gave a wall thickness/lumen diameter ratio of 9.2% for N and 9.05% for H arteries. In the large arteries, we were unable to measure medial thickness by the water-immersion-lens technique. In other measurements of

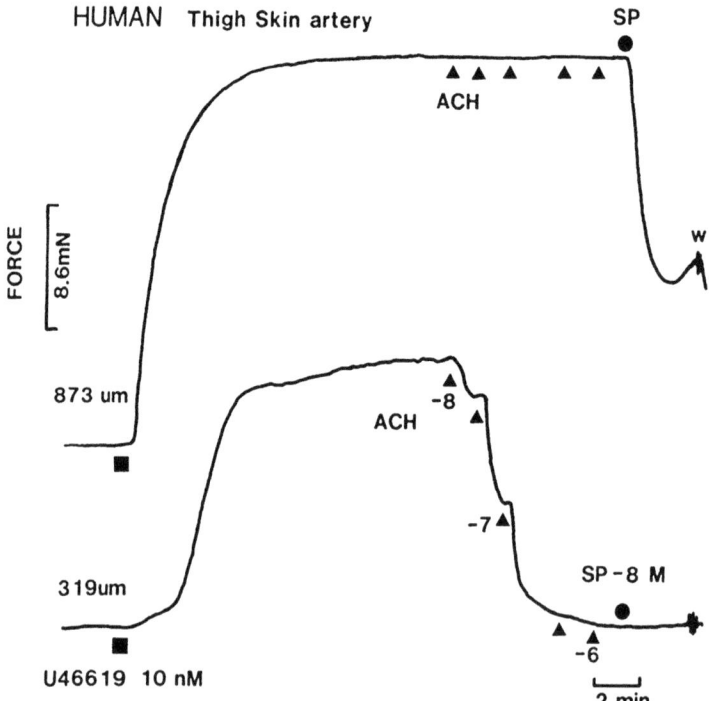

Fig. 13. Myograph chart records of two arteries removed from a buttock skin biopsy. The simultaneous records show that the submaximal contraction in response to U46619 was inhibited by acetylcholine ACH (log M) only in the smaller artery, which had a 319µm internal diameter. Substance P (*SP*, 10nM ●) relaxed the larger artery, indicating the presence of endothelial cells

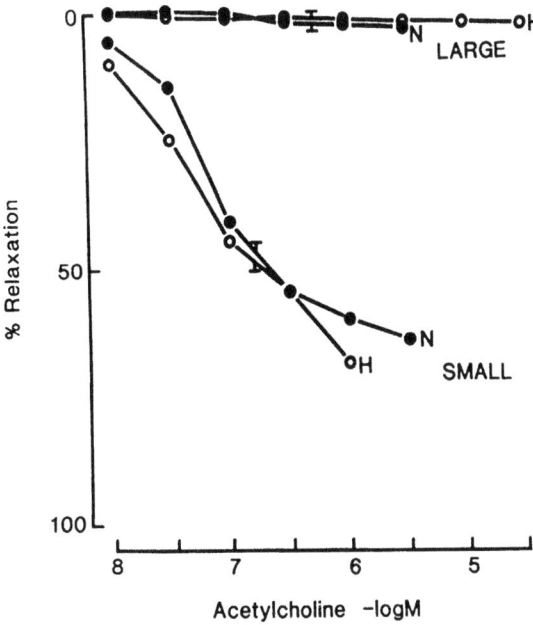

Fig. 14. Group data showing the relaxation of human isolated buttock skin resistance arteries in response to acetylcholine (–log M) in a Mulvany myograph. Six arteries from five normotensive (N, ●) and 12 arteries with internal diameters <500 µm (small) from eight patients with primary hypertension (H, ○) responded to acetylcholine, while larger arteries (>500 µm internal diameter; $n = 5$ for both the normal and hypertensive groups) failed to relax in response to the endothelium-dependent relaxant. Arteries were pre-contracted by U46619 (0.01–0.03 µM)

vasoconstrictor agents, we showed that the E_{max} (but not the EC_{50}) for NA, 5-HT and angiotensin II was significantly greater in H small arteries compared with N small arteries (ANGUS et al. 1991c).

These enhanced contractility responses were not observed in the larger arteries. Thus, the smaller subcutaneous resistance arteries may behave as pharmacological amplifiers in hypertensive circulation. This study again shows how a change in location of only one branch order in a resistance bed (250–500 µm i.d.) can apparently cause quite different pharmacology. To find *no* apparent functional receptors of acetylcholine on the smooth muscle or endothelium in the slightly larger arteries is particularly striking. Finally, in this study, as in some forearm plethysmography studies, we could find no evidence of endothelial dysfunction in human essential hypertension.

IV. Forearm Veins in Primary Hypertension

Venous distensibility measured by forearm plethysmography is decreased in essential hypertension (WALSH et al. 1969). The cause is not known, but we reasoned that vein biopsies and in vitro pharmacological studies could shed some insight into venous reactivity in hypertensive patients, where the veins would not have been exposed to increased pressure.

We recruited 15 volunteers with untreated primary hypertension and 14 normotensive subjects and removed a 1-cm-long segment of forearm vein close to the wrist under local anaesthetic (SUDHIR et al. 1990). In organ chambers, the passive length–tension relationship was determined by micrometer-controlled increments in circumference stretches of the vein ring segment while measuring wall tension. The veins from hypertensive subjects (BP = 156mmHg/103mmHg) were significantly stiffer, i.e. less compliant, than those from normotensive subjects (BP = 125mmHg/81mmHg). The internal diameters were 1.49 ± 0.42mm and 1.07 ± 0.31mm for N and H veins, respectively, and there was no significant difference in medial thickness or w/r_i ratio measured morphologically by fixing the tissue in the organ chamber under normalised stretch conditions.

On cumulative concentration–response curves, there was no difference between N and H veins with regard to sensitivity or E_{max} in response to K$^+$ or 5-HT, but H veins were significantly less sensitive to NA and the selective α_2-adrenoceptor agonist UK14204, but not methoxamine. However, the E_{max} resulting from all three α-adrenoceptor agonists was significantly decreased in H veins (Fig. 15). Despite this lower reactivity to α-adrenoceptor agonists, there was an enhanced contraction to transmural nerve stimulation in H veins and a suggestion of a decrease in neuronal uptake. Interestingly, there was a large increase in the range of contraction resulting from angiotensin II in H veins.

These in vitro studies have established (1) that forearm vein biopsies are a valuable aid to the detection of enhanced neural responses at the effector site and (2) that there are constrictor-selective changes in reactivity. It would be of great interest to repeat these studies in patients receiving treatment who experience normalisation of their hypertension.

V. Chronic Heart Failure

CHF of ischaemic origin is characterised by activation of various neurohumoral compensatory mechanisms to maintain tissue perfusion and peripheral vascular tone. NA is 163% higher in CHF-patient arterial plasma, due to an increase in spillover and reduced clearance (HASKING et al. 1986). In CHF patients, forearm plethysmographic studies show normal responses to hyperaemia and dilatation in response to the α_2-adrenoceptor antagonist yohimbine (KUBO et al. 1989). We set out to compare the in vitro reactivity and sensitivity of small resistance arteries from gluteal biopsies from patients with CHF (ANGUS et al. 1993). The six patients (mean age 65 years) had stable, long-term (>5 years) CHF of ischaemic origin with left and/or right ventricular ejection fractions less than 55% (New York Heart Association class II–III on treat-

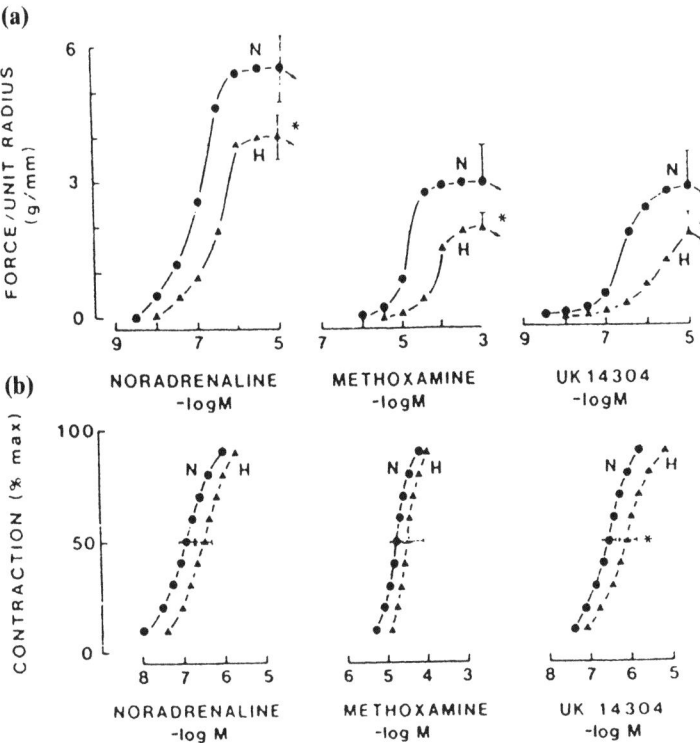

Fig. 15. a Average concentration-response curves to noradrenaline, methoxamine and UK14304 in vein segments from normal (N, ●) and hypertensive (H, ▲) subjects, showing a reduced maximum contractile force (F_{max}) in response to each α-adrenoceptor agonist. *Vertical error bars* are ± one standard error of the mean at normalised F_{max}. **b** Average concentrations corresponding to 10%–90% of the maximum response (EC_{50-90})

ment). All patients were brought into the hospital and had their medication (including digoxin, diuretics, ACEIs and nitrates) withdrawn 72 h prior to biopsy. For our control group, we had nine untreated, healthy volunteers (mean age 54 years). Myograph studies showed that the CHF arteries ($n = 6$; mean 294 ± 47 μm i.d.) contracted to only 65% of the maximum response to K^+ (124 mM), NA (1 μM) or both together, as measured in nine arteries from normals (N) (296 ± 28 μm i.d.; Fig. 16). This marked loss of E_{max} contractility in CHF arteries was observed for NA and angiotensin I and II concentration-response curves. Of great interest was the finding that the endothelium-dependent agonist acetylcholine was almost without effect in CHF arteries contracted by NA 1 μM, while N arteries relaxed to nearly 100%, with an EC_{50} of 7.55 plog M (Fig. 17). Calcitonin-gene-related peptide was less sensitive in CHF compared with N arteries, while the sensitivities and ranges of sodium nitroprusside (SNP) relaxation curves were superimposable in the two groups.

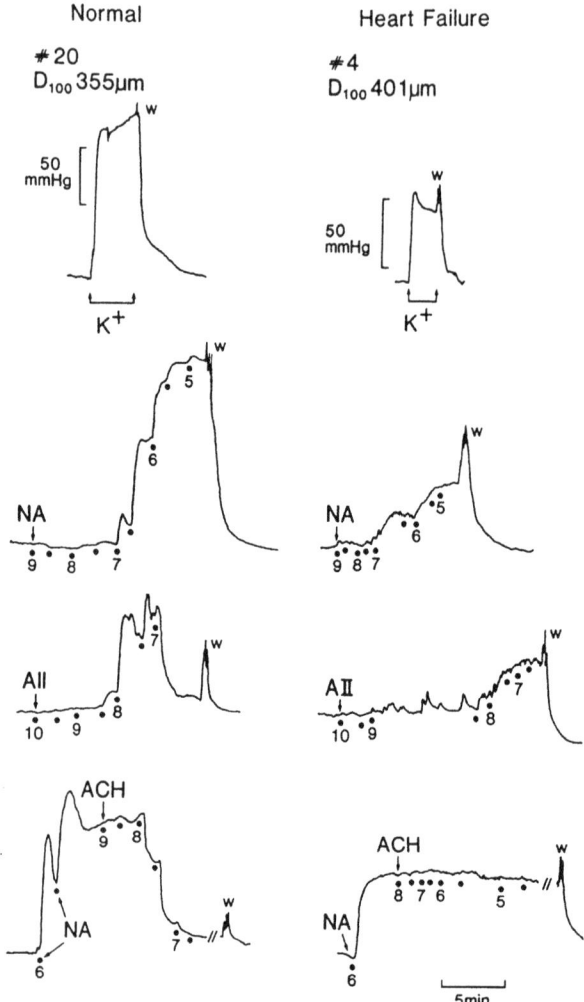

Fig. 16. Representative chart records of the isometric force from two skin small arteries from a normal volunteer (*left*) and a patient with chronic heart failure (*right*). Traces should be read in sequence from top to bottom; w, washout with drug free solution; D_{100}, the internal diameter of each vessel at the normalised transmural pressure of 100 mmHg. The isometric contraction is given in units of developed transmural pressure (mmHg). *Traces* indicate exposure to: K^+ (124 mM), noradrenaline (*NA*), angiotensin II (*AII*) and acetylcholine (*ACH*) in the presence of pre-contraction by 1 μM NA for the heart-failure vessel and 3 μM for the normal vessel. *Numbers* refer to –logM concentration, and *each unnumbered dot* indicates a 0.5-unit increase in concentration

These studies are limited by the difficulties associated with very sick patients and, thus, a low number of experiments. However, conclusions are that these gluteal resistance arteries contract poorly to a range of constrictor agents and may have an endothelial dysfunction. This "exhaustion" of contractile function could be caused by overexposure to circulating neurohumoral sub-

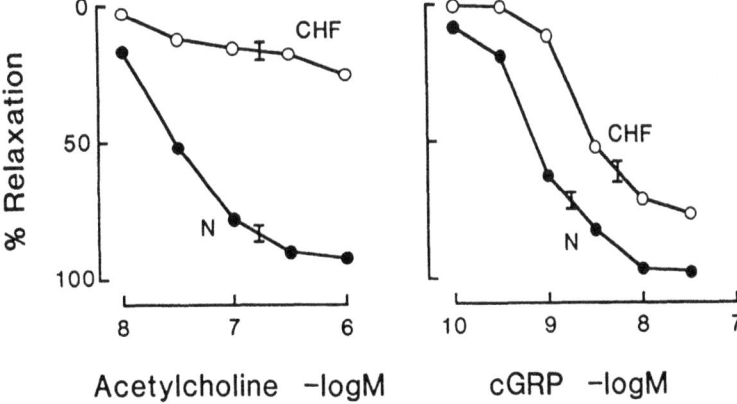

Fig. 17. Concentration–relaxation curves to acetylcholine (*left*) and calcitonin gene-related peptide (cGRP; *right*) in buttock skin small arteries. Relaxation was measured as a percent relaxation of the vessel pre-contracted by noradrenaline (1 μM). For acetylcholine, there were nine vessels in the normal (*N*) group and six in the chronic heart failure (*CHF*) group; for cGRP, there were nine N vessels and four vessels in the CHF group

stances, but we did not observe sensitivity changes. CHF could involve metabolic abnormalities in vascular smooth muscle and endothelium, reflecting hypoxia from low cardiac output – in vivo conditions quite different from the myograph chamber, where the pO_2 is high. The wire-mounting procedure destroys less than 30% of the endothelium and, as shown in the N arteries, the vessels can still relax to 100% of the maximum contraction. The smooth muscle reactivity to the NO donor SNP was normal in CHF, suggesting a loss of muscarinic receptors or that the NOS energy-requiring process is severely affected by CHF. The pattern of vascular dysfunction found in these skin small arteries from patients with CHF may point to a more general derangement in the peripheral circulation. It would be of great interest to examine biopsy vessels from the same patients after cardiac transplantation and recovery. The ethical issues of taking such vessels during ongoing immunosuppressive therapy, etc., and the risk of infection make such a study impossible to conduct.

D. Pharmacology of Vascular Conduits for Coronary-Bypass Graft Surgery

I. Introduction

Patients with severe angina caused by near-occlusive atherosclerotic plaques in large coronary arteries are now routinely grafted down stream of the occlusion by arterial grafts or venous conduits. In brief, the IMA is freed from the chest wall in its surrounding tissue, and the distal end is anastomosed (end to side) to the coronary artery. For vein grafts, lengths of the saphenous vein are harvested from the leg and anastomosed to the ascending aorta and coronary artery as a coronary jump graft. Newer arterial jump-graft procedures include the radial, gastroepiploic and inferior epigastric arteries. The arterial grafts have a longer patency rate than vein grafts. The occlusion rate of saphenous vein grafts in the first year is 10%–26% but, by 10 years after the procedure, 50% of the grafts are occluded (GRONDIN et al. 1984). In contrast, the IMA grafts spasm perioperatively, but the long-term patency rate is far superior to that for vein grafts (LYTLE et al. 1985). Saphenous veins spasm during harvesting because of surgical trauma, the tying off of side branches and testing for leaks using high pressure (up to 700 mmHg) from a syringe, which destroys much of the intima and media (RAMOS et al. 1976). In the longer term, these grafts take up lipid and remodel with severe intimal hypertrophy, with resultant reduction in lumen patency. The arterial grafts, however, spasm at the distal end at operation, probably because of spilt blood and surgical trauma. This makes anastomosis difficult and compromises coronary flow in the short term. We set out to develop a simple, pharmacological method to prevent spasm of both vein and arterial grafts prepared for bypass grafting.

II. Internal Mammary Artery

At surgery, we collected discarded distal end pieces of IMA and mounted them as 3-mm-long ring segments on wires in organ baths for isometric force recordings (HE et al. 1989). These segments were most sensitive and reactive to U46619, followed by NA, 5-HT and PE (Fig. 18). In submaximally contracted rings with K^+ or U46619, glyceryl trinitrate (GTN) and SNP caused full relaxation while, in K^+-depolarised rings, nifedipine, verapamil and diltiazem caused full relaxation (Fig. 19). If the vessels were pretreated with GTN, subsequent curves resulting from U46619 or K^+ were unaffected. In contrast, pretreatment with the calcium antagonist nifedipine abolished subsequent contractions in response to K^+ and markedly reduced the response curve to U46619.

These experiments suggested that treatment with GTN in the setting of spasm would be effective in relaxing the vessel by presumably (1) raising cyclic guanosine monophosphate release from NO inside the smooth muscle and (2) sequestering calcium. However, voltage-operated calcium-channel (VOCC) antagonists would be effective in reducing *subsequent* stimulation by K^+ or U46619. In the light of these laboratory studies, we developed a vasodilator solution (glyceryl trinitrate and verapamil; GV solution) containing verapamil (5 mg), GTN (2.5 mg), sodium bicarbonate (0.2 ml) and Ringers' solution (300 ml) to give a final concentration of both GTN and verapamil of $30\,\mu M$ at pH 7.4. These concentrations of GTN and verapamil were sufficient to maximally

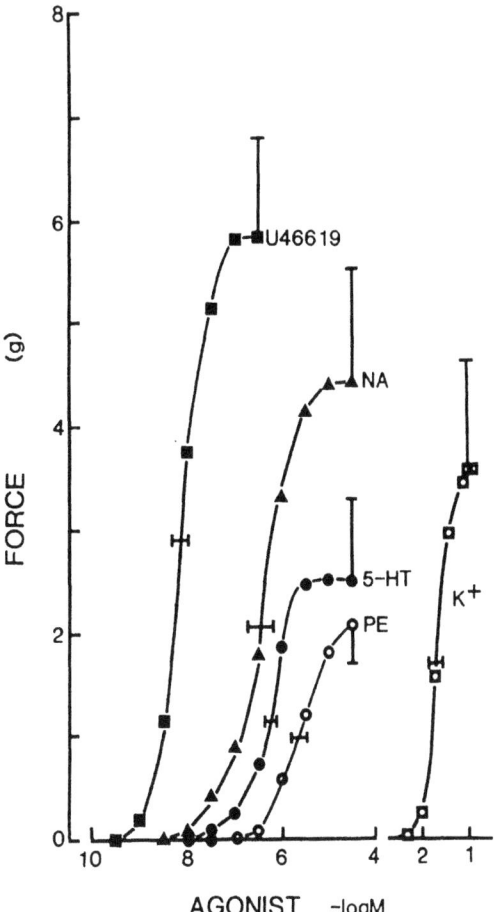

Fig. 18. Group data showing constrictor concentration (–logM)-response curves to thromboxane A_2 mimetic (*U46619*, ■), noradrenaline (*NA*, ▲), phenylephrine (*PE*, ○), serotonin (*5-HT*, ●), and K^+ (□) in human internal mammary artery segments. *Points represent mean increase in force (g) at fixed concentrations of agent. Horizontal error bars are one standard error of the mean for the effective concentration causing a 50% maximal response, averaged from logistic fitted curves from each ring. Vertical bars are error bars for the maximal response.* Numbers of rings are eight (from seven patients) for K^+ and six (from six patients) for all other constrictor agents

relax either K^+ (25 mM) or U46619 (15 nM) pre-contracted human IMA by more than 80% over 5 min for K^+-pre-contracted rings or over 25 min for U46619-pre-contracted IMA rings (Fig. 20). In a clinical trial in the operating theatre, intraluminal injection of the GV solution into one IMA caused an increase in flow of 95% above basal levels (by timed collection), compared with only a 53% increase following injection of Ringers' solution in the contralateral IMA. In a second study comparing papaverine (0.8 mM pH 4.4–4.8)

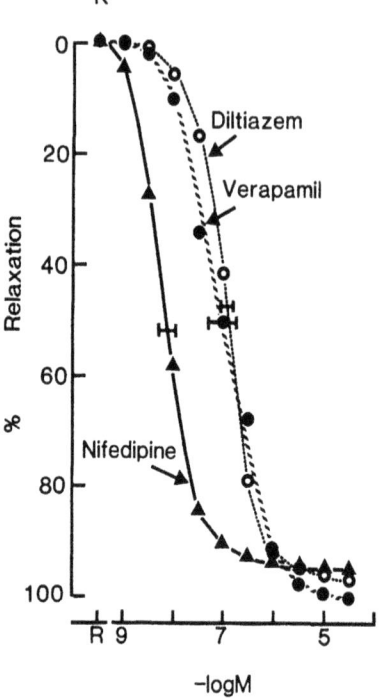

Fig. 19. Mean concentration (–logM)-response (percent relaxation) curves for nifedipine (▲), verapamil (●), and diltiazem (○) in internal mammary artery rings pre-contracted with 25 mM K⁺. *Symbols* represent data averaged from six rings (six patients). *Horizontal error bars* are one standard error of the mean for the effective concentration causing 50% of the maximal response. R, resting value

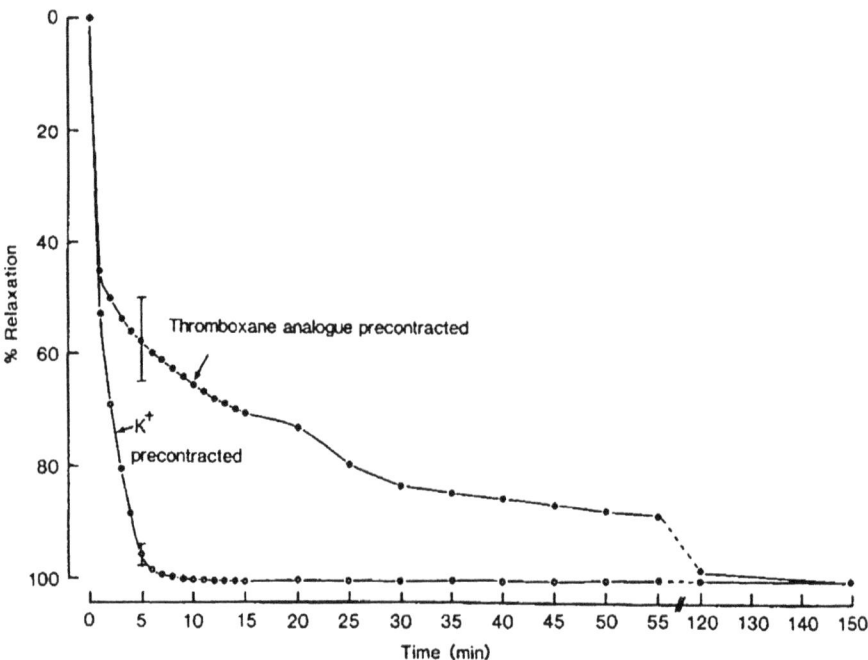

Fig. 20. Average relaxation responses to glyceryl trinitrate (30 μM) plus verapamil (30 μM) in human internal mammary artery segments pre-contracted by potassium (25 mM) or thromboxane A_2 mimetic U46619 (15 nM). *Points* represent average data from six vessels from six patients

with the GV solution, both treatments were equivalent, raising flow from baseline by 107% for GV and 80% for papaverine. Given that papaverine solution at acidic pH can destroy endothelium (CONSTANTINIDES and ROHMSON 1969) and a buffered, pH-7.4 solution of papaverine is unstable, we suggested that the GV irrigation solution was an effective perioperative relaxant solution and is the solution of choice for preventing or treating perioperative spasm of the IMA.

III. Saphenous Vein

Discarded, non-distended, 1- to 2-cm lengths of saphenous vein not required for grafting were suspended as 3-mm-long ring segments under isometric conditions in organ chambers. After normalisation of passive stretching to 20-mmHg transmural pressure, these veins contracted powerfully in response to 5-HT, with sensitivities in the order: U46619 > 5-HT > NA > PE (Fig. 21). In veins pre-contracted by K^+ or U46619, GTN caused rapid 100% relaxations, with EC_{50} values of $6.48\, p\log M$ and $6.07\, p\log M$, respectively. Verapamil caused a slower onset of relaxation to 100% in K^+-pre-contracted veins but only to 75% in U46619-pre-contracted veins.

As for the IMA, the combination of verapamil and GTN offered synergistic properties of sustained relaxation (verapamil) and rapid onset of action (GTN) for both the receptor-operated agonist (U46619) and K^+ depolarisation (HE et al. 1993). Similarly, when PE, the stable α_1-adrenoceptor agonist, was used to pre-contract the saphenous vein, the GV cocktail showed much efficacy (Fig. 22; ROSENFELDT et al. 1993).

In surgical practice, the GV solution is now perfused intraluminally and applied topically as the vein is cannulated from the distal end (Fig. 23). The cocktail should protect against spasm, lessening the need for use of damaging, highly distending pressures and subsequent failure of long-term patency. The precise nature of the cause of saphenous vein or IMA spasm is unknown. However, theoretically (and now in practice) we have shown that the GV solution is an effective prophylactic, spasmolytic treatment based on the known human vascular pharmacology of GTN and verapamil.

E. Human Vascular-to-Cardiac Tissue Selectivity of L- and T-Type VOCC Antagonists

A major issue in analytical pharmacology is how to measure drug action in two or more tissues that are highly dependent on assay conditions. For antagonists, the null-measure approach has stood the test of time. Here, the robust Schild plot (ARUNLAKSHANA and SCHILD 1959) or, more recently, the approach of global iterative fitting of a family of concentration–response curves is used to obtain the K_B value of the antagonist (STONE and ANGUS 1978; LEW and

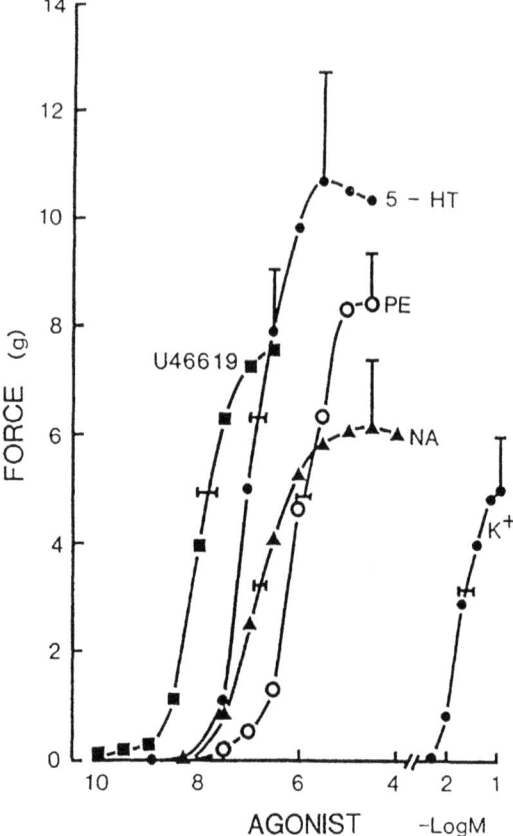

Fig. 21. Group data showing concentration–contraction curves to U46619, serotonin (*5-HT*), noradrenaline (*NA*), phenylephrine (*PE*) and potassium (*K⁺*) in human saphenous vein segments. *Horizontal error bars* are ± one standard error of the mean at the EC_{50} averaged from logistic fitted curves for each ring. Each *line* represents more than six vessels

Angus 1995); this is independent of agonist or tissue. It is a unique value of the concentration of antagonist; it will always right shift the agonist concentration–response curve with a concentration ratio of two. With VOCC antagonists, the option of displacing an agonist that opens the channel by increasing the concentration of an antagonist is not available. Therefore, scientists have relied upon measuring the concentrations of VOCC antagonists that decrease the response to channel activation by some stimulus in a particular tissue by 50% (EC_{50}). The key question is "what is the *best* starting response?"

We turned our attention to the problem of comparing a series of drugs classified as "L-" (long-) type VOCC antagonists that relax vascular tissue and decrease myocardial contractility. For the last 30 years, these long-opening

PRECONTRACTED BY PHENYLEPHRINE

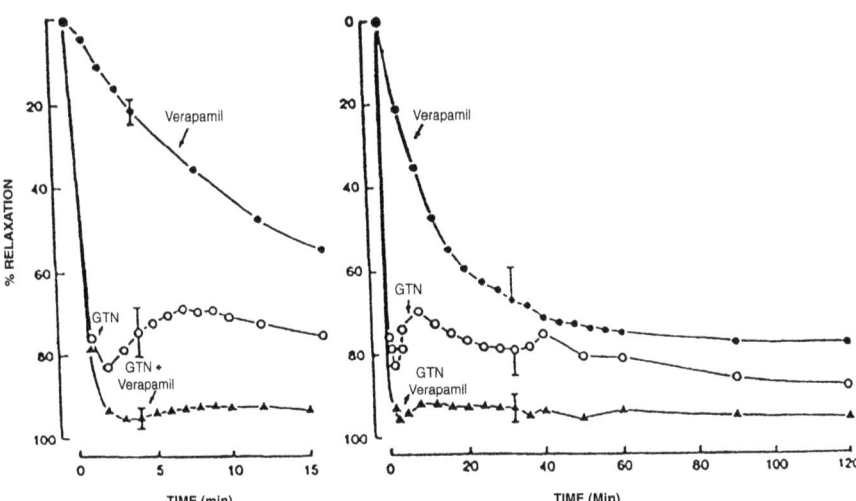

Fig. 22. Average relaxation response to glyceryl trinitrate (10 μM) and verapamil (10 μM), singly and in a combination in rings of saphenous vein pre-contracted with phenylephrine (5 μM). Relaxation was measured at 1-min intervals and expressed as a percentage of pre-contraction. *Left panel*: enlarged time scale over 15 min; *right panel*: responses up to 2 h

calcium channel antagonists (developed from three different chemical classes) have been used in the treatment of angina, hypertension, heart failure, supraventricular tachycardia and after myocardial infarction. Of recent interest were second-generation dihydropyridine derivatives from the parent molecule nifedipine; these were more vascular- than cardiac-selective, giving rise to the advantage of peripheral afterload reduction at doses that did not depress myocardial contractility. Most recently, a tetralol derivative, mibefradil (derived from the non-vascular- to cardiac-selective VOCC antagonist verapamil) was described; it is selective for T-type (rather than L-type) VOCC and displayed vascular to cardiac selectivity (CLOZEL et al. 1991).

A major difficulty in this area is trying to compare EC_{50} values for a few widely different VOCC antagonists in animal tissue and extrapolating the findings to man. We addressed this issue by taking four L-type VOCC antagonists and mibefradil and comparing the EC_{50} values of each of the five drugs in cardiac and vascular in vitro preparations removed from patients. We reasoned that the assays would give EC_{50} estimates that were at least internally robust and equivalent so that vascular- to cardiac-selectivity ratios could be compared over a wide range of compounds from the same laboratory.

Human right-atrial trabeculae muscle strips ($n = 1$–11) were prepared from each right-atrial appendage removed from patients undergoing heart–

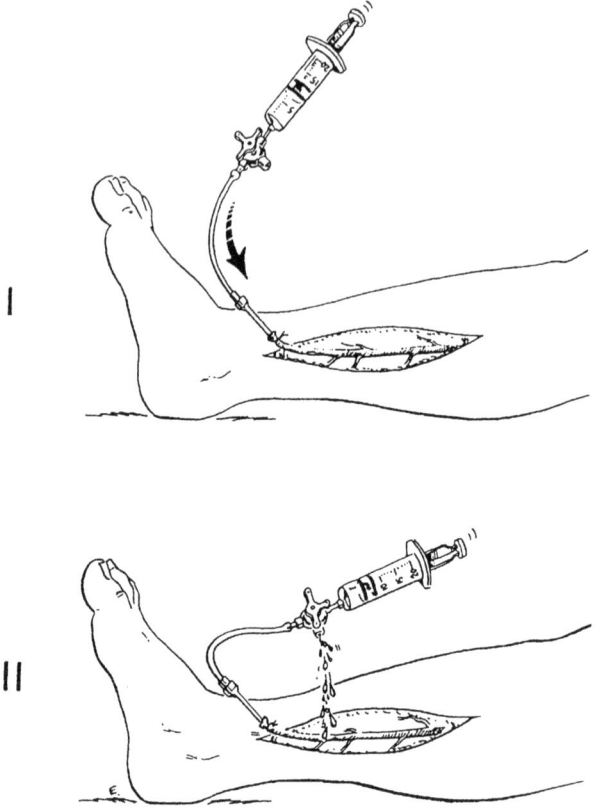

Fig. 23. Method of injecting the glyceryl trinitrate and verapamil solution in the lumen of the saphenous vein (*I*) and spraying it on to the surface of the vein (*II*)

lung bypass (SARSERO et al. 1998). These strips, less than 1 mm in diameter, were contracted at 1 Hz via suprathreshold stimulation from field electrodes. We had 16 organ chambers, allowing the advantage of testing only a single concentration of VOCC antagonist after allowing the submaximal inotropic response to isoprenaline (6 nM) to stabilise. Individual responses from 6–8 tissues at 6–8 concentrations from five different VOCC antagonists gave a family of concentration-negative inotropic response curves that were logistically fitted to provide estimates of EC_{50} values after 2-h incubation periods (Fig. 24).

For vascular responses, we chose a novel preparation of human aortic vasa vasorum. Arteries were dissected from the 1- to 2-cm-diameter patch of vasorum removed by the surgeons as they prepared the aorta for end-to-side anastomosis of the saphenous graft. From each patch of vasorum, we obtained from one to four 2-mm-long small resistance arteries; these were mounted in the Mulvany myograph. After normalisation, the vessels were activated by K^+ depolarisation by 50% K^+ Krebs' solution (KPSS; 62 mM K^+ with 62 mM Na^+)

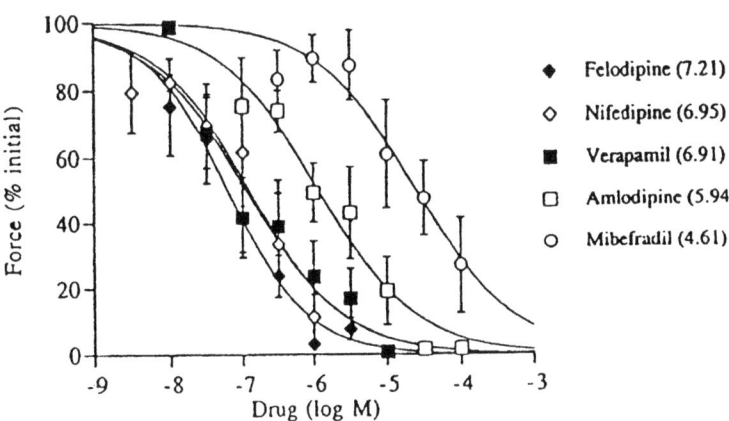

Fig. 24. Average inotropic responses to five calcium-channel antagonists in human isolated right atrial trabeculae. Each average value and standard error of the mean is the inotropic force at 2 h for 6–8 tissues in the presence of calcium antagonist. This force is calculated as the percentage of the initial response to a sub-maximal concentration of isoprenaline (6 nM). Note that only one concentration of calcium antagonist was applied per tissue. The values in *brackets* are the pIC_{50} values ($-\log IC_{50}$), which were calculated by fitting a logistic equation

for 2 min every 30 min. The contraction was terminated by washing with K^+-free solution. In this assay, we obtained EC_{50} values from cumulative addition of a VOCC antagonist as it decreased the peak contraction to sequential exposures to 50% KPSS (Fig. 25).

By taking the ratio of EC_{50} values for vasa vasorum (vascular) and negative inotropic responses (cardiac), we found the following ratios: mibefradil 41, felodipine 12, nifedipine 7, amlodipine 5 and verapamil 0.22. Thus, verapamil was fivefold more cardiac selective than it was at relaxing vascular smooth muscle. If this result is normalised to one, then mibefradil is 200 times more vascular selective than verapamil. These figures suggest that mibefradil is more vascular selective than the dihydropyridines felodipine, nifedipine and amlodipine and could be a result of T-type-VOCC selectivity.

These studies highlight the difficulties in assigning selectivity ratios to a range of drugs. The stimulus chosen to activate the smooth-muscle K^+ depolarisation, probably has no equivalent stimulus in vivo, and we may have measured different EC_{50}s (and thus ratios) if a receptor-operated agonist (such as methoxamine, endothelin-1, etc.) had been used to activate the tissue.

Similarly, for the cardiac tissue, we could have used the suprathreshold basal contraction as the starting point rather than β-adrenoceptor activation by isoprenaline. These choices raise the prospect of doing more extensive experiments. The scarcity of human tissue precludes comparative protocols, and one has to resort to animal tissue to test these questions (SARSERO et al. 1998). In a study similar to ours, Brixius et al. (1998) chose cumulative

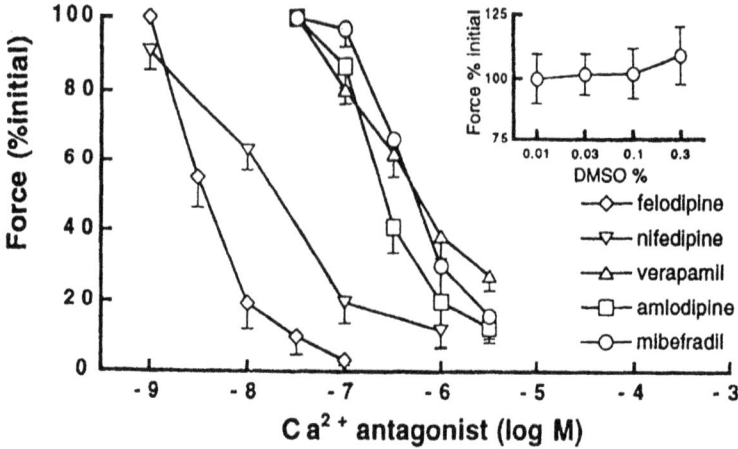

Fig. 25. Average contraction responses of human isolated arteries from aortic vasa vasorum exposed to K^+ (62 mM) in the presence of increasing concentrations of five calcium antagonists. *Symbols* are average responses (±one standard error of the mean) at specific concentrations generated from cumulative curves within artery. Responses were calculated as percentage of initial 2-min exposure to 62 mM K^+. *Inset* shows average effects of dimethyl sulfoxide cumulative concentrations on the contraction in response to 62 mM K^+ ($n = 5$ arteries)

concentration–response curves of mibefradil, nifedipine and diltiazem to inhibit contractions resulting from suprathreshold field stimulation of right-atria trabeculae muscle and left-ventricular papillary muscle from explant hearts with dilated cardiomyopathic heart failure. For their vascular tissue, they used large proximal left-anterior descending or circumflex coronary arteries from explant hearts pre-contracted by the receptor agonist prostaglandin $F_{2\alpha}$. They used concentrations of the three drugs that caused 25% relaxation of the coronary arteries and 25% negative inotropic response in atrial or ventricular tissue. They found that mibefradil was 316-fold vascular selective, compared with selectivities of 1.5 for nifedipine and 1.0 for diltiazem.

These studies serve to illustrate that the complexity of objective measurement of tissue selectivity is dependent on the choice of assay and tissue conditions. The incentive to using human tissue assays has to be balanced by the difficulties of robust quantitative analysis.

In summary, my experience with human in vitro tissue assays is that (1) the question needs to be very well defined before embracing clinical material; (2) sampling is never ideal, as factors of age, disease, premedication, site of biopsy, surgical trauma, transport to the laboratory, etc., must compromise to some degree the quality of the data; and (3) it is best, if one can, to run an experimental laboratory-animal assay in parallel with the human-tissue assay to ensure that the protocols are robust, with appropriate time or vehicle controls.

References

Aalkjaer C, Heagerty AM, Petersen K, Swales JD, Mulvany MJ (1987) Evidence for increased media thickness, increased neuronal amine uptake, and depressed excitation-contraction coupling in isolated resistance vessels from essential hypertensives. Circulation Research 61:181–186

Angus JA (1994) Arteriolar structure and its implications for function in health and disease. Current Opinion in Nephrology and Hypertension 3:99–106

Angus JA, Cocks TM (1996) Pharmacology of human isolated large and small coronary arteries. In The pharmacology of vascular smooth muscle. ed. C Garland and JA Angus. Oxford University Press, UK, pp 276–305

Angus JA, Lew MJ (1992) Interpretation of the acetylcholine test of endothelial cell dysfunction in hypertension. Journal of Hypertension 10 (Suppl 7):S179–S186

Angus JA, Wright CE, Cocks TM (1989) Vascular actions of serotonin in large and small arteries are amplified by loss of endothelium, atheroma and hypertension. In Serotonin: actions, receptors, pathophysiology. ed. E Mylecharane, JA Angus, I de la Lande and PPA Humphrey. Macmillan Press, UK, pp 225–232

Angus JA, Broughton A, Cocks TM, McPherson GA (1991a) The acetylcholine paradox: a constrictor of human small coronary arteries even in the presence of endothelium. Clin Exp Pharmacol Physiol 18:33–36

Angus JA, Broughton A, McPherson GA (1991b) Membrane potential and contractility responses to acetylcholine and other vasoconstrictor stimuli in human small coronary arteries. In: Resistance arteries, structure and function. ed. MJ Mulvany, C Aalkjaer, AM Heagerty, NCB Nyborg and S Trandgaard. Elsevier (Int. Congress Series), Amsterdam, pp 255–260

Angus JA, Jennings GL, Sudhir K (1991c) Enhanced contraction to noradrenaline, serotonin and nerve stimulation but normal endothelium-derived relaxing factor response in skin small arteries in human primary hypertension. Clinical and Experimental Pharmacology and Physiology 19 (Suppl1):39–47

Angus JA, Dyke DC, Jennings GL, Korner PI, Sudhir K, Ward JE, Wright CE (1992) Release of endothelium-derived relaxing factor from resistance arteries in hypertension. Kidney International 41:S73–S78

Angus JA, Ferrier CP, Sudhir K, Kaye D, Jennings GL (1993) Impaired contraction and relaxation in skin resistance arteries from patients with congestive heart failure. Cardiovascular Research 27:204–210

Arunlakshana O, Schild HO (1959) Some quantitative uses of drug antagonists. British Journal of Pharmacology 14:48–58

Black JW (1996) Receptors as pharmaceutical targets. In Textbook of receptor pharmacology. ed. JC Foreman and T Johansen. CRC Press, UK, pp 277–285

Brazenor RM, Angus JA (1981) Ergometrine contracts isolated canine coronary arteries by a serotonergic mechanism: no role for alpha adrenoceptors. Journal of Pharmacology and Experimental Therapeutics 218:530–536

Brixius K, Mohr V, Müller-Ehmsen J, Hoischen S, Münch G, Schwinger RHG (1998) Potent vasodilatory with minor cardiodepressant actions of mibefradil in human cardiac tissue. British Journal of Pharmacology 125:41–48

Chowienczyk PJ, Cockcroft JR, Ritter JM (1994) Blood pressure responses to intraarterial acetylcholine in man: effects of basal flow and conduit vessel length. Clinical Science 87:45–51

Clozel J-P, Osterrieder W, Kleinbloesem CH, Welker Ha, Schläppi B, Tudor R, Hefti F, Schmitt R, Eggers H (1991) Ro 40–5967: A new nondihydropyridine calcium channel antagonist. Cardiovascular Drug Research 9:4–17

Cockcroft JR, Chowienczyk PJ, Benjamin N, Ritter JM (1994) Preserved endothelium-dependent vasodilatation in patients with essential hypertension. New England Journal of Medicine 330:1036–1040

Cocks TM, Angus JA (1983) Endothelium-dependent relaxation of coronary arteries by noradrenaline and serotonin. Nature 305:627–630

Constantinides P, Rohmson M (1969) Ultrastructural injury of arterial endothelium: 1. Effects of pH, osmolarity, anoxia and temperature. Archives of Pathology 88: 99–105

Curry RC, Pepine CJ, Sabom MR, Feldman RL, Christie LG, Conti CR (1977) Effects of ergonovine in patients with and without coronary artery disease. Circulation 56:803–809

De Caterina R, Carpeggiani C, L'Abbate A (1984) Evidence against a role of serotonin in the genesis of coronary vasospasm. Circulation 69:889–894

Freedman SB, Chierchia S, Rodriguez-Plaza L, Bugiardini R, Smith G, Maseri A (1984) Ergonovine-induced myocardial ischemia: no role for serotonergic receptors? Circulation 70:178–183

Grondin CM, Campeau L, Lesperance J, Enjalbert M, Bourassa MG (1984) Comparison of late changes in internal mammary artery and saphenous vein grafts in two consecutive series of patients 10 years after operation. Circulation 70 (Suppl 1): 208–212

Hasking GL, Esler MD, Jennings GL, Burton J, Johns JA, Korner PI (1986) Noradrenaline spillover to plasma in congestive heart failure: evidence of increased cardiorenal and total sympathetic nerve activity. Circulation 73:615–621

He G-W, Rosenfeldt FL, Buxton BF, Angus JA (1989) Reactivity of human isolated internal mammary artery to constrictor and dilator agents. Implications for treatment of internal mammary artery spasm. Circulation 80:I141–I150

He G-W, Rosenfeldt FL, Angus JA (1993) Pharmacological relaxation of the saphenous vein during harvesting for coronary artery bypass grafting. Annals of Thoracic Surgery 55:1210–1217

Imaoka Y, Osanai T, Kamada T, Mio Y, Satoh K, Okumura K (1999) Nitric oxide-dependent vasodilator mechanism is not impaired by hypertension but is diminished with aging in the rat aorta. Journal of Cardiovascular Pharmacology 33:756–761

Kawachi Y, Tomoike H, Maruoka Y, Kikuchi Y, Araki H, Ishii Y, Tanaka K, Nakamura M (1984) Selective hypercontraction caused by ergonovine in the canine coronary artery under conditions of induced atherosclerosis. Circulation 69:441–450

Korner PI, Angus JA (1992) Structural determinants of vascular resistance properties in hypertension. Haemodynamic and model analysis. Journal of Vascular Research 29:293–312

Korner PI, Angus JA (1997) Vascular remodelling. Hypertension 29:1066–1067

Kubo SH, Rector TS, Heifetz SM, Cohn JN. (1989) α_2-adrenoceptor mediated vasconstriction in patients with congestive heart failure. Circulation 80:1660–1667

Kugiyama K, Ohgushi M, Motoyama T, Sugiyama S, Soejima H, Matsumura T, Yoshimura M, Ogawa H, Yasue H (1999) Enhancement of constrictor response of spastic coronary arteries to acetylcholine but not to phenylephrine in patients with coronary spastic angina. Journal of Cardiovascular Pharmacology 33:414–419

Langille BL (1993) Remodelling of developing and mature arteries: endothelium, smooth muscle and matrix. Journal of Cardiovascular Pharmacology 21:S11–S17

Lew MJ, Angus JA (1992) Wall thickness to lumen diameter ratios of arteries from SHR and WKY: comparison of pressurised and wire-mounted preparations. Journal of Vascular Research 29:435–442

Lew MJ, Angus JA (1995) Analysis of competitive agonist-antagonist interactions by nonlinear regression. Trends in Pharmacological Sciences 16:328–337

Lockette W, Otsuka Y, Carretero O (1986) The loss of endothelium-dependent vascular relaxation in hypertension. Hypertension 8 (Suppl II):II61–II66

Luscher TF and Vanhoutte PM. (1986) Endothelium-dependent contractions to acetylcholine in the aorta of the spontaneously hypertensive rat. Hypertension 8:344–348

Lytle BW, Loop FD, Cosgrove DM, Ratliff NB, Easley K, Taylor PC (1985) Long-term (5 to 12 years) serial studies of internal mammary artery and saphenous vein coronary bypass graft. Journal of Thoracic and Cardiovascular Surgery 89:248–258

Martin GR (1998) 5-Hydroxytryptamine Receptors. The IUPHAR Compendium of Receptor Characterization and Classification. ed. D Girdlestone. pp 167–184

Müller-Schweinitzer E (1980) The mechanism of ergometrine-induced coronary arterial spasm: in vitro studies on canine arteries. Journal of Cardiovascular Pharmacology 2:645–655

Mulvany MJ, Halpern W (1977) Contractile properties of small arterial resistance vessels in spontaneously hypertensive and normotensive rats. Circulation Research 41:19–26

Mulvany MJ, Baumbach GL, Aalkjaer C, Heagerty AM, Korsgaard N, Schiffrin EL, Heistad DD (1996) Vascular Remodeling. Hypertension 28:505–506

Panza JA, Quyyami AA, Brush JE Jr, Epstein SE (1990) Abnormal endothelium-dependent vascular relaxation in patients with essential hypertension. New England Journal of Medicine 323:22–27

Prinzmetal M, Kennamer R, Merliss R, Wada T, Bor N (1959) Angina pectoris. 1. A variant form of angina pectoris. American Journal of Medicine 27:375–388

Quillen JE, Sellke FW, Armstrong ML, Harrison DG (1991) Long-term cholesterol feeding alters the reactivity to primate coronary microvessels to platelet products. Arteriosclerosis Thrombosis 11:639–644

Ramos JR, Berger K, Mansfield PB, Sauvage LR (1976) Histologic fate and endothelial changes of distended and non-distended vein grafts. Annals of Surgery 183:205–228

Rosenfeldt FL, Angus JA, He G-W, Davis BB (1993) A new technique for relaxing the saphenous vein during harvesting for coronary bypass grafting. Australasian Journal of Cardiovascular and Thoracic Surgery 2:136–139

Sarsero D, Fujiwara T, Molenaar P, Angus JA (1998) Human vascular to cardiac tissue selectivity of L- and T-type calcium channel antagonists. British Journal of Pharmacology 125:109–119

Schiffrin EL, Deng LY, Larochelle P (1995) Progressive improvement in the structure of resistance arteries of hypertensive patients after 2 years of treatment with an angiotensin I-converting enzyme inhibitor: comparison with effects of a β-blocker. American Journal of Hypertension 8:229–236

Stone M, Angus JA (1978) Developments of computer-based estimation of pA_2 values and associated analysis. Journal of Pharmacology and Experimental Therapeutics 207:705–718

Stork AP, Cocks TM (1994) Pharmacological reactivity of human epicardial coronary arteries: characterization of relaxation responses to endothelium-derived relaxing factor. British Journal of Pharmacology 113:1099–1104

Sudhir K, Angus JA, Esler MD, Jennings GL, Lambert GW, Korner PI (1990) Altered venous responses to vasoconstrictor agonists and nerve stimulation in human primary hypertension. Journal of Hypertension 8:1119–1128

Thybo NK, Korsgaard N, Eriksen S, et al (1994) Dose-dependent effects of perindopril on blood pressure and small-artery structure. Hypertension 23:659–666

Treasure CB, Manoukian SV, Klein JL, et al (1992) Epicardial coronary artery responses to acetylcholine are impaired in hypertensive patients. Circulation Research 71:776–781

Van de Voorde J, Leusen I (1986) Endothelium-dependent and independent relaxation of aortic rings from hypertensive rats. American Journal of Physiology 250:H711–H717

Walsh JA, Hyman C, Maronde RF (1969) Venous distensibility in essential hypertension. Cardiovascular Research 3:338–349

Yang Z, Richard V, von Segesser L, Bauer E, Stulz P, Turina M, Luscher TF (1990) Threshold concentrations of endothelin-1 potentiate contractions to norepinephrine and serotonin in human arteries. Circulation 82:188–195

CHAPTER 2
Problems in Assigning Mechanisms: Reconciling the Molecular and Functional Pathways in α-Adrenoceptor-Mediated Vasoconstriction

M.J. LEW

A. Introduction

Advances in molecular and cellular biology have given us increased certainty in our descriptions of receptors and their associated transducer proteins, and help pharmacologists to document receptor stimulus-response pathways with increasing precision. However, in many cases, several concurrent interacting and branching pathways operate, resulting in a level of complexity that hinders understanding of the relative importance of those paths even in simple cellular systems. The inherent complexities of intact organs and whole animals compound those of the cellular level and, thus, it is doubly difficult to make a complete description of the overall stimulus-response pathways that might operate at those higher levels of organisation. Nonetheless, an attempt needs to be made because, from a therapeutic standpoint, the important pathways are those that operate in the relevant target organ under physiological and pathophysiological conditions. This chapter is a selective review of vascular stimulus-response mechanisms of α_1- and α_2-adrenoceptors and an attempt to integrate the observations from cellular and intact systems. α-Adrenoceptors are an ideal subject, because many studies explore and compare their coupling mechanisms, but this chapter is not intended to make a case for any special relationship between α_1- and α_2-adrenoceptors beyond their sharing of the physiological agonists norepinephrine and epinephrine. Most of the concepts aired in this review are also relevant to other receptor types in both vascular and non-vascular tissues, because the complications arising from intrinsic properties of vascular smooth muscle are likely to be analogous to complications that are important in many different tissues.

There are at least three subtypes of α_1-adrenoceptor and three subtypes of α_2-adrenoceptor currently accepted, but in order to make a comprehensible discussion of the overall stimulus–response coupling pathways of vascular α-adrenoceptors, repeated reference to issues of receptor subtypes has been avoided where possible. While there are undoubtedly differences in strengths and types of stimulus–response coupling between subclasses, we are fortunate in that the similarities are much more prominent than the differences. Vascular α-adrenoceptors are probably distributed as mixtures of subtypes anyway, so the simplification should not be detrimental to the overall picture of mechanisms that is developed. It is unlikely that all of the many different stimulus-response components associated with α_1- and α_2-adrenoceptors are equally important in mediating the final responses, and their relative importance must vary among tissues and with the pre-existing cellular state. No description of stimulus-response mechanisms can be sufficiently thorough in a complete description of the pathways and their roles.

B. Coupling Mechanisms at the Molecular and Cellular Levels

Vascular smooth muscle contracts when myosin is phosphorylated by myosin light-chain kinase (MLCK) and is thereby made to interact with actin filaments. Activation of MLCK is a result of the calcium-calmodulin complex and is therefore dependent on the cytoplasmic calcium concentration. The relationship between cytoplasmic calcium and myosin phosphorylation is not necessarily constant, but elevation of cytoplasmic calcium is nonetheless almost synonymous with activation of vascular smooth muscle contraction, and much of the work on stimulus–response coupling pathways in vascular smooth muscle is aimed at determining the sources and controls of cytoplasmic calcium.

I. α_1-Adrenoceptors

Activation of α_1-adrenoceptors leads [via the guanosine-triphosphate-binding protein (G protein) $G_{q/11}$] to the activation of phospholipase C (PLC), which hydrolyses phosphatidyl inositol (4,5) bisphosphate to inositol (1,4,5) trisphosphate (IP_3) and diacylglycerol (DAG). Several inositolphosphate species other than IP_3 are also formed after activation of PLC, but it is likely that IP_3 is the most important in initiating the responses of interest. IP_3 activates its receptors on the sarcoplasmic reticulum, stimulates the release of stored calcium into the cytoplasm and initiates smooth muscle contraction (which is relatively transient, because the stores can be rapidly depleted). Store depletion is a signal for capacitative calcium entry through "store-operated calcium channels", and that process may be important to the sustained phase of responses to α_1-adrenoceptor activation (GIBSON et al. 1988). DAG activates some isoforms of protein kinase C (PKC), which can then phosphorylate many different cellular regulatory proteins with generally excitatory effects in vascular smooth muscle. In particular, PKC can increase the sensitivity of contractile elements to activation by cytoplasmic calcium (NISHIMURA et al. 1990), probably by decreasing the activity of myosin light-chain phosphatase (BUUS et al. 1988). PKC can also enhance the flux of calcium through L-type calcium channels (Ca_L; FISH et al. 1988; LOIRAND et al. 1990; XIONG and SPERELAKIS 1995), but a depolarising stimulus is still needed to open the channels (PACAUD et al. 1991; XIONG and SPERELAKIS 1995). Following activation of α_1-adrenoceptors, depolarisation of both isolated vascular smooth muscle cells and vein segments appears to be the result of chloride currents (BYRNE and LARGE 1988; VAN HELDEN 1988) probably through Ca^{2+}-activated chloride channels (Cl_{Ca}; AMEDEE et al. 1990; PACAUD et al. 1991).

PACAUD et al. (1991) suggested that the sequence of events following activation of α_1-adrenoceptors in the portal vein is: (1) the receptor activates PLC to produce IP_3 and DAG; (2) the IP_3 induces release of calcium from the endoplasmic reticulum via the IP_3 receptor; (3) the calcium then causes depolarisation via Cl_{Ca} channels; (4) the depolarisation opens Ca_L channels to allow the influx of extracellular calcium; and (5) the influx of calcium is enhanced by the action of PKC, which is activated by the DAG. Potassium channels are also involved in vascular smooth muscle depolarisation following α_1-adrenoceptor activation, but their role is complex. Activation of PKC can decrease the flux of potassium through delayed rectifier K^+ channels (K_V; COLE et al. 1996) and adenosine triphosphate (ATP)-sensitive K^+ channels (K_{ATP}; BONEV and NELSON 1996), and one or both of those actions may be responsible for the slow depolarisation of venous smooth muscle, which is sometimes seen after activation of α_1-adrenoceptors (VAN HELDEN 1988). However, calcium released from intracellular stores has also been shown to open Ca^{2+}-activated K^+ channels, which can act as a feedback control of the depolarisation elicited by other channels (KNOT et al. 1998). It is clear that the overall balance of ion channel openings following activation of α_1-adrenoceptors leads to depolarisation, but the types of experiments that allow identification of individual channel types generally do not allow determination of the relative importance of those channels in the responses of intact vascular tissues.

Pertussis toxin-sensitive G proteins have been shown to be involved in responses to α_1-adrenoceptor stimulation in arteries and in neuronal tissues (WILSON and MINNEMAN 1990; ESBENSHADE et al. 1993; GURDAL et al. 1997). GURDAL et al. (1997) showed that α_1-adrenoceptor-mediated vasoconstriction responses and IP_3 accumulation in rat aortae were partly inhibited by pertussis toxin and suggested the involvement of G_o proteins, because immunoprecipitation experiments showed α_1-adrenoceptor coupling to both $G\alpha_q$ and $G\alpha_o$ proteins (predominantly the former) but not $G\alpha_i$ or $G\alpha_s$. It has also been shown, in cultured rat aortic smooth muscle cells, that the heterotrimeric G_i protein (i.e. $G\alpha_i\beta\gamma$) is associated with the IP_3 receptor on the sarcoplasmic reticulum, increasing the sensitivity of the sarcoplasmic reticulum to IP_3 (NEYLON et al. 1998). Thus, the effect of pertussis toxin on vasoconstriction mediated by α_1-adrenoceptors probably involves both decreased IP_3 formation and decreased IP_3 potency in releasing intracellular calcium stores. Fig. 1 is a schematic representation of the major stimulus-response pathways of α_1-adrenoceptors and should be viewed in conjunction with Fig. 2, which shows some of the important actions of PKC.

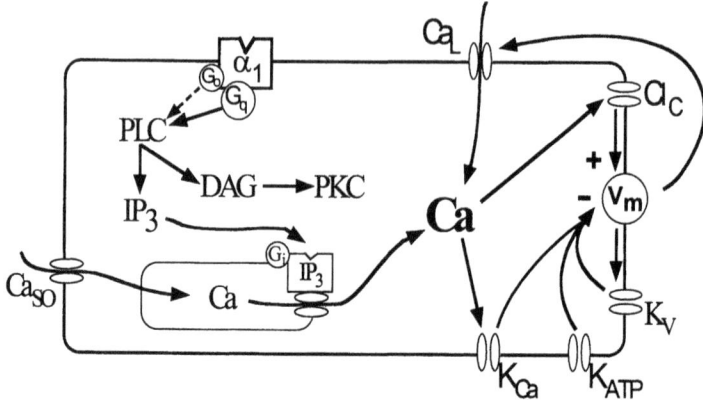

Fig. 1. Schematic diagram of some important features of α_1-adrenoceptor stimulus–response coupling pathways. The α_1-adrenoceptors activate phospholipase C (*PLC*) predominantly via G_q; the *PLC* makes inositol triphosphate (IP_3), which causes the release of sarcoplasmic Ca^{2+} stores via the IP_3 receptor. The calcium stores are replenished by a Ca^{2+} flux through store-operated calcium channels. Increased cytoplasmic Ca^{2+} activates chloride channels, which depolarise the cell (indicated by $+V_m$) and open L-type calcium channels to allow the influx of extracellular Ca^{2+}. The depolarisation is opposed by the Ca^{2+}-activated K^+ channels, which are opened by the elevation of cytoplasmic Ca^{2+}, and delayed rectifier K^+ channels, which are opened by the depolarisation. Diacylglycerol produced by *PLC* can activate protein kinase C, the effects of which are shown in Fig. 2. Abbreviations are as used in the text

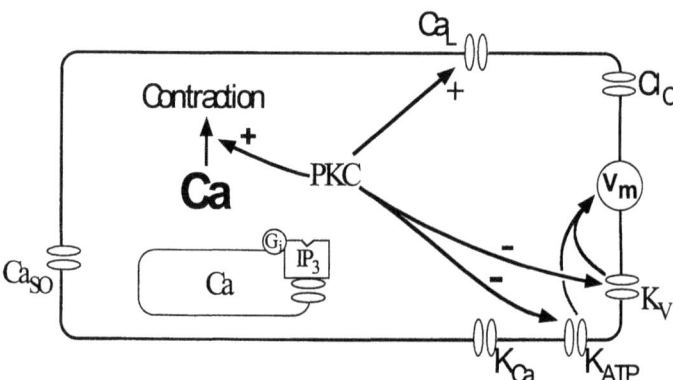

Fig. 2. Schematic diagram of some important effects of protein kinase C (*PKC*). *PKC* enhances the Ca^{2+} flux through open L-type calcium channels and decreases the opening of delayed rectifier K^+ channels and adenosine-triphosphate-sensitive K^+ channels to enhance any depolarisation. PKC activity also increases the amount of contraction that results from any level of cytoplasmic Ca^{2+}. Abbreviations are as used in the text

II. α_2-Adrenoceptors

α_2-Adrenoceptors are classically thought to evoke responses by inhibiting adenylate-cyclase activity to decrease the concentration of cyclic adenosine monophosphate (cAMP). Inhibition of adenylate cyclase is mediated by pertussis toxin-sensitive G proteins G_i and G_o but is not entirely straightforward, with most studies showing that activation of α_2-adrenoceptors differentially inhibits basal and stimulated adenylate-cyclase activity. For instance, WRIGHT et al. (1995) showed that α_2-adrenoceptor activation inhibited forskolin-stimulated cAMP production but not the basal rate of cAMP accumulation. That difference was not simply a matter of α_2-adrenoceptors reducing high (but not low) cAMP concentrations, because α_2-adrenoceptor activation also failed to inhibit cAMP accumulation in the presence of a phosphodiesterase inhibitor (rolipram), conditions of high cAMP concentration without activation of adenylate cyclase. The apparent inability to reduce basal adenylate-cyclase activity is not unique to α_2-adrenoceptors but is shared by other $G_{i/o}$-coupled receptors, such as neuropeptide Y receptors and melatonin receptors (FREDHOLM et al. 1985; CAPSONI et al. 1994; EBISAWA et al. 1994; WRIGHT et al. 1995). Thus, this inability could be a consequence of the way the $G_{i/o}$ protein subunits interact with adenylate cyclase. Another complication is that α_2-adrenoceptors have been shown to increase otherwise unstimulated adenylate-cyclase activity in several recombinant systems (NÄSMAN et al. 1997) and at least one physiological system, pancreatic islet cells (ULLRICH and WOLL-HEIM 1984). However, the stimulatory effect of α_2-adrenoceptors on adenylate cyclase is generally seen at high levels of activation and is both adrenoceptor-subtype and cell-line specific (DUZIC and LANIER 1992).

α_2-Adrenoceptors in vascular tissues have only been shown to inhibit forskolin- or receptor-stimulated cAMP accumulation in several vascular tissues (FREDHOLM et al. 1985; WRIGHT et al. 1995; ISHINE and LEE 1996) and, so far, there appears to be no evidence for α_2-adrenoceptor-mediated adenylate-cyclase activation in vascular tissues. The vasoconstrictor stimulus provided by inhibition of adenylate cyclase is really a reduction in the vasodilator effects of cAMP. Those effects are not completely characterised, but it is clear that many or most of the vasodilator effects of cAMP result from its activation of cAMP- and cyclic guanosine monophosphate (cGMP)-dependent protein kinases (PKA and PKG; JIANG et al. 1992). Phosphorylation of Ca_L channels in vascular smooth muscle cells by PKA or PKG inhibits calcium flux (XIONG and SPERELAKIS 1995; LIU et al. 1997), so a reduction in cAMP concentration following α_2-adrenoceptor activation would reduce PKA and PKG activation and disinhibit the influx of calcium through Ca_L channels. Phosphorylation (by PKA) of K_V channels in rabbit portal vein smooth muscle cells increases the channel-open probability (COLE et al. 1996), and K_{ATP} channels are opened following phosphorylation by PKA in rabbit mesenteric artery smooth muscle cells (QUAYLE et al. 1994). PKG activation seems responsible for the ability of cAMP to decrease intracellular Ca^{2+} levels in rat aortic

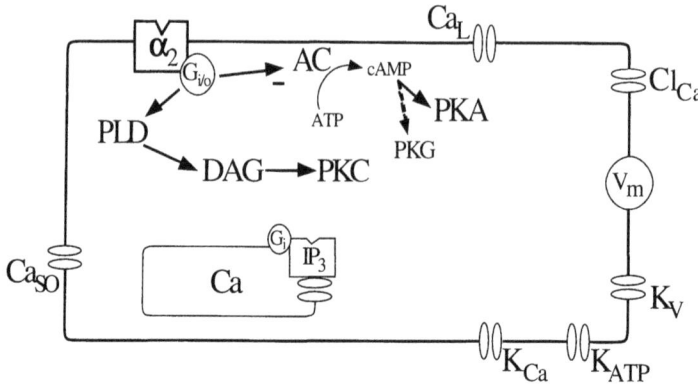

Fig. 3. Schematic diagram of some important features of α_2-adrenoceptor stimulus–response coupling pathways. The receptors inhibit adenylate cyclase via $G_{i/o}$, and this reduces the amount of cyclic adenosine monophosphate generated, thus reducing the activity of cyclic-adenosine-monophosphate-dependent protein kinase (*PKA*) and, to a lesser extent, cyclic-guanosine-monophosphate-dependent protein kinase (*PKG*). The α_2-adrenoceptors activate phospholipase D (*PLD*) via $G_{i/o}$, and the *PLD* makes diacylglycerol, which activates protein kinase C (*PKC*; Fig. 2 for the effects of PKC). The effects of *PKA* and *PKG* are shown in Fig. 4

smooth muscle cells (LINCOLN et al. 1990), and phosphorylation of the IP$_3$ receptor by PKG decreases the ability of IP$_3$ to release sarcoplasmic Ca^{2+} (KOMALAVILAS and LINCOLN 1994). Any decrease in the cellular cAMP concentration following α_2-adrenoceptor activation has the potential to reduce those vasodilator stimuli and effectively disinhibit any concomitant vasoconstrictor stimuli. α_2-Adrenoceptors can also activate phospholipase D, which makes DAG, which in turn activates PKC (ABURTO et al. 1995). Thus, the PKC-mediated vasoconstrictor effects discussed above are relevant to both α_1- and α_2-adrenoceptor-mediated vasoconstriction, particularly the enhancement of calcium fluxes through Ca$_L$ channels. Fig. 3 shows the major stimulus-response mechanisms of the α_2-adrenoceptor and should be viewed in conjunction with Fig. 2 and Fig. 4, which shows relevant effects of PKA and PKG.

C. Coupling Mechanisms in the Intact Animal

The conclusions that can be gleaned from functional data from *in vivo* preparations regarding stimulus–response coupling pathways are, of necessity, more broad than those from the cellular and molecular approaches. Nonetheless, they are essential in defining therapeutic targets and mechanisms. This section sets out those broad conclusions that have been obtained mostly from experiments involving measurement of pithed rat blood pressure. Many other experiments concerning α_1- and α_2-adrenoceptor stimulus–response coupling mechanisms have been performed using other *in vivo* preparations and intact blood vessels *in vitro*, but the pithed rat work was very influential and acted as a basis for interpretation of most of the other experiments. In 1981, VAN MEEL

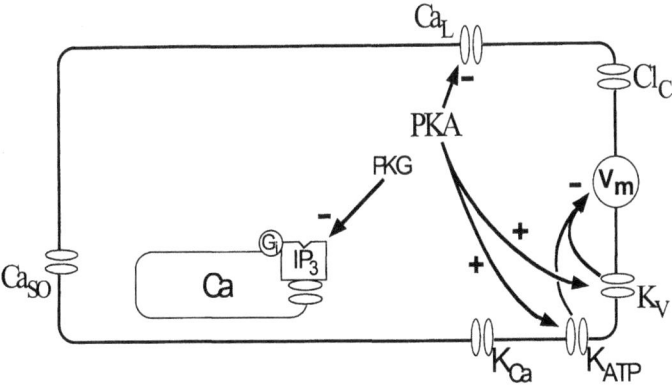

Fig. 4. Schematic diagram of some of the effects of cyclic-adenosine-monophosphate-dependent protein kinase (*PKA*) and cyclic-guanosine-monophosphate-dependent protein kinase (*PKG*) activation. Phosphorylation of L-type calcium channels by PKA decreases the flux of Ca^{2+}, but phosphorylation of adenosine-triphosphate-sensitive K^+ channels and delayed rectifier K^+ channels increases the K^+ flux, repolarising or hyperpolarising the cell. *PKG* can phosphorylate the inositol-triphosphate receptor and decrease the release of sarcoplasmic Ca^{2+} stores

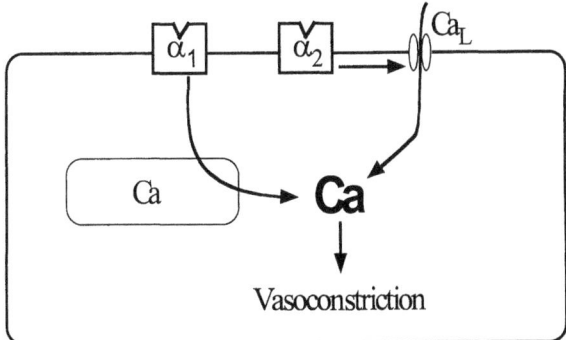

Fig. 5. Schematic diagram of vascular α-adrenoceptor stimulus–response coupling proposed by TIMMERMANS et al. α_1-Adrenoceptors mediate the release of intracellular Ca^{2+} stores, and α_2-adrenoceptors mediate the influx of extracellular Ca^{2+} through L-type calcium channels

and coworkers (1981a, 1981b) showed that pressor responses of pithed rats to α_1- and α_2-adrenoceptor stimulation differed in their susceptibility to calcium channel blockers. The responses to α_1-selective agonists were little affected by organic and inorganic calcium antagonists, but the responses to α_2-selective agonists were almost abolished. The authors concluded that the responses to α_1-adrenoceptors are dependent on the release of intracellular stores of calcium, whereas the responses to α_2-adrenoceptors are mediated exclusively by the influx of extracellular calcium (Fig. 5). That hypothesis found rapid support, but became more complicated when it was reported that removal of some of the α_1-adrenoceptors by phenoxybenzamine treatment greatly increased the sensitivity of α_1-adrenoceptor-mediated pressor responses to calcium channel blockers and that α_1-adrenoceptor partial agonists behaved like α_2-adrenoceptor agonists (RUFFOLO et al. 1984; COOKE et al. 1985; LEW et al. 1985). It was suggested that the coupling mechanisms of α_1- and α_2-adrenoceptors might be qualitatively similar, with the different susceptibilities to the L-channel blockers of the responses to α_1- and α_2-adrenoceptor stimulation being the predictable result of non-competitive antagonism of responses of systems with high (α_1) and low receptor reserve (α_2). That idea was not universally accepted (TIMMERMANS et al. 1984, 1987), and the fact that at least part of the response to α_1-adrenoceptor stimulation was refractory to inhibition by calcium channel blockers meant that coupling of α_1-adrenoceptors to vasoconstriction was through at least two distinct mechanisms. It had been suggested that the different coupling mechanisms were connected with different subtypes of α_1-adrenoceptor (MINNEMAN 1988). However, NICHOLS and RUFFOLO (RUFFOLO et al. 1991) proposed, in effect, that a single class of α_1-adrenoceptors coupled via two G protein transduction systems: one they called G_{PLC}, which stimulated phospholipase C to cause the release of intracellular calcium pools, and another they called G_{Ca}, which mediated influx of extracellular calcium. They suggested that the same G_{Ca} mediated the responses to α_2-adrenoceptor stimulation. Pertussis toxin mimicked Ca_L-channel blockers in its pattern of inhibition of α_1- and α_2-adrenoceptor-mediated pressor responses in the pithed rat, so G_{Ca} appeared to be a pertussis toxin-sensitive G protein. It had previously been proposed that α_2-adrenoceptors couple to vasoconstriction via G_i and inhibition of adenylate cyclase (BOYER et al. 1983), but NICHOLS and colleagues argued that G_{Ca} was distinct from G_i and G_o and that inhibition of adenylate cyclase played no role in the vasoconstriction responses to α_2-adrenoceptor stimulation (NICHOLS 1991; NICHOLS et al. 1988a, 1988b). The receptor reserve dependence of the effects of Ca_L-channel blockers on responses to α_1-adrenoceptor agonists was explained by the suggestion that the G_{Ca} system has high sensitivity but a relatively low capacity to mediate vasoconstriction and that the G_{PLC} system had a lower sensitivity but a higher capacity to mediate responses. Thus, low-efficacy agonists would primarily act via the high-sensitivity G_{Ca} system, giving responses that primarily involved influx of extracellular calcium. High-efficacy agonists could activate both the G_{Ca} and the lower-sensitivity G_{PLC} system, so the responses would involve both the influx of extracellular calcium and the release of intracellular calcium. Predictions from a simple mathematical model based on these ideas matched the experimental data well (NICHOLS and RUFFOLO 1988).

The main features of the stimulus–response coupling scheme that NICHOLS and RUFFOLO proposed based on pithed rat experiments (Fig. 6) are:

1. α_1-Adrenoceptors activate G_{PLC}, which activates PLC to make IP_3 and release intracellular calcium stores.
2. Both α_1- and α_2-adrenoceptors couple to opening of Ca_L channels via the same G protein, G_{Ca}.
3. G_{Ca} is inhibited by pertussis toxin.

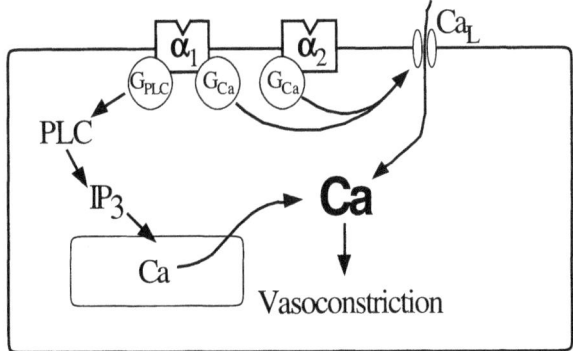

Fig. 6. Schematic diagram of vascular α-adrenoceptor stimulus–response coupling proposed by NICHOLS and RUFFOLO. α_1-Adrenoceptors mediate the release of intracellular Ca^{2+} stores via activation via G_{PLC} of phospholipase C and inositol-triphosphate generation. Both α_1- and α_2-adrenoceptors mediate (via G_{Ca}) the influx of calcium through L-type calcium channels

4. The relative reliance of α_1-adrenoceptor responses on G_{Ca} increases with decreasing receptor reserve.
5. Inhibition of adenylate cyclase by G_i or G_o is irrelevant to the effects of α_2-adrenoceptors.

D. Reconciliation

While it is not possible to confidently assert which are the most important of the many stimulus-response components implicated by the molecular and cellular studies, the form of the pithed rat data would not be easily predicted from the molecular and cellular studies. Both receptors can increase the flux of calcium though Ca_L channels, but the α_1-adrenoceptors appear much better able to depolarise the smooth muscle cells and, thus, to open those channels, so it is not easy to explain an almost complete reliance of α_2-adrenoceptor-mediated responses on the opening of Ca_L channels. The pertussis toxin sensitivity of the α_1-adrenoceptor-mediated vasoconstriction responses might be the result of the importance of pertussis toxin-sensitive G proteins in the formation and potency of IP_3, but that should give a pertussis toxin-sensitive component to the intracellular calcium-dependent response, not the pertussis toxin-sensitive opening of Ca_L channels that was proposed by NICHOLS and RUFFOLO. The cellular and molecular data point to inhibition of adenylate cyclase as being one of the important parts of α_2-adrenoceptor stimulus–response coupling, so we don't really have a clear single candidate for G_{Ca}. In the absence of a G_{Ca} molecule, it is difficult to assess the hypothesised reason for the agonist-efficacy-dependent reliance on calcium influx following activation of α_1-adrenoceptors, and it is also difficult to be confident that G_i and G_o are really irrelevant to the vasoconstrictor effects of α_2-adrenoceptor activation. Thus, there are substantial discrepancies between the descriptions of stimulus-response pathways obtained from the different types of experiment. This section attempts to find reasons for these discrepancies.

Some of the features of the stimulus-response scheme shown in Fig. 6 can be matched to the detailed molecular pathways: the G protein that NICHOLS and RUFFOLO called G_{PLC} is almost certainly G_q and mediates a calcium-influx-independent component of vasoconstriction responses by activating PLC to make IP_3 and release calcium from the sarcoplasmic-reticulum stores. However most of the components of the stimulus–response coupling scheme cannot easily be matched to molecular information. By itself, even equating G_{PLC} with G_q is not sufficient to make sense of the data because, if the calcium-influx-independent component of responses to α_1-adrenoceptor activation is mediated by G_q and IP_3, then one might expect it to be relatively transient or at least to peak rapidly and decline, forms of response that are not seen in the pithed rat data. This particular problem can be explained by a closer consideration of the time-courses of responses in the pithed rat. The cumulative bolus dose–response curves used in the relevant studies are rapid, with responses to each bolus reaching a peak or "plateau" within only a few seconds (LEW and ANGUS 1985). This will inevitably emphasise the mechanisms of the initial response over the later mechanisms and, thereby, will probably decrease the apparent role of the calcium influx in those responses to α_1-adrenoceptor activation. This idea was proposed by McGRATH and O'BRIEN (1987), who administered α_1- and α_2-adrenoceptor agonists to pithed rats by sustained infusion rather than by bolus injection. They found that nifedipine preferentially inhibited the late, sustained part of the responses to α_1-adrenoceptor agonists, leaving the initial, transient part relatively intact. The sensitivity of the sustained responses to α_1- and α_2-adrenoceptor activation to nifedipine were similar. Thus, it can be argued that the magnitude of the different sensitivities of α_1- and α_2-adrenoceptor-mediated pressor responses to Ca_L-channel blockers is affected by the rapid generation of agonist dose–response curves in the normal pithed rat assay method.

Intracellular mediators frequently affect more than one downstream target process or molecule, so stimulus-response pathways become branched and interconnected. That complication is widely known. It is probably less widely appreciated that, under many circumstances, activation of blood vessels by many different receptors can lead to recruitment of activation processes via physical and extracellular routes. Such recruited activation can occur in intact systems but not molecular or cellular systems, so they might be the reason for some of the discrepancies between the molecular mechanisms and intact functional responses to α-adrenoceptor activation. The most powerful of those recruited pathways is probably the smooth muscle myogenic response that can be recruited by increased smooth muscle cell tension, but activation by additional mediators released in response to α-adrenoceptor activation might also be an important component in the overall stimulus–response coupling mechanisms of α-adrenoceptors. Some of the remaining difficulties in

E. Myogenic Activation

Myogenic responses are intrinsic to the vascular smooth muscle and allow many blood vessels to respond to increased distending pressure or to stretch with increased vasoconstrictive effort. The most straightforward demonstrations of myogenic responses are experiments where small arteries or arterioles in vitro are exposed to increased distending pressure and the resulting change in diameter is tracked. The vessels are initially distended in a passive manner by the pressure step, but then they actively contract, often to a diameter smaller than they had before the pressure step. In general, it is found that small arteries and arterioles have a greater myogenic reactivity than large arteries. The proximate stimulus for myogenic responses has not been conclusively determined, but smooth muscle cell tension is very likely to be involved (MEININGER and DAVIS 1992), and basal active smooth muscle tone is probably necessary for arteries to sense and respond to increased distending pressure (SPEDEN and WARREN 1986). A role for the endothelium in these responses was also proposed but is now considered to be unlikely, as there is convincing evidence that endothelial cells are not required for myogenic responses in many arteries (MEININGER and DAVIS 1992).

The mechanisms involved in myogenic responses appear complex, and the many systems that have been proposed to play a role include stretch-activated cation channels, PKC, PLC and membrane depolarisation leading to opening of Ca_L channels (MEININGER and DAVIS 1992). Part of the reason that so many different systems have been proposed to play a part in the myogenic responses is that interventions reported to inhibit myogenic responses actually inhibited the active tone needed for myogenic activation to occur. Nonetheless, there is convincing evidence for smooth muscle cell depolarisation and the opening of Ca_L channels (BULOW 1996; WESSELMAN et al. 1996; KNOT and NELSON 1998; TANAKA and NAKAYAMA 1998).

In rat small cerebral arteries, increasing distending pressure from 10mmHg to 100mmHg caused constriction that was associated with smooth muscle cell depolarisation from -63 mV to -36 mV and a rise in intracellular calcium concentration from 119nM to 245nM. Inhibition of the Ca_L channels with nisoldipine decreased the intracellular calcium concentration and relaxed the arteries at 60mmHg but, importantly, the nisoldipine neither altered membrane potential nor reduced the depolarising effect of extracellular K^+ (KNOT and NELSON 1998). Thus, blockade of the Ca_L channels inhibited myogenic tone without inhibiting the proximal stimulus for the channel opening, and it cannot be argued that the effect was due to interference with the myogenic stimulus. Further evidence for a mechanistic antagonism of myogenic responses by Ca_L-channel blockade is provided by experiments in the rat femoral artery, where myogenic responses were enhanced following partial inhibition of active tone by acetylcholine or pinacidil but were almost abolished following an equivalent inhibition of the tone by Ca_L-channel blockers (BULOW 1996). Myogenic responses involve depolarisation and the influx of extracellular calcium through voltage-gated Ca_L channels and, therefore, Ca_L channels probably play a role in vasoconstriction elicited by any stimulus that recruits myogenic amplification.

Recruitment of myogenic responses during the responses initiated by receptors implies recruitment of the stimulus-response pathways of the myogenic response into the overall stimulus-response pathway of those receptors. Where the receptor and myogenic system have pathways in common, this would simply reinforce the activation of already active pathways but, otherwise, it would recruit pathways foreign to the receptor. Therefore, it seems inevitable that responses to α_1- and α_2-adrenoceptors will involve the mechanisms of the myogenic responses. Myogenic self-activation has been found to be a component in vasoconstriction responses initiated by α-adrenoceptors both in vivo and in vitro. MEININGER and TRZECIAKOWSKI (1988) showed that the effects of a phenylephrine infusion on the vascular resistance of rat intestine were reduced where the intestine was isolated from the pressor effect of the phenylephrine (i.e. when the intestinal vascular bed was held isobaric). They suggested that autoregulatory mechanisms in the intestinal bed were activated by the increased perfusion pressure elicited by the phenylephrine, implying that both the myogenic responses of the vasculature to the pressure per se and alterations in production of locally made vasoactive agents due to increased flow (i.e. overperfusion) in the vascular bed might be involved. However, the fact that myogenic responses can be elicited from arterial segments in vitro suggests that the myogenic response to pressure plays an important part and, thus, part of the primary response elicited by phenylephrine was amplified by a myogenic response. The interaction between receptor-mediated vasoconstriction responses and myogenic responses varies with the receptor type. Myogenic activation of small arteries in the rat cremaster are facilitated more by α_2-adrenoceptor activation than by α_1-adrenoceptor activation both in vivo and in vitro (FABER and MEININGER 1990; IKEOKA et al. 1992) and, thus, myogenic amplification of α_2-adrenoceptor-mediated vasoconstriction will be more important than myogenic amplification of α_1-adrenoceptor-mediated vasoconstriction. As the myogenic responses involve opening of Ca_L channels, the idea of myogenic recruitment helps to explain the importance of Ca_L channels in the response to α_2-adrenoceptor activation.

F. Autocrine and Paracrine Activation

Vasoactive purines and arachidonic-acid metabolites have been reported to be released from arterial smooth muscle or endothelium after stimulation of vascular α-adrenoceptors. Therefore, full consideration of α-adrenoceptor stimulus–response coupling in vascular tissues also requires attention to the activation of other types of receptor that might be activated in an autocrine or paracrine manner. Any autocrine effects of arachidonic acid released from smooth muscle cells will vary according to which prostaglandins are formed from the arachidonic acid. Significant amounts of prostacyclin have been reported to result from activation of both α_1- and α_2-adrenoceptors in rabbit aortic smooth muscle (NEBIGIL and MALIK 1993), so it is possible that activation of adenylate cyclase might result indirectly from activation of α-adrenoceptors. Presumably, however, vasoconstrictor pathways might also be stimulated by other types of prostaglandin. The release of arachidonic acid and prostacyclin are sensitive to pertussis toxin and appear to involve activation of phospholipase A and the influx of calcium through Ca_L channels (NEBIGIL and MALIK 1993; NISHIO et al. 1996).

Purines, including ATP and adenosine, can be released from arteries following stimulation of α_1-adrenoceptors, and those purines may act on smooth muscle cells in an autocrine or paracrine fashion. Purine release upon activation with noradrenaline was much greater from the rat tail artery than from the thoracic aorta, and the purines appeared to originate predominantly from the endothelium (SHINOZUKA et al. 1994). The location of the norepinephrine concentration–response curve for purine release from the tail artery was very similar to the concentration–response curve for vasoconstriction, so it is unlikely that the release of purines was simply a non-physiological artefact from high concentrations of noradrenaline or excessive vasoconstriction. Although the consequences of endothelial ATP and adenosine release for smooth muscle cell function were not tested, it was estimated that the amount of purines released from the endothelial cells was sufficient to affect vascular function (SHINOZUKA et al. 1994). A functional autocrine role for noradrenaline-induced ATP release has been shown in Chinese hamster ovary cells transfected with α_2-adrenoceptors. Those cells responded to noradrenaline with a rapid release of intracellular Ca stores, which was mimicked by exogenous ATP and inhibited by the ATP-hydrolysing enzyme apyrase and by the non-selective P2-receptor antagonist suramin (OKERMAN et al. 1998). Thus, activation of the α_2-adrenoceptor in those cells elicited a release of intracellular calcium stores via autocrine activation of P2Y receptors, which activate PLC to form IP_3. Because vascular smooth muscle cells express P2X rather than P2Y receptors, any autocrine activation by released ATP would presumably involve depolarisation via the P2X non-specific cation channel. Presumably, this would allow influx of calcium both through the P2X channel itself and through Ca_L channels opened as a consequence of depolarisation. If such paracrine activation was functionally significant, then one might expect vasoconstrictor responses to α-adrenoceptor activation to be partly sensitive to inhibition of purinoceptors. This has been reported to be the case in pithed rats, where responses to α_2-adrenoceptor stimulation were partially inhibited by desensitisation of P2X receptors with $\alpha\beta$-methylene-ATP (SCHLIKER et al. 1989; DALZIEL et al. 1990).

G. Functional Antagonism

Indirect recruitment of stimulus–response coupling mechanisms helps explain the importance of calcium influx in vasoconstriction responses to α-adrenoceptor stimulation but does not help to explain why α_1-adrenoceptor partial agonists behave more like α_2-adrenoceptor agonists than like full α_1-adrenoceptor agonists. We may be able to explain this by consideration of the efficacy-dependent features of functional interactions. Both PEDRINELLI et al. (1985) and LEW and ANGUS (1985) suggested that the depressor effect of calcium channel blockers may preferentially interfere with α_2-adrenoceptor agonists and partial α_1-adrenoceptor agonists by functional antagonism.

Before we can explore this issue in any depth, exact meanings of terms describing antagonism need to be agreed on. The term "functional antagonism" can mean either antagonism of an effect by activation of an opposite effect (a meaning that is sometimes called physiological antagonism) or antagonism by interference with events that follow receptor activation (JENKINSON et al. 1995). Because the meaning of functional antagonism that was intended by LEW and ANGUS and probably by PEDRINELLI et al. (1985) is the former but the antagonism of Ca_L-channel blockers for effects mediated by opening of Ca_L channels is the latter, we need to discriminate between them clearly. One might simply describe the former situation as physiological antagonism and thus allow functional antagonism to unambiguously mean the latter. However, because the term functional antagonism has been used to mean solely the former in many important works on the topic (VAN DEN BRINK 1973; BROADLEY and NICHOLSON 1979; MACKAY 1981; LEFF et al. 1985), it is preferable to reserve that meaning for functional antagonism and to coin a new term for the latter. It is suggested that "mechanistic antagonism" can be used to mean that type of antagonism that occurs when the antagonist specifically blocks an event that follows receptor activation. Thus, the antagonism of nifedipine for vasoconstriction responses that involve the opening of Ca_L channels is mechanistic antagonism, whereas any antagonism of vasoconstriction by vasodilatation per se is functional antagonism.

The exact patterns of functional antagonism interactions vary according to the systems in which they are observed and with the particular pathways used by the agonist and antagonist. However, where all else is equal, a full agonist will be less affected by a functional antagonist than will a partial agonist, and effects mediated by a strongly coupled receptor system will be less affected than responses mediated by a less strongly coupled receptor. This pattern results from the fact that the stronger systems have an excess capacity to mediate activation, so a larger amount of inactivation stimulus is needed to occlude their effects. This is sometimes explained in terms of receptor reserve, which is probably an adequate way to describe the differences between full and partial agonists acting at the same receptor. However, when dealing with drugs acting at different receptors, one cannot group all of the determinants of efficacy into receptor reserve; full agonists acting at a weakly coupled receptor system can have a receptor reserve but a low maximum effect. In other words, a full agonist at α_2-adrenoceptors may have receptor reserve even where the maximum effect of α_2-adrenoceptor activation is less than the effect of α_1-adrenoceptor activation, a situation that has been documented in the canine saphenous vein (RUFFOLO and ZEID 1985). Receptor reserve has predictive power when one is dealing with the effects of an irreversible competitive antagonist but not necessarily when one is dealing with functional antagonism. In that case, one has to look beyond receptor numbers (LEW 1995).

Because the maximum response to α_2-adrenoceptor activation in the pithed rat is less than that for α_1-adrenoceptor activation, the pattern of antagonism of α-adrenoceptor-mediated pressor responses by functional antagonism is qualitatively the same as that observed with Ca_L-channel blockers. Compared with responses to full agonists at α_1-adrenoceptors, responses to (1) α_2-adrenoceptor stimulation, (2) partial agonists of α_1-adrenoceptors and (3) full α_1-adrenoceptor agonists after partial alkylation of the receptors all have a lower ability to mediate vasoconstriction and preferentially inhibit vasodilatation in vitro (MARTIN et al. 1986; OHYANAGI et al. 1992) and in vivo (LEW and ANGUS 1985; PEDRINELLI and TARAZI 1985).

It is clear that at least part of the vasoconstriction responses mediated by both α_1- and α_2-adrenoceptors should be sensitive to mechanistic antagonism by Ca_L-channel blockers, so the question is not whether the effects of the blockers are the result of functional antagonism, but rather whether functional antagonism contributes to the effect of the Ca_L-channel blockers. Ca_L-channel blockers elicit a depressor effect in pithed rats and in anaesthetised rats with autonomic blockade. The basal blood pressure in such preparations may be as low as 40mmHg, probably low enough to make pressor responses in the pithed rat preparation susceptible to functional antagonism by any further decrease in pressure. That has been shown by the observation that while the blood pressure is reduced by sodium nitroprusside, the pressor responses to α_1-and α_2-adrenoceptor agonists are significantly inhibited by functional antagonism (LEW and ANGUS 1985; PEDRINELLI and TARAZI 1985). The pattern of inhibition is qualitatively the same as that caused by the Ca_L-channel blockers, with the responses to α_2-adrenoceptor agonists being almost abolished and the dose–response curve of full α_1-adrenoceptor agonists being merely shifted rightward.

TIMMERMANS et al. (1987) claimed to have excluded the possibility that the depressor effect of nifedipine was important to its effects in the pithed rat; however, they simply showed that prevention of the depressor effect by infusion of vasopressin did not completely prevent the antagonism of responses to phenylephrine by nifedipine in rats that had been treated with phenoxybenzamine. That experiment may have tested the hypothesis that functional antagonism was the sole mechanism of antagonism by nifedipine, but such a hypothesis had not been proposed. In fact, responses to phenylephrine in the presence of nifedipine were slightly larger in the vasopressin-infused rats than in the control rats, so the data actually support a role for functional antagonism. Functional antagonism is probably also important in explaining the apparent importance of the pertussis toxin-sensitive process in the coupling of α_1-adrenoceptors to calcium influx. In pithed rats, the pressor responses to α_1- and α_2-adrenoceptor agonists were substantially inhibited by pertussis toxin (RUFFOLO et al. 1991), with the magnitude and pattern of the effects of the toxin being similar to those of Ca_L-channel blockers (a more substantial inhibition of α_2-adrenoceptor agonists and partial α_1-adrenoceptor agonists than full α_1-adrenoceptor agonists). Thus, it was proposed that the adrenoceptors link to opening of Ca_L channels via a pertussis toxin-sensitive path, but the apparently critical experiment of testing for non-summation of the effects of pertussis toxin and Ca_L-channel blockers was not performed. Pertussis toxin causes a decrease in the basal blood pressure of the pithed rats sufficiently large that one would expect inhibition of pressor responses by functional antagonism. Presumably, mechanistic inhibition of both α_1-and α_2- adrenoceptor-mediated pressor responses also plays a role but, with the available data, it is not possible to compare the relative importance of the types of antagonism. Functional antagonism almost certainly plays a role in the antagonism of pressor responses of the pithed rat by agents which, like Ca_L-channel blockers and pertussis toxin, lower the blood pressure. Thus, the efficacy-dependent nature of functional antagonism can explain at least some of the selectivity of those agents for α_2-adrenoceptor and α_1-adrenoceptor partial-agonist-mediated pressor responses.

H. Adenylate Cyclase and α_2-Adrenoceptors

It was proposed that inhibition of adenylate cyclase plays little if any role in vasoconstriction responses to stimulation of α_2-adrenoceptors, despite the pertussis toxin-sensitive link between those receptors and vasoconstriction (NICHOLS et al. 1988; NICHOLS 1991). The logic of the arguments used is essentially that in order for inhibition of adenylate cyclase to mediate the large pressor response that can be obtained from stimulation of α_2-adrenoceptors (more than 80mmHg increase in blood pressure in the pithed rat), there would need to be a sufficient level of adenylate-cyclase activity to hold the pressure down by a similar amount. The authors suggested that there is little likelihood that there is a level of any circulating agent that activates adenylate cyclase in the pithed rat sufficient to account for such activation of the enzyme. That argument may be wrong on the grounds that it focuses solely on "circulating agents" and in the expectation that inhibition of adenylate cyclase has to mediate the whole response.

While there may be little circulating adrenaline to activate adenylate cyclase in the pithed rat, adenosine would be produced in tissues underperfused by the low basal blood pressure (~40mmHg diastolic). Adenosine A_2 receptors couple to vasodilatation via G_s and could easily provide an elevation of adenylate-cyclase activity and cAMP that can be inhibited by G_i activation following α_2-adrenoceptor activation. If part of the pressor effect is supplied by myogenic amplification, then relatively little adenylate-cyclase activation may be needed. One cannot say how much pressor stimulus is needed to initiate a pressor response of 80mmHg, but in vitro experiments suggest that it could be surprisingly little. Myogenic mechanisms can make the stimulus–response curve in pressurised isolated segments of rat mesenteric artery so steep that they respond to vasopressin in an almost all-or-none fashion when the diameter is held constant by feedback control of the distending pressure (VAN BAVEL and MULVANY 1994). Vasopressin concentrations sufficient to elicit any activation caused maximal contractile effort, leading the authors to conclude that even minor direct activation recruited myogenic activation sufficient for a maximal response. The conditions of that experiment may seem more extreme than those that would pertain to the pithed rat, but it should be remembered that at least some of the arteries and arterioles activated in the pithed rat would have experienced a marked increase in distending pressure as they contracted in response to the α-adrenoceptor agonists. Those vessels may have experienced the isometric conditions that were used in the in vitro experiments. Thus, the argument that inhibition of adenylate cyclase cannot be important for α_2-adrenoceptor-mediated vasoconstriction is based on an exaggerated assumption of the necessary initial level of adenylate-cyclase activity. Inhibition of a modest amount of adenylate-cyclase activity may be sufficient to allow the pressor responses that result from the combined direct and myogenic activation. The form of the interactions between α-adrenoceptors and both adenosine receptors and β-adrenoceptor stimulation is consistent with inhibition of adenylate cyclase as a primary stimulus-response mechanism of the α_2-adrenoceptors.

It has been shown that α_2-adrenoceptors are more sensitive than α_1-adrenoceptors to functional antagonism by the K_{ATP}-channel opener cromakalim (TATEISHI and FABER 1995) and by cGMP-elevating agents (LEW and ANGUS 1985; PEDRINELLI and TARAZI 1985; OHYANAGI et al. 1992). This is the expected pattern of functional antagonism because of the stronger stimulus–response coupling of the α_1-adrenoceptors. However, the opposite pattern has been shown in the rat cremaster, where the adenosine-receptor agonist 5'-N-ethylcarboxamidoadenosine is more potent at cancelling vasoconstriction mediated by α_1-adrenoceptors (norepinephrine in the presence of rauwolscine) than that mediated by α_2-adrenoceptors (norepinephrine in the presence of prazosin; NISHIGAKI et al. 1991). Similarly, in the pithed rat, salbutamol affected the concentration–response curves to the α_2-adrenoceptor agonist B-HT 933 and the full α_1-adrenoceptor agonist cirazoline equally (NICHOLS et al. 1989). In both of those cases, it is likely that there is a specific interaction between the activation of adenylate cyclase by both the adenosine receptors and β-adrenoceptors and the inhibition of adenylate cyclase by the α_2-adrenoceptors. In other words, where adenylate-cyclase activity is elevated above basal levels, α_2-adrenoceptors can couple to vasoconstriction via inhibition of adenylate cyclase. The inability of α_2-adrenoceptors to decrease basal levels of cAMP has been taken as evidence against a primary role of adenylyl-cyclase inhibition in α_2-adrenoceptor-mediated vasoconstriction (WRIGHT et al. 1995). However, it must be noted that the physiologically relevant state of adenylyl-cyclase activity may be well above the in vitro "basal" level because of the simultaneous presence of a wide variety of vasoactive agents. Thus, it seems likely that inhibition of adenylate cyclase is an important stimulus–response coupling mechanism for vascular α_2-adrenoceptors, not only in the cellular and molecular studies, but in vivo as well.

I. Conclusions

This review is a selective rather than comprehensive look at α_1- and α_2-adrenoceptor stimulus-response pathways and, thus, the resulting schemas might be both overly simple and biased. Important experiments that help complete our knowledge of stimulus-response pathways, particularly many in vitro studies of arteries and veins, have been omitted for the sake of clarity and brevity. However, that limitation does not affect the overall conclusions because, while there are many uncertainties in our picture of the stimulus–response coupling mechanisms of the α-adrenoceptors, it is clear that the pictures obtained from intact systems and the simpler cellular and molecular systems differ in important ways. Some of those differences come about because the increased relevance of more intact and complex experimental systems comes at the cost of decreased acuity of interpretation. For example, the role of pertussis toxin-sensitive G proteins in α_1-adrenoceptor-mediated Ca_L-channel opening and vasoconstriction is probably much smaller than the in vivo experiments suggested. Other important differences between intact and simple systems come about because the intact systems allow indirectly recruited mechanisms to come into the stimulus-response pathways, as exemplified by the recruitment of myogenic responses involving Ca_L-channel opening in α_2-adrenoceptor-mediated vasoconstriction. It might be argued that the mechanisms recruited indirectly are outside the proper or real stimulus-response pathway for a particular receptor, but indirectly recruited mechanisms appear to play a physiologically important role in the overall response of blood vessels to α-adrenoceptor activation and, therefore, cannot be ignored. The intrinsic complexity of the stimulus-response mechanisms and their interconnectedness means that we should expect that the roles of any particular component will vary not only among tissues but among conditions in a single tissue, and perhaps these roles will even vary from moment to moment. Which of the putative coupling pathways are actually involved in α-adrenoceptor stimulus–response coupling in blood vessels cannot be decided, in part because of the difficulty in equating the results from the widely differing experimental models and protocols used in signal transduction research. We must accept that many studies will document the possible rather than the actual and, as we gather more detailed information about stimulus–response coupling mechanisms, it is increasingly important to be cognisant of the scope of each type of experiment and to attempt to examine mechanisms at multiple levels of organisation and under differing conditions of cellular activation.

References

Aburto T, Jinsi A, Zhu Q, Deth RC (1995) Involvement of protein kinase C activation in α_2-adrenoceptor-mediated contractions of rabbit saphenous vein. Eur J Pharmacol 277:35–44

Akerman, KE, Nasman J, Lund PE, Shariatmadari R, Kukkonen JP (1998) Endogenous extracellular purine nucleotides redirect α_2-adrenoceptor signaling. FEBS Lett 430:209–212

Amedee T, Large WA, Wang Q (1990) Characteristics of chloride currents activated by noradrenaline in rabbit ear artery cells. J Physiol (Lond) 428:501–516

Bonev AD, Nelson MT (1996) Vasoconstrictors inhibit ATP-sensitive K^+ channels in arterial smooth muscle through protein kinase C. J Gen Physiol 108:315–323

Boyer JL, Cardenas C, Posadas C, Garcia-Sainz JA (1983) Pertussis toxin induces tachycardia and impairs the increase in blood pressure produced by $\alpha 2$-adrenergic agonists. Life Sci 33:2627–2633

Broadley KJ, Nicholson CD (1979) Functional antagonism as a means of determining dissociation constants and relative efficacies of sympathomimetic amines in guinea pig isolated atria. Br J Pharmacol 66:397–404

Bulow A (1996) Differentiated effects of vasodilators on myogenic reactivity during partial inhibition of myogenic tone in pressurized skeletal muscle small arteries of the rat. Acta Physiol Scand 157:419–426

Buus CL, Aalkjaer C, Nilsson H, Juul B, Moller JV, Mulvany MJ (1998) Mechanisms of Ca2+ sensitization of force production by noradrenaline in rat mesenteric small arteries. J Physiol (Lond) 510:577–590

Byrne NG, Large WA (1988) Membrane ionic mechanisms activated by noradrenaline in cells isolated from the rabbit portal vein. J Physiol (Lond) 404:557–573

Capsoni S, Viswanathan M, De Oliveria AM, Saavedra JM (1994) Characterization of melatonin receptors and signal transduction system in rat arteries forming the circle of Willis. Endorinology 135:373–378

Cole WC, Clement-Chomienne O, Aiello EA (1996) Regulation of 4-aminopyridine-sensitive, delayed rectifier K^+ channels in vascular smooth muscle by phosphorylation. Biochem Cell Biol 74:439–447

Cooke JP, Rimele TJ, Flavahan NA, Vanhoutte PM (1985) Nimodipine and inhibition of α-adrenergic activation of the isolated canine saphenous vein. J Pharmacol Exp Ther 234:598–602

Dalziel HH, Gray GA, Drummond RM, Furman BL, Sneddon P (1990) Investigation of the selectivity of α,β-methylene ATP in inhibiting vascular responses of the rat in vivo and in vitro. Br J Pharmacol 99:820–824

Duzic E, Lanier SM (1992) Factors determining the specificity of signal transduction by guanine-nucleotide-binding-protein-coupled receptors. III. Coupling of $\alpha 2$-adrenergic-receptor subtypes in a cell-type-specific manner. J Biol Chem 267:24045–24052

Ebisawa T, Karne S, Lerner MR, Reppert SM (1994) Expression cloning of a high-affinity melatonin receptor from *Xenopus* dermal melanophores. Proc Natl Acad Sci USA 91:6133–6137

Esbenshade TA, Han C, Murphy TJ, Minneman KP (1993) Comparison of $\alpha 1$-adrenergic receptor subtypes and signal transduction in SK-N-MC and NB41A3 neuronal cell lines. Mol Pharmacol 44:76–86

Faber JE, Meininger GA (1990) Selective interaction of α-adrenoceptors with myogenic regulation of microvascular smooth muscle. Am J Physiol 259:H1126–H1133

Fish RD, Sperti G, Colucci WS, Clapham DE (1988) Phorbol ester increases the dihydropyridine-sensitive calcium conductance in a vascular smooth muscle cell line. Circ Res 62:1049–1054

Fredholm BB, Jansen I, Edvinsson L (1985) Neuropeptide Y is a potent inhibitor of cyclic AMP accumulation in feline cerebral blood vessels. Acta Physiol Scand 124:467–469

Gibson A, McFadzean I, Wallace P, Wayman CP (1998) Capacitative Ca2+ entry and the regulation of smooth muscle tone. Trends Pharmacol Sci 19:266–269

Gurdal H, Seasholtz TM, Wang HY, Brown RD, Johnson MD, et al (1997) Role of Gαq or Gαo proteins in α1-adrenoceptor subtype-mediated responses in Fischer 344 rat aorta. Mol Pharmacol 52:1064–1070

Ikeoka K, Nishigaki K, Ohyanagi M, Faber JE (1992) In vitro analysis of α-adrenoceptor interactions with the myogenic response in resistance vessels. J Vasc Res 29:313–321

Ishine T, Lee TJ (1996) Norepinephrine attenuates serotonin inhibition of pial venous tone. Eur J Pharmacol 313:97–102

Jenkinson DH, Barnard EA, Hoyer D, Humphrey PP, Leff P, et al (1995) International Union of Pharmacology Committee on Receptor Nomenclature and Drug Classification. IX. Recommendations on terms and symbols in quantitative pharmacology. Pharmacol Rev 47:255–266

Jiang H, Colbran JL, Francis SH, Corbin JD (1992) Direct evidence for cross-activation of cGMP-dependent protein kinase by cAMP in pig coronary arteries. J Biol Chem 267:1015–1019

Knot HJ, Standen NB, Nelson MT (1998a) Ryanodine receptors regulate arterial diameter and wall [Ca^{2+}] in cerebral arteries of rat via Ca^{2+}-dependent K^+ channels. J Physiol (Lond) 508:211–221

Knot HJ, Nelson MT (1998b) Regulation of arterial diameter and wall [Ca^{2+}] in cerebral arteries of rat by membrane potential and intravascular pressure. J Physiol (Lond) 508:199–209

Komalavilas P, Lincoln TM (1994) Phosphorylation of the inositol 1,4,5-trisphosphate receptor by cyclic GMP-dependent protein kinase. J Biol Chem 269:8701–8707

Leff P, Martin GR, Morse JM (1985) Application of the operational model of agonism to establish conditions when functional antagonism may be used to estimate agonist dissociation constants. Br J Pharmacol 85:655–663

Lew MJ (1995) Extended concentration-response curves used to reflect full agonist efficacies and receptor occupancy–response coupling ranges. Br J Pharmacol 115:745–752

Lew MJ, Angus JA (1985) α_1- And α_2-adrenoceptor-mediated pressor responses: are they differentiated by calcium antagonists or by functional antagonism? J Cardiovasc Pharmacol 7:401–408

Lincoln TM, Cornwell TL, Taylor AE (1990) cGMP-dependent protein kinase mediates the reduction of Ca^{2+} by cAMP in vascular smooth muscle cells. Am J Physiol 258:C399–C407

Liu H, Xiong Z, Sperelakis N (1997) Cyclic nucleotides regulate the activity of L-type calcium channels in smooth muscle cells from rat portal vein. J Mol Cell Cardiol 29:1411–1421

Loirand G, Pacaud P, Mironneau C, Mironneau J (1990) GTP-binding proteins mediate noradrenaline effects on calcium and chloride currents in rat portal vein myocytes. J Physiol (Lond) 428:517–529

Mackay D (1981) An analysis of functional antagonism and synergy. Br J Pharmacol 73:127–134

Martin W, Furchgott RF, Villani GM, Jothianandan D (1986) Depression of contractile responses in rat aorta by spontaneously released endothelium-derived relaxing factor. J Pharmacol Exp Ther 237:529–538

McGrath JC, O'Brien JW (1987) Blockade by nifedipine of responses to intravenous bolus injection or infusion of α1- and α2-adrenoceptor agonists in the pithed rat. Br J Pharmacol 91:355–365

Meininger GA, Davis MJ (1992) Cellular mechanisms involved in the vascular myogenic response. Am J Physiol 263:H647–H659

Meininger GA, Trzeciakowski JP (1988) Vasoconstriction is amplified by autoregulation during vasoconstrictor-induced hypertension. Am J Physiol 254:H709–H718

Minneman KP (1988) α1-Adrenergic receptor subtypes, inositol phosphates, and sources of cell Ca2+. Pharmacol Rev 40:87–119

Nasman J, Jansson CC, Akerman KE (1997) The second intracellular loop of the α_2-adrenergic receptors determines subtype-specific coupling to cAMP production. J Biol Chem 272:9703–9708

Nebigil C, Malik KU (1993) α-Adrenergic receptor subtypes involved in prostaglandin synthesis are coupled to Ca^{++} channels through a pertussis toxin-sensitive guanine nucleotide-binding protein. J Pharmacol Exp Ther 266:1113–1124

Neylon CB, Nickashin A, Tkachuk VA, Bobik A (1998) Heterotrimeric G_i protein is associated with the inositol 1,4,5-trisphosphate receptor complex and modulates calcium flux. Cell Calcium 23:281–289

Nichols AJ (1991) α-Adrenoceptor signal-transduction mechanisms. In: α-Adrenoceptors: molecular biology, biochemistry and pharmacology, ed. Ruffolo RR Jr, pp 44–74. Basel: S. Karger AG

Nichols AJ, Ruffolo RR Jr (1988) The relationship between α-adrenoceptor reserve and agonist intrinsic efficacy to calcium utilization in the vasculature. Trends in Pharmacological Sciences 9:236–240

Nichols AJ, Motley ED, Ruffolo RR Jr (1988) Differential effect of pertussis toxin on pre- and post-junctional α2-adrenoceptors in the cardiovascular system of the pithed rat. Eur J Pharmacol 145:345–349

Nichols AJ, Motley ED, Ruffolo RR Jr (1989) Effect of pertussis toxin treatment on post-junctional α1- and α2-adrenoceptor function in the cardiovascular system of the pithed rat. J Pharmacol Exp Ther 249:203–209

Nishigaki K, Faber JE, Ohyanagi M (1991) Interactions between α-adrenoceptors and adenosine receptors on microvascular smooth muscle. Am J Physiol 260:H1655–H1666

Nishimura J, Khalil RA, Drenth JP, van Breemen C (1990) Evidence for increased myofilament Ca2+sensitivity in norepinephrine- activated vascular smooth muscle. Am J Physiol 259:H2–H8

Nishio E, Nakata H, Arimura S, Watanabe Y (1996) α_1-Adrenergic receptor stimulation causes arachidonic acid release through pertussis toxin-sensitive GTP-binding protein and JNK activation in rabbit aortic smooth muscle cells. Biochem Biophys Res Commun 219:277–282

Ohyanagi M, Nishigaki K, Faber JE (1992) Interaction between microvascular α1- and α2-adrenoceptors and endothelium-derived relaxing factor. Circ Res 71:188–200

Pacaud P, Loirand G, Baron A, Mironneau C, Mironneau J (1991) Ca^{2+} channel activation and membrane depolarization mediated by Cl^- channels in response to noradrenaline in vascular myocytes. Br J Pharmacol 104:1000–1006

Pedrinelli R, Tarazi RC (1985) Calcium entry blockade by nitrendipine and α-adrenergic responsiveness in vivo: comparison with non-calcium entry blocker vasodilators in absence and presence of phenoxybenzamine pre-treatment. J Pharmacol Exp Ther 233:636–642

Quayle JM, Bonev AD, Brayden JE, Nelson MT (1994) Calcitonin gene-related peptide activated ATP-sensitive K^+ currents in rabbit arterial smooth muscle via protein kinase A. J Physiol (Lond) 475:9–13

Ruffolo RR Jr, Zeid RL (1985) Relationship between α-adrenoceptor occupancy and response for the α1-adrenoceptor agonist, cirazoline, and the α2-adrenoceptor agonist, B-HT 933, in canine saphenous vein. J Pharmacol Exp Ther 235:636–643

Ruffolo RR Jr, Morgan EL, Messick K (1984) Possible relationship between receptor reserve and the differential antagonism of α_1- and α_2-adrenoceptor-mediated pressor responses by calcium channel antagonists in the pithed rat. J Pharmacol Exp Ther 230:587–594

Ruffolo RR Jr, Motley ED, Nichols AJ (1991) The effect of pertussis toxin on α_1-adrenoceptor-mediated vasoconstriction by the full agonist, cirazoline, and the partial agonist, (−)-dobutamine, in pithed rats. Fundam Clin Pharmacol 5:11–23

Schlicker E, Urbanek E, Gothert M (1989) ATP, α,β-methylene ATP and suramin as tools for characterization of vascular P_{2x} receptors in the pithed rat. J Auton Pharmacol 9:357–366

Shinozuka K, Hashimoto M, Masumura S, Bjur RA, Westfall DP, et al (1994) In vitro studies of release of adenine nucleotides andadenosine from rat vascular endothelium in response to α_1-adrenoceptor stimulation. Br J Pharmacol 113:1203–1208

Speden RN, Warren DM (1986) The interaction between noradrenaline activation and distension activation of the rabbit ear artery. J Physiol (Lond) 375:283–302

Tanaka Y, Nakayama K (1998) Measurement of intracellular Ca^{2+} concentration changes by use of fura-2 in the generation of myogenic contraction of dog cerebral artery in response to quick stretch. Res Commun Mol Pathol Pharmacol 99:169–186

Tateishi J, Faber JE (1995) ATP-sensitive K+ channels mediate α-2D-adrenergic receptor contraction of arteriolar smooth muscle and reversal of contraction by hypoxia. Circ Res 76:53–56

Timmermans PB, Mathy MJ, Thoolen MJ, de Jonge A, Wilffert B, et al (1984) Invariable susceptibility to blockade by nifedipine of vasoconstriction to various α_2-adrenoceptor agonists in pithed rats. J Pharm Pharmacol 36:772–775

Timmermans PB, Beckeringh JJ, Van Zwieten PA, Thoolen MJ (1987) Sensitivity of α_1-adrenoceptor-mediated pressor responses to inhibition by Ca^{++}-entry blockers in the pithed rat: arguments against the role of receptor reserve. J Pharmacol Exp Ther 240:864–870

Ullrich S, Wollheim CB (1984) Islet cyclic AMP levels are not lowered during $\alpha 2$-adrenergic inhibition of insulin release. J Biol Chem 259:4111–4115

Van Bavel E, Mulvany MJ (1994) Role of wall tension in the vasoconstrictor response of cannulated rat mesenteric small arteries. J Physiol (Lond) 477:103–115

Van den Brink FG (1973) The model of functional interaction. I. Development and first check of a new model of functional synergism and antagonism. Eur J Pharmacol 22:270–278

Van Helden DF (1988) An α-adrenoceptor-mediated chloride conductance in mesenteric veins of the guinea pig. J Physiol (Lond) 401:489–501

Van Meel JC, de Jonge A, Kalkman HO, Wilffert B, Timmermans PB, et al (1981a) Organic and inorganic calcium antagonists reduce vasoconstriction in vivo mediated by postsynaptic α_2-adrenoceptors. Naunyn Schmiedebergs Arch Pharmacol 316:288–293

Van Meel JC, De Jonge A, Kalkman HO, Wilffert B, Timmermans PB, et al (1981b) Vascular smooth muscle contraction initiated by postsynaptic α_2-adrenoceptor activation is induced by an influx of extracellular calcium. Eur J Pharmacol 69:205–208

Wesselman JP, VanBavel E, Pfaffendorf M, Spaan JA (1996) Voltage-operated calcium channels are essential for the myogenic responsiveness of cannulated rat mesenteric small arteries. J Vasc Res 33:32–41

Wilson KM, Minneman KP (1990) Pertussis toxin inhibits norepinephrine-stimulated inositol phosphate formation in primary brain cell cultures. Mol Pharmacol 38:274–281

Wright IK, Harling R, Kendall DA, Wilson VG (1995) Examination of the role of inhibition of cyclic AMP in $\alpha 2$-adrenoceptor mediated contractions of the porcine isolated palmar lateral vein. Br J Pharmacol 114:157–165

Xiong Z, Sperelakis N (1995) Regulation of L-type calcium channels of vascular smooth muscle cells. J Mol Cell Cardiol 27:75–91

CHAPTER 3
G_s Protein-Coupled Receptors in Human Heart

A.J. KAUMANN

A. Introduction

Endogenous amines, such as the neurotransmitter noradrenaline or humoral adrenaline, 5-hydroxytryptamine (5-HT) and histamine, enhance the rate and force of the heart beat. These amines act as agonists at specific membrane receptors that are usually coupled to the G_s protein which, in turn, usually uses adenylyl cyclase as an effector. Activation of some of these receptors can be beneficial or harmful to heart function. Some interesting properties of these cardiac receptors will be discussed, particularly their function in human heart. A vast array of cardiac tissues and cells from animals have been used as models of human heart function. Surprisingly, extrapolations and inferences about receptor-mediated modulation of human heart function from results obtained from animal cardiac tissues and cells can be misleading, resulting in the need for direct experimentation on human heart tissues and cells. The aim of this article is to concentrate on quantitative aspects of the function of coexisting G_s protein-coupled receptors in human heart to ascertain their relative importance. The value of choice of relevant animal models is critically stressed. The function of individual receptors will be compared first, followed by some indirect evidence for cross-talk among G_s-coupled receptors.

B. Receptor Subtypes

Three mammalian β-adrenoceptors, which usually couple to G_s protein, have been cloned so far (BYLUND et al. 1994). It has been known since 1972 (CARLSSON et al. 1972) that both β_1- and β_2-adrenoceptors can mediate cardiostimulant effects. β_3-adrenoceptors have been reported to mediate cardiodepressant effects (GAUTHIER et al. 1996, 1998) but, as discussed below, this is controversial. A third cardiostimulant β-adrenoceptor has been proposed (KAUMANN 1989), previously considered as a putative β_4-adrenoceptor (KAUMANN 1997), but now demonstrated to involve the β_1-adrenoceptor (KAUMANN et al. 2000).

It has been known since 1989 that myocardial receptors for 5-hydroxytryptamine (5-HT) are of the 5-HT$_4$ subtype and localised in the atrium (Kaumann et al. 1989 and 1990) and sinoatrial node (Kaumann 1990). (For a general classification of 5-HT receptors, see Hoyer et al. 1994 and Hartig et al. 1996). G_s protein-coupled receptors for histamine (Hill et al. 1997) have H_2 characteristics in heart (Black et al. 1972) and coexist with functional H_1 receptors in human atrium (Sanders et al. 1996).

C. β-Adrenoceptor Subtypes

I. Comparison of $β_1$- and $β_2$-Adrenoceptors

1. Localisation

Both $β_1$- and $β_2$-adrenoceptors have been found to coexist in mammalian hearts. Using autoradiographic techniques and combinations of a radioligand and selective antagonists for one or the other subtype, Molenaar and colleagues (Buxton et al. 1987; Elnatan et al. 1994; Summers and Molenaar 1995) mapped $β_1$- and $β_2$-adrenoceptors in human left and right atrium and ventricle, sinoatrial node, atrioventricular (AV) node, His bundle and Purkinje fibres. As found in the hearts of other species, the density of human cardiac $β_1$-adrenoceptors is higher than that of $β_2$-adrenoceptors. The density of $β_2$-adrenoceptors tends to be, however, higher in atrial than in ventricular myocardium and is higher in the AV conducting system than in the surrounding myocardium. The distribution of $β_1$- and $β_2$-adrenoceptors in the human AV conducting system resembles that of other species (Molenaar et al. 1990a, 1990b); the relative function of the two subtypes in this system is unknown.

2. Function of $β_1$- and $β_2$-Adrenoceptors

Noradrenaline and adrenaline can cause cardiostimulation through both $β_1$- and $β_2$-adrenoceptors. Contractility of isolated human atrial preparations is increased maximally by both noradrenaline and adrenaline through $β_1$- and $β_2$-adrenoceptors, respectively (Lemoine et al. 1988; Hall et al. 1990). Noradrenaline and adrenaline cause a similar incidence of arrhythmias mediated through $β_1$- and $β_2$-adrenoceptors, respectively, in an experimental model of arrhythmias in isolated human atrium (Kaumann and Sanders 1993). In isolated ventricular trabeculae obtained from human hearts in terminal failure, both noradrenaline and adrenaline enhanced contractility (through $β_1$- and $β_2$-adrenoceptors, respectively) to a similar extent (Kaumann et al. 1999). Furthermore, on these ventricular trabeculae, noradrenaline and adrenaline are also equipotent and equiefficacious in mediating hastened relaxation through $β_1$- and $β_2$-adrenoceptors, respectively (Kaumann et al. 1995a, 1999), as previously observed in human ventricular myocytes (Del Monte et al. 1993) and in human atria (Hall et al. 1990). Furthermore, in human atrium and ventricle both $β_1$- and $β_2$-adrenoceptors mediate similar PKA-dependent phosphorylations of phospholamban, troponin I and C protein, implicated in cardiac relaxation (Kaumann et al. 1996a; Kaumann et al. 1999). With the help of subtype-selective antagonists, the relaxant effect adrenaline in human ventricular trabeculae are found to be mediated to similar extents through $β_1$- and

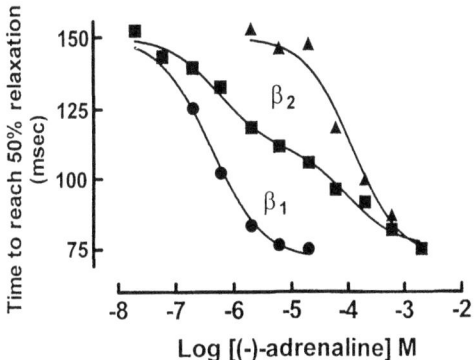

Fig. 1. Hastening of relaxation by (−)-adrenaline through both β_1- and β_2-adrenoceptors. Right-ventricular trabeculae, paced at 37°C, from a 54-year-male patient with ischaemic heart disease. The time to 50% relaxation, measured from 100-mm/s speed tracings, is shown as a function of (−)-adrenaline concentration. Cumulative concentration–effect curves for (−)-adrenaline were determined in the absence (*circles*) or presence of 300 nM CGP-20712 (*squares*) to block β_1-adrenoceptors or both 300 nM CGP-20712A and 50 nM ICI-118,551 (*triangles*) to block both β_1- and β_2-adrenoceptors. The concentration–effect curves were fitted with the equation

$$R = f_1 \frac{A}{A + K_{A1}(1 + ICI/K_{ICI} + CGP/K_{CGP1})} + f_2 \frac{A}{A + K_{A2}(1 + ICI/K_{ICI2} + CGP/K_{CGP2})}$$

where A, CGP and ICI are concentrations of (−)-adrenaline, CGP-20712 A and ICI-118,551, respectively, and K the corresponding equilibrium dissociation constants for β_1- and β_2-adrenoceptors. K values for ICI and CGP were taken from KAUMANN and LEMOINE (1987). pK_A values for (−)-adrenaline of 6.5 and 6.3 were estimated for β_1- and β_2-adrenoceptors, respectively. The fractions f_1 and f_2 of the relaxant response (R) mediated through β_1- and β_2-adrenoceptors were 0.44 and 0.56, respectively. The area between the *circles* and *squares* corresponds approximately to effects mediated through β_1-adrenoceptors. The area between the *squares* and *triangles* corresponds roughly to effects mediated through β_2-adrenoceptors (KAUMANN and SANDERS, unpublished)

β_2-adrenoceptors (Fig. 1). This evidence, taken together, is apparently puzzling because, in human heart, the β_1-adrenoceptor population is consistently greater than the β_2-adrenoceptor population, regardless of cardiac region (BUXTON et al. 1987; ELNATAN et al. 1994; SUMMERS and MOLENAAR 1995), and predominates in terminal heart failure (though less than in non-failing hearts; BRISTOW et al. 1986). Clearly, the contribution of β_1- and β_2-adrenoceptors to the effects of physiological catecholamines is not proportional to the size of the corresponding receptor populations. To account for this disproportion, a biochemical hypothesis has been advanced.

3. Selective Coupling of β_2-Adrenoceptors

A tighter coupling of β_2-adrenoceptors to the G_s protein–adenylyl cyclase system (compared with that of β_1-adrenoceptors) was proposed to occur in human atrial (GILLE et al. 1985) and ventricular membranes from non-failing (KAUMANN and LEMOINE 1987) and failing hearts (BRISTOW et al. 1989). The proposal came from the analysis of concentration–effect curves of adenylyl cyclase stimulation by noradrenaline and adrenaline in the absence and presence of antagonists selective for β_1- and/or β_2-adrenoceptors. Fractional participation of β_1- and β_2-adrenoceptors was estimated with non-linear analysis under the assumption of interaction of the catecholamine with the two receptor populations (GILLE et al. 1985; KAUMANN and LEMOINE 1987). Both adrenaline, which has similar affinity for human ventricular β_1- and β_2-adrenoceptors, and noradrenaline, which has an around twenty times higher affinity for β_1-adrenoceptors (KAUMANN et al. 1989a), stimulated adenylyl cyclase more through β_2- than through β_1-adrenoceptors (GILLE et al. 1985; KAUMANN and LEMOINE 1987; Fig. 2). Assessments of both membrane adenylyl cyclase stimulation and the density of membrane receptors yielded an estimate of the number of molecules of cyclic adenosine monophosphate (cAMP) produced through activation of one receptor per minute. This was obtained by dividing the agonist-evoked activity of the adenylyl cyclase mediated through a given receptor subtype by the corresponding subtype receptor density. When stimulated by noradrenaline and adrenaline, one β_2-adrenoceptor led to the production of around four and seven times more cAMP molecules, respectively, than did one β_1-adrenoceptor (Table 1).

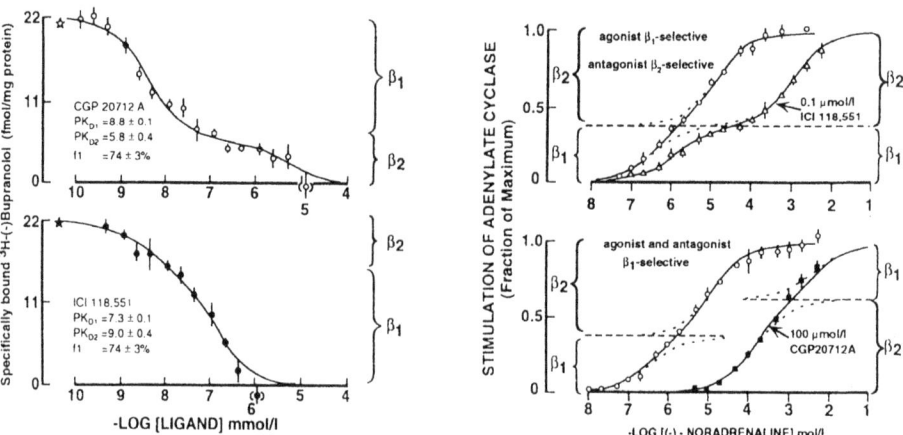

Fig. 2. Comparison of β_1- and β_2-adrenoceptor densities with the corresponding fractional responses of adenylyl cyclase in human ventricular membranes from patients without advanced heart failure. The receptors were labelled with [^{3}H]-(−)-bupranolol and β_1-selective CGP-20712A (*left upper panel*) and β_2-selective ICI-118551 (*left lower panel*) compete for binding. Concentration–effect curves for adenylyl (adenylate)-cyclase stimulation by (−)-noradrenaline and their shift by ICI-118551 and CGP-20712A are shown in the *right upper* and *lower panels*, respectively. Non-linear analysis was carried out with equations for two coexisting receptor populations. β_1- and β_2-adrenoceptor densities and fractional responses are depicted by *curly brackets*. Notice that fractional adenylyl cyclase responses of (−)-noradrenaline through β_2-adrenoceptors are greater than those through β_1-adrenoceptors, but the β_1-adrenoceptor density is greater (KAUMANN and LEMOINE 1987)

Table 1. Estimate of coupling of β_1- and β_2-adrenoceptors to the G_s protein–adenylyl cyclase system in membranes from non-failing human hearts. Calculated from atrial data of GILLE et al (1985) and ELNATAN et al. (1994) and ventricular data of KAUMANN and LEMOINE (1987)

Receptor	Tissue	Disease	Density fmol/mg	Adenylyl cyclase pmol/(mg × min)	Molecules cAMP min^{-1}
β_1	Ventricle	Mitral lesion	29.8	4.85 (adrenaline)	163
				6.85 (noradrenaline)	234
β_2	Ventricle	Mitral lesion	12.2	14.55 (adrenaline)	1195
				12.42 (noradrenaline)	1020
β_1	Atrium	Ischaemia	29.0	8.83 (noradrenaline)	304
β_2	Atrium	Ischaemia	15.2	17.92 (noradrenaline)	1179

cAMP, cyclic adenosine monophosphate.

Independent evidence obtained from human ventricle (Fig. 3) supports tighter coupling of β_2-adrenoceptors than of β_1-adrenoceptors. The ability of the receptor to form a high-affinity complex with hormone and G_s protein can be tested with guanosine triphosphate (GTP), guanosine diphosphate (GDP) or a non-hydrolysable analogue of GTP (GTPγS) known to cause dissociation of the hormone from the receptor (RODBELL et al. 1971; LONDOS et al. 1974) by decreasing its affinity. High- and low-affinity receptor states are conventionally described by the equilibrium dissociation constants K_H and K_L. The magnitude of the K_L/K_H ratio seems proportional to the tightness of coupling between receptor and G protein (KENT et al. 1980). Analysis of the experiment on human ventricular membranes with the non-hydrolysable GTP analogue GTPγS of Fig. 3 revealed that the K_L/K_H ratio is 13 for noradrenaline and 12 for adrenaline at β_1-adrenoceptors. In contrast, for β_2-adrenoceptors, the K_L/K_H ratio was considerably greater – 219 for noradrenaline and 387 for adrenaline. Interestingly, the K_L/K_H ratio appears to be independent of the affinity of the catecholamine for the high- and low-affinity states of the β_2-adrenoceptor (Fig. 3), suggesting that it is related to a similar conformational state evoked by the binding of noradrenaline ($K_H = 0.21\,\mu M$, $K_L = 46\,\mu M$) and adrenaline ($K_H = 0.015\,\mu M$, $K_L = 5.8\,\mu M$). The evidence is consistent with tighter coupling of β_2-adrenoceptors (compared with β_1-adrenoceptors) to the G_s protein–adenylyl cyclase system of human heart and appears to be tissue independent, because it has been confirmed with human recombinant transfected (GREEN et al. 1992) and co-transfected (LEVY et al. 1993) β_1- and β_2-adrenoceptors. There is β_1-adrenoceptor polymorphism. The more abundant arginine-389 variant appears to couple more firmly to the G_s protein–adenylyl cyclase system than the less abundant glycine-389 variant (MASON et al. 1999). GREEN et al. (1992) transfected the glycine-389 variant, while LEVY et al. (1993) co-transfected the arginine-389 variant (LEVY, personal

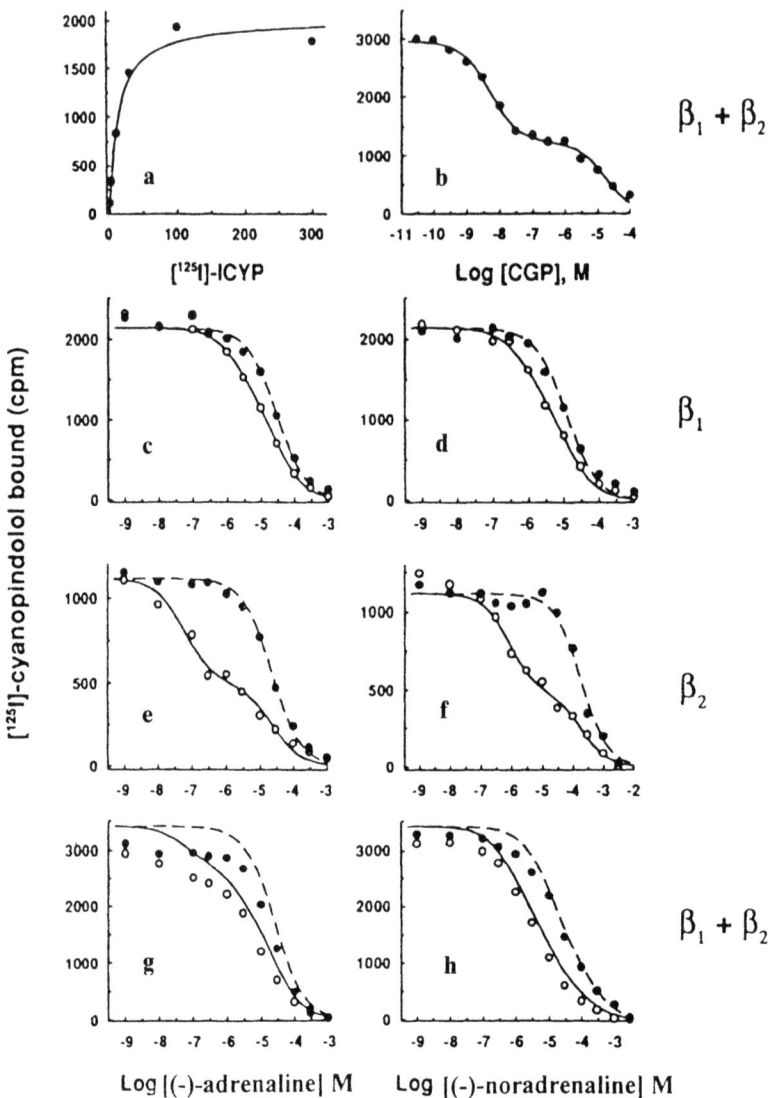

Fig. 3. Comparison of shifts of binding-inhibition curves of physiological catecholamines by guanosine triphosphate (GTP)γS (100 μM) through human ventricular $β_1$-adrenoceptors (**c,d**), $β_2$-adrenoceptors (**e,f**) and both $β_1$- and $β_2$-adrenoceptors (**g,h**). **c–h** depict binding-inhibition curves in the absence (○) and presence (●) of GTPγS. β-Adrenoceptors were labelled with (–)-[^{125}I]-cyanopindolol (**a**), and a binding-inhibition curve for CGP-20712A (**b**) was plotted to assess the proportion of $β_1$-adrenoceptors (58%) and $β_2$-adrenoceptors (42%). To block $β_2$-adrenoceptors, ICI-118551 (50 nM) was present in **c** and **d**. To block $β_1$-adrenoceptors, CGP-20712A (300 nM) was present in **e** and **f**. All assays were carried out on left-ventricular membranes from the heart of a 58-year-old male patient with dilated cardiomyopathy in terminal failure. Notice that GTPγS shifts were greater with $β_2$- than $β_1$-adrenoceptors. Data from **b–h** were fitted simultaneously with an equation for two receptor populations (KAUMANN et al. 1995a)

communication). Although both recombinant β_1-adrenoceptor variants couple less to the G_s protein–adenylyl cyclase system than the β_2-adrenoceptors do, it would be interesting to know whether it is possible to detect a difference in coupling with native cardiac β_1-adrenoceptors in homozygous individuals expressing either variant.

Selective coupling was not inferred from the joint analysis of experiments with receptor binding and adenylyl cyclase assays from cat heart, in which the magnitude of adenylyl cyclase stimulation is proportional to the corresponding densities of β_1- and β_2-adrenoceptors (KAUMANN et al. 1989a). In addition, feline cardiac β_1-adrenoceptors (but not β_2-adrenoceptors) hasten ventricular relaxation, suggesting that only the former but not the latter receptors use the G_s /cyclic AMP pathway (LEMOINE and KAUMANN 1991). Furthermore, the situation is even more disparate in murine heart compared with human heart. Rat β_2-adrenoceptors from adult rat and mouse ventricular myocytes appear to be coupled to G_i protein (XIAO et al. 1994, 1999) which inhibits adenylyl cyclase, with functional consequences unrelated to the human cardiac β_2-adrenoceptor coupled to the G_s protein–adenylyl cyclase pathway. Only when the G_i protein is inhibited by pertussis toxin can a coupling of rat β_2-adrenoceptors to the G_s protein–adenylyl cyclase pathway be demonstrated (XIAO et al. 1994, 1999). In mouse left atria β_2-adrenoceptors do not mediate positive inotropic effects of adrenaline, not even after pertussis toxin treatment (Oostendorp and Kaumann 2000), in sharp contrast to human atrium in which β_2-adrenoceptors mediate maximum inotropic effects of adrenaline and are coupled selectively to the G_s protein–adenylyl cyclase system (GILLE et al. 1985). Evidence from sheep (BOREA et al. 1992) and dog heart (ALTSCHULD et al. 1995) also suggests that only β_1-adrenoceptors (not β_2-adrenoceptors) couple effectively to a G_s protein–adenylyl cyclase pathway. These examples illustrate that it would be misleading to extrapolate conclusions from cardiac β_1- and β_2-adrenoceptors of several species to their function in human heart, stressing the need for direct experimentation on isolated preparations from human heart. In the hamster heart, however, β_2-adrenoceptors are more tightly coupled to the G_s protein–adenylyl cyclase system than β_1-adrenoceptors are (WITTE et al. 1995), thus resembling the situation in human heart and perhaps providing a relevant experimental model. Interestingly, in genetically cardiomyopathic hamsters, catecholamines appear to stimulate adenylyl cyclase entirely through β_2-adrenoceptors, while β_1-adrenoceptors cease to couple to the enzyme through G_s protein (WITTE et al. 1995, 1998).

The size of the functional G_s protein pool appears to be sufficient to permit complete coupling of both β_1- and β_2-adrenoceptors in human ventricle. This is demonstrated by the panels in Fig. 3g,h. When both β_1- and β_2-adrenoceptors are left to interact simultaneously with the G_s protein, the effect of GTPγS is roughly equivalent to the sum of the large effects on the smaller β_2-adrenoceptor population and the small effects on the larger β_1-adrenoceptor population. This finding is in contrast to the situation in some host cells, which greatly overexpress recombinant receptors at such high densities that the G_s protein pool is insufficient for complete coupling to the receptors (KENAKIN 1997).

Human β_2-adrenoceptors overexpressed approximately 50- to 200-fold in mouse heart couple spontaneously to the G_s protein–adenylyl cyclase system in the absence of agonist (MILANO et al. 1994; BOND et al. 1995). The hearts and atria of these transgenic mice contracted stronger and faster than those of littermates but were resistant to further contraction with (−)-isoprenaline. β-Adrenoceptor-blocking agents caused decreases in contractile force on atria of these transgenic mice, presumably by changing the β_2-adrenoceptors into a conformation that caused uncoupling from the G_s protein, i.e. by acting as inverse agonists (BOND et al. 1995). In human atrial myocytes, propranolol and atenolol can act as inverse agonists in the presence of forskolin by decreasing the L-type Ca^{2+} current (MEWES et al. 1993), perhaps through phosphorylation of the β_1-adrenoceptor. A puzzling observation is that ICI-118551, a selective β_2-adrenoceptor antagonist, failed to inhibit the L-type Ca^{2+} current (MEWES et al. 1993), suggesting that, under these conditions, native β_2-adrenoceptors are not pre-coupled to the G_s protein–adenylyl cyclase system. This is in contrast to agonist-stimulated β_2-adrenoceptors that mediated protein kinase A (PKA)-dependent increases of L-type Ca^{2+} current in human atrial myocytes (SKEBERDIS et al. 1997). Greatly overexpressed human β_2-adrenoceptors in mouse heart can also couple to G_i protein (Xiao et al. 1999). Native ventricular β_2-adrenoceptors from failing human hearts, however, couple to the G_s protein–adenylyl cyclase pathway (Kaumann et al. 1999), despite increased G_i protein levels (Feldman et al. 1988) compared with non-failing hearts. These profound functional discrepancies between native human cardiac β_2-adrenoceptors and recombinant β_2-adrenoceptors overexpressed in mouse heart preclude a variety of extrapolations from the latter to the former system.

D. Is There a Functional role for Cardiac β_3-Adrenoceptors?

I. Evidence Against Cardiostimulation

The recombinant β_3-adrenoceptor (EMORINE et al. 1989) and natively occurring β_3-adrenoceptors in adipocytes (ARCH and KAUMANN 1993) are usually (but not always) thought to couple to G_s protein. β_3-Adrenoceptor messenger RNA (mRNA) has been reported in human ventricle (KRIEF et al. 1993; GAUTHIER et al. 1996) and atrium (BERKOWITZ et al. 1995), but localisation in cardiac fat cells has not unambiguously been excluded. The advent of β_3-adrenoceptor-selective agonists (ARCH and KAUMANN 1993) has furnished tools to test the hypothesis, expected from a G_s protein coupled receptor, that myocardial β_3-adrenoceptors mediate cardiostimulant effects. However, four β_3-adrenoceptor-selective agonists failed to enhance contractions of human ventricular trabeculae in the presence of the β_1/β_2-adrenoceptor blocker nadolol (MOLENAAR et al. 1997a). This finding agrees with a similar lack of β_3-adrenoceptor-mediated cardiostimulant effects in rats in vivo (MALINOWSKA and SCHLICKER 1996) and in vitro (KAUMANN and MOLENAAR 1996) and ferrets in vitro (LOWE et al. 1998).

II. Evidence for Cardiostimulation

Recently, SKEBERDIS et al. (1999) reported that nanomolar concentrations of the β_3-adrenoceptor-selective agonists BRL-37344 and SR-58611 increased L-type Ca^{2+} current 1.7-fold and 2.2-fold in human atrial myocytes, respectively. The effects of BRL-37344 were resistant to blockade by $10\,\mu mol/l$ nadolol. These interesting results are puzzling because, in isolated human atrial preparations, both BRL-34377 and SR-58611 failed to elicit positive inotropic effects (KAUMANN et al. 1997), which would have been expected from the increases in L-type Ca^{2+} current.

III. Evidence for Cardiodepression

Under some conditions, however, β_3-adrenoceptors appear to be able to couple to inhibitory G_i protein in adipocytes (CHAUDRY et al. 1994; BEGIN-HEICK 1995). Recently, GAUTHIER et al. (1996) reported comprehensive evidence obtained from ventricular biopsies of transplanted human hearts that nanomolar concentrations of β_3-adrenoceptor agonists depress contractility and abbreviate the durations of action potentials. GAUTHIER et al. (1996, 1998) also claimed that high concentrations of (–)-isoprenaline and (–)-noradrenaline elicited cardiodepressant effects under a condition of blockade of both β_1- and β_2-adrenoceptors by nadolol, an antagonist that has low affinity for β_3-adrenoceptors (BOND and CLARKE 1988). As expected from β_3-adrenoceptors (ARCH and KAUMANN 1993), the cardiodepressive effects of a β_3-adrenoceptor-selective agonist, BRL-37344, were antagonised by bupranolol. GAUTHIER et al. (1996) found that the cardiodepressant effects of BRL-37344 were attenuated by pre-treatment of the ventricular tissue with pertussis toxin, and used this evidence to suggest coupling to a G protein (i.e. G_i) that inhibits adenylyl cyclase. Based on these experiments and observations (NANTEL et al. 1993) that β_3-adrenoceptors are more resistant to desensitisation than β_1- and β_2-adrenoceptors, GAUTHIER et al. (1996) proposed that the cardiodepressant effects of noradrenaline may further impair cardiac function in patients with heart failure.

More recently, GAUTHIER et al. (1998) went a step further and attributed the cardiodepressant effects of the β_3-adrenoceptor agonist BRL-37344 to release and action of nitric oxide (NO). The negative inotropic effects of BRL-37344 were reduced by inhibiting NO-evoked activation of guanylyl cyclase with methylene blue and inhibiting constitutively occurring NO synthase with the arginine analogues N^G-nitro-L-arginine methylester and N^G-monomethyl-L-arginine (L-NMMA). The inhibitory effect of L-NMMA was partially reversed by L-arginine. BRL-37344 increased NO production, and immunological staining of the NO synthase was demonstrated both in cardiomyocytes and in endothelial cells. Consistent with activation of the NO effector guanylyl cyclase, both BRL-34377 and (−)-isoprenaline caused a threefold increase in cyclic guanosine monophosphate (GMP) levels. The effect of BRL-37344 was abolished by L-NMMA, reduced by bupranolol and prevented by pertussis toxin, consistent with the NO pathway, mediation through β_3-adrenoceptors and coupling to G_i protein.

IV. Evidence Against Cardiodepression

The cardiodepressant effects reported by GAUTHIER et al. (1996) were not observed in isolated human cardiac preparations by others. Using micromolar concentrations of four β_3-adrenoceptor-selective agonists (including BRL-37344) in the presence of nadolol, KAUMANN and MOLENAAR (1997) failed to find cardiodepression in human atrium, and MOLENAAR et al. (1997a) did not observe cardiodepression in human ventricular trabeculae from failing hearts. HARDING (1997) did not detect cardiodepression with a β_3-adrenoceptor-selective agonist in human ventricular myocytes. Reasons for the lack of confirmation of β_3-adrenoceptor-mediated effects of GAUTHIER et al. (1996) are unknown. Their results are particularly puzzling, because it is the only evidence known for human β_3-adrenoceptors in which BRL-37344 is reputed to be an agonist with nanomolar potency. The agonist potency of BRL-37344 is, however, considerably lower for human than for murine β_3-adrenoceptors (ARCH and KAUMANN 1993). BRL-37344 has only micromolar affinity for human recombinant β_3-adrenoceptors and micromolar potency for the β_3-adrenoceptor of human adipocytes, for which it is only a partial agonist (SENNITT et al. 1998). BRL-37344 is not selective for human recombinant β_3-adrenoceptors, because its affinity for recombinant human β_2-adrenoceptors is slightly higher, and it is a partial agonist of similar intrinsic activity for adenylyl cyclase stimulation through these two receptors (SENNITT et al. 1998). A cardiostimulant effect of BRL-37344, mediated through β_2-adrenoceptors, has also been reported for human atrium (KAUMANN and SANDERS, cf in ARCH and KAUMANN 1993). Clearly, the interesting findings of GAUTHIER et al. (1996, 1998) – and especially the claim of mediation through β_3-adrenoceptors – require confirmation by other laboratories before achieving widespread acceptance.

E. Cardiostimulant Effects Through the Putative β_4-Adrenoceptor

I. Non-Conventional Partial Agonists

The existence of a third cardiostimulatory β-adrenoceptor was proposed in 1989 (KAUMANN 1989). The proposal was based on a class of β-adrenoceptor blocking agents that caused cardiostimulant effects at concentrations considerably higher than those that antagonised the cardiostimulant effects of catecholamines. These agents were designated *non-conventional partial agonists* to differentiate them from classical partial agonists in isolated cardiac muscle, which often (but not always; Fig. 9) exhibit concentration–effect curves that can be fitted closely by the corresponding β-adrenoceptor curve by using equilibrium dissociation constants estimated from antagonism (KAUMANN 1973, 1989). An important group of non-conventional partial agonists are the indoleamine pindolol and its analogues tert-butylpindolol, cyanopindolol, iodocyanopindolol, hydroxybenzylpindolol, iodohydroxybenzylpindolol, carazolol and tert-butylcarazolol (KAUMANN et al. 1979; BEARER et al. 1980; KAUMANN and BLINKS 1980; KAUMANN 1983). However, cardiostimulant effects of non-conventional partial agonists are not restricted to indoleamines, and include (in addition to pindolol) other clinically used β-adrenoceptor blocking agents, such as alprenolol and oxprenolol (KAUMANN and BLINKS 1980). The hydrophilic benzimidazolone CGP-12177 (STAEHELIN et al. 1983) is a non-conventional partial agonist that has become a particularly useful tool, both as an agonist (KAUMANN 1983, 1989, 1996; KAUMANN and MOLENAAR 1996, 1997; MALINOWSKA and SCHLICKER 1996; KAUMANN et al. 1998; LOWE et al. 1998) and as a radioligand (SARSERO et al. 1998a, 1998c, 1999) for researching the putative β_4-adrenoceptor.

The dissociation between the blockade of β_1/β_2-adrenoceptors and the cardiostimulant effects of CGP-12177 is illustrated in the experiments illustrated in Fig. 4. CGP-12177 enhanced sinoatrial beating rate and force of contraction of both paced left atrium and right-ventricular papillary muscle. The intrinsic activities of these cardiostimulant effects of CGP-12177 are smaller than those of the catecholamine (–)-isoprenaline and are tissue dependent. A maximally effective cardiostimulant concentration of CGP-12177 antagonised the effects of (–)-isoprenaline in a surmountable manner by causing around 3.5- to 4.5-log-unit shifts of the concentration–effect curve for (–)-isoprenaline. Using algebra and statistics (MARANO and KAUMANN 1976) based on classical stimulus theory (STEPHENSON 1956), equieffective concentrations of (–)-isoprenaline in the presence and absence of CGP-12177 allowed the calculation of an affinity estimate of CGP-12177 for the β-adrenoceptor populations with which (–)-isoprenaline interacted (i.e. mostly β_1- and β_2-adrenoceptors, for which CGP-12177 has nearly the same affinity; NANOFF et al. 1987). β_1/β_2-Adrenoceptor occupancy curves for CGP-12177 were then calculated and normalised to the maximum CGP-12177-evoked cardiostimulation. As seen in Fig. 4, the β_1/β_2-adrenoceptor occupancy curves are situated at CGP-12177 concentrations around 1.5–2.3 log units lower than those of the corresponding CGP-12177 cardiostimulant-effect curves, suggesting that the latter effects are not mediated through β_1- and β_2-adrenoceptors. The pattern of dissociation between stimulation and blockade for non-conventional partial agonists, as exemplified by the experiments of Fig. 4, led to the proposal of the existence of a third cardiostimulant β-adrenoceptor subtype (KAUMANN 1989).

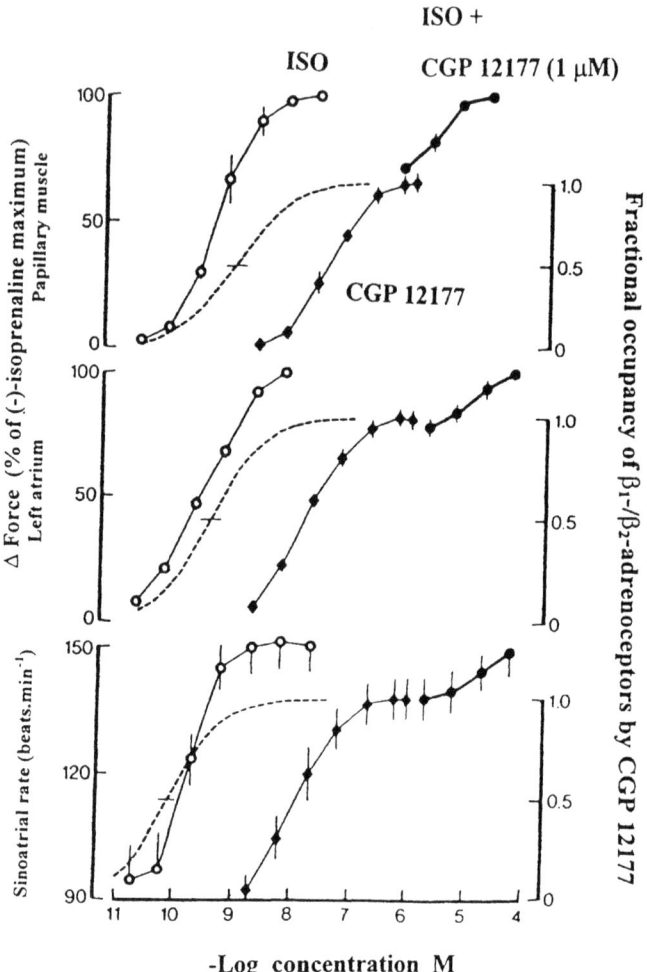

Fig. 4. CGP-12177 as non-conventional agonist. Comparison of the cardiostimulant effects of CGP-12177 with its antagonism of the positive chronotropic and inotropic effects of (−)-isoprenaline (ISO) on feline cardiac tissues. Three successive concentration–effect curves were carried out, the first for ISO (*open circles*) followed by washout, the second for CGP-12177 up to 1 µM (*diamonds*) and the third for ISO in the presence of 1 µM CGP-12177 (*closed circles*). Equilibrium dissociation constants for CGP-12177 were estimated (MARANO and KAUMANN 1976) and fractional receptor-occupancy curves calculated (*broken lines*); the logs of the errors of the constants are represented by *horizontal bars* through the midpoints of the occupancy curves. Notice the dissociation between blockade and stimulation by CGP-12177 (KAUMANN 1983)

II. The Putative β_4-Adrenoceptor Resembles – But Is Distinct from – the β_3-Adrenoceptor

At the time of the proposal of a third cardiostimulant β-adrenoceptor, it was noticed that its properties resembled those of the β_3-adrenoceptor (KAUMANN 1989). For example, several non-conventional partial agonists have agonist properties at adipocyte β_3-adrenoceptors, as verified on recombinant β_3-adrenoceptors (EMORINE et al. 1989). Both the third cardiostimulant β-adrenoceptor (KAUMANN and LYNHAM 1997) and β_3-adrenoceptor (EMORINE et al. 1989; SENNITT et al. 1998) mediate agonist effects consistent with activation of a G_s protein–cAMP-dependent pathway, and the effects are relatively resistant to blockade by antagonists of β_1- and β_2-adrenoceptors (propranolol; ARCH and KAUMANN 1993). Furthermore, after it was proposed that the β_1/β_2-adrenoceptor antagonist bupranolol also blocked the third cardiostimulant receptor (KAUMANN 1989), a variety of laboratories reported blockade of agonist effects mediated through β_3-adrenoceptors (ARCH and KAUMANN 1993). The assumption that the third cardiostimulant β-adrenoceptor was indeed a β_3-adrenoceptor prevailed (BOND and LEFKOWITZ 1996).

The development of β_3-adrenoceptor-selective agonists (ARCH and KAUMANN 1993) presented an opportunity to test the hypothesis as to whether or not the third cardiostimulant β_3-adrenoceptor was actually a β_3-adrenoceptor. The conclusion drawn from this approach was that β_3-adrenoceptor-selective agonists had neither agonist effects nor antagonist effects on the cardiostimulant effects of non-conventional partial agonists (KAUMANN and MOLENAAR 1996; MALINOWSKA and SCHLICKER 1996; MOLENAAR et al. 1997a). These experiments were inconsistent with the notion that the cardiostimulant effects of non-conventional partial agonists were mediated through β_3-adrenoceptors, leading to the designation of a "putative β_4-adrenoceptor" for what had previously been termed the third cardiostimulant β-adrenoceptor (KAUMANN 1989, 1997; KAUMANN and MOLENAAR 1997; MOLENAAR et al. 1997a). However, against this panorama, ARCH (1997) had made the interesting suggestion that the third cardiac β-adrenoceptor was actually a β_3-adrenoceptor that adopts a conformation that mediates the effects of non-conventional partial agonists but not of β_3-adrenoceptor-selective agonists. ARCH (1997) based his interpretation on association with different G proteins and/or different receptor conformations in a cell-dependent manner (KENAKIN 1995).

Conclusive experiments carried out on cardiac tissues of mice that had the β_3-adrenoceptor gene disrupted (i.e. β_3-adrenoceptor knockout mice) demonstrated that the cardiostimulant effects of (−)-CGP-12177 are unchanged compared with the effects in the hearts of wild-type mice (Fig. 5). Moreover, the cardiostimulant effects of (−)-CGP-12177 are relatively resistant to blockade by (−)-propranolol but are antagonised by (−)-bupranolol in cardiac tissues of both β_3-adrenoceptor knockout and wild-type mice (KAUMANN et al. 1998). The negative evidence with β_3-adrenoceptor-selective agonists and the persistence of (−)-CGP-12177-evoked cardiostimulant effects in the β_3-adrenoceptor knockout mice exclude the involvement of β_3-adrenoceptors. Consequently, it was thought at this stage that the putative β_4-adrenoceptor could be encoded by a gene that is distinct from the gene that encodes the β_3-adrenoceptor (KAUMANN et al. 1998).

Functional evidence for the cardiac putative β_4-adrenoceptor has been found in all mammalian species investigated so far, including man (KAUMANN

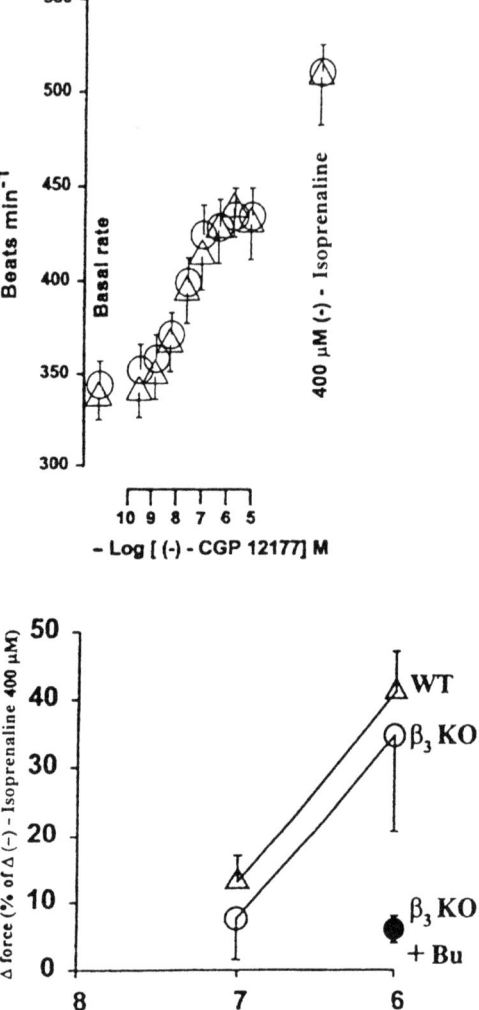

Fig. 5. Persistence of positive chronotropic (*upper panel*) and inotropic effects (*lower panel*) of (−)-CGP-12177 in atria from the hearts of mice lacking the functional β_3-adrenoceptor gene (*circles*). Comparison with atria from wild-type mice (*triangles*). The *closed circle* represents data from left atria incubated with (−)-bupranolol 1 μM (KAUMANN et al. 1998)

1996, 1997; KAUMANN and MOLENAAR 1997; SARSERO et al. 1998b). As expected for a cAMP-dependent pathway, the conventional partial agonist (−)-CGP-12177 not only increases contractile force but also hastens relaxation (i.e. a positive lusitropic effect), as demonstrated in isolated atrial and ventricular trabeculae from a failing human heart (Fig. 6) (SARSERO et al. 1998b) and also observed in rat ventricle (SARSERO et al. 1999). Similar cardiorelaxant effects of catecholamines and a β_2-adrenoceptor-selective ligand have been shown to

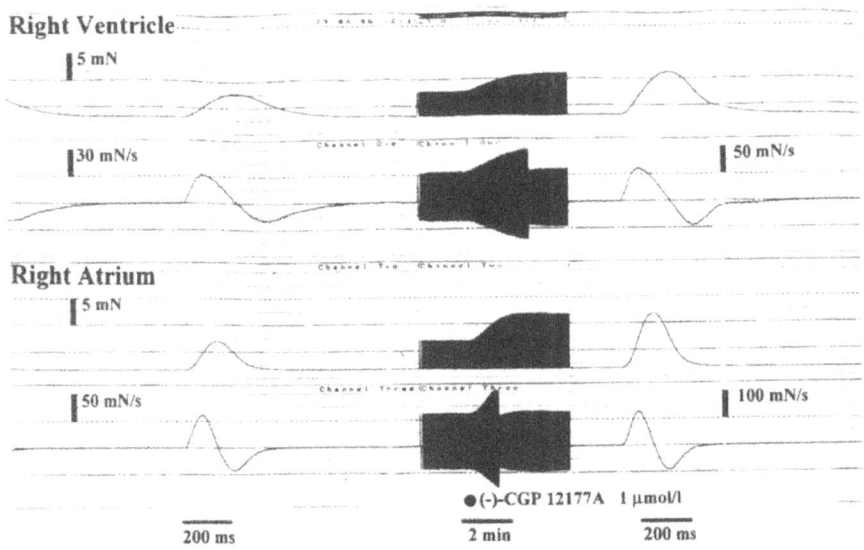

Fig. 6. Positive inotropic and lusitropic effects of (–)-CGP-12177 (added at *dot*) on a ventricular and atrial trabeculum obtained from a patient with adriamycin-induced cardiomyopathy undergoing transplant surgery. The tissues were set up in the same organ bath and paced at 1 Hz at 37°C in the presence of (–)-propranolol (200 nM) and 3-isobutyl-1-methylxanthine (60 μM). Fast and slow speed recordings of force measurement are shown together with the differentiated signal. Notice the change of calibration of the differentiated signals during the onset of action of (–)-CGP-12177 (KAUMANN and MOLENAAR 1997)

be mediated through β_1- and β_2-adrenoceptors in human ventricle (KAUMANN et al. 1999) and atrium (KAUMANN et al. 1996a) and are associated with cAMP-dependent protein-kinase-catalysed phosphorylation of proteins (phospholamban, troponin I, C protein) that are involved in the mediation of cardiac relaxation. However, whether or not these proteins are also phosphorylated under putative β_4-adrenoceptor stimulation remains an open question.

III. Which Endogenous Agonist for the Putative β_4-Adrenoceptor?

The identity of the endogenous agonist for the putative β_4-adrenoceptor is unknown. The provisional name of "putative β_4-adrenoceptor" is merely based on the high affinity of non-conventional partial agonists for β_1- and β_2-adrenoceptors, which hardly justifies its classification as a β-adrenoceptor subtype. If we are dealing with a β-adrenoceptor, one would expect catecholamines to activate this putative β_4-adrenoceptor. To investigate this, however, it is important to exclude the contribution of β_1-adrenoceptors in the mediation of catecholamine responses; this has proved frustrating, because β_1-adrenoceptor-selective antagonists are not selective enough to leave the population of β_4-adrenoceptors untouched. For example, although CGP-20712A is a highly selective antagonist for β_1-adrenoceptors, it is actually more potent as an antagonist of effects mediated through the putative β_4-adrenoceptor rather than through β_2-adrenoceptors (KAUMANN and MOLENAAR 1996). Conversely, although bupranolol blocks the putative β_4-adrenoceptor, its affinity for β_1- and β_2-adrenoceptors is considerably higher, thus making it more difficult to uncover function mediated through the putative β_4-adrenoceptor. The task of eliminating the contribution of coexisting β_1- and β_2-adrenoceptors with receptor-subtype-selective antagonists is particularly difficult in human cardiac systems, where the contribution of these two receptor populations is equally important and where the potency of the catecholamines is lower than in animal models. So far, from the analysis of blockade of cardiostimulant effects of catecholamines with antagonists selective for β_1- and β_2-adrenoceptors, no evidence has emerged for the participation of the putative β_4-adrenoceptor, suggesting that, if there is a role for catecholamines, their affinity for that receptor is probably low.

Taking advantage of the high hydrophilicity of CGP-12177 (which reduces non-specific binding), it has recently been possible to label a putative β_4-adrenoceptor population in the hearts of several species (MOLENAAR et al. 1997a; KAUMANN et al. 1998; SARSERO et al. 1998a, 1999), including man (SARSERO et al. 1998c). The density of the putative β_4-adrenoceptor population appears usually somewhat higher than the densities of coexisting β_1- and β_2-adrenoceptors. The binding affinity of (–)-CGP-12177 and other non-conventional partial agonists usually agrees with the corresponding cardiostimulant potencies. Interestingly, catecholamines also compete for binding in a stereoselective manner. However, the affinity of the catecholamines (–)-isoprenaline and (–)-noradrenaline is low, with dissociation equilibrium constants around 1 mmol/l. The low affinity of catecholamines is another property shared by the cardiac putative β_4-adrenoceptor (MOLENAAR et al. 1997a; SARSERO et al. 1998a) and the native β_3-adrenoceptor (GERMACK et al. 1997). The low affinity of catecholamines for the putative β_4-adrenoceptor may preclude detection of functional participation of catecholamines in cardiostimulant effects under current conditions. Clearly, highly selective β_1-adrenoceptor blockers of low toxicity are needed. Even more important, antagonists selective for the putative β_4-adrenoceptor are required for experiments to decide whether or not this receptor mediates cardiostimulant effects of endogenous catecholamines and, hence, can be classified as a β-adrenoceptor subtype. Recent evidence, however, indicates that the putative β_4-adrenoceptor is not a distinct catecholamine receptor but a special state of the β_1-adrenoceptor.

IV. The Putative β_4-Adrenoceptor is a Special State of the β_1-Adrenoceptor

The dissociation between blockade and stimulation caused by several β-blockers, i.e. non-conventional partial agonists (KAUMANN 1973), led to the proposal that their cardiostimulation was mediated through a receptor distinct from β_1/β_2-adrenoceptors (KAUMANN 1989): the putative β_4-adrenoceptor (KAUMANN 1997). Interestingly, however, the frequently used non-conventional partial agonist CGP-12177 exhibits agonist effects on human and rat recombinant β_1-adrenoceptors transfected into cell lines in which stimulation of adenylyl cyclase was assayed (PAK and FISHMAN 1996). The intrinsic activity of CGP-12177, compared with (−)-isoprenaline, increased from 0.21 to 0.94 as the density of transfected β_1-adrenoceptors increased from 130 fmol/mg protein receptor to 1570 fmol/mg protein receptor. Others have not detected stimulation of adenylyl cyclase for human recombinant β_1-adrenoceptors (SENNITT et al. 1998) transfected at around 200 fmol/mg and using up to 0.1 mmol/l CGP-12177 (ARCH, personal communication). Using (−)-[^{3}H]-CGP-12177, PAK and FISHMAN (1996) found two binding sites, one large (90%), with subnanomolar affinity, and another small (10%), with an approximately 100 times lower affinity; the latter was GTPγS sensitive.

Some binding and pharmacological properties of (−)-CGP-12177 in cardiac tissues differ from those found with recombinant β_1-adrenoceptors. Although the binding affinity of (−)-[^{3}H]-CGP-12177 for putative β_4-adrenoceptors is similar to that of the low-affinity site for recombinant β_1-adrenoceptors (PAK and FISHMAN 1996), the density of the putative β_4-adrenoceptor population appears consistently greater than that of the β_1-adrenoceptor population in cardiac membranes from rat, man and mouse (KAUMANN et al. 1998; SARSERO et al. 1998a, 1998c). In addition, unlike binding to the low-affinity state of the recombinant β_1-adrenoceptor, which is reduced by GTPγS (PAK and FISHMAN 1996), binding of (−)-[^{3}H]-CGP-12177 to cardiac putative β_4-adrenoceptor is not affected by GTP (SARSERO et al. 1998a). Although the positive inotropic potency of (−)-CGP-12177 is similar to the cardiac binding affinity of (−)-[^{3}H]-CGP-12177 at putative β_4-adrenoceptors, (−)-CGP-12177 is actually approximately 40 times more potent than (−)-isoprenaline in mouse ventricular myocytes in eliciting arrhythmic Ca^{2+} transients (FREESTONE et al. 1999). With these discrepancies it would appear unlikely that the putative β_4-adrenoceptor is a state or conformation of the β_1-adrenoceptors evoked by binding of non-conventional partial agonists to the β_1-adrenoceptor. Alternatively, three different conformations of the cardiac β_1-adrenoceptor mediate: (1) the classical cardiostimulant effects of (−)-isoprenaline and other catecholamines; (2) the arrhythmic effects of (−)-CGP 12177; and (3) the positive inotropic, lusitropic and chronotropic effects of (−)-CGP 12177 and other non-conventional partial agonists. Only conformations (1) and (3) but not (2) would be observed with the recombinant β_1-adrenoceptors of Pak and Fishman (1996).

CGP-12177 also has agonist properties for recombinant β_2-adrenoceptors, but with lower intrinsic activity than it has for recombinant β_1-adrenoceptors (PAK and FISHMAN 1996) and recombinant and native β_3-adrenoceptors (SENNITT et al. 1998). The effects of (−)-CGP 12177 were therefore investigated on cardiac tissues from mice lacking cardiac β_2-adrenoceptors (β_2-AR knockout – CHRUSCINSKI et al. 1999) and mice lacking both β_1- and β_2-adrenoceptors

(β_1-AR/β_2-AR double knockout – ROHRER et al. 1999). The cardiostimulant effects of (–)-CGP 12177 were present in wild-type and β_2-AR knockout mice but were absent in β_1-AR/β_2-AR double knockout mice, despite functional preservation of the post-receptor cAMP-dependent pathway (KAUMANN et al. 2000). A ventricular binding site with unaltered affinity for (–)-[^{3}H]-CGP 12177, previously attributed to represent part of the putative β_4-adrenoceptor (SARSERO et al. 1999), persisted in the β_1-AR/β_2-AR double knockout. Consequently, β_1-adrenoceptors have an obligatory role in the mediation of cardiostimulant effects of (–)-CGP 12177 and possibly other non-conventional partial agonists. The β_1-adrenoceptor may possess an allosteric binding site through which non-conventional partial agonists evoke or stabilise a conformation responsible for the pharmacology previously attributed to a putative β_4-adrenoceptor.

F. 5-HT$_4$ Receptors

Early work carried out in healthy volunteers demonstrated that intravenously administered 5-HT usually caused dose-dependent tachycardia accompanied by occasional chest pain (LE MESSURIER et al. 1959), sometimes preceded by bradycardia and hypotension (HOLLANDER et al. 1957). These observations summarise the main role of 5-HT in human heart. 5-HT$_{1B}$ and 5-HT$_{2A}$ receptors share the mediation of 5-HT-evoked contractions of human coronary artery (KAUMANN et al. 1994a) and may have mediated coronary spasm, thus accounting for chest pain in the volunteers. The bradycardia and early transient hypotension is probably (no direct evidence in man) elicited when 5-HT reaches vagal sensory nerve endings thereby initiating the Bezold-Jarisch reflex and simultaneously inhibits sympathetic nerve output, as observed experimentally in cats. The 5-HT-induced reflex and hypotension are prevented by application of a selective antagonist for 5-HT$_3$ receptors to the cardiac vagal nerve endings (MOHR et al. 1987).

I. Coupling to a cAMP Pathway

The main question was: which receptor mediates the 5-HT-evoked tachycardia in man? An indirect approach was to study in vitro the effects of 5-HT in human atrium. Although the author observed positive inotropic effects of 5-HT on paced isolated preparations of human atrium in 1983, the nature of the receptors involved remained obscure until 1989, because they were resistant to blockade by the 5-HT receptor antagonists available at that time. The effects of 5-HT were also resistant to blockade of α_1-, β_1- and β_2-adrenoceptors, ruling out indirect increases of contractile force through 5-HT-induced release of noradrenaline from cardiac nerve endings. Since it was known that, in mollusc hearts, 5-HT enhanced cAMP levels (SAWADA et al. 1984), cAMP was also measured in human atria and found to be increased with 5-HT; this was accompanied by an increase in cAMP-dependent protein kinase (PKA) activity (KAUMANN et al. 1989b, 1990). It was also observed that 5-HT hastened the relaxation of human atrium. It was then proposed that the 5-HT-induced increase in PKA activity causes the phosphorylation of the cardiac sarcolemmal L-type Ca^{2+} channel, phospholamban and troponin I (KAUMANN et al. 1990). Phosphorylation of the L-type Ca^{2+} channel enhances the Ca^{2+} current which, in turn, causes Ca^{2+}-induced Ca^{2+} release from the sarcoplasmic reticulum, thereby producing the positive inotropic effects of 5-HT. Phospholamban phosphorylation facilitates the activity of the Ca^{2+} pump of the sarcoplasmic reticulum by removing the inhibition caused by non-phosphorylated phospholamban, thus reducing Ca^{2+} concentrations in the vicinity of myofilaments and, consequently, hastening atrial relaxation. The phosphorylation of troponin I may also contribute to atrial relaxation by decreasing the affinity of troponin C for Ca^{2+}.

Electrophysiological data supported the proposal of a cAMP-dependent cascade for 5-HT responses of human atrium. 5-HT causes a reversible increase in L-type Ca^{2+} current that is as marked (about sixfold over basal current) as that produced by (-)-isoprenaline through β-adrenoceptors (JAHNEL et al. 1992, 1993; OUADID et al. 1992). The 5-HT-evoked increase in L-type Ca^{2+} current is not additive with the effects of intracellularly administered cAMP and is prevented by an inhibitor of PKA, consistent with an obligatory role of PKA (OUADID et al. 1992). 5-HT causes a greater availability of L-type Ca^{2+} channels (JAHNEL et al. 1993), presumably due to PKA-induced phosphorylation.

II. 5-HT$_4$-like Receptors

The proposal for the existence of cerebral 5-HT$_4$ receptors (DUMUIS et al. 1988, 1989) prompted the use of the 5-HT$_3$ receptor blockers tropisetron (ICS 205,930) and benzamide derivatives renzapride and cisapride as tools to define the nature of the human atrial 5-HT receptors. The positive inotropic effects of 5-HT were competitively blocked by tropisetron but not by the 5-HT$_3$ receptor-selective antagonists granisetron and MDL-72222 (KAUMANN et al. 1990). More recently, selective 5-HT$_4$-receptor antagonists were introduced and assayed on human atrium, including SB-203186 (PARKER et al. 1995), GR-113808 (KAUMANN 1993) and SB-207710 (KAUMANN et al. 1994b), which competitively blocked atrial 5-HT receptors (Table 2), supporting their 5-HT$_4$ nature. Radioiodinated SB-207710 was used to label atrial 5-HT receptors of piglet and man (KAUMANN et al. 1995b, 1996b). The density of human atrial 5-HT receptors is low, amounting to around 10% and 20% of the densities of β_1- and β_2-adrenoceptors, respectively (KAUMANN et al. 1996b). In piglet atria, the density of 5-HT$_4$ receptors is ten times lower than in human atria (KAUMANN et al. 1995b). The low densities of human and piglet atrial 5-HT$_4$ receptors may, in part, account for the smaller maximum positive inotropic

Table 2. Potency and affinity of ligands for human atrial 5-hydroxytryptamine (5-HT)$_4$ receptors. pEC$_{50}$ is –log(M) of the EC$_{50}$ for agonists and partial agonists. pK_P, pK_B and pK_D are equilibrium dissociation constants [–log(M)] estimated from antagonism of the effects of 5-HT by the partial agonist (P) or antagonist (B) and from binding (D) of ligands to 5-HT$_4$ receptors labelled with [^{125}I]SB-207710

	pEC$_{50}$	pK_P	pK_B	pK_D	References
5-HT	7.4–7.9			5.8	KAUMANN et al. (1990, 1991a, 1996); SANDERS et al. (1995)
5-CT	4.7			4.9	KAUMANN et al. (1991a, 1996b)
Renzapride	6.3	6.7		6.4	KAUMANN et al. (1991a, 1996b)
Cisapride	6.1	6.2		6.0	KAUMANN et al. (1991a, 1996b)
SB-207710			10.1	9.7	KAUMANN et al. (1994, 1996b)
[^{125}I] SB-207710				9.6	KAUMANN et al. (1996b)
GR-113808			8.8		KAUMANN (1993)
SB-203186			8.7	8.0	PARKER et al.(1995); KAUMANN et al. (1996b)
SDZ-205-557			7.7		ZERKOWSKI et al. (1993)
Tropisetron			6.7	6.1	KAUMANN et al. (1990, 1996)

5-CT, 5-carboxyamidotryptamine.

effects of 5-HT compared with those caused by catecholamines through human (KAUMANN et al. 1990, 1991a; SANDERS and KAUMANN 1992; SANDERS et al. 1995) and porcine (KAUMANN et al. 1991b; LORRAIN et al. 1992) β-adrenoceptors. In enzymatically disaggregated human atrial myocytes, however, 5-HT is more efficacious than in tissues, presumably due to additional 5-HT$_4$ receptors uncovered by the disaggregating enzymes (SANDERS et al. 1995).

The effects of 5-HT were also competitively antagonised by the substituted benzamides renzapride and cisapride (Fig. 7), which are partial agonists in human atrium (KAUMANN et al. 1991a) and piglet atrium (KAUMANN 1990; MEDHURST and KAUMANN 1993). By measuring increases in L-type Ca^{2+} current in atrial myocytes, OUADID et al. (1992) confirmed that, compared with 5-HT, renzapride is a partial agonist for human atrial 5-HT$_4$ receptors. As observed with 5-HT in brain (DUMUIS et al. 1988, 1989), 5-HT (and, to a lesser extent, renzapride and cisapride) increase human atrial cAMP levels and cAMP-dependent protein kinase activity (KAUMANN et al. 1990, 1991a; SANDERS et al. 1995). Unlike human and porcine atrium, however, cisapride and renzapride are actually more efficacious than (and at least as potent as) 5-HT in stimulating adenylyl cyclase in embryonic colliculi neurones in culture (DUMUIS et al. 1988). Furthermore, unlike brain 5-HT$_4$ receptors, where cisapride is slightly more efficacious and potent than renzapride, human atrial 5-HT$_4$ receptors have lower affinity and efficacy for cisapride than for renzapride, as confirmed both by drug-antagonism (Figs. 7, 8) and competition for binding with [^{125}I]-SB-207710 (Table 2). Due to these quantitative differences of the effects of the benzamides between atrial and brain 5-HT$_4$ receptors, atrial 5-HT$_4$ receptors were initially referred to as 5-HT$_4$-like receptors

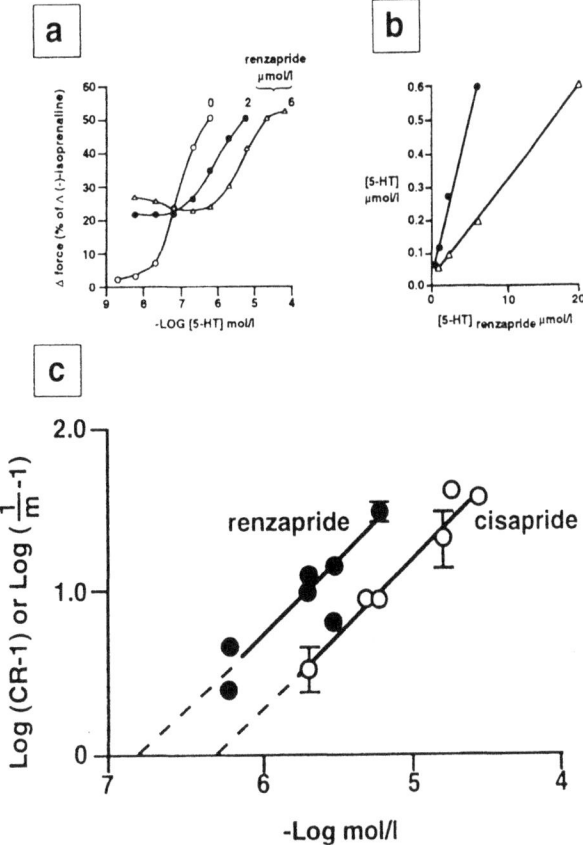

Fig. 7. Antagonism of the positive inotropic effects of 5-hydroxytryptamine (5-HT) in human atrium (**a,b**) by renzapride, and (**c**) comparison of affinity of renzapride and cisapride for 5-HT$_4$ receptors of human atrium. **a** Concentration–effect curves of 5-HT in the absence (○) and presence of 2 µM (●) and 6 µM (△) renzapride on three atrial strips obtained from a 63-year-old male patient with coronary-artery disease. **b** Plot of equieffective 5-HT concentrations in the absence and presence of the two renzapride concentrations used in **a** (MARANO and KAUMANN 1976). Fractional 5-HT$_4$ receptor-occupancy estimates calculated from the slopes of the plots in **b** were 0.91 and 0.97 with 2 µM and 6 µM renzapride, respectively; the corresponding equilibrium dissociation constants [–log(M)] were 6.70 and 6.76. **c** Schild-plots (CR, concentration ratios) or plots (m, slope of the regressions in **b**) according to LEMOINE and KAUMANN (1982). Notice that renzapride has an affinity for the 5-HT$_4$ receptors threefold higher than that of cisapride (KAUMANN et al. 1991a)

(KAUMANN 1990; KAUMANN et al. 1991a). It was also noted that human atrial 5-HT$_4$-like receptors closely resemble other peripheral 5-HT$_4$ receptors in various gut regions (KAUMANN et al. 1991a).

Since the cloning of the rat 5-HT$_4$ receptor [with its two splice variants: short (5-HT$_{4S}$) and long (5-HT$_{4L}$; GERALD et al. 1995)], it has become plausible that the pharmacological differences between brain and heart 5-HT$_4$ receptors can be attributed to the mediation by different splice variants. Despite the recent cloning of additional species homologues, the question of subtle dif-

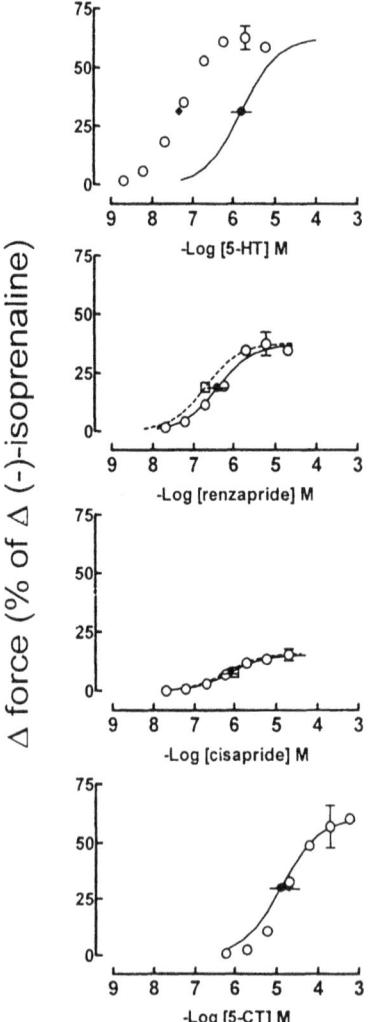

Fig. 8. Comparison of the concentration–effect curves (○) of the positive inotropic effects of 5-hydroxytryptamine (5-HT), renzapride, cisapride and 5-carboxyamidotryptamine with the corresponding fractional 5-HT$_4$ receptor occupancy-curves calculated from the binding-inhibition experiments of KAUMANN et al. (1996b; *solid lines*) and from antagonism of the positive inotropic effects of 5-HT by renzapride and cisapride (*broken lines*; KAUMANN et al. 1991a). 5-HT$_4$ receptors were labelled with [^{125}I]-SB-207710. Equilibrium dissociation constants and their standard errors are shown as ● and *horizontal bars*; for actual values, see Table 2. Notice that both the intrinsic activity and affinity are lower for cisapride than for renzapride

ferences between 5-HT$_4$ receptors in brain and heart has not yet been definitively solved. For example, both splice variants are expressed in the brains of rat (GERALD et al. 1995) and mouse (CLAEYSEN et al. 1996), but only 5-HT$_{4S}$ mRNA has been detected in rat atrium (GERALD et al. 1995) – a puzzling finding, because there is no functional evidence for atrial 5-HT$_4$ receptors in this species (KAUMANN 1991a). Both mouse recombinant and cerebral 5-HT$_{4L}$ receptors have higher binding affinity and efficacy for cisapride and renzapride than do human atrial 5-HT$_4$ receptors. Cisapride is even more potent (DUMUIS et al. 1988) and has a higher affinity than renzapride (CLAEYSEN et al. 1996), while the opposite is true in human atrium (Figs. 7, 8). Cisapride is a full agonist compared with 5-HT, producing a sevenfold increase in cAMP for a human cloned 5-HT$_4$ receptor (which resembles the rat 5-HT$_{4L}$ receptor) transfected into COS-7 cells at densities several hundreds times higher than the density of human atrial 5-HT$_4$ receptors (VAN DEN WYNGAERT et al. 1997). These results suggest that it is unlikely that the 5-HT$_{4L}$ splice variant plays an important role for the effects of cisapride in human atrium.

A human atrial 5-HT$_{4S}$ splice variant has recently been cloned (CLAEYSEN et al. 1997) and, when transfected at a density (10 fmol/mg) almost as low as that of native atrial 5-HT$_4$ receptors (4 fmol/mg; KAUMANN et al. 1996b) in COS-7 cells, has been shown to mediate weak partial agonist activity (cAMP) with renzapride. An atrial 5-HT$_4$ receptor cloned by BLONDEL et al. (1997) also exhibited only partial agonist activity with renzapride despite being transfected at an approximately 50-fold-higher density into COS-7 cells than its density in human atrium (KAUMANN et al. 1996b; CLAEYSEN et al. 1997). These studies suggest that the 5-HT$_{4S}$ splice variant – expressed in human atrium but not ventricle (BLONDEL et al. 1997) – plays an important role in the mediation of the effects of 5-HT. None of these studies with human recombinant 5-HT$_4$ receptors has systematically compared the effects of renzapride and cisapride on the same splice variant. VAN DEN WYNGAERT et al. (1997) only used cisapride, while both CLAEYSEN et al. (1997) and BLONDEL et al. (1997) only used renzapride. A systematic comparison of the pharmacology of the two recombinant human splice variants was made by BACH et al. (2000) who found that 5-HT$_{4S}$ and 5-HT$_{4L}$ receptors (expressed in COS-7 cells) have essentially identical pharmacology that, with the exception of cisapride, greatly resembled the pharmacology of human atrium. Cisapride and renzapride were partial agonists (adenylyl cyclase stimulation) at moderate receptor densities and became full agonists at high receptor densities. Unlike the situation in human atrium (KAUMANN et al. 1991a), cisapride showed both higher agonist potency and binding affinity than renzapride. mRNA for both splice variants was detected in both right and left human atrium, suggesting that the effects of 5-HT and renzapride (but perhaps not those of cisapride) are mediated through both splice variants (BACH et al. 2000). The possibility that other 5-HT$_4$ receptor splice variants function in human atrium cannot yet be discarded. In addition to the 5-HT$_{4S}$ and 5-HT$_{4L}$ splice variants (GERALD et al. 1995), another two cerebral splice variants of murine and human 5-HT$_4$ receptors have recently been cloned (BOCKAERT et al. 1998). The four splice variants are now designated 5-HT$_{4a}$ (5-HT$_{4S}$), 5-HT$_{4b}$ (5-HT$_{4L}$), 5-HT$_{4c}$ and 5-HT$_{4d}$ and appear to show pharmacological differences when transfected into host cells (BOCKAERT et al. 1998) and more splice variants are expected to be disclosed.

G. Cross-Talk Between Cardiac G$_s$-Coupled Receptors, as Revealed by Chronic Blockade of β_1-Adrenoceptors

An unsuspected inotropic hyperresponsiveness to (–)-adrenaline was observed in isolated right-atrial tissues obtained from British patients compared with atria from German patients. The discovery was prompted by comparing inotropic EC_{50} values for (−)-adrenaline in atrial tissues obtained from over 100 German patients with EC_{50} values from a dozen British patients. (−)-Adrenaline was ten times less potent on German atria than on British atria. It was soon found that the German surgeons did not operate on patients treated with β-adrenoceptor blocking agents, while the British surgeons did (KAUMANN et al. 1989c). The prevailing dogma of the 1980s was that chronic blockade of β_1-adrenoceptors enhanced their function due to receptor upregulation. The use of antagonists highly selective for β_1- and β_2-adrenoceptors as tools proved the dogma to be groundless for human atria when receptor density was assessed with two independent techniques. The hyperresponsiveness to (−)-adrenaline in atria from patients chronically treated with β_1-adrenoceptor-selective blockers was mediated through β_2-adrenoceptors (KAUMANN et al. 1989c; HALL et al. 1990). The densities of neither β_1- nor β_2-adrenoceptors, assessed with both cell autoradiography and membrane binding, differed from the densities in atria from non-β-adrenoceptor blocker-treated patients (KAUMANN et al. 1995a; MOLENAAR et al. 1997b). Binding affinities of ligands selective for either β_1- or β_2-adrenoceptors (including the affinity of a partial agonist of high intrinsic activity) in atrial membranes obtained from patients treated with β_1-adrenoceptor-selective blockers did not differ from affinities in membranes from patients who did not receive such treatment (MOLENAAR et al. 1997b).

In contrast, the cardiostimulant effects of salbutamol, mediated through human atrial β_2-adrenoceptors, are greatly enhanced in atria from patients chronically treated with β_1-adrenoceptor-selective blockers; however, the blocking potency of salbutamol for the β_2-adrenoceptors, estimated from the antagonism of (−)-adrenaline-evoked effects, is unchanged compared with the blocking potency in tissues from patients not treated with β-adrenoceptor blockers (Fig. 9). The concentration–effect curve of salbutamol on atria obtained from non-β-adrenoceptor blocker-treated patients can be fitted by the fractional occupancy of β_2-adrenoceptors while, in atria from β-adrenoceptor blocker-treated patients, the concentration–effect curve of salbutamol is situated at lower concentrations than its β_2-adrenoceptor-occupancy curve (Fig. 9). However, no β_1-adrenoceptor hyperresponsiveness was observed with (−)-noradrenaline (KAUMANN et al. 1989c; HALL et al. 1990) or (−)-adrenaline (KAUMANN 1991b). Even the maximal responses to a β_1-adrenoceptor-selective partial agonist, (−)-RO363, were only slightly enhanced (and inotropic potency was not changed) by chronic treatment of patients with β_1-adrenoceptor-selective blockers (MOLENAAR et al. 1997b). Unlike the β_2-adrenoceptor hyperresponsiveness to (−)-adrenaline and salbutamol, the positive inotropic effects of dibutyryl cAMP in atria (HALL et al. 1990) and of forskolin in atrial myocytes (SANDERS et al. 1995) are not changed by chronic β-adrenoceptor-blocker treatment, excluding modifications downstream of the adenylyl cyclase. Taken together, these results suggest that chronic β_1-adrenoceptor blockade improves the coupling of atrial β_2-adrenoceptors (HALL et al. 1990) and perhaps other atrial receptors coupled to G_s protein

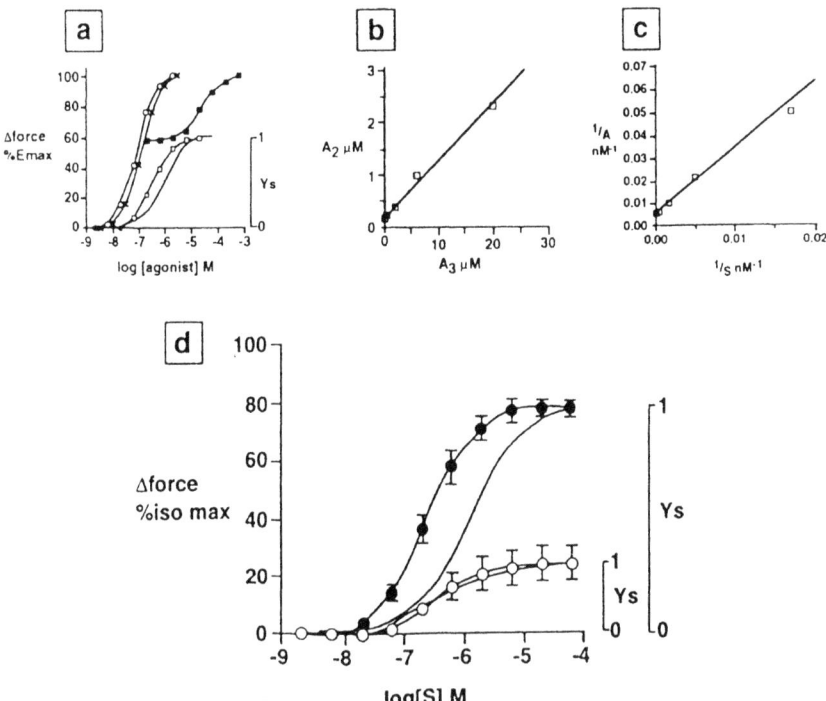

Fig. 9a–d. Inotropic hyperresponsiveness to salbutamol mediated through β_2-adrenoceptors without a change of affinity in atria from patients chronically treated with the β_1-adrenoceptor-selective antagonist atenolol. **a–c** Positive inotropic effects of salbutamol and antagonism of the positive inotropic effects of (–)-adrenaline by salbutamol in a human atrium obtained from an atenolol-treated patient. **a** A concentration–effect curve to (–)-adrenaline (○) followed by washout a curve for salbutamol was determined up to 20 μM (□) and a second curve for (–)-adrenaline was determined in the presence of salbutamol (■). The responses to (–)-adrenaline were corrected for desensitisation, as determined from successive curves for (–)-adrenaline on a paired atrial strip (✖). **b** Plot of equieffective concentrations of (–)-adrenaline in the absence (A_2) and presence (A_3) of salbutamol. The fractional β_2-adrenoceptor occupancy by salbutamol – $Y_S = [S]/([S] + K_S) = 1-m$, where K_S is the equilibrium dissociation constant), calculated from the slope m with the equation $A_2 = i + mA_3$ (MARANO and KAUMANN 1976) – is represented on the right side of the curve for salbutamol in **a**. **c** Plot of the inverse equieffective concentrations of (–)-adrenaline and salbutamol. K_S was estimated from the slope/intercept (WAUD 1969). The $-\log(K_S)$ values estimated from both **b** and **c** were 6.1. **d** Comparison of inotropic concentration–effect curves for salbutamol from 12 atenolol-treated patients (●) and ten non-atenolol-treated patients (○), with the corresponding fractional β_2-adrenoceptor occupancy curves. Notice that chronic atenolol treatment did not modify the affinity of salbutamol (i.e. the positions of the Y_S curves) and that the Y_S curves fitted the data for salbutamol in atria from non-atenolol-treated patients but not the salbutamol data from atenolol-treated patients (HALL et al. 1990)

(such as histamine H_2) and of 5-HT_4 (KAUMANN 1991b) coupled to effectors, including adenylyl cyclase.

Hitherto, the function of five receptor populations, all presumably mainly coupled to G_s protein, has been studied in atria from patients either not treated or treated with β-adrenoceptor blockers (usually blockers selective for β_1-adrenoceptors). The receptors are (in addition to the β_1- and β_2-adrenoceptors described above) 5-HT_4 receptors (KAUMANN and SANDERS 1994; SANDERS et al. 1995), histamine H_2 receptors (SANDERS et al. 1996) and putative β_4-adrenoceptors (KAUMANN 1996). The rank order of receptor hyperresponsiveness caused by chronic treatment of patients with β_1-adrenoceptor-selective blockers, as assessed with atrial positive inotropic responses, is:

$$\beta_2 > 5\text{-}HT_4 \approx H_2 \gg \beta_1 > \beta_4 \tag{1}$$

The human atrial hyperresponsiveness caused by chronic treatment with β_1-adrenoceptor blocking agents can be assumed to result from the elimination of a noradrenaline-evoked partial reduction of the function of some G_s protein-coupled receptors. Does this cross-talk occur in the same myocyte? If this were so, different receptor populations would have to coexist in the same myocyte. This has been shown to be the case for human ventricular β_1- and β_2-adrenoceptors (DEL MONTE et al. 1993), atrial β-adrenoceptors and 5-HT_4 receptors (SANDERS et al. 1995). The putative β_4-adrenoceptor has been shown to co-function with the β_1-adrenoceptor in the same murine atrial and ventricular myocytes (FREESTONE et al. 1999; SARSERO et al. 1999). These examples make it plausible that the β_1-adrenoceptor, activated chronically (on a scale of days) by noradrenaline, emits a signal that decreases the function of other coexisting G_s protein-coupled receptors in the same cardiomyocyte.

How would chronic β_1-adrenoceptor blockade improve coupling to β_2-adrenoceptors and perhaps other receptors of G_s protein? One possibility is that the inhibitory effects of G_i proteins on effectors, including adenylyl cyclase, are reduced by chronic blockade of β_1-adrenoceptors, thereby uncovering an otherwise partially repressed G_s function. This may be called the G_i hypothesis. High plasma levels of noradrenaline observed in heart-failure patients lead to an enhanced expression of inhibitory G_i proteins without changes in G_s protein expression (FELDMAN et al. 1988; NEUMANN et al. 1988). The hypothesis that G_i protein suppresses partially β-adrenoceptor-mediated contractility has received support by experiments of BROWN and HARDING (1992). They demonstrated that cardiomyocytes from failing hearts exhibit a blunted response to (–)-isoprenaline but that the (–)-isoprenaline-evoked contractions are increased after inactivating G_i protein with pertussis toxin. Experiments in rat cardiomyocytes (REITHMANN et al. 1989) and tissues (MÜLLER et al. 1993) demonstrate that high catecholamine concentrations increase the expression of G_i protein, probably due to a cAMP-dependent increase in $Gi\alpha_2$ mRNA transcription (MÜLLER et al. 1993).

Given that these effects are reputed to occur through β_1-adrenoceptor activation, one would expect chronic blockade of these receptors to eliminate the excess G_i levels and function. One apparent implication could perhaps be that when β_1-adrenoceptors are activated by noradrenaline they also manifest coupling to G_i protein. However, there is evidence for coupling of β_2-adrenoceptors (but not of β_1-adrenoceptors) to G_i protein in the rat and mouse heart (XIAO et al. 1994, 1999). On the other hand, in line with the G_i hypothesis, pigs injected with the β_1-adrenoceptor-selective antagonist bisoprolol for 35 days have decreased levels of atrial and ventricular $Gi\alpha_2$ mRNA and protein and also tend to have low G_s protein levels (PING et al. 1995). Another β_1-adrenoceptor-selective antagonist, metoprolol, reduced total G_i protein levels by one quarter in ventricular biopsies from patients with congestive heart failure (JAKOB et al. 1996). Although the latter two pieces of evidence agree with the G_i hypothesis, they do not explain why some G_s-coupled receptors show greater hyperresponsiveness than other G_s-coupled receptors after chronic β_1-adrenoceptor blockade. Positive inotropic responses of human atrial and ventricular preparations, mediated through β_1-, β_2- and putative β_4-adrenoceptors, are all reduced by adenosine and carbachol (MOLENAAR et al. 1998), consistent with functional antagonism through G_i protein activation. However, β_2-adrenoceptors – but neither β_1-adrenoceptors (KAUMANN et al. 1989c; HALL et al. 1990; MOLENAAR et al. 1997b) nor putative β_4-adrenoceptors (KAUMANN 1996) – exhibit hyperresponsiveness in human atria obtained from β-blocker-treated patients. From the G_i hypothesis, one would expect a general increase in the function of G_s-coupled receptors, which does not occur. Assuming that G_i protein function is reduced in atria from β-blocker-treated patients, the G_i hypothesis would have to be restricted to only the β_2-adrenoceptor compartment. Thus, the G_i hypothesis does not appear to generally account for receptor-dependent hyperresponsiveness after chronic blockade of β_1-adrenoceptors.

Human atrial β_2-adrenoceptor hyperresponsiveness after chronic β_1-adrenoceptor blockade could also be due to the suppression of (–)-noradrenaline-evoked phosphorylation of β_2-adrenoceptors by cAMP-dependent protein kinase (PKA; KAUMANN 1991b). The portion of the β_2-adrenoceptor population that is phosphorylated would not couple to G_s protein, thereby reducing inotropic responses. Indirect evidence is consistent with this mechanism (the PKA hypothesis). Under certain conditions, human coronary arteries are relaxed by adrenaline through β_2-adrenoceptors and by noradrenaline through β_1-adrenoceptors (FERRO et al. 1995). Prolonged incubation of the arteries with noradrenaline (16h) decreases the response to adrenaline, perhaps through PKA-dependent phosphorylation of β_2-adrenoceptors (FERRO et al. 1995). A limitation of the PKA hypothesis is that it does not account for the lack of PKA phosphorylation sites at the recombinant 5-HT$_4$ receptor (GERALD et al. 1995), whose native form exhibits marked atrial hyperresponsiveness after chronic β_1-adrenoceptor blockade (KAUMANN and SANDERS 1994; SANDERS et al. 1995).

Evidence with human atrial histamine receptors points to multiple factors involved in the hyperresponsiveness (caused by chronic β_1-adrenoceptor blockade) of some G_s protein-coupled receptors. Histamine H_2 receptors mediate increases in the contractile force of human atria, and there is hyperresponsiveness of this receptor in atria from β-blocker-treated patients (SANDERS et al. 1996). As found previously with 5-HT_4 receptors (SANDERS et al. 1995), H_2 receptors mediate enhanced increases of cAMP levels and more marked stimulation of PKA activity in atria from patients chronically treated with β-blockers selective for $β_1$-adrenoceptors compared with atria from non-β-blocker-treated patients (SANDERS et al. 1996). An assessment of the receptor specificity of the histamine responses led to the remarkable and unsuspected finding that the H_1-receptor blocker mepyramine attenuated both the inotropic and biochemical hyperresponsiveness to histamine in atria from $β_1$-adrenoceptor blocker-treated patients (SANDERS et al. 1996). Another related finding was that histamine produced tenfold and 25-fold increases in cyclic GMP levels that were blockable by mepyramine in atria from patients not treated and chronically treated with a $β_1$-adrenoceptor-selective blocker, respectively (SANDERS et al. 1996). These findings are consistent with hyperresponsiveness of both G_s protein-coupled H_2 receptors and G_q protein-coupled H_1 receptors induced by chronic blockade of $β_1$-adrenoceptors. It has been suggested that the extra cyclic GMP produced through H_1-receptor stimulation in atria from patients chronically treated with $β_1$-adrenoceptor blockers contributes to the inotropic and biochemical hyperresponsiveness to histamine mediated through H_2 receptors. It has been proposed (SANDERS et al. 1996) that the extra cyclic GMP inhibits phosphodiesterase III (PDE III), thereby preventing hydrolysis of cAMP, as illustrated below.

$$\begin{array}{l} \text{histamine} \to H_2 \to \text{cAMP} \to \text{PKA} \to \text{increased contractility} \\ \qquad\qquad\qquad\quad \uparrow \text{hydrolysis} \\ \qquad\qquad\qquad\quad \text{PDE III} \\ \qquad\qquad\qquad\quad \uparrow \text{inhibition} \\ \text{histamine} \to H_1 \to \text{cyclic GMP} \end{array} \qquad (2)$$

The example with histamine H_1 and H_2 receptors opens new questions about the heterogeneous consequences of chronic blockade of human atrial $β_1$-adrenoceptors. How is H_1 hyperresponsiveness initiated? Do other G_q protein-coupled receptors become hyperresponsive? Do PKA-catalysed phosphorylations of G_q-coupled receptors play a role? It has been reported in this context that endothelin-1, presumably acting through a G_q protein-coupled receptor, produces a higher incidence of experimental arrhythmias in atria obtained from β-blocker-treated patients than in atria from non-treated patients (BURRELL et al. 2000).

The discussed evidence from human atrial tissue and myocytes about receptor cross-talk points to diverse but, so far, mostly hidden mechanisms that need to be unravelled. Clearly, the cross-talk appears to occur mainly between receptors and effectors, i.e. at the level of G proteins. A simple candidate would be the G_s protein whose function could be enhanced by chronic $β_1$-adrenoceptor blockade. This G_s hypothesis has been supported by a moderately enhanced function of human $G_s α$ protein from $β_1$-adrenoceptor blocker-treated patients, with the $G_s\alpha$ protein reconstituted in S40 cyc⁻ cell membranes (WANG et al. 1999). However, this evidence does not account for the hyperresponsiveness of G_q protein-coupled receptors and is inconsistent with the near lack of β_1-adrenoceptor hyperresponsiveness in atria obtained from patients treated with β_1-adrenoceptor-selective blockers. Other candidates are G_i proteins that oppose the effects mediated through G_s protein-coupled receptors. The G_i proteins inhibit effectors, such as adenylyl cyclase and Ca^{2+} channels, but can activate both cyclic nucleotide phosphodiesterases and K^+ channels, and both the α unit and $\beta\gamma$ units of the G protein are involved. The G protein is tightly associated with GDP. Upon binding of an agonist, the GDP dissociates, and freely available GTP binds, causing a conformational change that results in dissociation of GTP-α from $\beta\gamma$, followed by activation of the above effectors. Activation is terminated by the GTPase activity of α. GTPase activity is considerably hastened by a family of GTPase-activating proteins called regulators of G protein signalling (RGS; BERMAN and GILMAN 1998). Two such RGS proteins, RGS4 and Gα-interacting protein, attenuate G_i-mediated inhibition of cAMP synthesis (HUANG et al. 1997). RGS4 does not bind to G_s protein, and the affinity of Giα1 is higher than that of Giα2 (BERMAN et al. 1996). Conceivably, during chronic blockade of cardiac β_1-adrenoceptors, the activity of an RGS protein is enhanced, thereby inhibiting G_i protein activity and allowing augmentation of the function of some (but not all; see above) G_s protein-coupled receptors. This speculation would imply some sort of inhibitory effect of noradrenaline (mediated through β_1-adrenoceptors) on the function or expression of the relevant RGS protein. Future research with human heart G_i proteins may provide clues as to whether RGS proteins and modulation by β_1-adrenoceptors contribute to receptor cross-talk.

H. Physiological, Pathophysiological and Therapeutic Relevance

Following agonist-evoked activation of G_s protein-coupled receptors, human heart functions are modified through a cAMP-dependent pathway leading to PKA activation. Activated PKA phosphorylates a number of target proteins, including L-type Ca^{2+} channels, phospholamban, troponin I, Na^+ channels (possibly) and other proteins involved in the physiology and pathology of human heart.

I. β_1- and β_2-Adrenoceptors

The G_s protein-coupled receptors that are known to mediate beneficial effects of humoral and neuronally released catecholamines are β_1- and β_2-adrenoceptors. The former is released mainly during exercise; both are released during stress. During exercise-induced tachycardia, noradrenaline activates not only sinoatrial but also ventricular β_1-adrenoceptors, which mediate beneficial hastening of relaxation, thereby causing a relative lengthening of diastole compared to systole, thus facilitating ventricular filling. A similar improvement of diastolic function may, in principle, occur during stress, when high plasma adrenaline acts to an important extent through sinoatrial and ventricular β_2-adrenoceptors. Furthermore, it has been suggested that endogenous surges of adrenaline could, in principle, improve diastolic function through activation of β_2-adrenoceptors in patients with heart failure under treatment with β_1-adrenoceptor-selective blockers (KAUMANN et al. 1999). However, this effect seems hardly exploitable therapeutically because of the arrhythmic effects of adrenaline.

Since the work of BRISTOW et al. (1982), evidence has accumulated that chronic exposure to high noradrenaline levels in advanced heart failure downregulates β_1-adrenoceptors. β_2-Adrenoceptors desensitise less than β_1-adrenoceptors in heart failure (BRISTOW et al. 1986), presumably by uncoupling partially from G_s protein (BRISTOW et al. 1989). β-Adrenoceptor uncoupling has been observed experimentally 8h after a 3-h exposure of human ventricular preparations to a high isoprenaline concentration (KAUMANN et al. 1989a). Under these conditions, the Vmax of the isoprenaline-evoked adenylyl cyclase stimulation is reduced to half, with a concomitant reduction of maximum isoprenaline-induced increase in contractility. Total β-adrenoceptor density, apparent affinity for (−)-isoprenaline and prostaglandin-1-evoked adenylyl cyclase stimulation were unchanged. These experiments suggest that a 3-h stress causes long-lasting uncoupling of β-adrenoceptors from the G_s protein–adenylyl cyclase system. It is not yet clear whether this is a physiological mechanism to reduce cardiac oxygen consumption after a stressful situation. Although it is likely that the uncoupling is mainly of β_2-adrenoceptors, this hypothesis still requires experimental verification. One would expect a blocker of β_2-adrenoceptors to prevent receptor uncoupling. Paradoxically, however, cardiac β_2-adrenoceptors of cardiomyopathic hamsters that (like human β_2-adrenoceptors) are coupled selectively to adenylyl cyclase, actually mediate a reduced enzyme stimulation without a change in G_s protein content after 4 weeks of drinking water containing the β_1/β_2-adrenoceptor blocker propranolol (WITTE et al. 1998). It appears, therefore, that the chronic influence of β-blockers on β_2-adrenoceptor function is different in human and cardiomyopathic-hamster heart.

An enzyme that causes uncoupling of agonist-bound β-adrenoceptors from G_s protein is β-adrenoceptor kinase 1 (βARK1), which interacts with membrane-bound $\beta\gamma$ subunits of G protein (PITCHER et al. 1992). βARK1 levels are enhanced in hearts in chronic failure (UNGERER et al. 1993), possibly contributing to reduced β-adrenoceptor responsiveness. Surprisingly, co-expression of a peptide cardiac-targeted (ct) inhibitor of βARK1, βARKct, normalises left-ventricular function and partially restores responsiveness to isoprenaline in a genetic model of murine dilated cardiomyopathy. Overexpression of βARKct prevents the appearance of heart failure in the murine model (ROCKMAN et al. 1998). These provocative experiments suggest that impediment of β-adrenoceptor coupling contributes to heart failure, at least in mouse models. A mutant of the human β_2-adrenoceptor, in which the threonine is switched to isoleucine at amino acid 164, exhibits decreased activation of the adenylyl cyclase by adrenaline due to defective coupling to G_s protein and reduced receptor sequestration when expressed in transgenic mice (TURKI et al. 1996). Furthermore, the isoleucine-164 β_2-adrenoceptor polymorphism adversely affects the prognosis of patients with congestive heart failure (LIGGETT et al. 1998). However, human β_2-adrenoceptors, overexpressed in the hearts of mice with genetic dilated cardiopathy, do not improve heart failure (ROCKMAN et al. 1998) despite being (presumably) maximally coupled to the G_s/cAMP pathway (MILANO et al. 1994; BOND et al. 1995). Thus, the beneficial effects of βARKct appear related, in part, to be due to an improvement of β_1-adrenoceptor coupling in mouse heart.

Although human β_2-adrenoceptors, overexpressed in mouse hearts, were proposed to represent a model of genomic treatment of heart failure (MILANO et al. 1994; KOCH et al. 1996), it is becoming increasingly clear that a high cardiac density of these receptors is harmful to heart function. The chronic maximum G_s protein-mediated signalling in mice overexpressing about 200 fold human cardiac β_2-adrenoceptors worsen experimental heart failure (ROCKMAN et al. 1998; DU et al. 2000). Relatively small overexpression of β_1-adrenoceptors (5–15 fold) produces cardiac hypertrophy followed by reduced cardiac function (ENGELHARDT et al. 1999) illustrating the deleterious effects of chronic hyperfunction of these receptors.

Major harmful cardiac effects mediated through β-adrenoceptors are enhanced cardiac oxygen consumption (against which β-blockers were successfully developed; BLACK 1989) and arrhythmias. Experimental arrhythmias, elicited by noradrenaline through β_1-adrenoceptors and by adrenaline through β_2-adrenoceptors, have been demonstrated in isolated human atrial preparations as a function of pacing rate and are interpreted as the result of Ca^{2+} overload (KAUMANN and SANDERS 1993). These arrhythmias are a model for transient atrial arrhythmias (including the triggering of atrial fibrillation) that occurs transiently in patients undergoing coronary-artery-bypass surgery. The incidence of transient post-surgical atrial fibrillation is reduced by propranolol (ORMEROD et al. 1984), probably by preventing the interaction of high plasma concentrations of noradrenaline and adrenaline with both β_1- and β_2-adrenoceptors.

β_2-Adrenoceptor-mediated arrhythmias of adrenaline have also been observed in human ventricular myocytes (DEL MONTE et al. 1993). More recently, β_2-adrenoceptor-mediated arrhythmias (ventricular fibrillation) have been confirmed to occur in dogs that have experimental infarcts at a healed stage concomitantly with brief coronary artery occlusions (BILLMAN et al. 1997). In agreement with the hypothesis of Ca^{2+} overload, canine ventricular myocytes made from hearts susceptible to ventricular fibrillation exhibit catecholamine-evoked increases of Ca^{2+} produced through β_2-adrenoceptor stimulation (BILLMAN et al. 1997). It has been suggested (KOCH et al. 1996; but see

Rockman et al. 1998; Du et al. 2000) that overexpression of β_2-adrenoceptors, which become constitutively active when expressed at high density (Milano et al. 1994; Bond et al. 1995), could be beneficial as gene therapy in heart failure. It has also been suggested that these receptors would not mediate arrhythmias (Altschuld et al. 1995). The experimental evidence for β_2-adrenoceptor-mediated arrhythmias would, however, preclude such an approach in heart failure.

Treatment of chronic heart failure with β_1-adrenoceptor blockers is becoming increasingly accepted, and there is evidence for an increased time before transplantation becomes necessary (with metoprolol; Waagstein et al. 1993). In addition to reducing myocardial oxygen consumption (an anti-ischaemic property), β-adrenoceptor blockers may also reduce the incidence of fatal arrhythmias. An important aspect in the treatment of heart failure is the question of whether β-blockers that block nearly non-selectively both β_1- and β_2-adrenoceptors are superior to β_1-adrenoceptor-selective blockers. At least three experimental observations favour the use of non-selective blockers. (1) Noradrenaline and adrenaline are nearly equieffective in mediating cardiostimulation and phosphorylation of proteins through β_1- and β_2-adrenoceptors, respectively, in human ventricular preparations from failing hearts (Kaumann et al. 1999), consistent with similar increases in oxygen consumption through both receptors. (2) Both β_1- and β_2-adrenoceptors mediate a similar incidence of experimental arrhythmias. (3) The incidence of experimental arrhythmias mediated through both β_1- and β_2-adrenoceptors is higher in atria from patients treated chronically with β_1-adrenoceptor-selective blockers (Kaumann and Sanders 1993). The latter results mimic the well-known arrhythmias (including atrial fibrillation) observed after acute withdrawal of β-adrenoceptor-blocking agents (Pritchard et al. 1983).

Several so-called non-selective β-blockers, such as propranolol (Gille et al. 1985), pindolol (Kaumann and Lobnig 1986) and timolol (Wang et al. 1996) have actually been found to be somewhat selective for β_2-adrenoceptors (two- to fivefold) compared with β_1-adrenoceptors of human heart. Interestingly, propranolol and timolol reduce the relative risk of death after myocardial infarction by 28% and 39%, respectively, while β_1-selective atenolol and metropolol only do so by 15% and 13%, respectively (Hennekens et al. 1996). A recently meta-analysis of double-blind, placebo-controlled, randomised trials from 3023 patients with chronic heart failure also concludes that the reduction of mortality risk is greater for β-blockers that are nearly non-selective for β_1- and β_2-adrenoceptors than for β_1-adrenoceptor-selective blockers (Lechat et al. 1998). Consequently, both experimental and clinical evidence favours the concept that the blockade of both β_1- and β_2-adrenoceptors is preferable to the blockade of only β_1-adrenoceptors, in agreement with the deleterious effects on heart function of mice overexpressing β_1- (Engelhardt et al. 1999) and β_2-adrenoceptors (Du et al. 2000). Pindolol does not reduce relative death risks (Soriano et al. 1997), possibly because it causes cardiostimulant effects presumably mediated through putative β_4-adrenoceptors, as discussed below.

II. Putative β_4-Adrenoceptors

It has not yet been demonstrated that catecholamine binding to a site of the β_1-adrenoceptor, responsible for putative β_4-adrenoceptor pharmacology, is of functional relevance (SARSERO et al. 1998a). However, several clinically used β-adrenoceptor blockers, including pindolol, have agonist activity mediated through this receptor (KAUMANN 1997). Pindolol may produce *beneficial* tachycardia in patients with orthostatic hypotension (MAN IN'T VELD and SCHALEKAMP 1981) and neurocardiogenic syncope (ISKOS et al. 1998), and it has been suggested that this effect is actually mediated through the putative β_4-adrenoceptor (KAUMANN 1989; KAUMANN and MOLENAAR 1997). The experimental arrhythmias (FREESTONE et al. 1999; SARSERO et al. 1999; LOWE et al 1998) observed with activation of the putative β_4-adrenoceptor make it plausible that β-blockers with intrinsic activity mediated through this receptor could potentially be harmful. A meta-analysis suggests that only β-blockers without cardiostimulant effects prolong the survival of patients with myocardial infarction. However, patients treated with β-blockers with cardiostimulant effects may not prolong survival or may even shorten survival (SORIANO et al. 1997). The deleterious effects of β-blockers with cardiostimulant effects mediated through the putative β_4-adrenoceptor could be due to the propensity of this receptor to mediate cardiac arrhythmias.

III. 5-HT$_4$ Receptors

5-HT can elicit arrhythmias in human atrial preparations and chronic β_1-adrenoceptor blockade enhances their incidence (KAUMANN and SANDERS 1994). These arrhythmias could be related to 5-HT$_4$ receptor-mediated increases of L-type calcium current (OUADID et al. 1992) and I_f pacemaker current (PINO et al. 1998) in human atrial myocytes. It has been proposed and argued that 5-HT released from platelets may contribute to the initiation and maintenance of atrial fibrillation, and may be involved in the production and mobilisation of emboli, leading to stroke (KAUMANN 1994). Besides platelets, there are other plausible sources of 5-HT, including heart tissue (SOLE et al. 1979) mast cells and nerve endings (COHEN 1985). The experiment illustrated in Fig. 10 demonstrates that nerve endings of human atrium can capture and release 5-HT. After loading nerve endings with 5-HT, neuronally released 5-HT can interact with the atrial 5-HT$_4$ receptors. This is demonstrated with field stimulation which, under blockade of β_1- and β_2-adrenoceptors with (–)-propranolol, elicits an increase in atrial contractility that is prevented by the highly selective 5-HT$_4$-receptor antagonist SB-207710 (Fig. 10).

Sinoatrial 5-HT$_4$-like receptors in piglet right atria, which greatly resemble 5-HT$_4$-like receptors of human atrium (KAUMANN 1990), mediate tachycardia evoked by 5-HT and partial agonists, such as renzapride and cisapride (Fig. 6). 5-HT and renzapride also elicit tachycardia in anaesthetised piglets (PARKER et al. 1995) and adult pigs (VILLALON et al. 1991) through 5-HT$_4$ receptors. Because cisapride causes significant tachycardia in man (BATEMAN 1986),

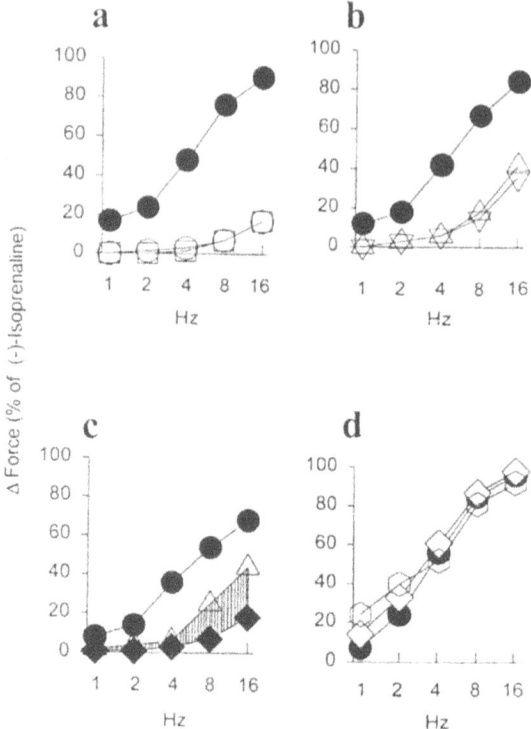

Fig. 10a–d. 5-Hydroxytryptamine (5-HT) release by field stimulation and interaction with 5-HT$_4$ receptors. Results from four trabeculae of a right-atrial appendage from a 50-year-old male patient undergoing coronary-artery surgery. The experiment was carried out at 37°C in the presence of 200 μM ascorbate. The trabeculae were bathed in Krebs solution and paced at 1 Hz through a punctiform electrode. To release neurotransmitters, the tissues were stimulated with field stimuli delivered through field electrodes into the absolute refractory period of the cardiac action potential (without re-exciting the trabeculum) at the indicated frequency (in Hz), as described by KAUMANN (1970). To avoid interaction of released acetylcholine with atrial muscarinic receptors, the experiments were carried out in the presence of 1 μM atropine. Three successive curves relating field-stimulation frequency (in Hz) to contractile force were expressed as percentages of the increase for each trabeculum. Increases in contractile force were expressed as percentages of the increase in peak contractile force caused by 200 μM (−)-isoproterenol administered at the end of the experiment. Each panel represents results from a single trabeculum. The first field-stimulation–force curve, determined in the absence of any antagonist, is shown by *filled circles*. In **a**, the second (*open circles*) and third (*open squares*) curves were determined in the presence of 200 nM (−)-propranolol incubated for 45 min before the second curve was begun. In **b**, the experiments for the second and third curves were carried out in the presence of 200 nM (−)-propranolol as in **a**, but the tissues were incubated for 30 min with 10 μM 5-HT followed by 10 min wash-out before determination of the second (*open triangles*) and third curves (*inverted open triangles*). The protocol of the experiment in **c** was as in **b**, except that the 5-HT$_4$-receptor-selective antagonist SB-207710 (100 nM; KAUMANN et al. 1994) was added after the second curve was finished and was present during the third curve (*filled diamonds*). In **d**, no 5-HT was present. The second curve (*open diamonds*) was determined in the absence of SB-207710 and the third curve (*open hexagons*) in the presence of SB-207710 (100 nM). The *hatched area* in **c** represents the increase in contractile force elicited from neuronally captured and released 5-HT interacting with 5-HT$_4$ receptors. Similar results were obtained from two other experiments. (KAUMANN and MOLENAAR, unpublished)

it is likely that the human sinoatrial 5-HT receptors that mediate tachycardia are of 5-HT_4 nature. Cisapride can also produce supraventricular arrhythmias in man (INMAN and KUNOTA 1992; OLSSEN and EDWARDS 1992). Patients with carcinoid heart disease and high levels of blood 5-HT exhibit occasional ectopic atrial rhythm and atrial fibrillation (LUNDIN et al. 1988). These clinical observations, taken together, support the hypothesis that human 5-HT_4 receptors mediate atrial arrhythmias (KAUMANN 1994) and are consistent with the arrhythmias observed experimentally in human atrium (KAUMANN and SANDERS 1994). The use of 5-HT_4-receptor-selective antagonists has been suggested for the prevention of atrial fibrillation and stroke in the elderly, especially when anticoagulants are contraindicated (KAUMANN 1994). Clinical studies with 5-HT_4-receptor antagonists will test the hypothesis of atrial 5-HT_4-receptor-mediated arrhythmias. Recent evidence, showing that a 5-HT_4-receptor-selective blocker terminates experimental fibrillation and atrial flutter in pig (RAHME et al. 1999), supports the concept of an involvement of 5-HT and 5-HT_4 receptors in these arrhythmias.

I. Epilogue

Some of the following questions are preludes to future research. Does inhibition of βARK improve human heart failure? Although this has been proposed as a target in humans (ROCKMAN et al. 1998), it would be expected to enhance catecholamine-evoked β-adrenoceptor function, including high myocardial oxygen consumption, which would be detrimental. This mitigates against the beneficial effects of β-adrenoceptor-blocking agents in heart failure. The introduction of βARK1 inhibition in the clinic will determine the outcome of this paradox.

Are there circumstances under which the β_1-adrenoceptor can elicit arrhythmias with putative β_4-adrenoceptor characteristics in human heart? Is it possible to develop selective antagonists of the β_1-adrenoceptor that mediates the putative β_4-adrenoceptor pharmacology, including arrhythmias? The introduction of selective and non-selective antagonists will contribute to clarify these questions.

At least two splice variants of the 5-HT_4 receptor (5-HT_{4a} and 5-HT_{4b}) of remarkably similar pharmacology are expressed in human atrium. Is there a functional role in human atrium for additional splice variants reported for brain? The development of antagonists selective for either splice variant could not only quench the curiosity of the researcher but could also have clinical advantages. For example, it may be desirable to selectively prevent atrial arrhythmias through one splice variant but leave unchanged the functions modified through other splice variants located in several brain regions.

Acknowledgements. I thank the British Heart Foundation for support. I am also grateful for the hospitality of Dr Peter Molenaar and Professor James Angus at the Department of Pharmacology of the University of Melbourne, where parts of this article were written.

References

Altschuld R, Starling RC, Hamlin RL, Billman GE, Hensley J, Castillo L, Fertel RH, Hohl CM, Robitaille P-ML, Jones L, Xiao RP, Lakatta EG (1995) Response of failing canine and human heart cells to β_2-adrenergic stimulation. Circulation 92: 1612–1618

Arch JRS (1997) β_3-adrenoceptors and other putative β-adrenoceptors. Pharmacol Res Commun 9:141–148

Arch JRS, Kaumann (1993) β_3- and atypical β-adrenoceptors. Med Res Rev 14:663–729

Bach T, Syversveen T, Kvingedal AM, Krobert KA, Brattelid T, Kaumann AJ, Levy FO (2000) 5-HT$_{4(a)}$ and 5-HT$_{4(b)}$ receptors have nearly identical pharmacology and are both expressed in human atrium and ventricle. Naunyn-Schmiedeberg's Arch Pharmacol (in press)

Bateman DN (1986) The action of cisapride on gastric emptying and the pharmacodynamics and pharmacokinetics of oral diazepam. Eur J Clin Pharmacol 30:205–208

Bearer CF, Knapp RD, Kaumann AJ, Swartz TL, Birnbaumer L (1980) Iodohydroxybenzylpindolol: preparation, localization of its iodine to the indole ring, and characterization as a partial agonist. Mol Pharmacol 17:328–338

Begin-Heick (1995) β_3-adrenoceptor activation of adenylyl cyclase in mouse white adipocytes: modulation by GTP and effect of obesity. J Cell Biochem 58:464–473

Berkowitz DE, Nardone NA, Smiley RM, Price DTY, Kreutter DK, Fremeau RT, Schwinn R. (1995) Distribution of β_3-adrenoceptor mRNA in human tissues. Eur J Pharmacol 289:223–228

Berman DM, Gilman AG (1998) Mammalian RGS proteins: barbarians at the gate. J Biol Chem 273:1269–1272

Berman DM, Kosaza T, Gilman AG (1996) The GTPase-activating protein RGS4 stabilizes the transition state for nucleotide hydrolysis. J Biol Chem 271:27209–27212

Billman GE, Castillo LC, Hensley J, Hohl CM, Altschuld RA (1997) β_2-adrenergic receptor antagonists protect against ventricular fibrillation. In vivo and in vitro evidence for enhanced sensitivity to β_2-adrenergic stimulation in animals susceptible to sudden death. Circulation 96:1914–1922

Black JW (1989) Drugs from emasculated hormones: the principle of synaptic antagonism. Science 245:486–493

Black JW, Duncan WAM, Durant CJ, Ganellin CR, Parsons EM (1972) Definition and antagonism of histamine H$_2$-receptors. Nature 236:385–390

Blondel O, Vandecasteele G, Gastineau M, Leclerc S, Dahmoune U, Langlois M, Fischmeister R (1997) Molecular and functional characterization of a 5-HT$_4$ receptor cloned from human atrium. FEBS Lett 412(3):465–474

Bockaert J, Claeysen S, Dumuis A (1998) Molecular biology, function and pharmacological role of 5-HT$_4$ receptors. Naunyn-Schmiedeberg's Arch Pharmacol 358:R 10

Bond RA, Clarke DE (1988) Agonist and antagonist characterization of a putative adrenoceptor with distinct pharmacological properties from the α- and β-subtypes. Br J Pharmacol 95:723–734

Bond RA, Lefkowitz RJ (1996) The third beta is not a charm. J Clin Invest 98:241

Bond RA, Leff P, Johnson TD, Milano CA, Rockman HA, McMinn TR, Apparsundaram S, Hyek MF, Kenakin TP, Allen LF, Lefkowitz RJ (1995) Physiological effects of inverse agonists in transgenic mice with myocardial overexpression of the β_2-adrenoceptor. Nature 374:272–276

Borea PA, Amerini S, Masini I, Cerbai E, Ledda F, Mantelli L, Varani K, Mugelli A (1992) β_1- and β_2-adrenoceptors in sheep cardiac ventricular muscle. J Mol Cell Cardiol 24:753–764

Bristow MR, Ginsburg R, Minobe W, Cubiciotti RS, Sageman WS, Lurie K, Billingham ME, Harrison DC, Stinson EB (1982) Decreased catecholamine sensitivity and β-adrenergic receptor density in failing human hearts. N Engl J Med 307:205–211

Bristow MR, Ginsburg R, Umans V, Fowler M, Minobe W, Rasmussen R, Zera P, Melove R, Shah P, Jamieson S, Stinson EB (1986) β_1- and β_2-adrenergic receptor subpopulations in nonfailing and failing human ventricular myocardium: coupling of both receptor subtypes to muscle contraction and selective β_1-receptor downregulation in heart failure. Circ Res 59:297–309

Bristow MR, Hershberger RE, Port JD, Rasmussen R (1989) β_1- and β_2-adrenergic receptor-mediated adenylate cyclase in non-failing and failing human ventricular myocardium. Mol Pharmacol 35:295–303

Brown LA, Harding SE (1992) The effect of pertussis toxin on β-adrenoceptor responses in isolated cardiac myocytes from noradrenaline-treated guinea pigs and patients with cardiac failure. Br J Pharmacol 106:115–122

Burrel KM, Molenaar P, Dawson PJ, Kaumann AJ (2000) Contractile and arrhythmic effects of endothelin receptor agonists in human heart in vitro: blockade by SB 209670. J Pharmacol Exp Ther 292:440–459

Buxton BF, Jones CR, Molenaar P, Summers RJ (1987) Characterization and autoradiographic localization of β-adrenoceptor subtypes in human cardiac tissues. Br J Pharmacol 92:299–310

Bylund DB, Eikenberg DC, Hieble JP, Langer SZ, Lefkowitz RJ, Minneman KP, Molinoff PB, Ruffolo RR Jr, Trendelenburg U (1994) IV International Union of Pharmacology Nomenclature of Adrenoceptors. Pharmacol Rev 46:121–136

Carlsson E, Åblad B, Brandström A, Carlsson B (1972) Differential blockade of the chronotropic effects of various adrenergic stimuli in cat heart. Life Sci II:953–958

Chaudry A, MacKenzie RG, Georgic LM, Granneman JG (1994) Differential interaction of β_1- and β_3-adrenoceptor with G_i in rat adipocytes. Cell Signalling 6:457–465

Chruscinski AJ, Rohrer DK, Schauble E, Desai KH, Bernstein D, Kobilka BK (1999) Targeted disruption of the β_2 adrenergic gene. J Biol Chem 274:16694-16700

Claeysen S, Seben M, Journot L, Bockaert J, Dumuis A (1996) Cloning, expression and pharmacology of the mouse 5-HT$_{4L}$ receptor. FEBS Lett 398:19–25

Claeysen S, Faye P, Seben M, Lemaire S, Bockaert J, Dumuis A (1997) Cloning and expression of human 5-HT$_{4S}$ receptors. Effect of receptor density on their coupling to adenylyl cyclase. Neuroreport 8:3189–3196

Cohen RA (1985) Platelet-induced neurogenic coronary contractions due to accumulation of the false neurotransmitter 5-hydroxytryptamine. J Clin Invest 75:286–292

Del Monte F, Kaumann AJ, Poole-Wilson PA, Wynne DG, Pepper J, Harding SE. (1993) Coexistence of functioning β_1- and β_2-adrenoceptors in single myocytes from human ventricle. Circulation 88:854–863

Du X-J, Autelitano DJ, Dilley RJ, Wang B, Dart AM, Woodcock EA (2000) β_2-adrenergic receptor overexpression exacerbates development of heart failure after aortic stenosis. Circulation 101:71–77

Dumuis A, Bouhelal R, Seben M, Cory R, Bockaert J (1988) A nonclassical 5-hydroxytryptamine receptor positively coupled with adenylate cyclase in the central nervous system. Mol Pharmacol 34:880–887

Dumuis A, Seben M, Bockaert J (1989) The gastrointestinal prokinetic benzamide derivatives are agonists at the non-classical 5-HT receptor (5-HT$_4$) positively coupled to adenylate cyclase in neurones. Naunyn-Schmiedeberg's Arch Pharmacol 340:403–410

Elnatan J, Molenaar P, Rosenfeldt FL, Summers RJ (1994) Autoradiographic localization and quantitation of β_1- and β_2-adrenoceptors in human atrioventricular conducting system: a comparison of patients with idiopathic dilated cardiomyopathy and ischemic heart disease. J Mol Cell Cardiol 26:313–323

Emorine LJ, Marullo S, Briend-Sutren MM, Patey G, Tate K, Delavier-Klutchko SE, Strosberg AD (1989) Molecular characterization of the human β_3-adrenoceptor. Science 245:1118–1121

Engelhardt S, Hein L, Wiesmann F, Lohse MJ (1999) Progressive hypertrophy and heart failure in β_1-adrenergic receptor transgenic mice. Proc Natl Acad Sci USA 96:7059–7064

Feldman AM, Gates AE, Veazey WB, Hershberger RE, Bristow MR, Baughman KL, Baumgartner WA, Van Dop C (1988) Increase of the 40,000 mol wt pertussis toxin substrate (G protein) in the failing human heart. J Clin Invest 82:189–197

Ferro A, Kaumann AJ, Brown MJ (1995) β-adrenoceptor subtypes in human coronary artery: desensitization of β_2-adrenergic stimulation in vitro. J Cardiovasc Pharmacol 25:134–141

Freestone NS, Heubach J, Wettwer E, Ravens U, Brown D, Kaumann AJ (1999) Putative β_4-adrenoceptors are more effective than β_1-adrenoceptors in mediating arrhythmic Ca^{2+} transients in mouse ventricular myocytes. Naunyn-Schmiedeberg's Arch Pharmacol 360:445–456

Gauthier C, Tavernier G, Charpentier F, Langin D, Le Marec H (1996) Functional β_3-adrenoceptor in the human heart. J Clin Invest 98:556–562

Gauthier C, Leblais V, Kobzig L, Trochu J-N, Khandoudi N, Bril A, Balligand J-L, Le Marec H (1998) The negative inotropic effect of β_3-adrenoceptor stimulation is mediated by activation of anitric oxide synthase in human ventricle. J Clin Invest 102:1377–1384

Gerald C, Adham N, Kao H-T, Olson MA, Laz TM, Schechter LE, Bard JA, Vaysee PJJ, Hartig PR, Branchek TA, Weinshank RL (1995) The 5-HT_4 receptor: molecular cloning and pharmacological characterization of two splice variants. EMBO J 14:2806–2815

Germack R, Starzec AB, Vassy R, Perret G (1997) β-Adrenoceptor subtype expression and function in rat white adipocytes. Br J Pharmacol 120:201–210

Gille R, Lemoine H, Ehle B, Kaumann AJ (1985) The affinity of (–)-propranolol for β_1- and β_2-adrenoceptors of human heart. Differential antagonism of the positive inotropic effects and adenylate cyclase stimulation by (–)-noradrenaline and (–)-adrenaline. Naunyn-Schmiedeberg's Arch Pharmacol 331:60–70

Green SA, Holt BD, Liggett SB (1992) β_1- and β_2-receptors display subtype-selective coupling to G_s. Mol Pharmacol 41:889–893

Hall JA, Kaumann AJ, Brown MJ (1990) Selective β_1-adrenoceptor blockade enhances positive inotropic effects of endogenous catecholamines through β_2-adrenoceptors in human atrium. Circ Res 66:1610–1623

Harding SE (1997) Lack of evidence for β_3-adrenoceptor modulation of contractile force in human ventricular myocytes. Circulation 95:I-53

Hartig PR, Hoyer D, Humphrey PPA, Martin GR (1996) Alignment of receptor nomenclature with the human genome: classification of 5-HT_{1B} and 5-HT_{1D} receptor subtypes. Trends Pharmacol Sci 17:103–105

Hennekens CH, Albert CM, Godfriend SL, Gaziano JM, Buring SE (1996) Adjunctive therapy of acute myocardial infarction – evidence from clinical trials. N Engl J Med 335:1660–1667

Hill SJ, Ganellin CR Timmerman H, Schwartz JC, Shankley NP, Young JM, Schunak W, Levi R, Haas HL. (1997) International Union of Pharmacology. XIII. Classification of histamine receptors. Pharmacol Rev 49:253–278

Hollander W, Michelson AL, Wilkins RW. (1957) Serotonin and antiserotonins. I. Their circulatory, respiratory and renal effects in man. Circulation 16:246–255

Hoyer D, Clarke DE, Fozard JR, Hartig PR, Martin GR, Mylecharane EJ, Saxena PR, Humphrey PPA (1994) VII International Union of Pharmacology Classification of receptors for 5-hydroxytryptamine (Serotonin). Pharmacol Rev 46:157–203

Huang C, Hepler JR, Gilman AG, Mumby SM (1997) Attenuation of G_i- and G_q-mediated signaling by expression of RGS4 or GAIP in mammalian cells. Proc Natl Acad Sci USA 94:6159–6163

Inman W, Kunota K (1992) Tachycardia during cisapride treatment. Br med J 305:1019

Iskos D, Dutton J, Scheinman MM, Lurie KG (1998) Usefulness of pindolol in neurocardiogenic syncope. Am J Cardiol 82:1121–1124

Jahnel U, Rupp J, Ertl R, Nawrath H (1992) Positive inotropic responses to 5-HT in human atrial but not in ventricular muscle. Naunyn-Schmiedeberg's Arch Pharmacol 346:482–485

Jahnel U, Nawrath H, Rupp J, Ochi R (1993) L-type calcium channel activity in human atrial myocytes as influenced by 5-HT. Naunyn-Schmiedeberg's Arch Pharmacol 348:396–402

Jakob SM, Becker H, Hanrath P, Schumacher C, Eschenhagen T, Schmitz W, Scholz H, Steinfath M (1996) Effects of metoprolol on myocardial beta-adrenoceptors and G_i alpha-proteins in patients with congestive heart failure. Eur J Clin Pharmacol 51(2):127–132

Kaumann AJ (1970) Adrenergic receptors in heart muscle: relations among factors influencing the sensitivity of the cat papillary muscle to catecholamines. J Pharmacol Exp Ther 173:383–398

Kaumann AJ (1973) Adrenergic receptors in cardiac muscle. Two different mechanisms of β-blockers as partial agonists. Int. Union of Biochemistry, Symposium Nr 52, Acta Physiol Latamer 23:235–236

Kaumann AJ (1983) Cardiac β-adrenoceptors – experimental viewpoints. Z Kardiol 72:63–82

Kaumann AJ (1989) Is there a third heart β-adrenoceptor? Trends Pharmacol Sci 10:316–320

Kaumann AJ (1990) Piglet sinoatrial 5-HT receptors resemble human atrial $5\text{-}HT_4$-like receptors. Naunyn-Schmiedeberg's Arch Pharmacol 342:619–622

Kaumann AJ (1991a) $5\text{-}HT_4$-like receptors in mammalian atria. J Neural Transm [Suppl] 34:195–201

Kaumann AJ (1991b) Some aspects of heart beta adrenoceptor function. Cardiovasc Drugs Ther 5:549–560

Kaumann AJ (1993) Blockade of human atrial $5\text{-}HT_4$ receptors by GR 113808. Br J Pharmacol 110:1172–1174

Kaumann AJ (1994) Do human atrial $5\text{-}HT_4$ receptors mediate arrhythmias? Trends Pharmacol Sci 15:451–455

Kaumann AJ (1996) (–)-CGP-12177-induced increase of human atrial contraction through a putative third β-adrenoceptor. Br J Pharmacol 117:93–98

Kaumann AJ (1997) Four β-adrenoceptor subtypes in mammalian heart. Trends Pharmacol Sci 18:70–76

Kaumann AJ, Blinks JR (1980) β-adrenoceptor blocking agents as partial agonists in isolated heart muscle: dissociation of stimulation and blockade. Naunyn-Schmiedeberg's Arch Pharmacol 311:205–218

Kaumann AJ, Lemoine H (1987) β_2-adrenoceptor-mediated positive inotropic effects of adrenaline in human ventricular myocardium. Quantitative discrepancies between binding and adenylate cyclase stimulation. Naunyn-Schmiedeberg's Arch Pharmacol 335:403–411

Kaumann AJ, Lobnig BM (1986) Mode of action of (–)-pindolol on feline and human myocardium. Br J Pharmacol 89:207–218

Kaumann AJ, Lynham JA (1997) (–)-CGP-12177 stimulates cyclic AMP-dependent protein kinase in rat atria through an atypical β-adrenoceptor. Br J Pharmacol 120:1187–1189

Kaumann AJ, Molenaar P (1996) Differences between the third cardiac β-adrenoceptor and the colonic β_3-adrenoceptor in the rat. Br J Pharmacol 118:2085–2098

Kaumann AJ, Molenaar P (1997) Modulation of human cardiac function through 4 β-adrenoceptor populations. Naunyn-Schmiedeberg's Arch Pharmacol 355:667–681

Kaumann AJ, Sanders L (1993) Both β_1- and β_2-adrenoceptors mediate catecholamine-evoked arrhythmias in isolated human atrium. Naunyn-Schmiedeberg's Arch Pharmacol 348:536–540

Kaumann AJ, Sanders L (1994) 5-Hydroxytryptamine causes rate-dependent arrhythmias through $5\text{-}HT_4$ receptors in human atrium: facilitation by chronic β-adrenoceptor blockade. Naunyn-Schmiedeberg's Arch Pharmacol 349:331–337

Kaumann AJ, Morris TH, Birnbaumer L (1979) A comparison of the influence of N-isopropyl and N-tert butyl substituents on the affinity of ligands for sinoatrial β-adrenoceptors in rat atria and β-adrenoceptor-coupled adenylyl cyclase in kitten ventricle. Naunyn-Schmiedeberg's Arch Pharmacol 307:1-8

Kaumann AJ, Lemoine H, Schwederski-Menke U, Ehle B (1989a) Relations between β-adrenoceptor occupancy and increases of contractile force and adenylate cyclase activity induced by catecholamines in human ventricular myocardium. Acute desensitization and comparison with feline myocardium. Naunyn-Schmiedeberg's Arch Pharmacol 339:99–112

Kaumann AJ, Murray KJ, Brown AM, Sanders L, Brown MJ (1989b) A receptor for 5-HT in human atrium. Br J Pharmacol 98:664P

Kaumann AJ, Hall JA, Murray KJ, Wells FC, Brown MJ (1989c) A comparison of the effects of adrenaline and noradrenaline on human heart: the role of adenylate cyclase and contractile force. Eur Heart J 10 [Suppl B]:29–37

Kaumann AJ, Sanders L, Brown AM, Murray KJ, Brown MJ (1990) A 5-hydroxytryptamine receptor in human atrium. Br J Pharmacol 100:879–885

Kaumann AJ, Sanders L, Brown AM, Murray KJ, Brown MJ (1991a) A 5-HT$_{4\text{-like}}$ receptor in human right atrium. Naunyn-Schmiedeberg's Arch Pharmacol 344:150–159

Kaumann AJ, Brown AM, Raval P (1991b) Putative 5-HT$_{4\text{-like}}$ receptors in piglet left atrium. Br J Pharmacol 101:98P

Kaumann AJ, Frenken M, Posival H, Brown AM (1994a) Variable participation of 5-HT$_{1\text{-like}}$ receptors and 5-HT$_2$ receptors in serotonin-induced contraction of human isolated coronary arteries. 5-HT$_{1\text{-like}}$ receptors resemble cloned 5-HT$_{1D\beta}$ receptors. Circulation 90:1141–1153

Kaumann AJ, Gaster LM, King FD, Brown AM (1994b) Blockade of human atrial 5-HT$_4$ receptors by SB-207710, a selective and high affinity 5-HT$_4$ receptor antagonist. Naunyn-Schmiedeberg's Arch Pharmacol 349:546–548

Kaumann AJ, Lynham JA, Sanders L, Brown AM, Molenaar P (1995a) Contribution of differential efficacy to the pharmacology of human β_1- and β_2-adrenoceptors. Pharmacol Comm 6:215–222

Kaumann AJ, Lynham JA, Brown AM (1995b) Labelling with [^{125}I]-SB-207710 of a small 5-HT$_4$ receptor population in piglet right atrium: functional relevance. Br J Pharmacol 115:933–936

Kaumann AJ, Sanders L, Lynham JA, Bartel S, Kuschel M, Karczweski P, Krause EG (1996a) β_2-adrenoceptor activation by zinterol causes protein phosphorylation, contractile effects and relaxant effects through a cAMP pathway in human atrium. Mol Cell Biochem 163/164:113–123

Kaumann AJ, Lynham JA, Brown AM (1996b) Comparison of the densities of 5-HT$_4$ receptors, β_1- and β_2-adrenoceptors in human atrium: functional implications. Naunyn-Schmiedeberg's Arch Pharmacol 353: 592–595

Kaumann AJ, Lynham JA, Sarsero D, Molenaar P (1997) The atypical cardiac cardiostimulant β-adrenoceptor is distinct from β_3-adrenoceptor and is coupled to a cyclic AMP-dependent pathway in rat and human myocardium. Br J Pharmacol 120:102 P

Kaumann AJ, Preitner F, Sarsero D, Molenaar P, Revelli J-P, Giacobino JP (1998) (–)-CGP-12177 causes cardiostimulation and binds to cardiac putative β_4-adrenoceptors in both wild-type and β_3-adrenoceptor knockout mice. Mol Pharmacol 53:670–675

Kaumann AJ, Bartel S, Molenaar P, Sanders L, Burrell K, Vetter D, Hempel P, Karczewski P, Krause EG (1999) Activation of β_2-adrenergic receptors hastens relaxation and mediates phosphorylation of phospholamban, troponin I and C protein in ventricular myocardium from patients with terminal heart failure. Circulation 99:65–72

Kaumann AJ, Engelhardt S, Molenaar P, Lohse M (2000) (–)-CGP 12177-evoked cardiostimulation is abolished in double β_1/β_2-adrenoceptor (AR) knockout mice. Proc Aust Soc Clin Exp Pharmacol Toxicol 7:70

Kenakin T (1995) Agonist receptor efficacy II: agonist trafficking of receptor signals. Trends Pharmacol Sci 16:232–238

Kenakin T (1997) Differences between natural and recombinant G protein-coupled receptor systems with varying receptor/G protein stoichiometry. Trends Pharmacol Sci 18:456–464

Kent RS, De Lean A, Lefkowitz RJ (1980) A quantitative analysis of beta-adrenergic receptor interactions: resolution of high and low affinity states of the receptor by computer modelling of ligand binding data. Mol Pharmacol 17:14–23

Koch WJ, Milano CA, Lefkowitz RJ (1966) Transgenic manipulation of myocardial G protein-coupled receptors and receptor kinases. Circ Res 78:511–516

Krief S, Lönnqvist F, Raimbault S, Baude B, van Spronson A, Arner P, Strosberg AD, Riquier D, Emorine LJ (1993) Tissue distribution of β_3-adrenoceptor mRNA in man. J Clin Invest 91:344–349

Le Messurier DH, Schwartz CJ, Whelan R (1959) Cardiovascular effects of intravenous 5-hydroxytryptamine in man. Br J Pharmacol 14:246–250

Lechat P, Packer M, Chalon S, Cucherat M, Arab T, Boissel (1998) Clinical effects of β-adrenergic blockade in chronic heart failure. A meta-analysis of double-blind, placebo-controlled, randomized trials. Circulation 98:1184–1191

Lemoine H, Kaumann AJ (1982) A novel analysis of concentration-dependence of partial agonism. Naunyn-Schmiedeberg's Arch Pharmacol 320:130–144

Lemoine H, Kaumann AJ (1991) Regional differences of β_1- and β_2-adrenoceptor-mediated functions in feline heart. A β_2-adrenoceptor-mediated positive inotropic effect possibly unrelated to cyclic AMP. Naunyn-Schmiedeberg's Arch Pharmacol 344:56–69

Lemoine H, Schönell H, Kaumann AJ (1988) Contribution of β_1- and β_2-adrenoceptors of human atrium and ventricle to the effects of noradrenaline and adrenaline as assessed with (−)-atenolol. Br J Pharmacol 95:55–66

Levy FO, Zhu X, Kaumann AJ, Birnbaumer L (1993) Efficacy of β_1-adrenergic receptors is lower than that of β_2-adrenergic receptors. Proc Natl Acad Sci USA 90: 10798–10802

Liggett SB, Wagoner LE, Craft LL, Hornung RW, Hoit BD, McIntosh TC, Walsh RA (1998) The Ile 164 β_2-adrenergic receptor polymorphism adversely affects the outcome of congestive heart failure. J Clin Invest 102:1534–1539

Londos C, Salomon Y, Lin MC, Harwood JP, Schramm M, Wolff J, Rodbell M (1974) 5-guanylylimidodiphosphate, a potent activator of adenylate cyclase system in eukaryotic cells. Proc Nat Acad Sci 71:3087–3090

Lorraine J, Grosset A, O'Connor E (1992) 5-HT$_4$ receptors, present in piglet atria and sensitive to SDZ 205-557, are absent in papillary muscle. Eur J Pharmacol 229: 105–108

Lowe MD, Grace AA, Vandenberg JI, Kaumann AJ (1998) Action potential shortening through the putative β_4-adrenoceptor in ferret ventricle: comparison with β_1- and β_2-adrenoceptor-mediated effects. Br J Pharmacol 124:1341–1344

Lundin L, Nordheim I, Landelius J, Oberg K, Theodorsson-Nordheim E (1988) Carcinoid heart disease: relationship of circulating vasoactive substances to ultrasound-detectable cardiac abnormalities. Circulation 77:264–269

Malinowska B, Schlicker E (1996) Mediation of the positive chronotropic effect of CGP-12177 and cyanopindolol in the pithed rat by atypical β-adrenoceptors, different from β_3-adrenoceptors. Br J Pharmacol 117: 943–949

Man'tInt Veld AJ, Schalekamp MADH (1981) Pindolol acts as a beta-adrenoceptor agonist in orthostatic hypotension: therapeutic implications. Br Med J 282:929–931

Marano M, Kaumann AJ (1976) On the statistics of drug-receptor constants for partial agonists. J Pharmacol Exp Ther 198:518–526

Mason DA, Moore JD, Green SA, Liggett SB (1999) A gain-of-function polymorphism in a G-protein coupling domain of the human β_1-adrenergic receptor. J Biol Chem 274:12670–12674

Medhurst A, Kaumann AJ (1993) Characterisation of the 5-HT$_4$ receptor mediating tachycardia in isolated piglet right atrium. Br J Pharmacol 110:1023–1030

Mewes T, Dutz S, Ravens U, Jakob KH (1993) Activation of calcium currents in myocytes by empty β-adrenoceptors. Circulation 88:2916–2922

Milano CA, Allen LF, Rockman HA, Dolber PC, McMinn TR, Chien KR, Johnson TD, Bond RA, Lefkowitz RJ (1994) Enhanced myocardial function in transgenic mice overexpressing the β_2-adrenergic receptor. Science 264:582–586

Mohr B, Bom AH, Kaumann AJ, Thämer V (1987) Reflex inhibition of the efferent renal sympathetic activity by 5-hydroxytryptamine and nicotine elicited by different epicardial receptors. Pflügers Arch 409:145–151

Molenaar P, Smolich JJ, Russell FD, McMartin LR, Summers RJ (1990a) Differential regulation of beta-1 and beta-2 adrenoceptors in guinea pig atrioventricular conducting system after (−)-isoproterenol infusion. J Pharmacol Exp Ther 255:393–400

Molenaar P, Canale E, Summers RJ (1990b) Densitometric analysis of β_1-and β_2-adrenoceptors in guinea-pig atrioventricular conducting system. J Mol Cell Cardiol 22:483–495

Molenaar P, Sarsero D, Kaumann AJ (1997a) Proposal for the interaction of non-conventional partial agonists and catecholamines with the 'putative β_4-adrenoceptor' in mammalian heart. Clin Exp Pharmacol Physiol 24:647–656

Molenaar P Sarsero D, Arch JRS, Kelly J, Henson SM, Kaumann AJ (1997b) Effects of (−)-RO363 at human atrial β-adrenoceptor subtypes, the human cloned β_3-adrenoceptor and rodent intestinal β_3-adrenoceptors. Br J Pharmacol 120:165–176

Molenaar P, Sarsero D, Lynham JA, Kaumann AJ (1998) The putative β_4-adrenoceptor activates the cyclic AMP pathway in human heart: inhibition of positive inotropic responses by adenosine and carbachol. Naunyn-Schmiedeberg's Arch Pharmacol 358:R629

Müller FU, Boheler KR, Eschenhagen T, Schmitz W, Scholz H (1993) Isoprenaline stimulates gene transcription of the inhibitory G-protein Giα_2 in rat heart. Circ Res 72:696–700

Nanoff C, Freissmuth M, Schütz W(1987) The role of a low β_1-adrenoceptor selectivity of [^{3}H]-CGP-12177 for resolving subtype-selectivity of competitive ligands. Naunyn-Schmiedeberg's Arch Pharmacol 336:519–525

Nantel F, Bonin H, Emorine LJ, Zilberfarb V, Strosberg AD, Bouvier M, Marullo S (1993) The human β_3-adrenoceptor is resistant to short-term agonist-promoted desensitization. Mol Pharmacol 43:548–555

Neumann J, Schmitz W, Scholz H, Meyerink LV, Doring V, Kalma P (1988) Increase in myocardial G$_i$-proteins in heart failure. Lancet II:936–937

Olssen S, Edwards IR (1992) Tachycardia during cisapride treatment. Br med J 305:748–749

Oostendorp J, Kaumann AJ (2000) Pertussis toxin suppresses carbachol-evoked cardiodepression but does not modify cardiostimulation mediated through β_1- and putative β_4-adrenoceptors in mouse left atria: no evidence for β_2- and β_3-adrenoceptor function. Naunyn-Schmiedeberg's Arch Pharmacol 361:134–145

Ormerod OJM, McGregor CGA, Stone DL, Wisbey C, Petch MC (1984) Arrhythmias after coronary bypass surgery. Br Heart J 51:618–621

Ouadid H, Seguin J, Dumuis A, Bockaert J, Nargeot J (1992) Serotonin increases calcium current in human atrial myocytes via the newly described 5-hydroxytryptamine$_4$ receptor. Mol Pharmacol 41:346–351

Pak MD, Fishman PH (1996) Anomalous behavior of CGP-12177 A on β_1-adrenergic receptors. J Receptor Signal Transduction Research 16 (1 and 2):1–23

Parker SG, Taylor EM, Hamburger SA, Vimal M, Kaumann AJ (1995) Blockade of human and porcine myocardial 5-HT$_4$ receptors by SB 203186. Naunyn-Schmiedeberg's Arch Pharmacol 335:28–35

Ping P, Gelzer-Bell R, Roth DA, Kiel D, Insel PA, Hammond HK (1995) Reduced β-adrenergic receptor kinase activity in porcine heart. J Clin Invest 95:1271–1280

Pino R, Cerbai E, Calamai G, Alajmo F, Borgioli A, Braconi L, Cassai M, Montesi GF, Mugelli A (1998) Effect of 5-HT$_4$ receptor stimulation on the pacemaker current If in human isolated atrial myocytes. Cardiovasc Res 40: 516–522

Pitcher JA, Inglese J, Higgins JB, Arriza JL, Casey PJ, Kim C, Benovic JL, Kwatra MM, Caron MG, Lefkowitz RJ (1992) Role of $\beta\gamma$ subunits of G proteins in targeting the β-adrenergic receptor kinase to membrane-bound receptors. Science 257:1264–1267

Pritchard BNC, Tomlinson B, Walden RJ, Bhattacharjee P (1983) The β-adrenergic blockade withdrawal phenomenon. J Cardiovasc Pharmacol 5 [Suppl 1]:56–62

Rahme MM, Cotter B, Leistad E, Wadha MK, Mohabir R, Ford APDW, Eglen RM, Feld GK (1999) Electrophysiological and antiarrhythmic effects of the atrial selective 5-HT$_4$ receptor antagonist RS-100302 in experimental atrial flutter and fibrillation. Circulation 100:2010–2017

Reithmann C, Gierschik P, Sidiropoulos D, Werdan K, Jakobs KH (1989) Mechanism of noradrenaline-induced heterologous desensitization of adenylate cyclase stimulation in rat heart muscle cells: increase in the level of inhibitory G-protein α-subunits. Eur J Pharmacol 172:211–221

Rockman HA, Chien KE, Choi D-J, Iaccarino G, Hunter JJ, Ross Jr J, Lefkowitz RJ, Koch WJ (1998) Expression of a β-adrenergic receptor kinase 1 inhibitor prevents the development of myocardial failure in gene-targeted mice. Proc Natl Acad Sci USA 95:7000–7005

Rodbell M, Krans HMJ, Pohl SL, Birnbaumer L (1971) The glucagon-sensitive adenyl cyclase system in plasma membranes of rat liver. IV. Effects of guanyl nucleotides on binding of ^{125}I-glucagon. J Biol Chem 246:1872–1876

Rohrer DK, Chruscinski A, Schauble EH, Bernstein D, Kobilka BK (1999) Cardiovascular and metabolic alterations in mice lacking both β_1- and β_2-adrenergic receptors. J Biol Chem 274:16701–16708

Sanders L, Kaumann AJ (1992) A 5-HT$_{4-like}$ receptor in human left atrium. Naunyn-Schmiedeberg's Arch Pharmacol 345:382–386

Sanders L, Lynham JA, Bond B, del Monte F, Harding SE, Kaumann AJ (1995) Sensitization of human atrial 5-HT$_4$ receptors by chronic β blocker treatment. Circulation 92:2526–2639

Sanders L, Lynham JA, Kaumann AJ (1996) Chronic β_1-adrenoceptor blockade sensitises the H$_1$ and H$_2$ receptor systems in human atrium. Naunyn-Schmiedeberg's Arch Pharmacol 353:661–670

Sarsero D, Molenaar P, Kaumann AJ (1998a) Validity of (–)-[^{3}H]-CGP-12177 as a radioligand for the "putative β_4-adrenoceptor" in rat atrium. Br J Pharmacol 123:371–380

Sarsero D, Molenaar P, Kaumann AJ (1998b) The "putative β_4-adrenoceptor" mediates positive inotropic responses and hastens relaxation through a cAMP pathway in human heart. Aust NZ J Med 28:147

Sarsero D, Molenaar P, Kaumann AJ (1998c) (–)-[^{3}H]-CGP-12177 radiolabels β_1-, β_2- and putative β_4-adrenoceptors in human atrium and ventricle. Naunyn-Schmiedeberg's Arch Pharmacol 358:R 629

Sarsero D, Molenaar P, Kaumann AJ, Freestone NS (1999) Putative β_4-adrenoceptors mediate increases in contractile force and cell Ca^{2+}: comparison with atrial responses and relationship to (–)-[^{3}H]-CGP 12177 binding. Br J Pharmacol 128:1445–1460

Sawada M, Ichinose M, Ito I, Maeno T, Mcadoo DJ (1984) Effects of 5-hydroxytryptamine on membrane potential, contractility, accumulation of cyclic AMP, and Ca^{2+} movements in anterior aorta and ventricle of Aplysia. J Neurophysiol 51:361–374

Sennitt MV, Kaumann AJ, Molenaar P, Beeley LJ, Young PW, Kelly K, Chapman H, Henson SM, Berge JM, Dean DK, Kotecha NR, Morgan HKA, Rami HK, Ward RW, Thompson M, Wilson S, Smith SA, Cawthorne MA, Stock MJ, Arch JRS (1998) The contribution of classical ($\beta_{1/2}$-) and atypical β-adrenoceptors to the stimulation of human white adipocyte lipolysis and right atrial appendage contraction by novel β_3-adrenoceptor agonists of different selectivities. J Pharmacol Exp Ther 285:1084–1095

Skeberdis VA, Jurevicius J, Fischmeister R (1997) β_2-Adrenergic activation of L-type Ca^{++} current in cardiac myocytes. J Pharmacol Exp Ther 282:452–461

Skeberdis VA, Jurevicius J, Fischmeister R (1999) β_3-Adrenergic regulation of L-type Ca^{2+} current in human atrial myocytes. Abstract to the 43rd Annual Meeting of the Biophysical Society, February 13–17, Baltimore, Maryland, USA

Sole MJ, Shum A, VanLoon GR (1979) Serotonin: metabolism in the normal and failing heart. Circ Res 45:629–634

Soriano JB, Hoes AW, Meems L, Grobee DE (1997) Increased survival with β blockers: importance of ancillary properties. Progress in Cardiovascular Diseases 39: 445–456

Staehelin M, Simons P, Jaeggi K, Wigger H (1983) CGP-12177. A hydrophilic β-adrenergic receptor radioligand reveals high affinity binding of agonists to intact cells. J Biol Chem 258:3496–3502

Stephenson RP (1956) A modification of receptor theory. Br J Pharmacol 11:379–393

Summers RJ, Molenaar P (1995) Autoradiography of β_1- and β_2-adrenoceptors. In: Methods in Molecular Biology 41: Signal transductions protocols. Kendall DA, Hill SJ, eds, Humana Press Inc Totowa NJ, pp 25–39

Turki J, Lorenz JN, Green SA, Donnelly ET, Jacinto M, Liggett SB (1996) Myocardial signalling defects and impaired cardiac function of a human β_2-adrenergic receptor polymorphism expressed in transgenic mice. Proc Natl Acad Sci USA 93:10483–10488

Ungerer M, Böhm M, Elce JS, Erdmann E, Lohse MJ (1993) Altered expression of β-adrenergic receptor kinase and β_1-adrenergic receptors in failing human heart. Circulation 87:454–463

Van den Wyngaert I, Gommeren W, Verhasselt P, Jurzak M, Leysen J, Luyten W, Bender E (1997) Cloning and expression of a human serotonin 5-HT$_4$ receptor cDNA. J Neurochem 69:1810–1819

Villalon CM, Den Boer MO, Heiligers JPC, Saxena PR (1991) Further characterization, by use of tryptamine and benzamide derivatives, of the putative 5-HT$_4$ receptor mediating tachycardia in the pig. Br J Pharmacol 102:107–112

Waagstein F, Bristow MR, Swedberg K, Camerini F, Fowler MB, Silver MA, Gilbert EM, Johnson MR, Gross FG, Hjalmarson A (1993) Beneficial effects of metoprolol in idiopathic dilated cardiomyopathy. Lancet 342:1441–1446

Wang T, Kaumann AJ, Brown MJ (1996) (–)-Timolol is a more potent antagonist of the positive inotropic effects of (–)-adrenaline than those of (–)-noradrenaline in human atrium. Br J Clin Pharmacol 42:217–223

Wang T, Plumpton C, Brown MJ (1999) Selective β_1-adrenoceptor blockade enhances the activity of the stimulatory G-protein in human atrial myocardium. Br J Pharmacol 128:135–141

Waud DR (1969) On the measurement of the affinity of partial agonists for receptors. J Pharmacol Exp Ther 170:117–122

Witte K, Schnecko A, Olbrich HG, Lemmer B (1995) Efficiency of beta-adrenoceptor subtype coupling to cardiac adenylyl cyclase in cardiomyopathic and control hamsters. Eur J Pharmacol 290:1–10

Witte K, Schnecko A, Hauth D, Wirzius S, Lemmer B (1998) Effects of chronic application of propranolol on β-adrenergic signal transduction in heart ventricles from myopathic BIO TO2 and control hamsters. Br J Pharmacol 125:1033–1041

Xiao RP, Ji X, Lakatta EG (1994) Functional coupling of the β_2-adrenoceptor to a pertussis toxin-sensitive G protein in cardiac myocytes. Mol Pharmacol 47:322–329

Xiao R-P, Avdonin P, Zhou Y-Y, Cheng H, Akhter SA, Eschenhagen T, Lefkowitz RJ, Koch WJ, Lakatta EG (1999) Coupling of β_2-adrenoceptor to G$_i$ proteins and its physiological relevance in murine cardiac myocytes. Circ Res 84:43–52

Zerkowski H-R, Broede A, Kunde K, Hillemann S, Schäfer E, Vogelsang M, Michel MC, Brodde O-E (1993) Comparison of the positive inotropic effects of serotonin, histamine, angiotensin II, endothelin and isoprenaline in the isolated human right atrium. Naunyn-Schmiedeberg's Arch Pharmacol 346:347–352

Section II
New Theoretical Concepts and Molecular Mechanisms of Receptor Function

CHAPTER 4
Kinetic Modeling Approaches to Understanding Ligand Efficacy

J.J. LINDERMAN

A. Introduction

The responses of cells are governed to a large extent by the binding of ligands to cell surface receptors and the signal transduction steps that follow. Although these steps are being studied in detail to determine the identities of the molecules participating and the reactions that occur during signal transduction, there is, by comparison, little attention paid to quantitative aspects of the signal transduction process. However, quantitative studies are necessary to predict how specific numbers of bound receptors trigger particular levels of responses and thus obtain the ability to carefully control or manipulate the extent of cellular responses. An understanding of the kinetics of these processes is vital to such quantification. The goal of this chapter is to describe some kinetic modeling approaches that, in combination with experimental data, suggest key parameters relating ligand–receptor binding to the cellular response. A particular focus is on the ligand-specific parameters which underlie differences in ligand efficacy.

The generation of a response following the binding of ligand to cell surface receptors – the underlying basis of ligand efficacy – is not simply a function of the number of bound receptors. The ability of bound receptors to transduce a response may depend on their "state" – e.g., active, inactive, desensitized, or internalized – and this state may, in part, depend on the ligand used to bind the receptor. Furthermore, receptors in a fully active state may be unable to transduce a signal if they are unable to find and interact with the appropriate effector in the signal transduction cascade. Thus, any quantitative understanding of the link between receptor–ligand binding and cellular responses or efficacy must entail a detailed grasp of receptor states, the kinetics of receptor processing (or conversion between states), and the ability of receptors in these different states to transmit the message of ligand binding. In this chapter, four ligand-dependent parameters that relate to receptor states and may contribute to ligand efficacy are discussed. Two of these focus on G protein-coupled receptors.

B. Background

I. Efficacy

It is well known that different ligands binding to the same number of receptors of a particular type on the same cell type may elicit different levels of a particular response; these ligands are said to have different efficacies. As reviewed in KENAKIN (1993), there have been several attempts to link receptor occupancy with response for different ligands (occupation theory). In a classical but still popular approach, the term "efficacy" means the intrinsic efficacy (ε), as defined by FURCHGOTT (1966), and the stimulus (S) is defined by

$$S = \frac{\varepsilon[A][R_{tot}]}{[A]+K_D}, \quad (1)$$

where [A] is the agonist concentration, [R_{tot}] is the total number of receptors available for binding ligand, and K_D is the equilibrium dissociation constant. The response is then related to the stimulus by

$$\text{Response} = f(S) \quad (2)$$

where f is an unknown (generally assumed hyperbolic) function (STEPHENSON 1956). The efficacy ε is ligand-specific, whereas the values of [R_{tot}] and the function f are assumed to be tissue-specific. Efficacy is positive for agonists, negative for inverse agonists, and zero for neutral antagonists. Thus, in this model, ligands are assumed to have two key properties: affinity and efficacy. In other words, when the cellular response [increase in cell number, cyclic adenosine monophosphate (cAMP) production, increase in intracellular free calcium concentration, etc.] is plotted as a function of the scaled agonist concentration [A]/K_D, differences in the responses induced by different ligands are said to be a result of different ligand efficacies. One example of this is shown in Fig. 1. Equilibrium binding experiments can presumably be used to determine a value for the affinity, but efficacy is a parameter that is often used in a qualitative and relative sense (e.g. one ligand has a greater efficacy than another). Although Eqs. 1 and 2 do indeed describe many observed dose–response characteristics, the parameter ε and the function f lack any mechanistic underpinnings. Both are simply fit to the data.

Further, as pointed out by COLQUHOUN (1987), the above model cannot be correct; binding must influence efficacy and vice versa due to thermodynamic considerations. For example, in G protein-coupled systems, the affinity of agonist for receptor in the absence of G proteins is now believed to be determined by its affinities for both an active and inactive state of the receptor (SAMAMA et al. 1993; WEISS et al. 1996b), and the ability of the ligand to selectively bind to the receptor in the active state is believed to contribute to its efficacy as an agonist. Therefore, the two properties – affinity and efficacy – are necessarily intertwined.

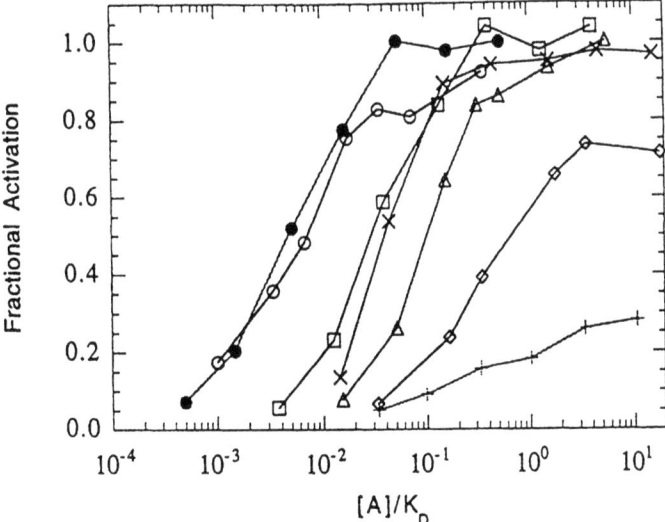

Fig. 1. Differences in ligand efficacy among ligands for the β-adrenergic receptor. Dose–response curves for seven different agonists – epinephrine (*filled circle*), isoproterenol (*open circle*), salbutamol (*open square*), metaproterenol (×), zinterol (*open triangle*), dobutamine (*open diamond*), and ephedrine (+) – binding to the β-adrenergic receptor on S49 cells are shown. Fractional activation represents 10 min of cAMP accumulation normalized by the maximum response obtained using epinephrine (STICKLE and BARBER 1991)

In this chapter, I will use the qualitative definition of efficacy as the property of a ligand that affects responses and is not simply related to the ability of ligand to occupy cell surface receptors. The premise of this chapter, consistent with all of the above, is that ligand efficacy is, in fact, dependent on a variety of physical phenomena and thus depends on the parameters that describe those phenomena. A combination of theoretical and experimental work may enable us to determine some of the physical parameters that underlie efficacy and, ultimately, may allow us to predict efficacy. In this context, physical parameters mean quantities that, in principle, are measurable, e.g., diffusivities, concentrations, and – notably – kinetic rate constants. The work described below is by no means complete; indeed, only a few of the physical parameters that may affect efficacy and cellular responses in general are discussed.

II. Modeling

In this chapter, the focus is on models that link receptor–ligand binding with the generation of cellular responses. Developing models requires that assumptions be made about the system to be described. Some of those assumptions are very obvious, such as including one reaction but leaving another out. Some are less obvious and can be hidden in the framework of the model; it is the modeler's job to understand and make those assumptions explicit. For example, the decision to treat all receptors in one particular state (e.g. ligand-bound) as equivalent in a model is routinely made, and then only the concentration of that particular receptor population is tracked in time. In Sect. C.I, we will see that this assumption may be limiting, and a different model framework that allows tracking of individual receptors will be used.

1. Equilibrium, Steady State, and Kinetic Models

It is useful to briefly define some of the terms and discuss some of the principles that will be used to describe the models in this chapter. The models under consideration here are ones that predict the concentrations of various molecular species (free receptors, ligand-bound receptors, desensitized receptors, activated G proteins, etc.). Consider the reversible reaction

$$A \underset{k_r}{\overset{k_f}{\rightleftarrows}} B \tag{3}$$

which describes the conversion of species A into species B with first-order rate constants k_f and k_r. If you begin with only species A, initially only the forward reaction occurs. Once some of species B is present, the reverse reaction also occurs. As the concentration of A decreases, the rate of the forward reaction decreases and, as the concentration of B increases, the rate of the reverse reaction increases. Obviously, eventually a point is reached at which the rates of the forward and reverse reactions are equal, and

$$k_f[A] = k_r[B] \tag{4}$$

or

$$\frac{[A]}{[B]} = K_D \tag{5}$$

where K_D, the equilibrium dissociation constant, is equal to k_r/k_f. At this point, no further net change in the amounts of A and B takes place, and the reaction mixture is said to be in chemical equilibrium (FELDER and ROUSSEAU 1986). The equilibrium solution to this model is given by Eq. 5. This, then, is an *equilibrium* model. There are no dynamics; one does not know how long it will take A to be converted into a mixture of A and B. To answer that question, one must consider not equilibrium thermodynamics but chemical kinetics.

One can also describe the model above with the kinetic equations

$$d[A]/dt = -k_f[A] + k_r[B] \tag{6a}$$

and

$$d[B]/dt = k_f[A] - k_r[B] \tag{6b}$$

the solutions of which describe the change in the concentrations of A and B with time. This is a *dynamic* or *kinetic model*. Eventually, a point at which the concentrations of A and B do not change any further with time will be reached, and the process is said to be at a *steady state*; before this point, the process is at an *unsteady state*. The solution to this model depends on the rate constants and not simply the equilibrium constant. For example, if one increases both k_f and k_r by the same factor, the equilibrium solution (which depends only on their ratio) is unchanged, but the approach to that state is more rapid.

For the model just described (Eq. 6a, b), the equilibrium and steady state solutions to the model are identical. However, this is not always the case. For example, in some systems, there may be a net flow of material and an accompanying steady state but no equilibrium (WYMAN 1975). Consider the situation of internalization and recycling of receptors via the endocytic pathway (for simplicity, imagine that all internalized receptors are recycled). If the internalization pathway is suddenly "turned on" (perhaps by ligand binding), receptors initially found on the surface will begin to appear inside the cell and, eventually, these receptors will be returned to the cell surface. After some time, the rates of internalization and recycling will be equal, and the numbers of receptors on the cell surface and inside the cell will not change with time. This steady state does not represent an equilibrium; there is a constant consumption of energy (this is an "open" system, not a closed one) and a constant flow of receptors around the loop. An analogous situation would be the cycling of G proteins between inactive [$\alpha\beta\gamma$-guanosine diphosphate (GDP)] and active [α-guanosine triphosphate (GTP) and $\beta\gamma$] states. Even if a steady state in the concentration of active and inactive G proteins was reached, there would still be a continuing hydrolysis of GTP to GDP (and thus a continuing need for GTP) and no thermodynamic equilibrium.

Of course, signal transduction is likely to be an inherently non-equilibrium process. The process is initiated when agonist is added, and may then desensitize in the continued presence of agonist. In the body (and in some in vitro experiments), the added complication of agonist concentrations that change with time is present. In this chapter, it is shown that kinetic modeling is a valuable tool for understanding ligand efficacy and signal transduction in general.

2. Diffusion-Versus Reaction-Controlled Events

In the simple example reaction scheme above (Eq. 3), rate constants are used to describe the rate at which an event occurs. For example, in Eq. 6a above, species A is converted to species B at a rate given by the product of the kinetic rate constant k_f and the concentration of A. The physics behind the rate constant describing the conversion of A to B has not yet been specified.

Consider, for example, the reaction of receptor (R) with ligand (L) to form a receptor–ligand complex (C) according to

$$R + L \underset{k_r}{\overset{k_f}{\rightleftharpoons}} C \tag{7}$$

Although only a single arrow is shown for the forward (association) process, the binding of receptor to ligand is, in actuality, a two-step process (reviewed in LAUFFENBURGER and LINDERMAN 1993). Receptor and ligand must first be transported near to each other (typically by diffusion), and then the two may react with intrinsic kinetics dictated by the properties of the molecules themselves. If the diffusion step is much faster than the intrinsic reaction step, then the overall process is termed *reaction-controlled* or reaction-limited. In this case, the value of k_f is determined by the value of the intrinsic forward reaction rate constant. However, if the intrinsic reaction step is much faster than the diffusion step, then the overall process is termed *diffusion-controlled* or diffusion-limited. In such a case, reactions occur with essentially 100% probability once two molecules capable of reacting actually collide. Finally, a reaction may be partially diffusion-controlled (or equivalently, partially reaction-controlled), and such a reaction would occur with less than 100% probability once the molecules collide.

The binding of a ligand to a cell surface receptor is typically (though not always) reaction-controlled (LAUFFENBURGER and LINDERMAN 1993 and references therein). It is reasonable to assume that hydrolysis of GTP by the α subunit of a G protein is also reaction-controlled [although the role of regulators of G protein signaling (RGS) proteins in regulating the hydrolysis rate may complicate that picture; BERMAN and GILMAN 1998]. However, the conversion of inactive G proteins to active G proteins, occurring upon collision of a G protein with a ligand-bound receptor (assuming adequate GTP is available), is likely to be either diffusion-controlled or partially diffusion-controlled (JANS 1992; MAHAMA and LINDERMAN 1994b, 1995). The interaction of two receptors, one bound by ligand and one not, to form two cross-linked receptors is also believed to be partially or completely diffusion-controlled (DEMBO et al. 1982). Modeling treatments in these situations often take the diffusion step explicitly into account.

3. Model Structures and Dose–Response Curves

Two types of models will be described in this chapter. *Ordinary differential-equation models (time-dependent)*, such as those given by Eq. 6a and b, can be solved (by analytical or numerical techniques) to give the concentrations of the various molecular species (ligand-bound receptors, desensitized receptors, internalized receptors) as a function of time. Note that there is no ability to distinguish between individual members of a particular species. Nor is there the ability to distinguish between molecules based on their spatial position, because time is the only independent variable. By identifying the concentration of one of the molecular species in the model (e.g. cAMP or α-GTP) as indicative of a response, a dose–response curve can be generated.

Kinetic Modeling Approaches to Understanding Ligand Efficacy

Alternatively, models that also consider the positions of molecules may also be developed. In this chapter, *Monte Carlo models* which track individual molecules in time and space are described. Although there are many types of Monte Carlo models (the name "Monte Carlo" simply means that there is some random element in the model), the particular type described here can be shown schematically in Fig. 2a. In this numerical-simulation scheme, molecules associated with the cell membrane are placed on a lattice (they can also be restricted to subregions or domains, as in Fig. 2b). Individual molecules are chosen at random and allowed to move in a random direction a distance in accordance with a specified diffusion coefficient and the time step of the simulation. Molecules are also allowed to undergo a conversion to a different molecular species. Some of these conversions are completely reaction-controlled, and they occur with a probability related to the intrinsic kinetic rate constant for that reaction. An example of such a conversion would be the conversion of a free receptor to a ligand-bound receptor, which is (in most cases) not

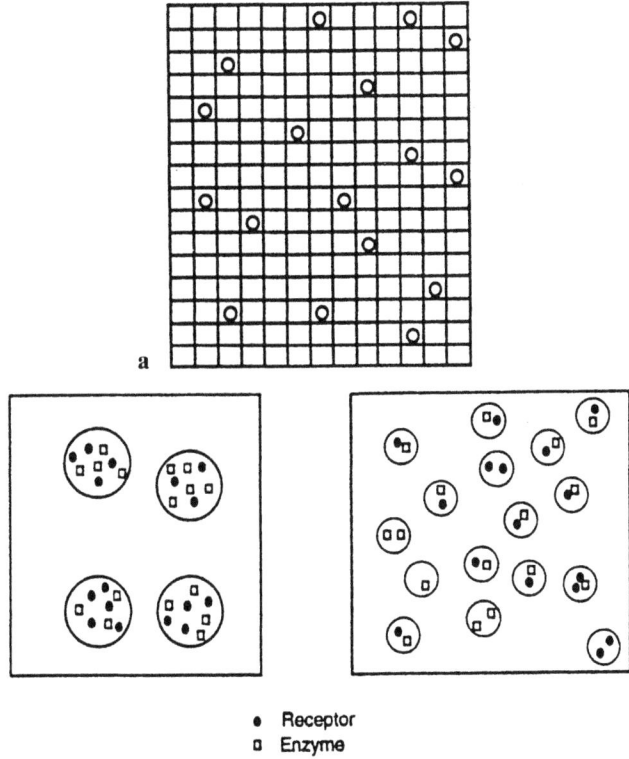

Fig. 2a,b. Monte Carlo simulations. **a** Monte Carlo model lattice with particles. Molecules (*circles*) are placed on a cell membrane (*grid*) and allowed to move among grid locations (diffusion), collide with other molecules, and react. **b** Restriction of molecules into specific subregions or domains. For clarity, the grid detail is not shown

influenced by diffusion (LAUFFENBURGER and LINDERMAN 1993). Diffusion-controlled conversions are set to occur with 100% probability once two molecules capable of reacting collide, and partly diffusion-controlled conversions are set to occur with some intermediate probability once the molecules collide. Overall, the simulation scheme simply keeps track of the individual molecules as they move and react on the two-dimensional surface of the plasma membrane (MAHAMA and LINDERMAN 1994). One can then count the number of each particular species for comparison with the ordinary differential-equation-model solution described above. Again, the identification of the concentration of one particular species as a response (e.g. the number of α-GTP) allows a dose–response curve to be generated.

C. Parameters Contributing to Ligand Efficacy

In the sections below, four categories of ligand-dependent kinetic rate constants that are related to receptor states and may contribute to ligand efficacy are discussed. Although the parameters and underlying phenomena are treated somewhat individually here, all would act simultaneously in a cell to modulate the cellular response.

I. The Lifetime of an Individual Receptor–Ligand Complex ($1/k_r$)

Consider a G protein-coupled receptor system in which receptors exist in only one of two states, ligand-bound or unbound, and in which only bound receptors are capable of activating G proteins. Are all bound receptors equivalent? Two situations of equal receptor occupancy with agonist are shown in Fig. 3. In the left panel of the figure, occupancy of receptor by agonist ligand is shared among all of the receptors. In the right panel, a very tightly binding antagonist is used to block some of the receptors. The remaining receptors are bound by agonist. Note that the agonist concentrations in the two panels must be different in order to achieve equivalent occupancies. Is there a difference in terms of expected G protein activation between the two situations shown?

Experimental data suggest that the equivalent receptor occupancies shown in Fig. 3 may give unequal G protein activation. STICKLE and BARBER (1989) measured the amount of cAMP production induced by epinephrine binding to β-adrenergic receptors on S49 murine lymphoma cells. Cells were incubated with agonist with or without preincubation with the antagonist propranolol to block some of the receptor sites. As shown in Fig. 4a, the fractional response declined with increasing concentrations of the agonist despite equivalent receptor occupancy by agonist. MAHAMA and LINDERMAN (1994a, 1995) measured the increase in cytosolic free-calcium concentrations in individual BC3H1 cells stimulated with the α1-adrenergic receptor agonist phenylephrine. These cells were found to exhibit an all-or-none response, and the frac-

Kinetic Modeling Approaches to Understanding Ligand Efficacy

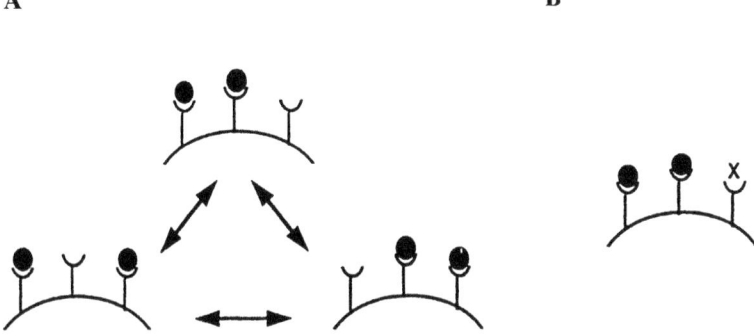

Fig. 3A,B. Ligand movement among cell surface receptors. Ligand is assumed to bind and dissociate from receptors with probabilities proportional to $k_f[A]$ and k_r, respectively. **A** Two-thirds of the receptors are occupied by agonist at any time, and each receptor is occupied by agonist 66% of the time. **B** One-third of the receptors have been blocked with a tightly binding antagonist. The remaining two-thirds of the receptors are occupied by agonist 100% of the time

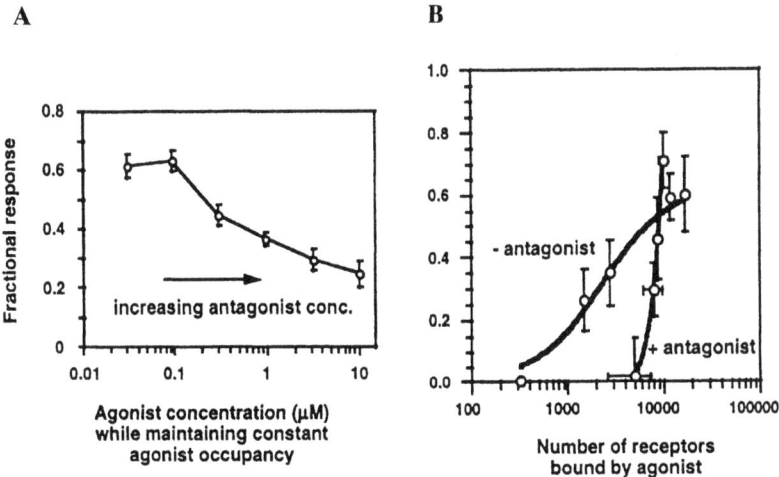

Fig. 4A,B. Impairing the movement of ligand among cell surface receptors may inhibit G protein activation. **A** Fractional response (cAMP production) as a function of agonist concentration (STICKLE and BARBER 1989). **B** Fractional response (fraction of cells exhibiting a calcium rise) as a function of the number of receptors bound by agonist and in the presence or absence of antagonist (MAHAMA and LINDERMAN 1995)

tion of cells responding was found to be an increasing function of the ligand concentration. In Fig. 4b, the fractional response (percentage of cells responding) as a function of the number of receptors bound by agonist is shown. Cells were either incubated with agonist alone or were pre-incubated with the tightly binding antagonist prazosin. These data suggest that movement of ligand among receptors (Fig. 3a) allows a greater cellular response than if such movement is inhibited (Fig. 3b).

The development of models for G protein activation offer an explanation of the data of Fig. 4 and suggest a parameter that may contribute to ligand efficacy. The hypothesized basis for the differences seen in Fig. 4 is the inability of diffusion to act as an effective mixing mechanism in two dimensions. Receptors and G proteins are believed to find each other by diffusion, as shown in Fig. 5a. A receptor that remains ligand-bound for a long time (shown schematically in Fig. 3b) may activate all nearby G proteins and have a difficult time accessing G proteins further away and, thus, a depletion zone in which few inactive G proteins are available may develop around that receptor. However, when the occupancy of receptors by ligand is effectively shared among the entire receptor population (shown schematically in Fig. 3a), such depletion zones are expected to be minimized, and greater G protein activation is anticipated. To test this scenario, the Monte Carlo models described earlier are used to follow the interaction of *individual* receptors and G proteins to produce the response of G protein activation (MAHAMA and LINDERMAN 1994b, 1995; SHEA and LINDERMAN 1997). The events included in the model are: binding and dissociation of ligand and receptor, collision of ligand-bound receptors with $\alpha\beta\gamma$-GDP to form α-GTP and $\beta\gamma$ (the exchange of GTP for GDP is assumed to be rapid), the hydrolysis of GTP by the α subunit to form α-GDP, the collision of α-GDP and $\beta\gamma$ to re-form $\alpha\beta\gamma$-GDP, and the diffusion of receptors and G proteins (all forms) in the membrane. A model framework that explicitly follows the spatial position of individual molecules is convenient, because one can then observe whether the hypothesized depletion zone does indeed develop. The interaction between a ligand-bound

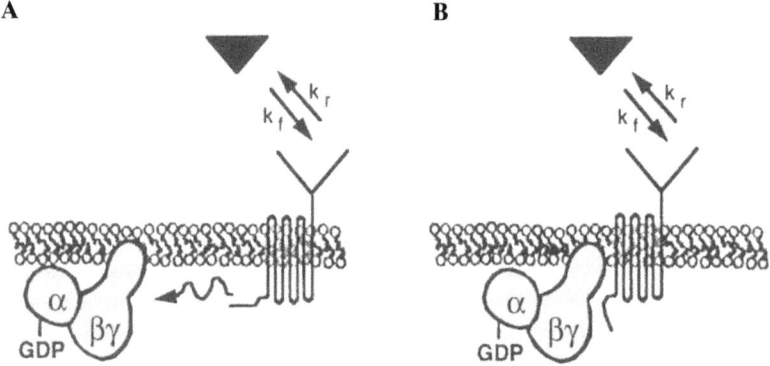

Fig. 5A,B. Role of diffusion in receptor–G protein coupling. **A** Receptors and G proteins find each other by diffusion in the plane of the membrane. **B** Receptors precoupled to G proteins need not diffuse to activate a G protein

receptor and inactive G protein to produce α-GTP and $\beta\gamma$ subunits (leaving the ligand-bound receptor unchanged) may be simulated as completely or partially diffusion-controlled. This is a non-equilibrium model; there is a consumption of GTP (assumed to be present in excess) due to the constant cycling of G proteins between activated and inactivated states.

The prediction of this model can be plotted in the form of a dose–response curve giving the amount of α-GTP at steady state as a function of the scaled ligand concentration, as shown in Fig. 6a. Note that the dose–response curve shifts to the left as the ligand dissociation rate constant k_r increases (i.e., as the mean receptor-ligand complex lifetime $1/k_r$ decreases). This occurs despite correction for the differing K_{DS} of the ligands. As hypothesized above, the reason for this shift in the dose–response curve is the development of larger depletion zones around individual receptors bound for long periods of time as compared with those bound for relatively short periods of time. A depletion zone is shown schematically in Fig. 7 and the size of the predicted depletion zone has been demonstrated to vary in a predictable fashion with both k_r and the diffusion coefficient (SHEA et al. 1997). The larger values of k_r (Fig. 7a) give a smaller depletion zone and are representative of the situation shown in Fig. 3a; the smaller values of k_r (Fig. 7b) give a larger depletion zone and are representative of the situation shown in Fig. 3b. It is interesting to note that the shape of the curve is hyperbolic – this is not an assumption as with the STEPHENSON model (1956) noted earlier but rather a consequence of the reactions specified in the simulation.

Two additional factors can influence the degree to which the mean lifetime of a receptor-ligand complex ($1/k_r$) affects the efficacy. First, pre-coupled

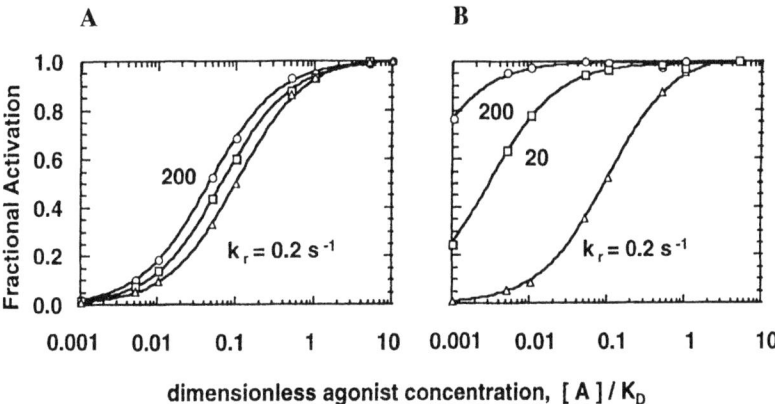

Fig. 6A,B. Predicted effect of the mean receptor–ligand complex lifetime $1/k_r$ on G protein activation. Results for k_r equal to 200, 20, and $0.2\,s^{-1}$ are shown in each plot. **A** Activation in the absence of any receptor–G protein pre-coupling. **B** Activation in the ongoing presence of receptor–G protein pre-coupling. For this simulation, coupling parameters are set such that 30% of receptors are pre-coupled to G protein prior to ligand addition (SHEA and LINDERMAN 1997)

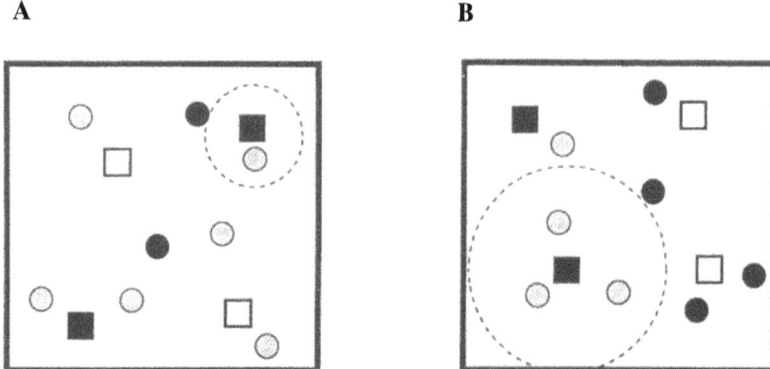

Fig. 7A,B. Schematic of the depletion zones that may develop around agonist-bound receptors. Bound receptors (*filled squares*), free receptors (*open squares*), activated G proteins (*shaded circles*), and inactive G proteins (*filled circles*) are shown. **A** Situation existing when the mean receptor–ligand complex lifetime $1/k_r$ is short. **B** Situation existing when the mean receptor ligand complex lifetime $1/k_r$ is long. In each case, there are the same number of agonist-bound receptors. More activated G proteins are produced when k_r is large, and those activated G proteins are more evenly distributed in the membrane. Hypothetical depletion zones, regions near receptors that are devoid of activatable G proteins, are shown by *dotted lines*. When k_r is small, receptors remain bound long enough to significantly deplete the region of inactive G proteins

receptor–G protein complexes (Fig. 5b) may be present. In the simulation of this scenario, one allows colliding free receptors and inactive G proteins ($\alpha\beta\gamma$-GDP) to complex with some set probability throughout the simulation, and for those complexes to break apart unproductively with a particular rate constant. When these pre-coupled receptor–G protein complexes are present, the need for a receptor-ligand complex to diffuse to find a G protein is reduced, at least for finding the initial G protein to activate. The rapid sharing of receptor occupancy among all surface receptors now conveys an even greater advantage, and the dose-response curve in the presence of pre-coupling is dramatically influenced by the mean lifetime of a receptor-ligand complex, as shown in Fig. 6b. The magnitude of this effect depends on both the equilibrium constant describing receptor–G protein coupling and uncoupling (not shown) and the kinetics of receptor–G protein pre-coupling (Fig. 8).

Second, receptors and G proteins may be present only within discrete microdomains on the cell surface. Such domains have been widely postulated (NEUBIG 1994; NEER 1995), although the size of the microdomains and the factors regulating them are unknown. One can use the Monte Carlo model described above (with the grid set up as in Fig. 2b) to determine the effect of domains on G protein activation. Generally, restricting receptors and G proteins to a fraction of the membrane increases activation relative to the case of no such restriction, as expected. When comparing cases of equivalent area coverage of domains but different numbers of domains (i.e., many small domains

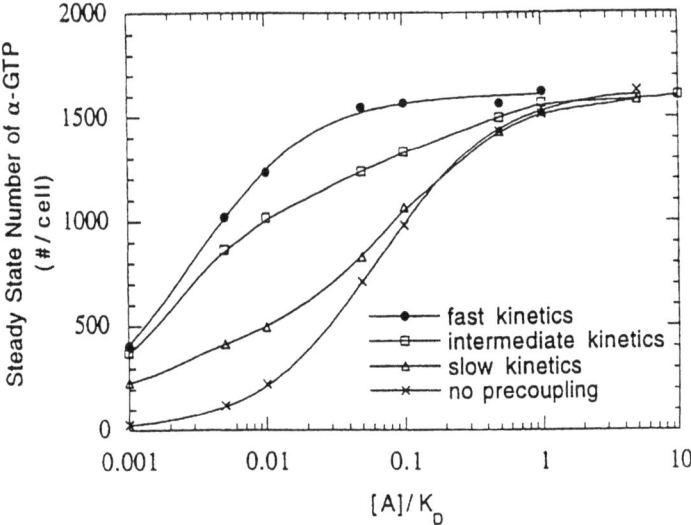

Fig. 8. Predicted effect of pre-coupling kinetics on activation. For simulations with pre-coupling of receptors and G proteins, 30% of the receptors were pre-coupled prior to ligand addition, thus setting the equilibrium constant describing receptor coupling and uncoupling. The kinetic rate constants making up that equilibrium constant were then varied. Larger values of those rate constants are predicted to shift the dose–response curve to the left (SHEA and LINDERMAN 1997)

versus a few large domains), however, one finds that the magnitude of the increase caused by compartmentalization to domains is dependent upon additional physical parameters, including the ligand-dissociation rate constant k_r. When there are small domains, there is a distinct advantage to the sharing of receptor occupancy among all receptors in all domains, as measured by the mean lifetime of a receptor–ligand complex ($1/k_r$). This effect on efficacy is shown in Fig. 9.

Thus, the conclusion from this work is that ligand efficacy in G protein-coupled systems is a function of the mean lifetime of the receptor–agonist complex ($1/k_r$). The value of k_r contributes not only to the determination of the number of bound receptors but also to the ability of those receptors to signal. A number of presumably cell-dependent parameters, such as the number of receptors pre-coupled to G proteins in the absence of ligand, the kinetics of that pre-coupling (and uncoupling) reaction, the translational-diffusion coefficients of the membrane molecules, and the presence and size of microdomains, influence the degree to which the value of k_r affects the dose–response curve.

II. The Receptor Desensitization Rate Constant k_x

In the previous section, we allowed receptors to be in ligand-bound, G protein-coupled, or free (unbound) states. Now consider the possibility that receptors

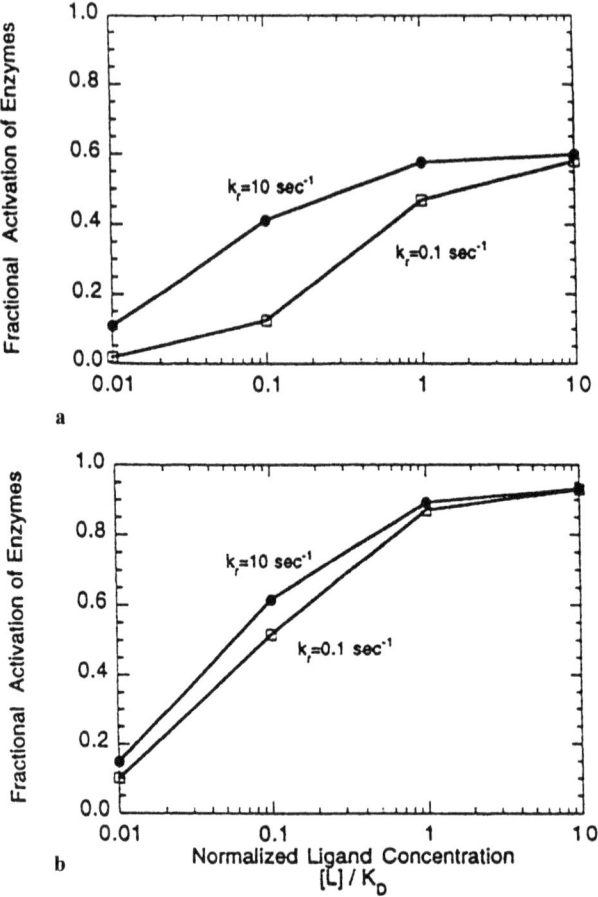

Fig. 9a,b. Predicted dose–response curves for varying values of the ligand dissociation rate constant k_r. **a** Receptors and G proteins are localized to 504 domains of diameter 154 nm. **b** Receptors and G proteins are localized to one domain of diameter 3612 nm. For each case, 20% of the cell surface is covered with domains. Ligand movement among surface receptors has a more significant effect on activation in the case of small domains (SHEA and LINDERMAN 1998)

can have yet another state, a desensitized state, as shown in Fig. 10. The ability of bound receptors to signal is clearly dependent on the value of the desensitization rate constant k_x. If the desensitization rate constant varies with the identity of the ligand, then k_x also contributes to ligand efficacy.

Quantitative investigations into the value of the desensitization rate constants k_x for different ligands binding to the same receptor type on the same cell type are few. One of the complicating factors is that a number of receptor-processing events (e.g. desensitization, internalization, recycling, upregulation) can take place simultaneously and, thus, must be considered when designing or analyzing experiments.

Fig. 10. Receptor desensitization. Receptors may be converted from a signaling-capable to a signaling-incapable form with rate constant k_x

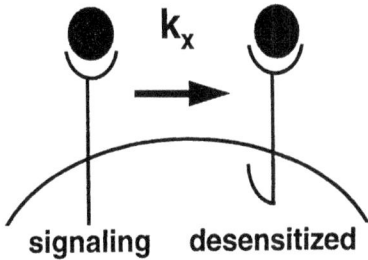

$$L + R_s \underset{k_r}{\overset{k_f}{\rightleftarrows}} LR_s \overset{k_x}{\longrightarrow} LR_n \underset{k_{fn}}{\overset{k_{rn}}{\rightleftarrows}} L + R_n$$

Fig. 11. Model scheme used by HOFFMAN et al. (1996) to analyze binding data and determine the desensitization rate constant k_x for ligands binding to the *N*-formyl peptide receptor on human neutrophils

HOFFMAN et al. (1996) followed the binding of several ligands to the *N*-formyl peptide receptor (a G protein-coupled receptor) on human neutrophils at 4°C using flow-cytometric techniques. This receptor is known to convert from a low-affinity state to a higher-affinity state (SKLAR 1987), a step that may involve or occur in series with phosphorylation of the receptor (ALI et al. 1993; TARDIF et al. 1993; PROSSNITZ 1997) and/or interaction of the receptors with cytoskeletal elements (KLOTZ and JESAITIS 1994) and which correlates with receptor desensitization (SKLAR et al. 1985). Fluorescent ligands, or non-fluorescent ligands in competition with fluorescent ligands, were used. At 4°C, receptor internalization and upregulation are inhibited; however, desensitization and ligand binding to and dissociation from the signaling and desensitized forms of the receptor occur. A schematic of the model used to analyze these events is shown in Fig. 11; the equations describing this model are time-dependent ordinary differential equations. In order to determine the rate constants of the model, binding data were analyzed using a statistical package to fit the model of Fig. 11 to the data. The values of k_x for three different agonist ligands of different (but positive) efficacy at 4°C are given in Table 1. Note that the value of the receptor-desensitization rate constant was found to be ligand dependent.[1]

These data suggest, then, that differences in ligand efficacy may, in part, be due to differences in desensitization rates for receptors bound by these different ligands. The underlying reason for the differences in desensitization rate constants is not known. For G protein-coupled receptors, recent

---

[1] There are also differences in the ligand-dissociation rate constant k_r among these ligands (HOFFMAN et al. 1996). This may contribute to differences in efficacy, as described in the previous section.

Table 1. Desensitization rate constants for three N-formyl peptide receptor agonists at 4°C (HOFFMAN et al. 1996)

Agonist	Desensitization rate constant k_x (s^{-1})
CHO-NLFNTK-fl[a]	$1.3 \pm 0.4 \times 10^{-2}$
CHO-MLF[b]	$1.1 \pm 0.1 \times 10^{-4}$
FNLP[c]	$1.0 \pm 0.4 \times 10^{-4}$

[a]N - formyl - norleucyl - leucyl - phenylalanyl - norleucyl - tyrosyl-lysine fluorescein.
[b]N-formyl-methionyl-leucyl-phenylalanine.
[c]N-formyl-norleucyl-leucyl-phenylalanine.

data and models (SAMAMA et al. 1993; WEISS et al. 1996a) have suggested that ligand-bound, non-desensitized receptors may exist in active (R*) and inactive (R) forms and that the distribution between these forms may depend on the ligand identity (Sect. C.4). If active and inactive forms of the receptor are desensitized at different rates, then the overall observed differences in desensitization rates of receptors bound by different ligands may be due partly or entirely to the underlying differences in the distribution of active versus inactive receptors (RICCOBENE et al. 1999).

III. The Ligand Binding and Dissociation Rate Constants at Endosomal pH

Let us now consider the trafficking of receptors to the interior of the cell and the possible recycling of those receptors back to the cell surface. This adds at least one additional receptor "state" (that of the internalized receptor) and presumably more (internalized receptors may be ligand-bound, desensitized, etc.). Although signal transduction and membrane trafficking have previously been considered as separate subdisciplines within cell biology, this clearly cannot continue (SEAMAN et al. 1996). For example, internalization and recycling of receptors may allow for resensitization of G protein-coupled receptors (KRUEGER et al. 1997), and signals generated by receptor–agonist binding (which may be ligand-dependent) may influence internalization (COLOMBO et al. 1994). In this section, however, the other side of that picture is discussed; the outcome of endocytosis – i.e., whether receptors are recycled or not – may influence signaling and may be the result of ligand-dependent properties.

The number of receptors on the cell surface that are available for initiating signal transduction is modulated by endocytosis. As shown in Fig. 12, internalized receptors are sorted in endosomes (which have a pH of about 5.0) into at least two pathways: the recycling pathway returns receptors to the cell surface, and the degradative pathway sends receptors to lysosomes. A similar choice of pathways is available to internalized ligands. If ligand properties

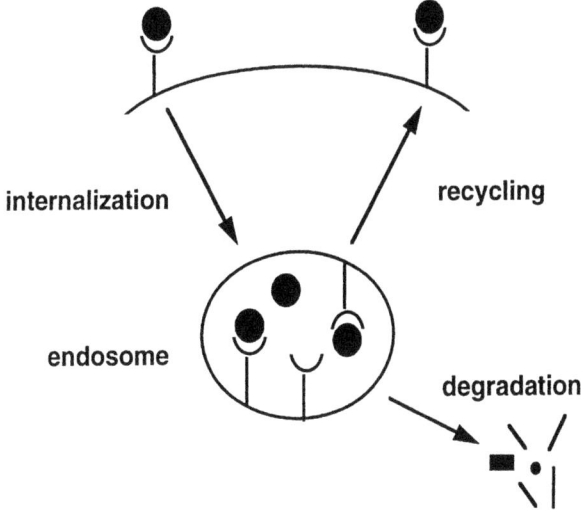

Fig. 12. Receptor and ligand sorting in endosomes. Receptors and ligands are internalized, delivered to acidic endosomes, and there targeted for either recycling or degradation

influence the outcome of the sorting process, then these properties would also be expected to contribute to ligand efficacy.

A model for the endosomal sorting process, developed by LINDERMAN and LAUFFENBURGER (1988) and extended by FRENCH and LAUFFENBURGER (1996), is based on the hypothesis that receptors found in endosomes may be sorted differently based on their state in the endosome (bound, free, or even cross-linked by a multivalent ligand). The expected scenario is that receptors that are unbound by ligand are capable of diffusing into the tubular portion of the endosome that is involved in recycling to the cell surface. Receptors that are bound by ligand can be trapped by an interaction with a "retention component" and thus are forced to remain behind in the central vesicular portion of the endosome and are targeted for degradation. This scheme is shown in Fig. 13. Although, when originally proposed, the "retention components" were simply hypothesized to exist, a molecule that may function as a sorting nexin for the epidermal growth factor (EGF)-receptor system has recently been identified (KURTEN et al. 1996).

Because different ligands are likely to have different abilities to bind at the low pH of the endosome, the outcome of the sorting process is expected to vary with the identity of the ligand. A ligand which readily dissociates from its receptor at low pH will leave many receptors free (unbound by ligand), unlikely to interact with retention components, and able to be recycled. Conversely, a ligand that does not readily dissociate will leave many receptors bound, likely to interact with retention components, and unable to recycle.

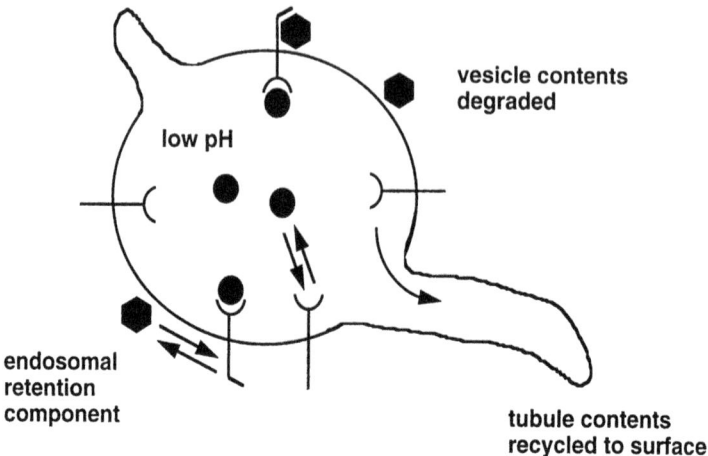

Fig. 13. Proposed model for sorting of receptors and ligands inside endosomes. Receptors and ligands that enter tubules are recycled, while those that do not are later degraded. Ligand-bound receptors may be trapped by endosomal retention components and prohibited from diffusing into tubules. Thus, the ability of ligand to bind receptor in the endosome influences the outcome of the sorting process

Thus, the different abilities of ligands to dissociate at low pH will contribute to different efficacies.

For this model (Fig. 13), time-dependent ordinary differential equations are written to describe the number of each species (bound receptors in vesicles, bound receptors in tubules, free receptors in tubules, free ligand in tubules, receptors trapped by the retention components, etc.). Receptors (bound and free) that are not trapped by retention components are free to diffuse into tubules. Receptors can bind and dissociate ligand with rate constants evaluated at endosomal pH and can interact with retention components with forward and reverse rate constants.

A major prediction of the model for the outcome of the sorting process at steady state is shown in Fig. 14. In this figure, the outcome of the sorting process for a particular ligand is shown as a function of loading by the ligand, i.e., the intracellular ligand concentration. At low ligand concentrations (region I in the figure), few of the ligands are bound and, thus, the fraction of ligand that is recycled corresponds to the fraction of the endosomal volume found in tubules (here estimated as ~60%). As the ligand concentration is increased, a greater fraction of the ligand is bound to receptors, and those receptors remain behind in the vesicular portion of the endosome. Thus, the amount of ligand recycled drops (region II). Finally, as the ligand concentration is increased still further, the endosomal retention components are saturated, and bound receptors are found in the recycling tubules. In this region (region III), both free ligand and much of the bound ligand is recycled, and the fraction of ligand recycled increases. This is the qualitative behavior expected for every ligand; however, the locations and sizes of the different

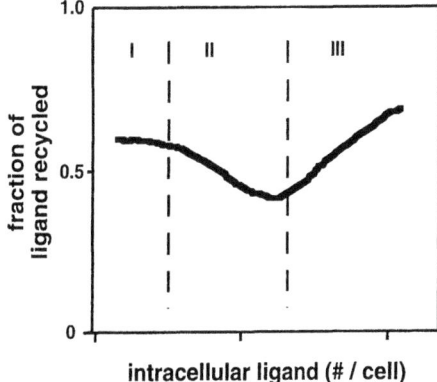

Fig.14. Predicted qualitative behavior of the sorting model of Fig. 13. The fraction of internalized ligand that is recycled to the cell surface depends on the intracellular ligand loading. At low ligand loading (*region I*), much of the ligand is free. At higher loading (*region II*), more ligand is bound, and those bound receptors are trapped by endosomal retention components. At the highest ligand loading (*region III*), endosomal retention components are saturated and ligand recycling increases. Plotting the sorting outcome as a function of the intracellular ligand loading rather than the extracellular ligand concentration allows differences in binding of ligands at the cell surface to be normalized (FRENCH and LAUFFENBURGER 1996)

Table 2. Association (k_f) and dissociation (k_r) rate constants for three epidermal growth factor (EGF)-receptor ligands at pH 6 (FRENCH et al. 1995)

Ligand	k_f min^{-1}M^{-1}	k_r min^{-1}
Mouse EGF	$2.7 \pm 1.0 \times 10^7$	0.75 ± 0.16
Human EGF	$8.5 \pm 1.0 \times 10^6$	0.66 ± 0.12
TGFα	$5.7 \pm 2.0 \times 10^6$	2.30 ± 0.91

TGF, transforming growth factor.

regions will depend on ligand- and cell-dependent parameters such as the association and dissociation rate constants for ligand binding to receptor at endosome pH and the number of retention components.

Quantitative data to support the model of Fig. 13 are difficult to obtain, because it is difficult to isolate the outcome of the sorting process from other receptor- and ligand-processing events. Recently, however, elegant experiments by FRENCH and co-workers (1995) have measured different outcomes of sorting that are qualitatively consistent with model predictions. The investigators used three ligands – mouse EGF, human EGF, and transforming growth factor-α (TGFα) – which bind to the EGF receptor and have different pH sensitivities at low pH (data is only available to pH 6), as shown in Table 2. To determine the outcome of sorting, B82 fibroblasts were transfected with wild-type EGF receptors. Cells were incubated at 37°C for 2 h in various concentrations of labeled ligand, were washed at pH 3.0 and 4°C to remove surface-bound ligand, and were chased at 37°C with an excess of unlabelled ligand. Gel filtration was used to separate degraded from intact (recycled) radiolabeled ligands in the medium. The three ligands were found to have different sorting outcomes, as shown in Fig. 15. In this figure, differences in binding at the cell surface are normalized by plotting the amount of ligand recycled versus the amount of *intracellular* ligand. Note that TGFα behaves as in region II of Fig. 14. There is enough intracellular ligand nearby to bind some of the internalized receptors, and these receptors, along with the TGFα, are presumably trapped by the retention components and ultimately degraded. Mouse and human EGF, which are more likely to remain bound to receptors at endosome pH than are TGFα (Table 2), behave as in region III of Fig. 14. Because many of the receptors remain bound by ligand, the endosomal retention apparatus is saturated, and many of the receptors, along with the EGF, are recycled.

Thus, the conclusion drawn from this work is that ligand efficacy is a function of the ligand association rate constant k_f and the dissociation rate constant k_r at endosomal pH. Two ligands acting identically at the cell surface (same association and dissociation rate constants, same ability of receptor–ligand complexes to transmit a signal, same rate of desensitization and/or internalization, etc.) may have different efficacies if, for example, ligand A is likely to stay bound to receptors within the endosomes and ligand B is likely to dissociate. Ligand A will direct more receptors to the degradative pathway, thus reducing further signal transduction, while ligand B will allow receptors

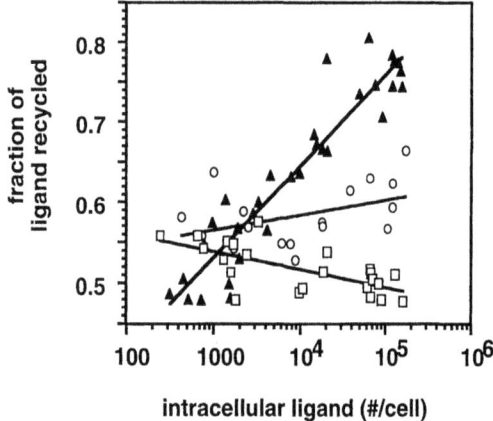

Fig. 15. Measured endosomal sorting outcomes for several epidermal growth factor (EGF)-receptor ligands. The fraction of internalized ligand that is recycled to the cell surface as a function of the amount of intracellular ligand is shown for mouse EGF (*filled triangle*), human EGF (*open circle*), and transforming growth factor α (*open square*) (FRENCH et al. 1995)

Kinetic Modeling Approaches to Understanding Ligand Efficacy

to recycle and again signal at the cell surface. Furthermore, cell-dependent parameters (such as the number of endosomal retention components and the kinetics of their interaction with ligand-bound receptors) may contribute to the abilities of these ligands to induce cellular responses.

IV. Rate Constants in a Ternary Complex Model

Recently, new experimental data have suggested that there may be multiple interconverting states for G protein-coupled receptors on the cell surface. Receptors potentially able to interact with G proteins are hypothesized to exist in active (R^*) and inactive (R) conformations. These may actually represent families of conformations (KENAKIN 1996a), and the minimum number of such "families" that will be needed in a model is not known. Thus, the ternary (i.e., ligand/receptor/G protein) complex model (TCM) proposed by JACOBS and CUATRECASAS (1976) and DELEAN et al. (1980) and extensively studied in its various forms (KENAKIN 1996b and references therein) has been modified to include these additional states. Two such modifications have been proposed: the extended ternary complex model (eTCM) shown in Fig. 16a (SAMAMA et al. 1993) and the more thermodynamically complete cubic ternary complex model (cTCM) shown in Fig. 16b (Chap. 2.2; WEISS et al. 1996a). The eTCM and cTCM account for, among other data, the observation that the increase in the affinity of ligands for a constitutively active mutant (CAM) receptor is correlated with their intrinsic activity at the wild-type receptor (SAMAMA et al. 1993). They also allow ligands to behave as inverse agonists.

The models shown in Fig. 16, like the original TCM, are equilibrium models. They have, however, been used to predict activation in the presence

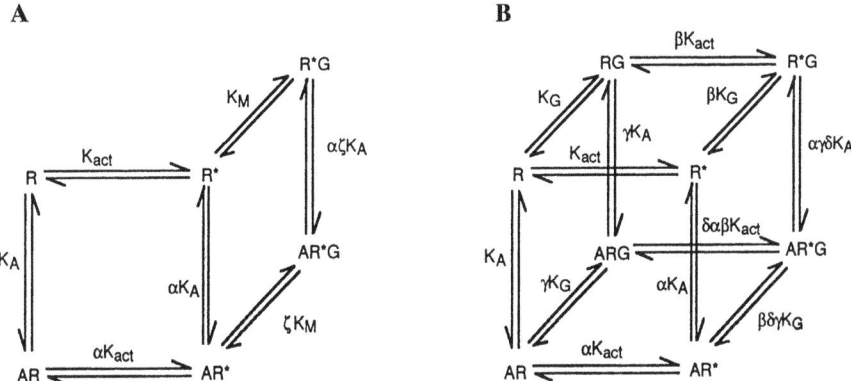

Fig. 16. Equilibrium ternary complex models that include active (R^*) and inactive (R) receptor states. **A** Extended ternary complex model (eTCM). **B** Cubic ternary complex model (cTCM). The cTCM reduces to the eTCM when the concentrations of RG and ARG are small. Equilibrium constants for each reaction are shown

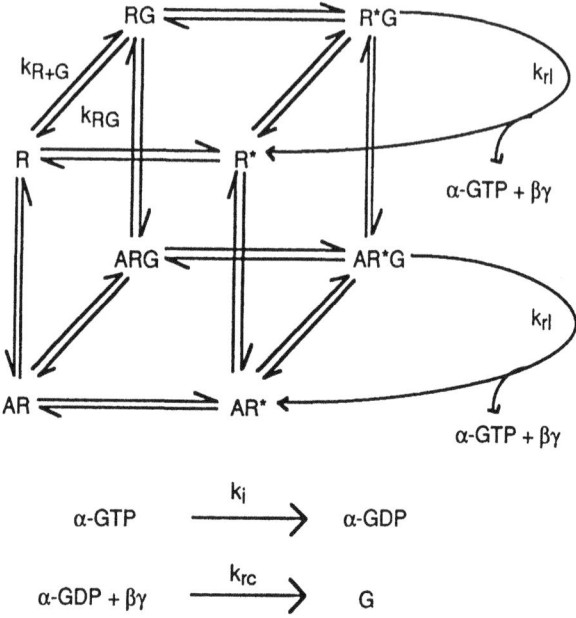

Fig. 17. The ternary complex activation model (TCAM). The steps of G protein activation, hydrolysis of guanosine triphosphate by the α subunit, and recombination of α and $\beta\gamma$ subunits are explicitly included (rate constants k_{rl}, k_i, and k_{rc}, respectively). Rate constants that make up each equilibrium reaction in the cubic ternary complex model (cTCM) are specified. For example, k_{R+G}/k_{RG} is seen to be equal to the equilibrium constant K_G of the cTCM. For clarity, other rate constants are not shown, but see SHEA (1997)

of guanine nucleotides by making the assumption that activation is proportional to [R*G] + [AR*G] (SAMAMA et al. 1993; KENAKIN 1996b, 1997). In these models, in other words, the "active" receptor coupled to G protein is assumed to produce activated G proteins in proportion to the receptor concentration calculated from the model, although this step is not explicitly included. Alternatively, one can think of these models as most appropriate when applied to data obtained in the absence of GTP.

We have examined an extension to these models that also accounts for the activation of G proteins in the presence of GTP (SHEA 1997; SHEA et al. 1998). This model – a non-equilibrium model termed the ternary complex activation model (TCAM) – is shown in Fig. 17 and is based, in part, on the cTCM of Fig. 16b.[2] Transitions between receptor states that are part of the "cube" are described by forward and reverse rate constants (their ratios give the equilib-

---

[2] An analogous model based instead on the eTCM would give qualitatively similar results, as the eTCM is a subset of the cTCM.

rium constants of Fig. 16b). Of these, for simplicity, only two are shown in Fig. 17. Activation of G proteins by the AR*G (or R*G) complex includes the exchange of GTP for GDP and the breakup of the complex into AR* (or R*), α-GTP, and $\beta\gamma$. THOMSEN and NEUBIG (1989) found these events to be adequately described by a single rate-limiting step (the breakup of the complex) and, thus, only a single step with rate constant k_{rl} is used in the TCAM. Using the kinetic model shown in Fig. 17 (and the time-dependent ordinary differential equations that describe it) gives new insights into ligand efficacy.

It is worth noting first that the TCAM, which reduces to the cTCM at equilibrium in the absence of guanine nucleotides, is capable of describing the ligand-binding behavior that originally motivated the development of the eTCM, as described in SAMAMA et al. (1993). Furthermore, if one uses the α-GTP concentration as a measure of the cellular response (one could also use the concentration of $\beta\gamma$ subunits or, by adding other elements to the model, the concentration of another downstream molecule), one can describe positive, neutral, and negative agonism (Fig. 18a) and the greater activity of CAMs compared with wild-type receptors (Fig. 18b) using the TCAM.

KENAKIN (1996b, 1997, 1998) and WEISS et al. (1996c) have discussed how ligand-dependent (α) and system-dependent (K_{act}) equilibrium parameters of the cTCM contribute to efficacy and the dose–response curve. Below, it is shown that, in addition, the individual kinetic rate constants of the TCAM contribute to ligand efficacy.

In Fig. 19, sample predictions of the kinetic model (TCAM) and equilibrium model (cTCM) are compared for identical values of the nine constants the two models have in common (equilibrium constants K_G, K_{act}, and K_A, constants α, β, γ, and δ, and the total numbers of receptors and G proteins; Fig. 16b). For the kinetic model, additional parameters (the rate constants) must also be specified, and the prediction shown represents only one of many. The response plotted for the TCAM is the amount of α-GTP at steady state; the response plotted for the cTCM is the number of receptors in the active state that are coupled to G proteins ([R*G]+[AR*G]).

Note first that activation (as measured by α-GTP in the kinetic model) is *not* predicted from the value of [R*G] + [AR*G] in the equilibrium model (SHEA 1997). In the particular case shown, the equilibrium and kinetic models predict positive and negative agonism, respectively. By comparing the kinetic model to the equilibrium model, one can identify three factors that lead to different predictions (WOOLF and LINDERMAN 2000). First, the addition of a second pathway (the activation pathway) from R*G to R* and from AR*G to AR* in the kinetic model alters the distribution of receptor states so that the non-G protein-coupled states are more favored than when the activation pathway is not present. Second, the number of G proteins available for coupling to receptors is reduced in the kinetic model, because some of the G proteins are now present as α-GTP and α-GDP. Finally, the models are qualitatively different – the kinetic model is used to determine a steady state, which

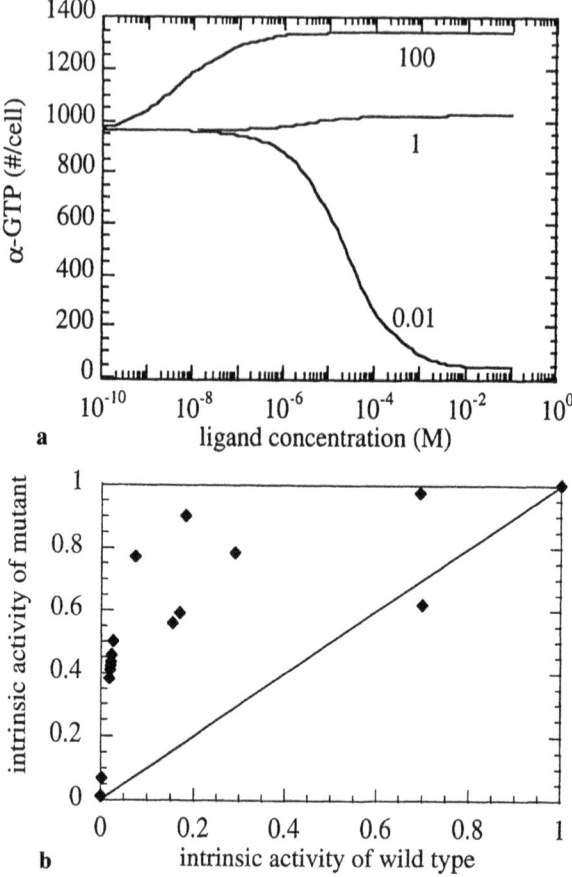

Fig. 18a,b. Sample predictions of the ternary complex activation model. **a** Positive, neutral, and inverse agonism are found as α is varied as shown and all remaining parameters (rate constants and total numbers of receptors and G proteins) are held constant. Responses are assumed to be proportional to the amount of α-guanosine triphosphate (GTP) at steady state. **b** Intrinsic activity (normalized amount of α-GTP) of a constitutively active mutant as a function of the activity of the wild type. Each *point* represents a different ligand simulated by choosing different values of α and β (all other parameters were held constant). Mutant receptors were assumed to have a 100-fold larger value of K_{act} than wild-type receptors. The behavior shown is reminiscent of that reported in SAMAMA et al. (1993)

is not at all the same as equilibrium. These factors conspire – for the parameter set chosen here – to cause negative agonism in the activation model.[3] Thus, knowledge of the parameters in the equilibrium model does not uniquely determine activation and, in fact, the equilibrium model may give qualitatively and quantitatively different predictions than the related kinetic model. This is

---

[3] Even when the two models predict the same type of behavior, such as inverse agonism, the magnitudes of the responses predicted are often very different.

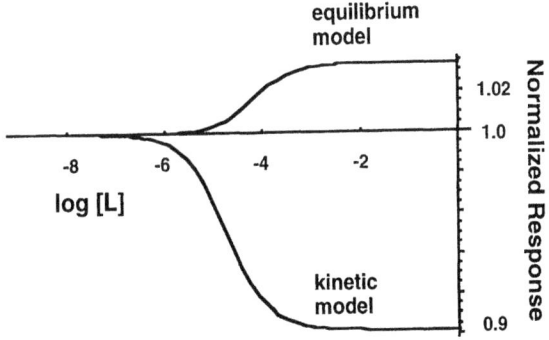

Fig. 19. Comparison of the predictions of the equilibrium (cubic ternary complex model; cTCM) and kinetic (ternary complex activation model; TCAM) models. Response in the cTCM is given by the sum of the concentrations of AR*G and R*G. Response in the TCAM is given by the steady state amount of αGTP. All parameters common to the two models are identical; in addition, the kinetic model requires that individual rate constants be specified. For the particular parameters used here, the TCAM predicts inverse-agonist behavior, while the cTCM predicts positive agonism

not only true for predictions of "responses" (Fig. 19) but also for predictions of concentrations of various species, such as receptors pre-coupled to G proteins (SHEA et al. 1998).

Thus, the conclusion from this work is that efficacy in G protein-coupled systems may be a function of the kinetic parameters describing conversion between receptor states – for example, the ligand-dependent rate constants describing the interconversion of inactive ligand-bound receptor (AR) and active ligand-bound receptor (AR*) and the interconversion of AR* and AR*G. It is also possible that the ability of RGS proteins to regulate GTP hydrolysis by the α subunit of the G protein may be influenced by the ligand (BERMAN and GILMAN 1988), suggesting that the value of the rate constant for hydrolysis (k_i in Fig. 17) may also contribute to efficacy.

D. Concluding Remarks

The premise of this chapter is that kinetic models of receptor signaling offer insights into the determinants of ligand efficacy. Several kinetic models are described here to suggest that ligand efficacy is a function of (1) the ligand-dissociation rate constant k_r when the receptor must diffuse to find the next component – here a G protein – in the signal transduction pathway, (2) the receptor desensitization rate constant k_x, and (3) the many rate constants describing receptor state (e.g. active/inactive, coupled/uncoupled to G protein) interconversions in G protein-coupled systems. In addition to these parameters, other potentially ligand-dependent kinetic rate constants can also contribute to ligand efficacy. Examples of such kinetic parameters include the receptor internalization rate constant and the rate constant for receptor resensitization. The influence that each of these parameters may have on the dose–response curve may be modulated by a number of cell-dependent parameters including, for example, the numbers of receptors, G proteins, receptor kinases (involved in desensitization), and endosomal retention components, and the rate constants describing, for example, interconversion between active and inactive forms of a G protein-coupled receptor in the absence of ligand.

Although not the focus of this chapter, model validation through experiments is essential. With the growing ability of molecular biologists to alter the structures of receptors, ligands, and effectors, we may be able to directly test many of these models by intentionally modifying key kinetic rate constants and other system parameters. At that point, we may be able to further identify the molecular structures that underlie differences in the measured values of kinetic rate constants and use that knowledge to rationally design drugs with a desired efficacy.

Acknowledgements. The author gratefully acknowledges helpful conversations with Drs. Terry Kenakin, Robert Lefkowitz, Lonnie Shea, Rick Neubig, and Geneva Omann, the hospitality of R. Lefkowitz and his group during the 1996–1997 academic year, and the assistance of Peter Woolf and Lonnie Shea with some of the figures. The support of National Science Foundation grants BES-9410403 and BES-9713856 is also acknowledged.

References

Ali H, Richardson RM, Tomhave ED, Didsbury JR, Snyderman R (1993) Differences in phosphorylation of formyl peptide and C5a chemoattractant receptors correlates with differences in desensitization. J Biol Chem 268:24247–24254

Berman DM, Gilman AG (1998) Mammalian RGS proteins: barbarians at the gate. J Biol Chem 273:1269–1272

Colombo MI, Mayorga LS, Nishimoto I, Ross, EM, Stahl PD (1994) G_s regulation of endosome fusion suggests a role for signal-transduction pathways in endocytosis. J Biol Chem 269:14919–14923

Colquhoun D (1987) Affinity, efficacy, and receptor classification: is the classical theory still useful? In: Black JW, Jenkinson DH, Gerskowitch VP (eds) Perspectives on receptor classification, Vol. 6, Alan Liss, Inc., pp 103–114

DeLean A, Stadel JM, Lefkowitz RJ (1980) A ternary-complex model explains the agoinst-specific biding properties of the adenylate-cyclase-coupled β-adrenergic receptor. J Biol Chem 255:7108–7177

Dembo M, Kagey-Sobotka A, Lichtenstein L, Goldstein B (1982) Kinetic analysis of histamine release due to covalently linked Ig dimers. Mol Immunol 19:421–434

Felder RM, Rousseau RW (1986) Elementary principles of chemical processes. Second edition. John Wiley & Sons, Inc., New York

French AR, Lauffenburger DA (1996) Intracellular receptor/ligand sorting based on endosomal retention components. Biotech and Bioeng 51:281–297

French AR, Tadaki DK, Niyogi SK, Lauffenburger DA (1995) Intracellular trafficking of epidermal-growth-factor family ligands is directly influenced by the pH sensitivity of the receptor/ligand interaction. J Biol Chem 270:4334–4340

Furchgott RF (1966) The use of β-haloalkylamines in the differentiation of receptors and in the determination of dissociation constants of receptor–agonist complexes. In: Advances in drug research, Vol. 3, edited by NJ Harper and AB Simmonds, pp 21–55. Academic Press, New York

Hoffman J, Keil ML, Riccobene TA, Omann GM, Linderman JJ (1996) Interconverting receptor states at 4°C for the neutrophil N-formyl peptide receptor. Biochemistry 35:13047–13055

Jacobs S, Cuatrecasas P (1976) The mobile-receptor hypothesis and "cooperativity" of hormone binding. Biochim Biophys Acta 433:482–495

Jans DA (1992) The mobile receptor hypothesis revisited: a mechanistic role for hormone receptor lateral mobility in signal transduction. Biochim Biophys Acta 1113:271–276

Kenakin TP (1993) Pharmacologic analysis of drug–receptor interaction. Second edition. Raven Press, Ltd., New York

Kenakin TP (1996a) Receptor conformational induction vs. selection: All part of the same energy landscape. Trends Pharm Sci 17:190–191

Kenakin TP (1996b) The classification of seven transmembrane receptors in recombinant expression systems. Pharmacol Rev 48:413–463

Kenakin TP (1997) Protean agonists: keys to receptor active states? Ann NY Acad Sci 812:116–125

Kenakin TP (1998) Differences between natural and recombinant G protein-coupled receptor systems with varying receptor/G protein stoichiometry. Trends Pharm Sci 18:456–464

Klotz K-N, Jesaitis AJ (1994) Neutrophil chemoattractant receptors and the membrane skeleton. Bioessays 16:193–198

Krueger KM, Daaka Y, Pitcher JA, Lefkowitz RJ (1997) The role of sequestration in G protein-coupled receptor resensitization. Regulation of β2-adrenergic receptor dephosphorylation by vesicular acidification. J Biol Chem 272:5–8

Kurten RC, Cadena DL, Gill GN (1996) Enhanced degradation of EGF receptors by a sorting nexin, SNX1. Science 272:1008–1010

Lauffenburger DA, Linderman JJ (1993) Receptors: models for binding, trafficking, and signaling. Oxford University Press, New York

Linderman JJ, Lauffenburger DA (1988) Analysis of intracellular receptor/ligand sorting in endosomes. J Theor Biol 132:203–245

Mahama PA, Linderman JJ (1994a) Calcium signaling in individual BC$_3$H1 muscle cells: speed of calcium mobilization and heterogeneity. Biotechnology Prog 10:45–54

Mahama PA, Linderman JJ (1994b) A Monte Carlo study of the dynamics of G-protein activation. Biophys J 67:1345–1357

Mahama PA, Linderman JJ (1995) Monte Carlo simulations of membrane signal transduction events: effect of receptor blockers on G-protein activation. Ann Biomed Eng 23:299–307

Neer E (1995) Heterotrimeric G proteins: organizers of transmembrane signals. Cell 80:249–257

Neubig RR (1994) Membrane organization in G-protein mechanisms. FASEB 8:939–946

Prossnitz ER (1997) Desensitization of N-formylpeptide receptor-mediated activation is dependent upon receptor phosphorylation. J Biol Chem 272:15213–15219

Riccobene TA, Omann GM, Linderman JJ (1999) Modeling activation and desensitization of G-protein coupled receptors provides insight into ligand efficacy. J Theor Biol 200:207–222

Samama P, Cotecchia S, Costa T, Lefkowitz RJ (1993) A mutation-induced activated state of the β2-adrenergic receptor: extending the ternary-complex model. J Biol Chem 268:4625–4636

Seaman MN, Burd CG, Emr SD (1996) Receptor signalling and the regulation of endocytic membrane transport. Curr Opin Cell Biol 8:549–556

Shea LD (1997) Kinetics of ligand, G-protein, and receptor interaction for signal transduction: a modeling study. University of Michigan

Shea LD, Linderman JJ (1997) Mechanistic model of G-protein signal transduction: Determinants of efficacy and the effect of pre-coupled receptors. Biochem Pharm 53:519–530

Shea LD, Linderman JJ (1998) Compartmentalization of receptor and enzymes affects activation for a collision coupling mechanism. J Theor Biol 191:249–258

Shea LD, Omann GM, Linderman JJ (1997) Calculation of diffusion-limited kinetics for the reactions in collision coupling and receptor cross-linking. Biophys J 73: 2949–2959

Shea LD, Neubig RR, Linderman JJ (1998) The critical role of kinetics in G-protein activation. In preparation

Sklar LA (1987) Real-time spectroscopic analysis of ligand-receptor dynamics. Ann Rev Biophys Biophys Chem 16:479–506

Sklar LA, Hyslop PA, Oades ZG, Omann GM, Jesaitis AFG, Cochrane CG (1985) Signal transduction and ligand-receptor dynamics in the human neutrophil. Transient responses and occupancy–response relations at the formyl peptide receptor. J Biol Chem 260:11461–11467

Stephenson RP (1956) A modification of receptor theory. Br J Pharmacol 11:379–393

Stickle D, Barber R (1989) Evidence for the role of epinephrine binding frequency in activation of adenylate cyclase. Mol Pharm 36:437–445

Stickle D, Barber R (1991) Comparisons of the combined contributions of agonist-binding frequency and intrinsic efficiency to receptor-mediated activation of adenylate cyclase. Mol Pharm 40:276–288

Tardif M, Mery L, Brouchon L, Boulay F (1993) Agonist-dependent phosphorylation of N-formyl peptide and activation peptide from the fifth complement component of C5a chemoattractant receptors in differentiated HL60 cells. J Immunol 150:3534–3545

Thomsen WJ, Neubig RR (1989) Rapid kinetics of $\alpha 2$-adrenergic inhibition of adenylate cyclase. Evidence for a distal rate-limiting step. Biochem 28:8778–8786

Weiss JM, Morgan PH, Lutz MW, Kenakin TP (1996a) The cubic ternary-complex receptor-occupancy model. I. Model description. J Theor Biol 178:151–167

Weiss JM, Morgan PH, Lutz MW, Kenakin TP (1996b) The cubic ternary-complex receptor-occupancy model. II. Understanding apparent affinity. J Theor Biol 178:169–182

Weiss JM, Morgan PH, Lutz MW, Kenakin TP (1996c) The cubic ternary-complex receptor-occupancy model. III. Resurrecting efficacy. J Theor Biol 181:381–397

Woolf P, Linderman JJ (2000) From the static to the dynamic. Three models of signal transduction in G-protein coupled receptors. In: Biomedical applications of computer modeling, edited by A Christopoolos, CRC Press. In press

Wyman J (1975) The turning wheel: a study in steady states. Proc Natl Acad Sci USA 72:3983–3987

CHAPTER 5
The Evolution of Drug-Receptor Models: The Cubic Ternary Complex Model for G Protein-Coupled Receptors

T. KENAKIN, P. MORGAN, M. LUTZ, and J. WEISS

A. Receptor Models

One of the earliest explicit mathematical representations of drug–receptor kinetics was given by CLARK (1933, 1937). It was known as occupancy theory, a name that suggested that occupation of a receptor by a drug could evoke a response. To differentiate drugs that simply occupied the receptor from those that occupied and presumably changed the receptor (to produce a physiological response), a proportionality factor (intrinsic activity) was added to receptor occupancy for these latter drugs (ARIENS 1954). These ideas were later extended and made more applicable to experimental pharmacology by STEPHENSON (1956) with the introduction of "efficacy".

This chapter will describe the evolution of the ternary complex model for G protein-coupled receptors (GPCR). As a preface, two important ideas in receptor pharmacology, which led to the development of the ternary complex model, will be discussed. The first is the description of allosterism in receptor proteins from ideas describing the behavior of ion channels (KATZ and THESLEFF 1957). The ability to monitor the flux of ions through open and shut ion channels afforded a method to study the various conformations of protein ion channels and the development of two-state theory. Open and closed ion channels allowed such quantification by means of measurement of current flow through the open channel. The other idea crucial to the thinking about seven transmembrane (7TM) receptors was one that described receptors as mobile entities in the lipid of the cell membrane, i.e., the "mobile receptor" hypothesis by CUATRECASAS (1974). This suggested a new realm of interaction of receptors and other membrane-bound proteins, specifically G proteins.

These ideas came together in the ternary complex model, as published by DE LEAN and colleagues (DE LEAN et al. 1980). This model considered the activation of a receptor by an agonist to produce a ternary complex with a G protein, thereby activating the G protein to initiate cellular response. A modified, so-called extended ternary complex model has been described by SAMAMA et al. (1993); it incorporates constitutive receptor activity. Since its publication in 1980, the ternary complex model and its extension have become the standard by which drug interaction with 7TM receptors has been studied.

A recent further extension of this 7TM model, referred to as the cubic ternary complex model (CTC model), has been presented; it differs from the extended ternary complex model by allowing the G protein to interact with the inactivated receptor (WEISS et al. 1996a, 1996b). The thermodynamic reasons for allowing this interaction are discussed later.

This chapter will derive the various equations used to describe these models and show relationships that can be used to experimentally verify the specific predictions of each of the models. In particular, some predictions of the extended ternary complex and the CTC models will be contrasted. As a preface to the description of these complex models, it is useful to review the behaviors of receptors and drugs in terms of the root two-state model.

B. The Ternary Complex Model of Receptor Function

In view of data suggesting that receptors translocate within the two-dimensional space of cell membranes, CUATRECASAS (1974) proposed a model whereby receptors translocate within the lipid bilayer of membranes to interact with other membrane-bound components. The first specific model to incorporate this idea into receptor function was given by DE LEAN, STADEL, and LEFKOWITZ (1980). It was termed the ternary complex of receptors. The components of this system are receptors ([R]), drugs ([A]), and membrane-bound protein couplers (in the case of 7TM receptors, these are G proteins, denoted [G]). Response is produced by activation of G protein by the active state receptor. The system is:

$$
\begin{array}{ccc}
G & & G \\
+ & K_a & + \\
A + R & \rightleftharpoons & AR \\
K_g \updownarrow & & \updownarrow \gamma K_g \\
A + RG & \rightleftharpoons & ARG \\
& \gamma K_a &
\end{array}
\qquad (1)
$$

where γ represents the differential affinity of the G protein for the receptor when it is ligand bound. Alternatively, γ represents the differential affinity of the ligand for the G protein-bound (versus the unbound or R) state of the receptor. This model introduced the G protein into the receptor system and made it an integral part of ligand and receptor behavior patterns. In general, during the development of the ternary complex model, a synoptic view of a receptor system (as opposed to a receptor in isolation) was taken. The definitions of the various components of this model are shown in Table 1.

While the ternary complex model was a radical step forward in terms of defining the mechanism of action of GPCRs, it still viewed agonist activation of the receptor as a prerequisite to G protein activation. As in occupancy theory, the ternary complex model also viewed the mechanism by which receptor activation takes place as a conformational change brought on by agonist

Table 1. Parameters for receptor models

Symbol	Definition
[R]	Concentration of receptor in the system
[R$_i$]	Concentration of receptor in the inactive state (i.e., this species does not activate G proteins
[R$_a$]	Concentration of receptor in the active state (this species activates G proteins)
[G]	Concentration of G protein in the system
K_a	Equilibrium association constant for agonist and receptor
K_A	Equilibrium dissociation constant of the agonist–receptor complex ($K_A = 1/K_a$)
K_g	Equilibrium association constant for receptor and G protein
K_G	Equilibrium dissociation constant of the receptor/G protein complex ($K_G = 1/K_g$)
γ	Factor defining the differential affinity of the receptor for G proteins when the receptor is ligand bound. In the cubic ternary complex model, it defines the effect of ligand binding on the interaction of the receptor with G protein when the receptor is in the inactive state
L	Allosteric constant denoting the ratio of receptor in the active vs. inactive state (L = [R$_a$]/[R$_i$])
α	Factor defining the differential affinity of an agonist for the active vs. the inactive state. Also, the effect of ligand binding on receptor activation
β	Factor defining the differential affinity of the receptor for G protein when the receptor is in the active state
δ	Factor defining the synergy produced by simultaneous ligand binding and receptor activation on the interaction of the G protein with receptor

binding. To understand the major conceptual changes in this model leading to the extended ternary complex model years later (SAMAMA et al. 1993), discussion of two-state theory is useful. Two-state theory was introduced to describe the behavior of ion channels (KATZ and THESLEFF 1957) and was later applied to autonomic and neurotransmitter receptors (KARLIN 1967; COLQUHOUN 1973; THRON 1973).

C. Two-State Theory

The two-state model describes an equilibrium between two conformations of a receptor (denoted R$_a$ for active state and R$_i$ for inactive state) controlled by an allosteric constant L. The following system is described:

$$R_i \underset{}{\overset{L}{\rightleftharpoons}} R_a \qquad (2)$$

The affinity of a ligand for the inactive state (R$_i$) is denoted by the association constant K_a. Since the second receptor state is a different conformation, the drug may have a different affinity for that state (to be denoted αK_a, where α is the difference in affinity caused by the change in receptor state). The scheme in which drug A interacts with a two-state receptor system is given by:

$$\begin{array}{ccc} A & & A \\ + & & + \\ R_i & \rightleftharpoons & R_a \\ K_a \updownarrow & & \updownarrow \alpha K_a \\ & \alpha L & \\ AR_i & \rightleftharpoons & AR_a \end{array} \qquad (3)$$

The various parameters describing the model are given in Table 1. Two-state theory introduces the idea that, because of an intrinsic property of the receptor, the receptor system can have basal constitutive activity in the absence of an agonist. Thus, the basal ligand-independent activity of the receptor system is given by:

$$\text{Constitutive Activity} = \frac{L}{1+L} \qquad (4)$$

From this equation, it can be seen that the allosteric constant controls the magnitude of the constitutive activity of the system. This same constitutive activity also affects the observed affinity of a ligand according to the equation:

$$K_{obs} = K_A \frac{1+L}{1+\alpha L} \qquad (5)$$

From Eq. 5, it can be seen that the factor α and the set point of the receptor system (the magnitude of L) affect the observed affinities of ligands. Two-state theory offers a molecular mechanism for efficacy in terms of selective affinity for the two receptor states. If the affinity of the agonist is greater for the active state of the receptor ($\alpha > 1$), then the selective binding of the agonist to that species will shift the equilibrium toward the production of more R_a, and response will result. All of these features laid the foundation for the extended ternary complex model.

Two-state receptor theory also introduced a new concept to receptor theory, namely the idea that drugs can not only produce excitation of receptor systems (possess positive efficacy) but can also could reverse spontaneous activity of receptor system (have negative efficacy). A drug having a higher affinity for the inactive state of the receptor (i.e., $\alpha < 1$) would shift the prevailing equilibrium toward R_i, the inactive state. If there is a measurable spontaneous effect due to spontaneously formed R_a, then this type of drug (called an inverse agonist) will block the spontaneous basal response.

Certain experimental observations with recombinant receptor systems could not be accommodated with the simple ternary complex model. For example, it had been shown, in numerous experimental receptor systems, that receptors could spontaneously form active states and go on to activate G proteins in the absence of agonists. Thus, spontaneous activation of the receptor, as predicted by two-state theory, became a requirement for the behavior of 7TM receptor systems. In 1993, SAMAMA and colleagues introduced a formal modification of the ternary complex model to account for different spontaneously formed receptor conformations within the ternary complex model.

D. The Extended Ternary Complex Model

The components of GPCR systems, according to the extended ternary complex model, are:

$$
\begin{array}{ccc}
A + R_i & \xrightleftharpoons{K_a} & AR_i \\
L \,\big\updownarrow & & \big\updownarrow\, \alpha L \\
A + R_a & \xrightleftharpoons{\alpha K_a} & AR_a \\
+ & & + \\
G & & G \\
\beta K_a \,\big\updownarrow & & \big\updownarrow\, \beta\gamma K_g \\
R_a G & \xrightleftharpoons{\alpha\gamma K_a} & AR_a G
\end{array}
\tag{6}
$$

In this model, it can be seen that receptors can exist spontaneously in either the active (R_a) or inactive (R_i) form, ligands can interact with either form, and G proteins can interact with the active form whether it is occupied or unoccupied by ligand. At this point, it is useful to develop the equilibrium equations describing this model and contrast them with the statistically complete version to be described later. The definitions of the various components of the system are given in Table 1. The expressions for certain experimentally observable features of receptor systems (according to the extended ternary complex model) are given in Table 2.

The extended ternary complex model cannot be regarded as a simple joining of two-state theory and the ternary complex model. This is because the insertion of the factor β formally indicates that there are thermodynamic differences between the unbound receptor (activated and inactivated), the ligand-bound receptor, and the G protein-bound receptor. This distinction prevents the model from qualifying as a two-state receptor system.

From a thermodynamic point of view, the extended ternary complex model for receptor activation is incomplete. This is because the inactive state of the receptor, either ligand bound ([AR_i]) or unbound [R_i], coexists with G protein in the system, and these species must have some constant of interaction (albeit possibly very small). In terms of thermodynamic closure (WYMAN 1975), there must also be a thermodynamic energy pathway between all species in the system. While all of these interactions may not take place to an appreciable extent at equilibrium, the mechanism for their formation must exist. These considerations led to the development of the statistically complete

Table 2. Extended ternary complex model response as a fraction of total receptor

$$\rho = \frac{\beta L[G]/K_G (1+\alpha\gamma[A]/K_A)}{A/K_A(1+\alpha L(1+\gamma\beta[G]/K_G))+L(1+\beta[G]/K_G)+1}$$

where ρ = response-producing species [RaG] and [ARaG] expressed as a fraction of the total receptor

Maximal agonist response	$\dfrac{\alpha\beta\gamma L[G]/K_G}{1+\alpha L(1+\gamma\beta[G]/K_G)}$
Observed agonist affinity	$\dfrac{K_A(1+L(1+\beta[G])/K_G)}{1+\alpha L(1+\gamma\beta[G]/K_G)}$
GTP shift	$\dfrac{(1+L)(1+\alpha L(1+\gamma\beta[G]/K_G))}{(1+\alpha L)(1+L(1+\beta[G]/K_G))}$
Constitutive activity	$\dfrac{\beta L[G]/K_G}{1+L(1+\beta[G]/K_G)}$
Constitutive/maximum response	$\dfrac{(1+\alpha L(1+\gamma\beta[G]/K_G))}{\alpha\gamma(1+L(1+\beta[G]/K_G))}$

GTP, guanosine triphosphate.

ternary complex model, named the CTC model because of the conceptual arrangement of the components in Euclidean space.

E. The CTC Model

In the thermodynamically complete version of the ternary complex model, it can be seen that the principle of microscopic reversibility dictates that the G protein must be able to interact with the inactive form of the receptor. This type of model is considerably more complex, because the effect of each element on the interaction of the other two elements must be accounted for (WEISS et al. 1996a, 1996b). This is done by modifying interactions between any two components by introducing a third.

To understand the implications of the CTC model and the various constants employed, it is useful to build the cube from various sides. The first is the effect of ligand activation of receptors. Ligands may or may not recognize the two states of the receptor ($[R_a]$ and $[R_i]$) with different affinities, but this possibility is accounted for by the inclusion of the factor α (as with two-state theory and the extended ternary complex model). Thus, the standard two-state receptor scheme for this situation is given by:

$$\begin{array}{ccc} R_i & \xrightleftharpoons{L} & R_a \\ \| K_a & & \| \alpha K_a \\ AR_i & & AR_a \end{array} \quad (7)$$

The equilibrium constant of the reaction from R_i to R_a is the product of the equilibrium constants of the component reactions. Thus, the thermodynamically derived equilibrium association constant for the production of AR_a from R_i is $\alpha K_a L$. The free energy for the production of R_i from AR_a must be independent of the route chosen (i.e., either via AR_i or R_a); this concept of thermodynamic closure is generally referred to as the principle of microscopic balance or microscopic reversibility (WYMAN 1975). Under these circumstances, $\alpha K_a / K_a = \alpha L / L$, and the scheme becomes:

$$
\begin{array}{ccc}
R_i & \xrightleftharpoons{L} & R_a \\
\| K_a & & \| \alpha K_a \\
AR_i & \xrightleftharpoons[\alpha L]{} & AR_a
\end{array}
\tag{8}
$$

The term α has two interpretations. First, it measures the differential propensity of bound and unbound receptors to convert to their active forms (the effect of ligand binding on activation). Second, it measures the differential affinity of the ligand for the activated receptor (the effect of receptor activation on ligand binding).

In the extended ternary complex model, only the active form of the receptor couples with G protein. In this model, the concept of thermodynamic closure must allow the inactive receptor R_i to interact with the G protein. This is not a unique constraint on the model since all proteins have an affinity constant (albeit possibly small) for interaction. The interaction of receptors with G proteins can be viewed by the following scheme:

$$
\begin{array}{ccc}
R_i & \xrightleftharpoons{L} & R_a \\
\| K_G & & \| \beta K_G \\
R_i G & & R_a G
\end{array}
\tag{9}
$$

where the equilibrium between R_i and G proteins is controlled by K_G. Again, thermodynamic closure dictates that $\beta K_G / K_G = \beta L / L$; therefore, the scheme can be written:

$$
\begin{array}{ccc}
R_i & \xrightleftharpoons{L} & R_a \\
\| K_G & & \| \beta K_G \\
R_i G & \xrightleftharpoons[\beta L]{} & R_a G
\end{array}
\tag{10}
$$

Under these circumstances, the term β can be thought of as the effect of G protein coupling on receptor activation or, alternatively, the differential affinity that the activated receptor has for G proteins.

The effect of ligand on the receptor–G protein interaction is considered a differential affinity of the ligand for the $R_i G$ complex versus R_i; this is accommodated by a factor denoted by γ. Under these circumstances,

$$\begin{array}{ccc} AR_i & & AR_iG \\ \| \; K_a & & \| \; \gamma K_a \\ R_i & \xrightleftharpoons[K_G]{} & R_iG \end{array} \qquad (11)$$

Again, by thermodynamic closure, $\gamma K_G/K_G = \gamma K_a/K_a$, making the complete scheme:

$$\begin{array}{ccc} AR_i & \xrightleftharpoons[]{\gamma K_G} & AR_iG \\ \| \; K_a & & \| \; \gamma K_a \\ R_i & \xrightleftharpoons[K_G]{} & R_iG \end{array} \qquad (12)$$

The factor γ measures both the extent to which G protein binding to the receptor alters ligand binding and the effect of ligand binding on receptor–G protein binding.

At this point, all but one of the vertices of the cube are constructed, and all that remains is to arrive at the AR_aG complex. From R_i, there are three processes: the binding of ligand to the receptor, the activation of the receptor, and the binding of G protein. The order for these is immaterial, and there are six (3!) pathways. If one of the pathways is defined, then thermodynamic closure will necessitate what constants control the other two.

It is useful to define one of the edges in terms of the effect of ligand binding on G protein binding to both active and inactive receptors. Thus, a factor γ that differentiates the binding of G protein to bare receptor R_i (defined by the equilibrium association constant K_g) and ligand-bound R_i (equlibrium association constant γK_g) can be defined. If the activation state of the receptor is not relevant to the binding of G protein, then it is assumed that the state of the receptor (active or inactive) is not relevant to the binding of G protein. Thus, γK_g would also control the interaction between R_a and G. Since the binding of R_a to G protein to form R_aG is βK_g, the ligand effect on G protein binding would then be $\gamma \beta K_g$. However, if the activation state of the receptor is important, then the ligand effect will differ; this will be denoted by the factor δ. Under these circumstances, the complete set of factors for the reaction $AR_a + G = AR_aG$ will be $\delta \gamma \beta K_g$. The remaining constants are set by thermodynamic closure to yield the complete model shown in Fig. 1.

The front face of the cube shows the interaction of ligand with the active and inactive states of the receptor and, moving from the front face to the back, the interaction with G protein. Moving from the left-hand face to the right-hand face represents the effect of receptor activation, while moving from the bottom to the top represents ligand binding. These movements are shown in Fig. 2.

The definitions of the various components of GPCR systems, arranged according to the cubic ternary complex model, are given in Table 1. The equi-

Fig. 1. The cubic ternary complex model for G protein-coupled receptors. The receptor is allowed to exist in one of two conformations, both of which can interact with G protein and ligand

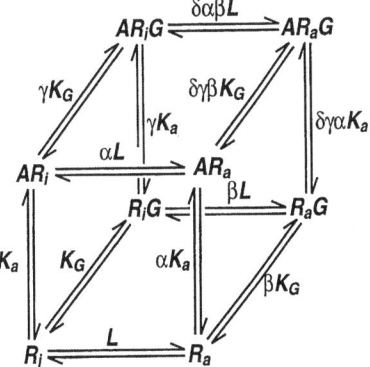

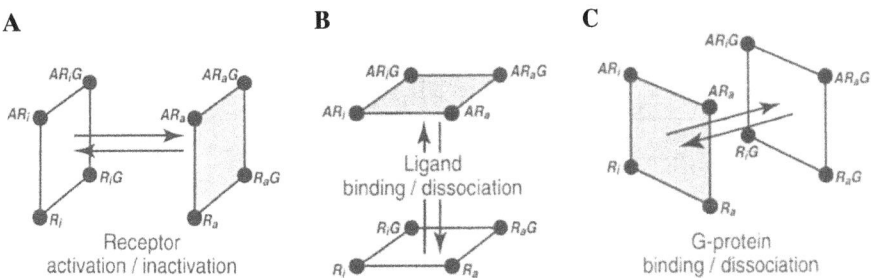

Fig. 2A–C. Transitions within the cubic ternary complex model. **A** The activation of receptors (either through binding to agonist or spontaneously) causes a shift of species from the left to the right face. **B** The binding of ligand shifts species from the bottom to the top. **C** The binding of G protein shifts species from the back face to the front face

librium equations describing some observable features of the cubic ternary complex model are shown in Table 3.

F. General Application of the Cubic Model

The cubic model is a statistically complete model that subsumes a great many previously described models of drug–receptor interaction. The relationships of these models and the ways they relate to the CTC model are shown in Fig. 3. The earliest described model is the classical occupancy model, which assumed that the binding of the agonist activates the receptor. The simple two-state model described the agonist-bound receptor existing in one of two states (active and inactive). The full two-state model extended this idea to the receptor, which existed in one of the two states whether or not it was ligand bound. With the addition of G protein to the system, agonism was described in terms of either receptor activation through agonist binding with subsequent G

Table 3. Cubic ternary complex model response as a fraction of total receptor

$$\rho = \frac{\beta L[G]/K_G(1+\delta\alpha\gamma[A]/K_A)}{[A]/K_A(1+\alpha L+\gamma[G]/K_G(1+\delta\alpha\beta L))+G/K_G(1+\beta L)+1+L}$$

where ρ = response-producing species [RaG] and [ARaG] expressed as a fraction of the total receptor

Maximal agonist response	$\dfrac{\delta\alpha\beta\gamma L[G]/K_G}{(1+\alpha L+\gamma[G]/K_G(1+\delta\alpha\beta L))}$
Observed agonist affinity	$\dfrac{K_A(1+L+([G])/K_G(1+\beta L))}{(1+\alpha L+\gamma[G]/K_G(1+\delta\alpha\beta L))}$
GTP shift	$\dfrac{(1+L+[G]/K_G)(1+\alpha L+\gamma[G]/K_G(1+\delta\alpha\beta L))}{(1+\alpha L+\gamma[G]/K_G)(1+L+[G]/K_G(1+\beta L))}$
Constitutive activity	$\dfrac{\beta L[G]/K_G}{1+L+[G]/K_G(1+bL)}$
Constitutive/maximum response	$\dfrac{1+\alpha L+\gamma[G]/K_G(1+\delta\alpha\beta L)}{\delta\alpha\beta(1+L+[G]/K_G(1+\beta L))}$

GTP, guanosine triphosphate.

protein coupling (simple ternary complex model I) or receptor activation through G protein coupling to agonist-bound receptor (simple ternary complex model II). This latter model is directly related to the complete ternary complex model, in which receptor activation occurs through G protein binding and agonist binding promotes G protein coupling. The simple ternary complex model I, the complete ternary complex model, and the full two-state model culminate in the extended ternary complex model, in which G protein coupling and receptor activation are separate steps. Binding of the agonist promotes receptor activation and G protein coupling, but only to the activated receptor. The extension introduced in the CTC model is that the inactive receptor also can bind G protein (Fig. 3).

At this point, it is worth considering the advantages and disadvantages of using the more complex CTC model rather than the more simple and frugal partial models, such as the ternary or extended ternary complex models. One drawback of the cubic model is the large number of terms and the corresponding lack of estimatability of constants. This relegates the model to a descriptive and predictive role, one not amenable to the fitting of real data. However, aside from the statistical completeness of the cubic model, the existence of a receptor in two states that can interact with a ligand and a G protein compels the thermodynamic existence of eight species. An important consideration at this point is the relevance of the AR_iG state; i.e., what is the evidence that the inactive receptor interacts and complexes with the G protein without activating it?

The Cubic Ternary Complex Model for G Protein-Coupled Receptors

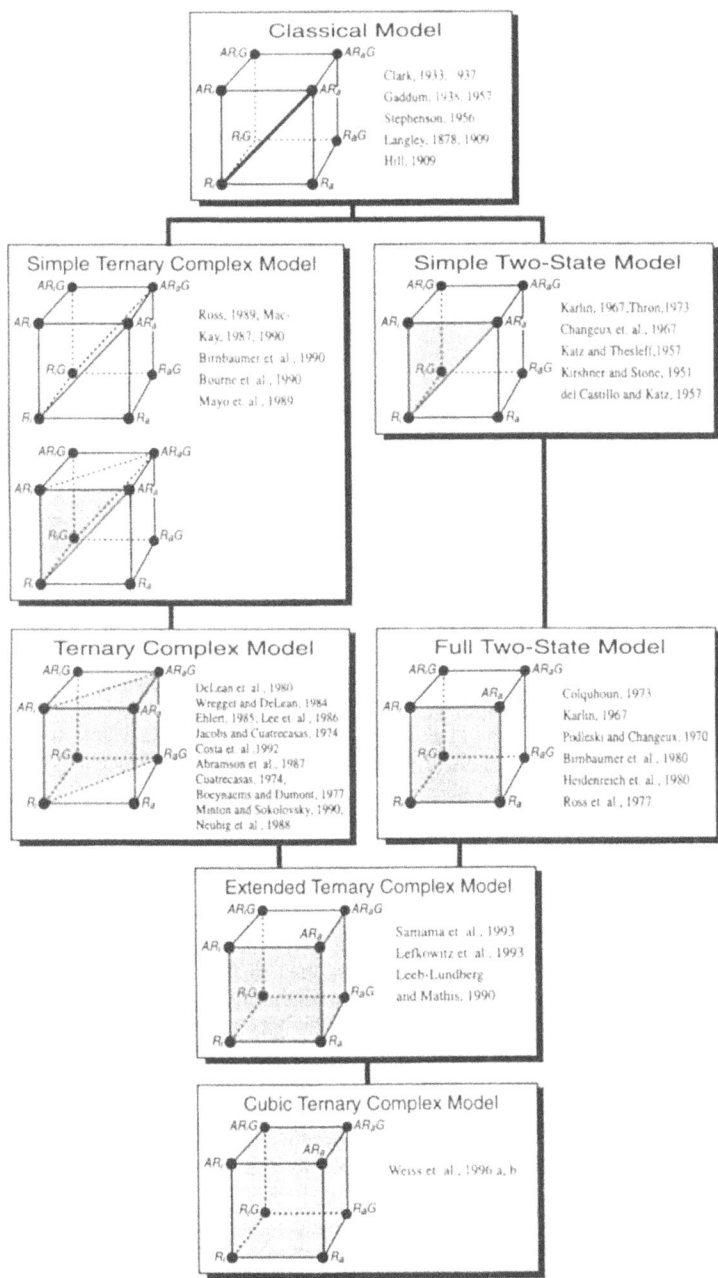

Fig. 3. Interrelationships between pharmacologic receptor models

I. Evidence for the AriG Complex

There is a theoretical approach that may be useful in the evaluation of the CTC model. There are substantially different predictions made by the extended ternary complex model and the CTC model regarding the maximal amount of constitutive receptor activity that can be obtained by receptor overexpression. The expression for the amount of constitutive activity produced by spontaneous receptor activation (expressed as a fraction of the maximal response to a full agonist) is given as:

$$\frac{\text{Constitutive Activity}}{\text{Maximal Response}} = \frac{\beta L[R]/K_G}{1+\beta L[R]/K_G} \qquad (13)$$

It can be seen that, as [R] approaches infinity, the constitutive receptor activity, expressed as a function of the maximal response obtainable with a full agonist, will approach unity. Thus, if an effectively infinite stoichiometric amount of G protein can be applied to the system then, theoretically, *all* of the receptor can be converted from the inactive form to the signaling RaG complex.

An interesting contrast to this is predicted by the CTC model. With this model, the constitutive receptor activity, expressed as a fraction of the maximal response to a full agonist, is given as a function of receptor density, with:

$$\frac{\text{Constitutive Activity}}{\text{Maximal Response}} = \frac{[R]/K_G(1+\delta\alpha\beta L)}{\delta\alpha(1+[R]/K_G(1+\beta L))} \qquad (14)$$

Here, it can be seen that, as [R] approaches infinity, constitutive activity due to receptor over-expression does not equal unity, but

$$\frac{\text{Constitutive Activity}}{\text{Maximal Response}} = \frac{(1/\delta\alpha)+\beta L}{(1+\beta L)} \qquad (15)$$

If a reasonably efficacious agonist is used to measure the maximal response of the receptor system, then the term $1/\delta\alpha$ approaches zero, and the maximal constitutive activity reduces to:

$$\frac{\text{Constitutive Activity}}{\text{Maximal Response}} = \frac{\beta L}{1+\beta L} \qquad (16)$$

an expression entirely dependent on the receptor system and not the agonist. From Eq. 16, it can be seen that, even in the presence of an apparently infinite amount of receptor and G protein, the constitutive activity may remain sub-maximal (depending on the intrinsic reactivity of the receptor with G protein and the spontaneous formation of the active state). This is because of the potential for the inactive state of the receptor to interact with G protein, thus reducing the possibility of complete G protein activation. This is shown schematically in Fig. 4.

The respective equations describing constitutive activity in terms of the extended ternary complex model and the CTC model were used to simulate

Extended Ternary Complex Model

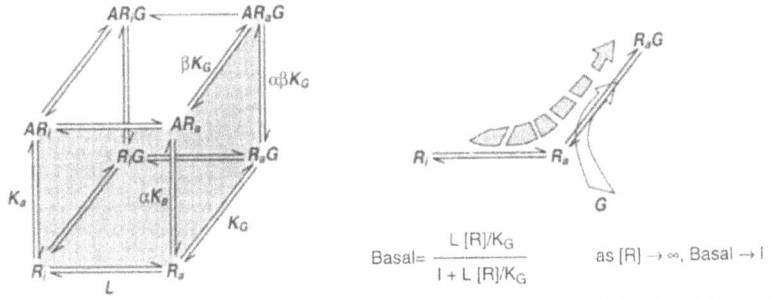

Cubic Ternary Complex Model

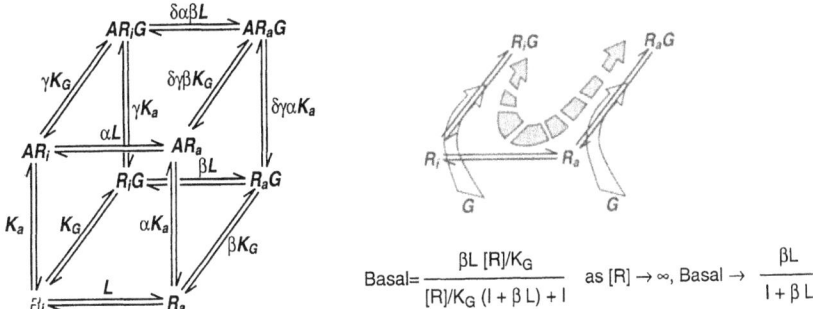

Fig. 4. Schematic description of constitutive production of the activated receptor–G protein complex ([RaG]) with unlimited G protein. In terms of the extended ternary complex model, all of the receptor species will eventually be converted to [RaG]. In contrast, the cubic ternary complex predicts a possible sub-maximal formation of [RaG] if K_G allows a substantial amount of [RiG] species to be formed. Since [RiG] does not produce physiological response, a submaximal response (in terms of [RaG]) could result

constitutive activity in Fig. 5A for a receptor with a relatively high energy barrier to formation of the active state. It can be seen that the extended ternary complex model (Eq. 13) predicts that, when the G protein concentration is not limiting, then a high receptor density will lead to a maximal constitutive response (as given by spontaneous formation of the [RaG] complex). The same parameters were used in Eq. 14 to simulate the spontaneous production of the [RaG] species, and it can be seen from Fig. 5A that the maximal constitutive activity falls far short of the maximum attainable by a full agonist.

While, in theory, this distinction is capable of differentiating which receptors adhere to either the extended ternary complex model or the CTC model, in practice, there are a number of caveats that preclude this. The first is that, if a receptor had an intrinsically low energy barrier for activation (i.e., if L was large), then the constitutive activity may well approach the maximal response achieved with a full agonist (Eq. 16). There is another complication encoun-

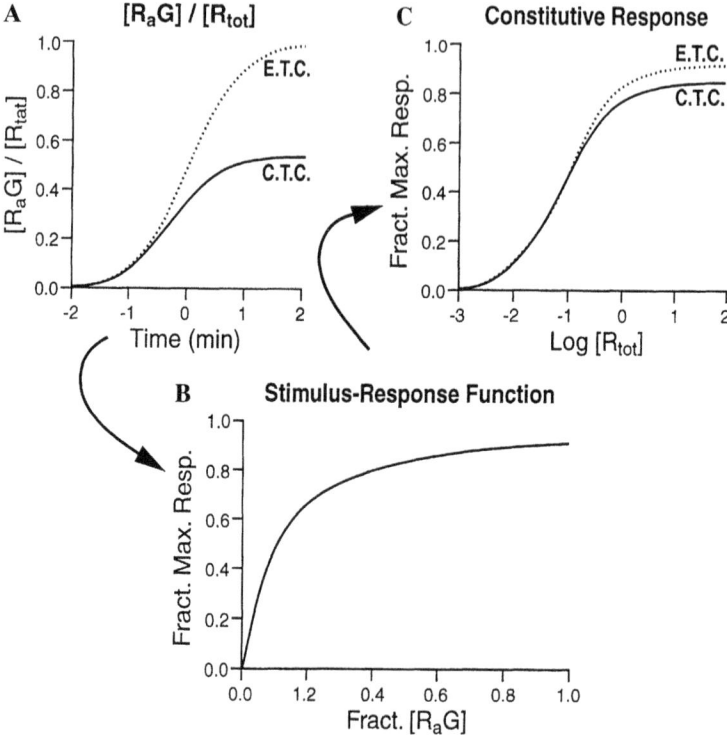

Fig. 5A–C. Constitutive response with unlimited G protein. **A** Formation of [RaG] complex in terms of the extended ternary complex model (ETC) and the cubic ternary complex model (CTC; $\alpha = \beta = 10$, $\gamma = \delta = 1$; $L = 0.1$). **B** Stimulus-response function (according to Eq. 17) with $\phi = 0.01$. The amplification of the [RaG] species allows near-maximal response to result from submaximal formation of [RaG]. **C** The response output from **B** for the [RaG] produced in **A**. Note how the CTC model allows observation of maximal constitutive response, as does the ETC model

tered when attempting to use observed physiological responses to determine maximal constitutive receptor activation. The complication arises from the translation of receptor stimulus into tissue response. In general, tissues amplify receptor signals, and the effects of stimulus response mechanisms can be modeled by the insertion of a hyperbolic function between receptor occupancy and tissue response. Thus, response can be depicted as a function of the response producing receptor species [RaG] and [ARaG] according to:

$$\text{Response} = \frac{([RaG]+[ARaG])}{([RaG]+[ARaG])+\phi} \tag{17}$$

where ϕ represents a stimulus–response-coupling constant (KENAKIN and BEEK 1980). Under these circumstances, the constitutive receptor activity, as measured by physiological response, is given by the CTC model as:

$$\frac{\text{Constitutive Activity}}{\text{Maximal Response}} = \frac{\beta L(R)/K_G}{[R]/K_G(\beta L + \phi(1+\beta L)) + \phi} \quad (18)$$

It can be seen from this equation that, as [R] approaches infinity, the maximal constitutive receptor activity becomes:

$$\frac{\text{Constitutive Activity}}{\text{Maximal Response}} = \frac{\beta L}{\beta L(1+\phi) + \phi} \quad (19)$$

It can be seen from this equation that, if ϕ is much less than 1, then constitutive receptor activity can approach the maximum observed with a full agonist, i.e., the same prediction as that of the extended ternary complex model. The instance of a low (relative to unity) value for ϕ is not uncommon, since many tissues have high degrees of stimulus amplification (i.e., low amounts of receptor occupancy by an agonist lead to tissue maximal response).

The spontaneously formed amounts of the [RaG] complex calculated in Fig. 5A were amplified by a stimulus-response cascade function (Eq. 17; Fig. 5B) with a receptor coupling efficiency factor of $\phi = 0.01$. As can be seen from Fig. 5C, the dramatic difference in the maximal constitutive receptor activity disappears and is replaced by a difference in the location parameter along the receptor concentration axis. As there is no experimental method to verify where this location parameter should be in terms of either model, the ability of this approach to differentiate between the two models is lost. In general, while the proclivity of receptor systems to produce constitutive receptor activity may help differentiate which systems are represented by the extended versus the cubic ternary complex models, it cannot be relied upon to do so.

Perhaps a more useful approach would be to explore whether or not the [ARiG] complex can be detected, either by direct biochemical means or through inference, in pharmacological effects. There is provocative data with cannabinoid receptors that may be relevant to this latter point.

In Chinese hamster ovary (CHO) cells transfected with human central cannabinoid receptors (CB1), the inverse agonist SR-141716A [*N*-(piperidino-1-yl)-5-(4-chlorophenyl)-1-(2,4-dichlorophenyl)-4-methyl-pyrazole-3-carboxamide] decreased constitutive CB1 receptor activity (as measured by activation of mitogen-activated protein kinase, MAPK) as expected, but also unexpectedly blocked the pertussis-sensitive activation of the same kinase by insulin and insulin-like growth factor 1 (IGF-1) receptors also (BOUABOULA et al. 1997). Figure 6 shows the blockade of insulin and IGF-1 effects by SR-141716A and the lack of effect of this inverse agonist on MAPK activation through fibroblast growth factor receptors. This latter receptor does not utilize a pertussis-sensitive G protein. Figure 6D–F show the dependence of the effect on CB1 receptors (no effects are observed in wild type CHO cells not transfected with CB1 receptors). This crossover inhibition was also seen when Mas-7 (a mastoparan analogue) was used to directly activate G_i protein. These data clearly show that the inverse agonist can block G_i protein-mediated effects of

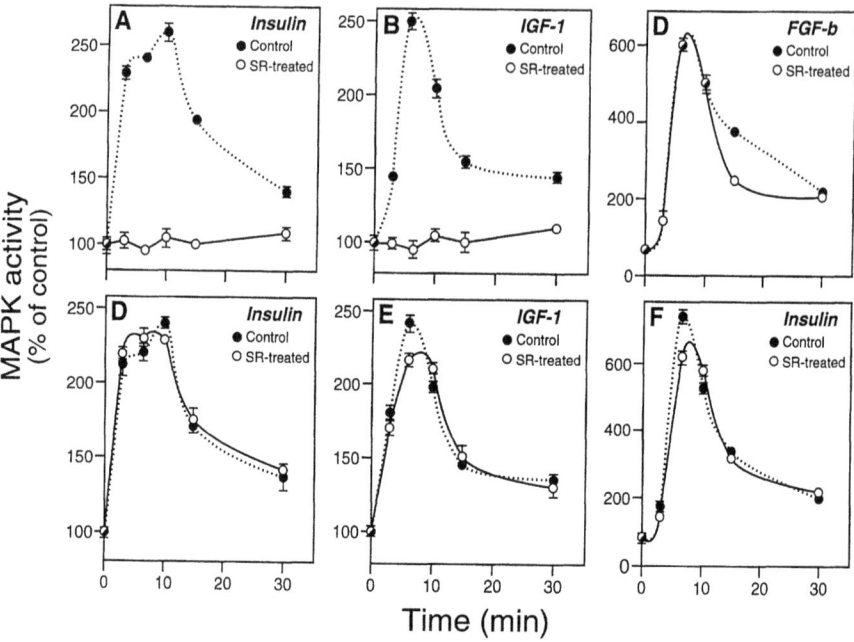

Fig. 6. The effect of inverse agonism for CB1 receptors on responses to ligands for insulin (**A,D**), insulin-like growth factor-1 (**B,E**) and fibroblast growth factor b (FGF-b; **C,F**). Chinese hamster ovary cells were transfected with CB1 receptors pre-treated (*open circles*) and not pre-treated (*closed circles*) with the CB1 inverse agonist SR-141716A (see text for details). Cells were then stimulated with agonist shown, and mitogen-activated protein kinase activity was measured. Note how the inverse agonist blocks the G_i protein-mediated CB1 and insulin response but not the G_i protein-independent FGF-b response (BOUABOULA et al. 1997)

other receptors, thereby indicating circumstantial evidence for the formation of the [ARiG] species in this system.

Another finding in CHO cells stably transfected with μ-opioid receptors indicates the formation of an inactive ligand–receptor–G protein complex for opioid receptors (BROWN and PASTERNAK 1998). Thus, the potent μ-opioid receptor antagonist naloxone benzoylhydrazone (NalBzoH) blocks agonist-stimulated cyclic adenosine monophosphate responses with no accompanying stimulation; in all functional assays, NalBzoH is an antagonist. Equilibrium binding studies indicate that there is a threefold enhancement of affinity for NalBzoH in the presence of the GTP-stable analogue guanylylimido diphosphate [Gpp(NH)p], indicating a low level of negative efficacy. This is consistent with the proposal that NalBzoH has a preferential affinity for the inactive state of the receptor. Surprisingly, ^{3}H-NalBzoH demonstrated biphasic kinetics, indicative of two affinity states. The elimination of the high-affinity state by Gpp(NH)p indicated an association with G protein (with no concomitant signaling; Fig. 7A). The production of this same effect with pertussis toxin

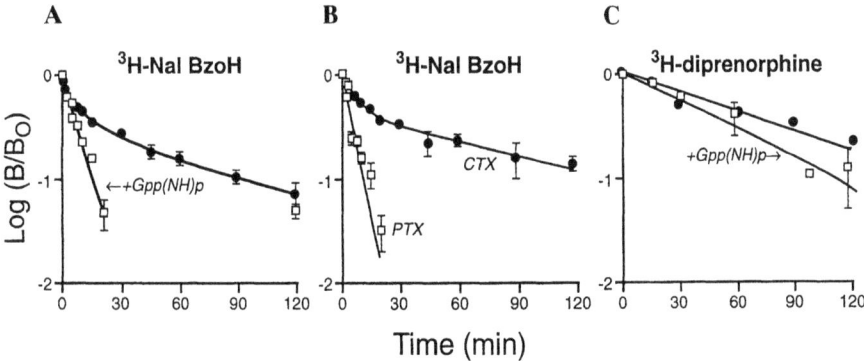

Fig. 7A–C. G protein-dependent antagonist binding; evidence for an [AriG] complex. *Ordinate axes*: logarithms of specific binding at various times as a fraction of specific binding at time zero. *Abscissae*: time (min). **A** Effect of guanylylimido diphosphate [Gpp(NH)p; 0.1 mM] on offset kinetics of bound ^{3}H-naloxone benzoylhydrazone. **B** Effect of pertussis toxin inactivation of G_i protein on offset kinetics. Inactivation of G_s with cholera toxin had no effect on the rate of offset. **C** Lack of effect of Gpp(NH)p on offset kinetics of the antagonist diprenorphine (BROWN and PASTERNAK 1998)

treatment (Fig. 7B) indicated that the high-affinity component was a ligand-associated receptor complexed with G_i protein. This was not observed with the μ-opioid antagonist diprenorphine (Fig. 7C). The kinetic data and the fact that NalBzoH affinity was enhanced with Gpp(NH)p are consistent with the notion that NalBzoH forms an inactive receptor complex with GDP-associated G_i protein; this complex (AriG complex) does not signal.

G. Conclusion

Presently, it is not clear whether G protein-coupled receptors adhere generally to the CTC model or the more simple extended ternary complex model. It may be that the behavior of some receptors requires one model and that of others requires another.

References

Ariens EJ (1954) Affinity and intrinsic activity in the theory of competitive inhibition. Arch Int Pharmacodyn Ther 99:32–49

Birnbaumer LG, Yatani A, VanDongen AMJ, Graf R, Codina J, Okabe K, Mattera R, Brown AM (1990) G protein-coupling of receptors to ionic channels and other effector systems. In: Nathanson NM, Harden TK (eds) G proteins and signal transduction. Rockerfeller University Press, New York, p 169

Boeynaems JM, Dumont JE (1977) The two-step model of ligand–receptor interaction. Mol Cell Endocrinol 7:33–47

Bouaboula M, Perrachon S, Milligan L, Canatt X, Rinaldi-Carmona M, Portier M, Barth F, Calandra B, Pecceu F, Lupker J, Maffrand J-P, Le Fur G, Casellas P (1997) A selective inverse agonist for central cannabinoid receptor inhibits mitogen-

activated protein-kinase activation stimulated by insulin or insulin-like growth factor. J Biol Chem 272: 22330–22339
Bourne HR, Sanders DA, McCormick F (1990) The GTPase superfamily: a conserved switch for diverse cell functions. Nature 348:125–132
Brown GP, Pasternak GW (1998) ^{3}H-Naloxone benzoylhydrazone binding in MOR-1-transfected chinese hamster ovary cells: evidence for G-protein-dependent antagonist binding. J Pharmacol Exp Ther 286:376–381
Clark AJ (1933) The mode of action of drugs on cells London. Edward Arnold
Clark AJ (1937) General pharmacology. In: Heffner's Handbuch d. exp. Pharmacol. Suppl. vol. 4. Springer, Berlin Heidelberg New York
Colquhoun D (1973) The relationship between classical and cooperative models for drug action. In: Rang HP (ed) Drug receptors. University Park Press, Baltimore. pp 149–182
Costa T, Ogino Y, Munson PJ, Onaran HO, Rodbard D (1992) Drug efficacy at guanine-nucleotide-binding regulatory protein linked receptors: thermodynamic interpretation of negative antagonism and of receptor activity in the absence of ligand. Mol Pharmacol 41:549–560
Cuatrecasas P (1974) Membrane receptors. Annu Rev Biochem 43:169–214
De Lean A, Stadel JM, Lefkowitz RJ (1980) A ternary complex model explains the agonist-specific binding properties of adenylate-cyclase-coupled β-adrenergicreceptor. J Biol Chem 255:7108–7117
Del Castillo J, Katx B (1957) Interaction at end-plate receptors between different choline derivatives. Proc R Soc London B 146:369–381
Ehlert FJ (1985) The relationship between muscarinic receptor occupancy and adenylate-cyclase inhibition in the rabbit myocardium. Mol Pharmacol 28:410–421
Heidenreich KA, Weiland GA, Molinoff PB (1980) Characterization of radiolabeled agonist binding to β-adrenergic receptors in mammalian tissues. J Cyclic Nucleotide Res 6:217–230
Hill AV (1909) The modes of action of nicotine and curari, determined by the form of the contraction curve and the method of temperature coefficients. J Physiol Lond 39:361–373
Iyengar R, Abramowitz J, Bordelon-Riser M, Birnbaumer L (1980) Hormone receptor-mediated stimulation of adenylyl-cyclase systems. Nucleotide effects and analysis in terms of a simple two-state model for the basic receptor-affected enzyme. J Biol Chem 255:3558–3564
Jacobs S, Cuatrecasas P (1976) The mobile receptor hypothesis and "cooperativity" of hormone binding: applications to insulin. Biochim Biophys Acta 433:482–495
Karlin A (1967) On the application of a "plausible model" of allosteric proteins to the receptor for acetylcholine. J Theor Biol 16:306–320
Katz B, Thesleff S (1957) A study of the "desensitization" produced by acetylcholine at the motor end-plate. J Physiol (Lond.) 138:63–80
Kenakin TP, Beek D (1980) Is prenalterol (HI33/80) really a selective β1-adrenoceptor agonist? Tissue selectivity resulting from differences in stimulus-response relationships. J Pharmacol Exp Ther 213:406–412
Lee TWT, Sole MJ, Wells JW (1986) Assessment of a ternary model for the binding of agonists to neurohumoral receptors. Biochemistry 25:7009–7020
Lefkowitz RJ, Cotecchia S, Samama P, Costa T (1993) Constitutive activity of receptors coupled to guanine-nucleotide-regulatory proteins. Trends Pharmacol Sci 14:303–307
MacKay D (1987) Use of null equations, based on classical receptor and ternary models of drug action, to classify receptors and receptor-effector systems even if some of the agonists and antagonists available have additional actions on the tissues. In: Black JW, Jenkinson DH, Gerskowitch VP (eds) Perspectives on receptor classification. A.R. Liss, New York, pp 193–206
MacKay D (1990) Agonist potency and apparent affinity: interpretation using classical and steady-state ternary complex models. Trends Pharmacol Sci 11:17–22

Mayo KH, Nunez M, Burke C, Starbuck C, Lauffenberger D, Savage CR Jr (1989) Epidermal-growth-factor-receptor binding is not a simple, one-step process. J Biol Chem 264:17838–17844

Minton AP, Sokolovsky M (1990) A model for the interaction of muscarinic receptors, agonists, and two distinct effector substances. Biochemistry 29:1586–1593

Neubig RR, Gantzos RD, Thomsen WJ (1988) Mechanism of agonist and antagonist binding to $\alpha 2$-adrenergic receptors: evidence for a pre-coupled receptor–guanine-nucleotide protein complex. Biochemistry 27:2374–2384

Podleski TR, Changeux J-P (1970) On the excitability and cooperativity of electroplax membrane. In: Danielli JF, Moran JF, Triggle DJ (eds) Fundamental concepts in drug-receptor interaction. Academic Press, New York, pp 93–119

Ross EM (1989) Signal sorting and amplification through G protein-coupled receptors. Neuron 3:141–152

Ross EM, MaGuire ME, Sturgill TW, Biltonen RL, Gilman AG (1977) Relationship between the β-adrenergic receptor and adenylate cyclase. J Biol Chem 252:5761–5775

Samama P, Cotecchia S, Costa T, Lefkowitz RJ (1993) A mutation-induced activated state of the β_2-adrenergic receptor: extending the ternary complex model. J Biol Chem 268:4625–4636

Stephenson RP (1956) A modification of receptor theory. Br J Pharmacol 11:379–393

Thron CD (1973) On the analysis of pharmacological experiments in terms of an allosteric receptor model. Mol Pharmacol 9:1–9

Weiss JM, Morgan PH, Lutz MW, Kenakin TP (1996a) The cubic ternary complex receptor occupancy model. I Model description. J Theoret Biol 178:151–167

Weiss JM, Morgan PH, Lutz MW, Kenakin TP (1996b) The cubic ternary complex receptor-occupancy model. II Understanding apparent affinity. J Theoret Biol 178:169–182

Wreggett KA, De Lean A (1984) The ternary complex model: its properties and application to ligand interactions with the D2-dopamine receptor and the anterior pituitary gland. Mol Pharmacol 26:214–227

Wyman J (1975) The turning wheel: a study in steady states. Proc Natl Acad Sci USA 72:3983–3987

CHAPTER 6
Inverse Agonism

R. BOND, G. MILLIGAN, and M. BOUVIER

A. Background

For the superfamily of guanine nucleotide-binding regulatory protein (G protein)-coupled receptors, traditional receptor theory has postulated that agonist binding to the receptor induces a conformational change of the receptor. As a result, the agonist–receptor complex possesses increased affinity for the G protein. This increase in affinity has been proposed to define the ligand's intrinsic efficacy (KENAKIN 1988; MACKAY 1988; LEFF and HARPER 1989). Unoccupied receptors are believed to be in a single, quiescent state until they are activated by agonist binding. According to this theory, antagonists are believed to bind to the receptor and to prevent agonist binding, but not believed to alter the affinity of the complex for the G protein.

However, in the last few years, numerous reports have documented the ability of G protein-coupled receptors (GPCRs) to couple with G proteins and signal a cellular response in the absence of hormones or agonists (MILLIGAN et al. 1995a). Simultaneous with the discovery of the spontaneous activity of unliganded receptors was the discovery of compounds that could decrease this activity.

The first studies implicating the spontaneous activity of GPCRs were performed using radioligand-binding assays and reconstituted or transfected cell systems that monitored guanosine triphosphatase (GTPase) activity (COSTA and HERZ 1989; COSTA et al. 1990). Perhaps the strongest evidence in these earlier studies was a report by COSTA and HERZ (1989) using NG-108 cells endogenously expressing the δ-opioid receptor. By substituting potassium for sodium in the media, these authors demonstrated an increase in unstimulated "baseline" GTPase activity. With this increased baseline, the authors were able to show that several compounds previously classified as opioid-competitive antagonists were able to decrease the spontaneously increased GTPase activity. These compounds were referred to as "negative antagonists", and subsequent papers have also used the term "negative agonists", but the term "inverse agonist" is now commonly used. Furthermore, the authors were able to show that a neutral antagonist, a compound which does not affect baseline in the absence of hormone, was able to block the responses to both inverse agonists and agonists.

The report by COSTA and HERZ (1989) also demonstrated the importance of raising the baseline as a means of detecting the inverse-agonist properties of drugs. While there are reports of inverse agonist activity in native systems (see below), the general rule is that one has to increase the number of spontaneously active receptors to reveal inverse-agonist efficacy. The two most commonly used methods today are: overexpression of wild-type receptors and receptor mutations which increase the level of spontaneous activity.

The list of wild-type and mutant GPCRs demonstrating constitutive activity that can be inhibited by inverse agonists is now extensive. It includes β_2-, α_1- and α_2-adrenoceptors (CHIDIAC et al. 1994; TIAN et al. 1994; BOND et al. 1995; LEE et al. 1997); several subtypes of 5-hydroxytryptamine (5-HT) receptors [5-HT$_{2C}$ (CHIDIAC et al. 1994; LABRECQUE et al. 1995); 5-HT$_{1A}$ (BARR and MANNING 1997; NEWMAN-TANCREDI et al. 1997)], dopamine receptors [D$_5$ (TIBERI and CARON 1994); D$_2$ (HALL and STRANGE 1997); D$_3$ (MALMBERG et al. 1998)], muscarinic receptors (M$_1$, M$_2$, M$_3$, and M$_5$; MEWES et al. 1993; BURSTEIN et al. 1997), δ-opioid receptors (COSTA and HERZ 1989; MULLANEY et al. 1996; SZERKERES and TRAYNOR 1997), canabinoid receptors (CB1; BOUABOULA et al. 1997; COUTS and PERTWEE 1997; LANDSMEN et al. 1997), bradykinin B2 receptors (LEEB-LUNDBERG 1994), histamine H$_2$ receptors (SMIT et al. 1996), calcitonin receptors (POZVEK et al. 1997) and adenosine A$_1$ receptors (SHYROCK et al. 1998), among others. The number of papers about or discussing inverse agonists for GPCRs has gone from about four in calendar year 1994 to about 35 in calendar year 1997.

This chapter will review in more detail some of the papers that helped establish the concepts of spontaneously active GPCRs, and inverse agonism and will discuss the underlying mechanisms and the potential functional consequences of this change in pharmacological paradigm.

B. Overexpression, Spontaneous Activity and Inverse Agonism

As mentioned above, overexpression systems have been instrumental in demonstrating ligand-independent cellular signalling through GPCRs. The signalling responses being measured ranged from the aforementioned GTPase activity to responses involving effector mechanisms, such as adenylate-cyclase activity (CHIDIAC et al. 1994) or phosphoinositide (PI) hydrolysis (BARKER et al. 1994), and even in vivo measurements, such as cardiac contractility (BOND et al. 1995).

Demonstration that wild-type GPCRs could activate their effector systems in the absence of agonist stimulation came from studies in Sf9 cells overexpressing the human β_2-adrenoceptor following use of the baculovirus-infection expression system. Baseline adenylate-cyclase activity was shown to increase proportionately with receptor density, and several compounds previously classified as β-adrenoceptor antagonists were shown to be inverse agonists, as they inhibited baseline adenylate cyclase activity in the apparent absence of agonist stimulation (CHIDIAC et al. 1994). To rule out competition with contaminating endogenous hormone as the cause of the inhibition, these authors presented four lines of evidence.

1. The increase in the baseline and inverse-agonist activities of some ligands could be observed even in the absence of serum from the medium, eliminating the most likely source of contaminating catecholamines.
2. The magnitude of the inverse-agonist effect, E_{inv}, was not correlated with the binding affinity of the ligands for the β_2-adrenoceptor, a correlation that would have been predicted if competition with endogenous hormone were the mechanism for inhibition of baseline activity.
3. β-Adrenoceptor antagonists that had little effect on baseline activity (herein termed neutral antagonists) were able to block the inhibitory effect of inverse agonists. Again, if competition with endogenous catecholamines were the primary cause of inhibitory effects on baseline activity, then a combination of two "antagonists" should have augmented the response by producing greater competition with the hormone.
4. Pretreatment of the cells for 24 h with high concentrations of alprenolol, a treatment that would displace any possibly contaminating agonist from the receptor, followed by extensive washing, did not significantly decrease baseline activity or inhibition by inverse agonists (CHIDIAC et al. 1994).

Using PI hydrolysis as the functional endpoint, similar results were obtained for the 5-HT$_{2C}$ receptor expressed in NIH-3T3 fibroblasts (BARKER et al. 1994) and Sf9 cells LABRECQUE et al. 1995). Cells overexpressing the receptor showed an increase in baseline inositol-monophosphate (IP) formation relative to non-transfected cells, and some ligands previously classified as 5-HT antagonists functioned as inverse agonists and decreased baseline IP formation. In one of these studies (BARKER et al. 1994), the authors ruled out competition with endogenous 5-HT as the mechanism for inhibition of baseline activity. They measured the amount of 5-HT in the culture medium by high-performance liquid chromatography (HPLC) and concluded that the amount of 5-HT present (0.1 nM) could not solely account for the increase in baseline. Furthermore, they used a neutral antagonist, 2-bromolysergic acid diethylamide, to block the effects of the inverse agonists mianserin and mesulergine (BARKER et al. 1994).

The use of receptor overexpression to detect spontaneous activity and inverse agonism was not limited to cell-culture systems. Indeed, spontaneous activity of the β_2-adrenoceptor has also been detected in transgenic mice specifically overexpressing the human receptor in cardiomyocytes (MILANO et al. 1994). Three such lines of transgenic mice have been developed: TG35, TG33 and TG4. These have, respectively, 50, 100 and 200 times the β-adrenoceptor density of control mice (MILANO et al. 1994). Basal adenylate-cyclase activity in cardiac membranes was significantly greater in the TG4 and TG33 lines relative to control mice. In fact, the basal level of activity measured in the TG4 line was comparable to the maximal agonist-stimulated (isoproterenol) response observed in control mice (MILANO et al. 1994). Nevertheless, this elevated adenylate-cyclase activity could be further stimulated on β-adrenoceptor-agonist addition. Direct evidence of receptor and G protein pre-coupling ($G_{\alpha s}$ subunit) in the absence of apparent agonist stimulation was also obtained. Following immunoprecipitation of $G_{\alpha s}$ from cardiac membranes of TG4 and control mice, radioligand binding with the β-adrenoceptor ligand [^{125}I]-iodocyanopindolol (^{125}I-CYP) was performed on the immunoprecipitate. The density of β-adrenoceptors in the immunoprecipitate of the TG4 line was approximately 50-fold higher than in the control line (GURDAL et al. 1997).

In addition, to demonstrate that spontaneous activity can be biochemically detected in cardiac tissues, the transgenic mice enabled assessment of the functional consequences of such spontaneous activity both in vitro and in vivo. In vitro, isometric contractility was measured in the isolated, paced left atria of control and TG4 mice. Similar to the results obtained for adenylate-cyclase activity, baseline atrial contractility of TG4 mice was found to be significantly elevated when compared with controls, and the baseline of the TG4 line was comparable to controls following maximal β-adrenoceptor stimulation with isoproterenol (10 nM). However, unlike the adenylate-cyclase data, in which further increases in enzyme activity were observed after isoproterenol administration, the TG4 atria did not respond to isoproterenol. Thus, atrial contractility was maximal in the TG4 line and could not be further enhanced by β-adrenoceptor activation (MILANO et al. 1994).

In the three lines of transgenic mice (TG35, TG33, and TG4), but not in the control line, the β_2-adrenoceptor ligand ICI-118,551 produced concentration-dependent inhibition of basal left-atrial tension. The maximal inhibition correlated with the receptor density, suggesting a receptor-mediated event (BOND et al. 1995). To again rule out competition with endogenous hormone as the mechanism of inhibition, three lines of evidence were provided. First, if the inhibition produced by the antagonist were due to displacement of endogenous catecholamines, then the inhibition produced by various antagonists should be equal if the amount of such displacement were equal. However, at variance with predictions of the "competition of endogenous hormones" explanation, four different β_2-adrenoceptor antagonists all produced varying amounts of inhibition when used at concentrations that were 300 times their respective K_B values for the β_2-adrenoceptor (BOND et al. 1995). Second, animals were treated with an intraperitoneal (i.p.) dose of reserpine (0.3 mg/kg 24 h prior to sacrifice), which produced greater than 98% depletion of cardiac catecholamines, as measured by HPLC coupled to electrochemical detection. This treatment with reserpine failed to alter the concentration–response curve (CRC) obtained with ICI-118,551, again indicating that displacement of endogenous catecholamines was not the mechanism of inhibition by ICI-118,551 (BOND et al. 1995). Third, a neutral antagonist, alprenolol, was used to block the inhibitory effects of ICI-118,551. Alprenolol (100nM) produced an approximately 30-fold shift to the right of the ICI-118,551 CRC. This shift is appropriate for alprenolol's affinity for the β_2-adrenoceptor and is indicative of a competitive interaction between the two compounds for a single receptor (BOND et al. 1995). These data also rule out the possibility that the inhibitory effect of ICI-118,551 is due to a non-specific effect of the drug. If a non-specific effect were the cause, the combination of two blockers should have exacerbated the effect or, at least, left the larger effect intact. Thus, the ICI-118,551 effect was exerted via the β-adrenoceptor but not by competition with endogenous agonist.

It was also possible to demonstrate the spontaneous activity of the β_2-adrenoceptor and the inverse agonist activity of ICI-118,551 in vivo. Cardiac catheterisation and haemodynamic measurements were performed in anaesthetised TG4 and control mice. The maximum first derivative of the left-ventricular pressure (LV dP/dt_{max}) was used as an index of in vivo cardiac contractility. The TG4 mice had significantly higher LV dP/dt_{max} values compared with controls. Again, the inverse agonist ICI-118,551 produced inhibition of this parameter, and this inhibitory effect could be blocked by prior administration of the neutral antagonist alprenolol (BOND et al. 1995).

Although spontaneous GPCR activity can be more easily demonstrated and characterised in overexpression systems, evidence for constitutively active receptors and compounds functioning as inverse agonists have also been obtained in native, untransfected cells. For example, such evidence exists for the β-adrenoceptor of turkey erythrocytes, using adenylate-cyclase activity as the endpoint (GOTZE and JAKOBS 1994), and for β-adrenoceptors and muscarinic receptors of cardiac myocytes from various species in which electrophysiologic measurements of Ca^{2+} and K^+ currents were assessed (HANF et al. 1993; MEWES et al. 1993). There is also evidence (in rat myometrial cells) for the spontaneous activity of the bradykinin B2 receptor and for the inverse activity of ligands previously classified as B2-receptor antagonists (LEEB-LUNDBERG et al. 1994). In this study, contamination by endogenously produced bradykinin was ruled out by measuring bradykinin levels. Bradykinin was not detected in an assay with a sensitivity of 1×10^{-12} M (LEEB-LUNDBERG et al. 1994).

Thus, the phenomenon of spontaneous receptor activity and the existence of inverse agonists has been demonstrated for a variety of GPCRs that can respond to chemically distinct neurotransmitters (acetylcholine, epinephrine, bradykinin, 5-HT, etc.). The evidence has been obtained not only from a wide variety of receptors but also from various methodologies, including second-messenger assays, isolated cells and tissues and in vivo studies. At the present time, there is no overt reason to suggest that this phenomenon could not apply to all of the hundreds of GPCRs. However, the questions of what percentage of the receptors is in the active state(s) and whether the spontaneous isomerisation has any (patho)physiological relevance remain largely unanswered.

However, evidence that spontaneous activity may have a functional significance under normal physiological conditions is accumulating. One argument often used to dispute the importance of spontaneous activity and inverse agonism is that artificial overexpression is required to observe the phenomenon. As discussed above, this is not the case. In addition to being observed in native tissues, spontaneous activity and inverse agonism have been readily observed in cell lines expressing as little as 200 fmol of β2-adrenoceptor/mg membrane protein (CHIDIAC et al. 1994), a density very close to that observed in human lungs (~150 fmol/mg protein) and certainly lower than the local density found at synapses. The observation that closely related receptors display markedly different levels of spontaneous activity also lends support to the potential physiological importance of the phenomenon. Such differences have been seen for the dopamine D_A vs D_{1B} (TIBERI and CARON 1994), the prostaglandin EP3α vs EP3β (HASEGAWA et al. 1998) and the adrenoceptor β2 vs β1 (NOUET and BOUVIER, unpublished). In regard to the D_{1A} and D_{1B} dopamine receptors, the authors found that, despite displaying a very similar pharmacology (as determined by the affinities of various compounds for both receptor subtypes), the D_{1B} receptor produced greater increases in basal adenylate-cyclase activity at comparable levels of expression (TIBERI and CARON 1994). This led the authors to propose that such difference could account for the basis of heterogeneity within a given class of neurotransmitter receptors. They also raised the interesting possibility that some psychotropic antagonist drugs used in the management of certain brain disorders may have beneficial effects as inverse agonists.

The more recent discovery that some endogenous ligands can act as inverse agonists for specific receptors further supports a physiological role for spontaneous activity. The first demonstration of an endogenous inverse agonist was obtained for the central melanocortin receptor MC1 in mouse B16 melanoma cells (SIEGRIST et al. 1997). Indeed, the peptide agouti, which is the product of the coat-colour gene *Agouti* in mice, acts as an efficacious inverse agonist on the spontaneous adenylate-cyclase stimulation promoted by the melanocortin receptor. Since then, the concept of endogenous ligands with intrinsic negative activity has been extended to virally encoded GPCRs. Recently, a virally encoded chemokine receptor, the Kaposi's-sarcoma-associated herpes virus (also called ORF74) GPCR, which acts as an oncogene and angiogenesis activator in acquired immune deficiency syndrome-related malignancy, was found to have very high spontaneous activity when compared with its mammalian homologue, the interleukin-8 receptor. In very recent reports (GERAS-RAAKA et al. 1998; ROSENKILDE et al., in press), the interferon-γ-inducible protein 10 and stromal cell-derived factor-1 were found to inhibit ORF74 spontaneous activity both in inositol triphosphate and cell-proliferation assays, thus behaving as endogenous inverse agonists.

C. Mutations and Diseases of Spontaneous Receptor Activity

The above discussion indicates that, although mounting evidence suggests the physiological importance of GPCRs spontaneous activity in normal conditions, further studies are required before the generality of this phenomenon can be clearly established. However, a large body of data clearly demonstrates that agonist-independent receptor activity, at least in the case of constitutively activating mutations (CAMs), contributes to various pathophysiological processes.

The first indications that certain mutations could lead to the constitutive activation of GPCRs came from studies carried out by COTECCHIA et al. (1990), demonstrating that substitutive mutations in the third cytoplasmic loop of the α1b-adrenoceptor leads to constitutive activation of the receptor even in the absence of agonist. Equivalent mutations in the β-adrenoceptor also promoted the appearance of high levels of agonist-independent activity (SAMAMA et al. 1993). Interestingly, in the case of this last receptor, it was shown that the same inverse agonists that could inhibit the spontaneous activity of the overexpressed wild-type receptor (CHIDIAC et al. 1994) also inhibited the constitutive activity of the mutant receptor (SAMAMA et al. 1994). Thus, inverse agonists can inhibit spontaneous activity whether it is revealed by overexpression or promoted by mutations. Following these early studies, mutations of an increasing number of GPCRs were found to promote various levels of constitutive activity. CAMs were even found for receptors that had no detectable spontaneous activity in overexpression systems (PREZEAU et al. 1996; GROBLEWSKI et al. 1997).

Soon after the first reports that specific site-directed mutation of GPCRs can lead to constitutive activity, naturally occurring mutations that also lead to constitutive activity were identified in tissues obtained from patients afflicted with rare congenital diseases. In fact, it has been shown that several human diseases result from mutations of GPCRs that promote their constitutive activity. The first published example was a type of male precocious puberty resulting from a mutation of the leutinising-hormone (LH) receptor in the testes; this mutation produces spontaneous function of the receptor in the absence of circulating LH, thus leading to an exaggerated testosterone secretion and a precocious onset of puberty (SHENKER et al. 1993). Additional examples include: the Jansen-type metaphyseal chondrodysplasia, which results from a CAM of the parathyroid-hormone receptor (SCHIPANI et al. 1995); congenital and somatic hyperthyroidism caused by CAMs of the thyrothropin receptor (PARMA et al. 1993; KOPP et al. 1995); and retinitis pigmentosa and stationary night blindness as a consequence of CAMs of rhodopsin (RIM and OPRIAN 1995). One can conclude from these examples that alterations of GPCR spontaneous-activity levels can underlie pathological conditions and, therefore, that it represents an important physiological parameter. Based on the observation (previously reviewed) that inverse β-

adrenoceptor agonists can inhibit the spontaneous activity of constitutively activated β_2-adrenoceptor, it could be argued that an inverse agonist would be of greater therapeutic utility than neutral antagonists in the diseases cited above, since the problem is spontaneous activity of the receptor, not excess hormonal stimulation. Until inverse agonists for the culprit receptors are identified or developed, this clinically important hypothesis will remain untested. Whether inverse agonists could also represent better drugs than neutral antagonists in pathologies resulting from the overexpression of a given receptor subtype also remains an open question.

D. Modulation of Receptor Function by Agonists and Inverse Agonists

The topic of modulation of receptor function by agonists and inverse agonists has recently been reviewed, and the following discussion contains excerpts of a previously published review, reprinted with permission (MILIGAN and BOND 1997). As GPCRs are the primary recognition point for the presence, variation and intensity of hormonal and neurotransmitter-encoded information, their regulation at transcriptional, translational, post-translational and degradative stages is to be anticipated (HADDOCK and MALBON 1991; COLLINS et al. 1992; MILLIGAN et al. 1995b). One facet of this regulation that has been widely explored is the capacity of agonist ligands to cause desensitisation, internalisation or sequestration of GPCRs over a relatively short period of exposure. This can be followed by a combination of resensitisation and recycling to the cell surface or, if the stimulus is both prolonged and intense, degradation of the GPCR, resulting in an overall reduction in cellular levels, a process called downregulation. These processes have been studied at both biochemical and cell-biological levels and are becoming well understood (VON ZASTROW and KOBILKA 1992; BARAK et al. 1997; MOLINO et al. 1997; RUIZ-GOMEZ and MAYOR 1997).

There is no reason a priori to assume that the conformational state enhanced by agonist binding to produce G protein coupling is the same conformation that is a substrate for phosphorylation and downregulation. For example, although D-Tyr-Gly-[(norleucine 28,31, D-Trp30)cholecystokinin 26–32]-phenethyl ester functions as a cholecystokinin antagonist with no evidence of agonist function, it is able to cause the internalisation of the cholecystokinin-A receptor expressed in Chinese-hamster ovary (CHO) cells (ROETTGER et al. 1997). This is also true of a non-peptide antagonist at this receptor. There are also studies (in primary cultures of epithelial cells derived from the choroid plexus, which express the 5-HT$_{2C}$ receptor endogenously) showing downregulation of receptors is not always caused by agonists. In this system, the inverse agonist mianserin causes downregulation of the receptor whereas an antagonist ligand fails to reproduce this effect (BARKER et al. 1994). Furthermore, following expression in insect Sf9 cells, a range of antagonist/inverse-agonist ligands produced receptor downregulation. However, there was no strong correlation between inverse efficacy and the degree of effect, indicating these two features are not intrinsically linked (LABRECQUE et al. 1995). Conversely, in HEK 293 cells expressing δ- and μ-opioid receptors, morphine was shown to activate opioid receptors without producing internalisation (KEITH et al. 1996).

However, for most receptor systems, downregulation of receptors is associated with ligands that function to increase G protein coupling (agonists). Thus, the receptor conformation that interacts with G proteins is often also a substrate for desensitisation and downregulation. For example, in a recent study using HEK-293 cells transfected with the β_2-adrenoceptor, the same rank order of agonist potency was observed for adenylate-cyclase activation, desensitisation, and receptor internalisation (JANUARY et al. 1997), thus suggesting that the conformational requirements are similar for all three processes. According to the potentially overly simplistic complementary view of agonism and inverse agonism developed above based on the selective stabilisation of inactive and active GPCR conformations, if prolonged exposure to agonist ligands frequently results in GPCR downregulation, equivalent treatment with inverse agonists might be anticipated to result in their upregulation. A large literature exists on such effects produced in vivo by administration of "antagonist" ligands. However, it is often difficult to discern whether these effects are simply antagonist effects that prevent a degree of GPCR downregulation due to the presence of circulating endogenous agonist or are truly a reflection of the inverse-agonist nature of the ligand employed. This issue has recently been addressed in cell lines stably transfected to express GPCRs. Following expression of the rat histamine-H_2 receptor in CHO cells, SMIT et al. recorded a time- and concentration-dependent upregulation in levels of this receptor when the cells were exposed to cimetidine or ranitidine. This was not replicated by treatment with burimamide or by ligands with good selectivity for the histamine-H_1 or -H_3 receptors (SMIT et al. 1996). Both cimetidine and ranitidine were able to inhibit basal cyclic-adenosine-monophosphate production in H_2-receptor-expressing cells but not in untransfected CHO cells, demonstrating the function of these compounds as inverse agonists. By contrast, burimamide functioned as a neutral antagonist (SMIT et al. 1996). Interestingly, the inverse agonist cimetidine did not promote any upregulation of a mutant form of the H_2-histamine receptor (Leu^{124}Ala) having limited basal and histamine-stimulated activity, thus suggesting that the capacity of an inverse agonist to reverse GPCR-produced elevated basal second-messenger levels is central to upregulation.

The authors further suggested that these findings may yield a plausible explanation for the observed development of tolerance with the use of cimetidine and ranitidine. Their chronic use may produce upregulation of the H_2 receptor, which would oppose its therapeutic value of preventing histamine activation of the H_2 receptor and subsequent release of hydrochloric acid. It follows that, in this case, a neutral antagonist may be a preferred therapeutic agent. Very similar results were observed for the wild-type human β_2-adrenoceptor following stable expression in NG108-15 neuroblastoma × glioma hybrid cells. Sustained exposure to either betaxolol or sotalol, which function as inverse agonists, resulted in approximately a doubling in membrane levels of the receptor, whereas equivalent treatment with the antagonist alprenolol was unable to produce this effect (MACEWAN and MILLIGAN 1996). Furthermore, treatment of patients with β-blocker drugs (many of which display inverse-agonist properties) is known to produce a degree of β-adrenoceptor upregulation. If a clear distinction develops between inverse agonists and antagonists in their capacity to upregulate wild-type GPCRs, and if this trend is extended following more detailed examination of a range of GPCRs that display relatively high degrees of constitutive activity then, to avoid potential onset of subsequent short-term supersensitivity to endogenous agonist, therapeutic use of such ligands may require careful monitoring on their removal until the cells and tissues have re-established the initial steady-state levels. As such, these concerns may be more theoretical than practical, but they clearly require further consideration and the development of robust assays to allow discrimination among ligands possessing antagonist and inverse-agonist properties.

The preceding discussion clearly indicates that, in addition to having immediate effects on receptor signalling, inverse agonists can have long-term effects on the responsiveness of the systems. These effects appear, in most cases, to be the mirror images of the effects classically promoted by sustained agonist treatments. In most instances, investigations have considered the consequences of either agonist or inverse-agonist treatment on the subsequent responsiveness to agonists. Only one study assessed the effect of sustained agonist stimulation on the efficacy of inverse agonists. In Sf9 cells overexpressing the human β_2-adrenoceptor, treatment for 30 min with isoproterenol, which promotes desensitisation to further agonist stimulation, significantly increased the negative efficacy of the inverse agonists (CHIDIAC et al. 1996). Pre-treatment with agonist, therefore, had reciprocal effects on the responsiveness of the system to agonist and inverse agonist (i.e., reducing the efficacy of the agonists and increasing the negative efficacy of the inverse agonists). Overall, however, pre-treatment with agonist pre-set the system so that, in all cases, lower levels of activity are observed. This study also illustrated that signalling efficacy is not exclusively a ligand-dependent parameter. Indeed, dichloroisoproterenol (which behaves, under control conditions, as a weak partial agonist) became a weak inverse agonist when tested in membrane preparations derived from desensitised cells. This dramatic change in apparent efficacy clearly indicates that some compounds can act as either agonists or inverse agonists, depending on the initial conditions of the system. Ligands that posses such dual signalling properties are now referred to as protean ligands (KENAKIN 1995; JANSSON et al. 1998). The theoretical basis that could underlie the dual behaviour of ligands has recently been reviewed (KENAKIN 1997), and some of the models (such as the cubic model; see below and Chap. 2.2) that have been developed to explain receptor activation and can accommodate the existence of protean ligands.

E. Models

Overall, the findings obtained in the last few years concerning spontaneous activity and inverse agonism at GPCRs are difficult to reconcile with the classical induced-fit model of receptor activation, which considers that the ternary complex agonist–receptor–G protein is the only active form of the receptor (L+R↔R*L↔GR*L). However, the findings are entirely consistent with the "two-state" receptor model, whereby the receptor exists in at least two-states – R (inactive) and R* (active) – even in the absence of ligands. This spontaneous isomerisation to an active conformer would be responsible for the spontaneous activity observed in systems expressing high levels of receptor. Under basal conditions, the inactive state would largely predominate for most receptors, and spontaneous activity could be detected only if the absolute number of receptors in the active conformation is sufficient to promote a sizeable signal. In such a model, mutations leading to constitutive activation would be envisioned as stabilising or favouring the transition towards the active isomer, most likely by relieving a molecular constraint that increases the transition energy. However, the conformational changes permitted by the mutations can not be seen as rigid or irreversible since, as mentioned above, inverse agonists can inhibit the spontaneous activity of the constitutively activated receptor mutants.

In a two-state model, the actions of agonists and inverse agonists can then be described in terms of their preferential affinities for a given state; inverse agonists and agonists preferentially bind to and stabilise the inactive and the active states, respectively. A neutral antagonist would not discriminate between the two conformers and would thus act as a competitive antagonist towards both agonists and inverse agonists. Interestingly, and similarly to the phenomenon of partial agonism, various levels of inverse efficacy were found. In fact, a continuum going from full inverse agonist to neutral antagonist was observed for many receptors. These different efficacies can be rationalised by assuming that different compounds will have distinct relative preferences for the active and inactive conformers. In a first effort to formalise a two-state model in the context of receptor–G protein interactions, SAMANA et al. (1993) proposed the extended ternary-complex model, which allows both ligand-unoccupied and ligand-bound receptors to adopt the R* conformation and, thus, to bind to the G protein. It follows that, in contrast to the classical ternary-complex model, at least two active complexes exist (R*G and LR*G). One theoretical limitation of this model is that G proteins are allowed to bind to R* but not R. However, there is no a priori reason to rule out either the process by which GR* relaxes to GR or the binding of R to G (albeit with a lower affinity than R*). A more general model termed the cubic model has thus been proposed (WEISS et al. 1996; Chap. 2.2). This model, although more complex, has the advantage of being thermodynamically complete. According to both the extended ternary-complex and the cubic models, ligand efficacy may be viewed as conformational *selection* of pre-existing states. Although this selected-fit model offers an easy explanation for spontaneous activity and for the mode of action of inverse agonists, various lines of evidence suggest that different ligands may promote or stabilise ligand-specific conformational changes. If the existence of such multiple active states of the receptor (which may have distinct signalling efficacies and isomerisation constants) is experimentally confirmed, models that pretend to describe receptor-activation processes would need to incorporate them.

F. Summary and Conclusions

Recent evidence has now demonstrated beyond question that numerous GPCRs exhibit spontaneous activity and signal cellular responses in the absence of agonist binding. At the present time, there is no reason to assume that the spontaneous activity of GPCRs is not applicable to all of the hundreds of GPCRs though, for several neurotransmitters and hormones, the percentage of receptors in the spontaneously active form can vary within the receptor subtypes. For these spontaneously active GPCRs, ligands termed inverse agonists have been shown to decrease the level of spontaneous activity.

Several diseases where mutations of the receptor increased the level of spontaneous activity have been identified in humans. It would seem that the development of inverse agonists for these mutated receptors would be a rational form of treatment. Furthermore, inverse agonists and antagonists can have differential effects on receptor upregulation, thereby suggesting that chronic treatments may also reveal differences that are not acutely observable between inverse agonists and antagonists. For these reasons, it appears that the drug industry must develop appropriate screens to determine whether their compounds function as inverse agonists or antagonists (BLACK and SHANKLEY 1995).

Finally, the classic receptor-theory model, based on a single quiescent state of GPCRs, must be modified. At present, most of the observed experimental results can be satisfactorily modelled based on an equilibrium between two receptor conformations: an inactive and an active conformation. However, the existence of ligand-specific selection of different conformational states of a receptor has necessitated an expansion beyond the two-state model.

References

Barak LS, Ferguson SSG, Zhang J, Martenson C, Meyer T, Caron MG (1997) Internal trafficking and surface mobility of a functionally intact β_2-adrenergic green fluorescent protein conjugate. Mol Pharmacol 51:177–184

Barker EL, Westphal RS, Schmidt D, Sanders-Bush E (1994) Constitutively active 5-hydroxytryptamine$_{2c}$ receptors reveal novel inverse agonist activity of receptor ligands. J Biol Chem 269:11687–11690

Barr AJ, Manning DR (1997) Agonist-independent activation of Gz by the 5-hydroxytryptamine$_{1A}$ receptor co-expressed in *Spodoptera frugiperda* cells. J Biol Chem 272:32979–32987

Black JW, Shankley NP (1995) Inverse agonists exposed. Nature 374:214–215

Bond RA, Johnson TD, Milano CA, Leff P, Rockman HA, McMinn T, Apparsunduram S, Hyek MF, Kenakin TP, Allen LF, Lefkowitz RJ (1995) Physiologic effects of inverse agonists in transgenic mice with myocardial overexpression of the β_2-adrenoceptor. Nature 374:272–276

Bouaboula M, Perrachon S, Milligan L, Canat X, Rinaldi-Carmona M, Portier M, Barth F, Calandra B, Pecceu F, Lupkers J, Maffrand J-P, Le Fur G, Cassellas P (1997) A selective inverse agonist for central cannabinoid receptor inhibits mitogen-activated protein kinase activation stimulated by insulin or insulin-like growth factor 1. J Biol Chem 272:22330–22339

Burstein ES, Spalding TA, Brann MR (1997) Pharmacology of muscarinic receptor subtypes constitutively activated by G proteins. Mol Pharm 51:312–319

Chidiac P, Hebert TE, Valiquette M, Dennis M, Bouvier M (1994) Inverse agonist activity of β-adrenergic antagonists. Mol Pharmacol 45:490–499

Chidiac P, Nouet S, Bouvier M (1998) Agonist-induced modulation of inverse agonist efficacy at the β_2-adrenergic receptor. Mol Pharmacol 49:in press

Collins S, Caron MG, Lefkowitz RJ (1992) From ligand binding to gene expression: new insights into the regulation of G-protein-coupled receptors. Tends Biochem Sci 17:37

Costa T, Herz A (1989) Antagonists with negative intrinsic activity at opioid receptors coupled to GTP-binding proteins. Proc Natl Acad Sci USA 86:7321–7325

Costa T, Lang J, Gless C, Herz A (1990) Spontaneous association between opioid receptors and GTP-binding regulatory proteins in native membranes: specific regulation by antagonists and sodium ions. Mol Pharmacol 37:383–394

Cotecchia S, Exum S, Caron MG, Lefkowitz RJ (1990) Regions of the a1-adrenergic receptor involved in coupling to phosphatidyl-inositol hydrolysis and enhanced sensitivity of biological function. Proc Natl Acad Sci USA 87:2896–2900

Coutts AA, Pertwee RG (1997) Inhibition by cannabinoid receptor agonists of acetylcholine release from the guinea-pig myenteric plexus. Br J Pharmacol 121:1557–1566

Geras-Raaka E, Varma A, Ho H, Clark-Lewis I, Gershengorn MC (1998) Human interferon-gamma-inducible protein 10 inhibits constitutive signaling of Kaposi's sarcoma-associated herpes virus G protein-coupled receptor. J Exp Med 188:405–408

Gotze K, Jakobs KH (1994) Unoccupied β-adrenoceptor-induced adenylyl cyclase stimulation in turkey erythrocyte membranes. Eur J Pharmacol 268:151–158

Groblewski T, Maigret B, Larguier R, Lombard C, Bonnafous J-C, Marie J (1997) Mutation of Asn-111 in the third transmembrane domain of the AT1A angiotensin II receptor induces its constitutive activation. J Biol Chem 272:1822–1826

Gurdal HM, Bond RA, Johnson MD, Friedman E, Onaran HO (1997) An efficacy-dependent effect of cardiac overexpression of β2-adrenoceptor on ligand affinity in transgenic mice. Mol Pharm 52:187–194

Haddock JR, Malbon CC (1991) Regulation of receptor expression by agonists: transcriptional and post-transcriptional controls. Trends Neurosci 14:242–247

Hall DA, Strange PG (1997) Evidence that antipsychotic drugs are inverse agonists at D$_2$ dopamine receptors. Br J Pharmacol 121:731–736

Hanf R, Li Y, Szabo G, Fischmeister R (1993) Agonist-independent effects of muscarinic antagonists on Ca2+ and K+ currents in frog and rat cardiac cells. J Physiol (Lond) 461:743–765

Hasegawa H, Negishi M, Ichikawa A (1998) Two isoforms of the prostaglandin E receptor EP3subtype different in agonist-independent constitutive activity. J Biol Chem 271:1857–1860

Jansson CC, Kukkonen JP, Nasman J, Huifang G, Wurster S, Virtanen R, et al (1998) Protean agonism at α-2A-adrenoceptors. Mol Pharmacol 53:963–968

January B, et al (1997) β_2-adrenergic-receptor desensitization, internalization and phosphorylation in response to full and partial agonists. J Biol Chem 272: 23871–23879

Keith DE, Murray SR, Zaki PA, Chu PC, Lissin DV, Kang L, Evans CJ, von Zastrow M (1996) Morphine activates opioid receptors without causing their rapid internalization. J Biol Chem 271:19021–19024

Kenakin TP (1988) Are receptors promiscuous? Intrinsic efficacy as a transduction phenomenon. Life Sci 43:1095–1101

Kenakin TP (1995) Pharmacological proteus? Trends Pharmacol Sci 16:356–358

Kenakin TP (1997) Protean agonists. Keys to receptor active states? Ann NY Acad Sci 812:116–125

Kopp P, Van Sande J, Parma J, Duprez L, Gerber H, Joss E, et al (1995) Congenital hyperthyroidism caused by a mutation in the thyrotropin-receptor gene. New Engl J Med 332:150–154

LaBrecque J, Fargin A, Bouvier M, Chidiac P, Dennis M (1995) Serotonergic antagonists differentially inhibit spontaneous activity and decrease ligand binding capacity of the rat 5-hydroxytryptamine type 2C receptor in Sf9 cells. Mol Pharmacol 48:150–159

Landsman RS, Burkey TH, Consroe P, Roeske WR, Yamamura HI (1997) SR141716A is an inverse agonist at the human cannabinoid CB_1 receptor. Eur J Pharmacol 334:R1–R2

Lee TW, Cotecchia S, Milligan G (1997) Up-regulation of the levels of expression and function of a constitutively active mutant of the hamster alpha1B-adrenoceptor by ligands that act as inverse agonists. Biochem J 325:733–739

Leeb-Lundberg LM, Mathis SA, Herzig MC (1994) Antagonists of bradykinin that stabilize a G-protein-uncoupled state of the B2 receptor act as inverse agonists in rat myometrial cells. J Biol Chem 269:25970–25973

Leff P, Harper D (1989) Do pharmacological methods for the quantification of agonists work when the ternary complex mechanism operates? J Theor Biol 140: 381–397

MacEwan DJ, Milligan G (1996) Inverse-agonist-induced upregulation of the human β_2-adrenoceptor in transfected neuroblastoma × glioma hybrid cells. Mol Pharmacol 50:1479–1486

Mackay D (1988) Continuous variation of agonist affinity constants. Trends Pharmacol Sci 9:156–157

Malmberg A, Mikaels A, Mohell N (1998) Agonist and inverse agonist activity at the dopamine D_3 receptor measured by guanosine 5'-[γ-thio]triphosphate-[^{35}S] binding. J Pharmacol Exp Ther 285:119–126

Mewes T, Dutz S, Ravens U, Jakobs KH (1993) Activation of calcium currents in cardiac myocyte by empty β-adrenergic receptors. Circulation 88:2916–2922

Milano CA, Allen LF, Dolber P, Rockman H, Chien KR, Johnson TD, Bond RA, Lefkowitz RJ (1994) Enhanced myocardial function in transgenic mice with cardiac overexpression of the human β_2-adrenergic receptor. Science 264:582–586

Milligan G, Bond RA (1997) Inverse agonism and the regulation of receptor number. Trends Pharmacol Sci 18:468–47450

Milligan G, Bond RA, Lee M (1995a) Inverse agonism: Pharmacological curiosity or potential therapeutic strategy? Trends Pharmacol Sci 16:10–13

Milligan G, Parenti M, Magee AI (1995b) The dynamic role of palmitoylation in signal transduction. Trends Biochem Sci 20:181–186

Molino M, et al (1997) Thrombin receptors on human platelets. Initial localization and subsequent redistribution during platelet activation. J Biol Chem 272:6011–6017

Mullaney I, Carr IC, Milligan G (1996) Analysis of inverse agonism at the δ-opioid receptor after expression in Rat-1 fibroblasts. Biochem J 315:227–234

Newman-Tancredi A, Conte C, Chaput C, Spedding M, Millan MJ (1997) Inhibition of the constitutive activity of human 5-HT1A receptors by the inverse agonist, spiperone but not the neutral antagonist. WAY 100,635

Parma J, Duprez L, Van Sande J, Cochaux P, Gervy C, Mockel J, et al (1993) Somatic mutations in the thyrotropin receptor gene cause hyperfunctioning thyroid adenomas. Nature 365:649–651

Pozvek G, Hilton JM, Quiza M, Houssami S, Sexton PM (1997) Structure/function relationships of calcitonin analogues as agonists, antagonists, or inverse agonists in a constitutively activated receptor cell system. Mol Pharmacol 51:658–665

Prezeau L, Gomzea J, Ahern S, Mary S, Galvez T, Bockaert J, et al (1996) Changes in the carboxyl-terminal domain of metabotropic glutamate receptor by alternative splicing generate receptors with differing agonist-independent activity. Mol Pharmacol 49:422–429

Rim J, Oprian DD (1995) Constitutive activation of opsin: interaction of mutants with rhodopsin kinase and arrestin. Biochemistry 34:11938–11945

Roettger BF, et al (1997) Antagonist-stimulated internalization of the G-protein-coupled cholecystokinin receptor. Mol Pharmacol 51:357–362

Rosenkilde MM, Kledal TN, Brauner-Osborne H, Schwartz TW (1998) Agonists and inverse agonists for the herpes-virus-8-encoded constitutively active 7TM-oncogene product ORF74. J Biol Chem (in press)

Ruiz-Gomez A, Mayor F Jr (1997) β-adrenergic receptor kinase (GRK2) colocalizes with β-adrenergic receptors during agonist-induced receptor internalization. J Biol Chem 272:9601–9604

Samama P, Cotecchia S, Costa T, Lefkowitz RJ (1993) A mutation-induced activated state of the β_2-adrenergic receptor. J Biol Chem 268:4625–4636

Samama P, Pei G, Costa T, Cotecchia S, Lefkowitz RJ (1994) Negative antagonists promote an inactive conformation of the β_2-adrenergic receptor. Mol Pharmacol 45:390–394

Schipani E, Kruse K, Juppner HA (1995) Constitutively active mutant PTH-PTHrP receptor in Jansen-type metaphyseal chondrodysplasia. Science 268:98–100

Shenker A, Laue L, Kosug S, Merendino JJJ, Minegishi T, Cutler GBJ (1993) A constitutively activating mutation of the luteinizing hormone receptor in familial male precocious puberty. Nature 365:652–654

Shryock JC, Ozeck MJ, Belardinelli L (1998) Inverse agonists and neutral antagonists of recombinant human A1-adenosine receptors stably expressed in Chinese-hamster ovary cells. Mol Pharmacol 53:886–893

Siegrist W, Drozdz R, Cotti R, Willard DH, Wilkinson WO, Eberle AN (1997) Interactions of α-melanotropin and agouti on B16 melanoma cells: evidence for inverse agonism of agouti. J Receptor Sign Transduc Res 17:75–98

Smit MJ, et al (1996) Inverse agonism of histamine H_2 antagonists accounts for upregulation of spontaneously active histamine H_2 receptors. Proc Natl Acad Sci USA 93:6802–6807

Szekeres PG, Traynor JR (1997) δ-Opioid modulation of the binding of guanosine-5'-O-(3-[^{35}S]thio)triphosphate to NG108-15 cell membranes: characterization of agonist and inverse agonist effects. J Pharmacol Exp Ther 283:1276–1284

Tian WN, Duzic E, Lanier SM, Deth RC (1994) Determinants of α_2-adrenergic receptor activation of G-proteins: evidence for a pre-coupled receptor/G-protein state. Mol Pharmacol 45:524–531

Tiberi M, Caron MG (1994) High agonist-independent activity is a distinguishing feature of the dopamine-D_{1B}-receptor subtype. J Biol Chem 269:27925–27931

Von Zastrow M, Kobilka BK (1992) Ligand-regulated internalization and recycling of human β_2-adrenergic receptors between the plasma membrane and endosomes containing transferrin receptors. J Biol Chem 267:3530–3538

Weiss JM, Morgan PH, Lutz MW, Kenakin TP (1996) The cubic ternary complex receptor-occupancy model I. Model description. J Theor Biol 178:151–167

CHAPTER 7
Efficacy: Molecular Mechanisms and Operational Methods of Measurement. A New Algorithm for the Prediction of Side Effects

T. KENAKIN

A. Introduction

This chapter will consider efficacy, the property of a molecule that changes the behavior of a receptor towards its host. It will be seen that efficacy can be positive or negative and that the manifestation of efficacy can change in accordance with the set point of the receptor system. The chapter will be divided into two main themes. The first will review the molecular mechanisms of efficacy. The second will consider the operational methods available to measure the relative efficacy of agonists and will describe a new method whereby relative efficacy estimates can be extended to predict agonist selectivity in vivo.

B. The Molecular Nature of Efficacy

Membrane receptors are extraordinary in that they repeatedly react to agonist messenger molecules to translate the chemical information they bear without changing that messenger. Unlike enzymes, receptors do not change their "substrates" but rather read the information encoded in the chemical structure of the agonist and change their behavior toward the cell accordingly. There has been a great deal of study into the mechanisms by which receptors cause physiological responses. In many cases, the specific mechanisms relate to the specific receptor types, but there are general global mechanisms by which receptors can translate information. These involve the ways in which proteins change tertiary conformation.

In early formulations of receptor theory, the agonist molecule was thought to bind to the receptor and thus change the receptor conformation to make it reactive toward other cellular components. Work on the allosteric nature of enzymes provided the rationale for thinking that the binding of an agonist at one part of the receptor could impart a change in conformation in another part of the receptor, thus initiating physiological responses (KOSHLAND 1960; MONOD et al. 1965). A pivotal discussion of how these ideas relate to agonism and drug receptors was given by BURGEN (1966), who coined terms for two apparently divergent mechanisms. One he called *conformational selection*, whereby agonists selectively bind to one of two (or more) receptor conformations and thus bias the population towards that conformation. The rationale for this mechanism comes from mass action kinetics.

It is known that G protein-coupled receptors (GPCRs) can adopt active (with respect to interaction with G proteins) conformations spontaneously. The equilibrium between an inactive state and an active state of the receptor can be described by an intrinsic equilibrium constant (termed an allosteric constant, denoted L). Within this theoretical framework, inactive ($[R_i]$) and active ($[R_a]$) states of the receptor can be depicted as:

$$[R_i] \rightleftharpoons [R_a] \tag{1}$$

where $L = [R_a]/[R_i]$.

If a ligand binds selectively to one of the conformations, it effectively removes that conformation from the equilibrium formerly controlled by the allosteric constant L.

$$[R_i] \overset{L}{\rightleftharpoons} [R_a]+[A] \overset{K_A}{\rightleftharpoons} [AR_a] \tag{2}$$

In this scheme, the binding of the agonist to the receptor removes R_a from the equilibrium controlled by the allosteric constant L; therefore, more R_a is formed to make up the deficit left by the conversion of R_a to AR_a. If response is a function of the amount of receptor in the active state (R_a plus AR_a), then the selective binding of the agonist to the R_a species will lead to physiological response.

BURGEN (1966) outlined another, apparently different, mechanism for agonism; he termed it *conformational induction*. This idea resembled the historical view of efficacy in that agonists were thought to bind to and, by that binding, to deform the receptor into an active state that then initiates physiological response. As presented, these ideas appeared to be divergent and representative of separate mechanisms of agonism. However, in thermodynamic terms, a substantial energy of binding would be required for a small molecule to cause a receptor to adopt a different tertiary conformation. It is more likely that molecules stabilize pre-existing conformations of the receptor protein. Proteins are thought to adopt a number of tertiary conformations and variably adopt these according to levels of thermal energy. The concept of an "energy landscape" can be used to depict the various conformations present at any instant in a population of proteins (FRAUNFELDER et al. 1988, 1991). Thus, at any instant in time, the numbers of receptors in any given conformation could be depicted as a spectrum of frequency histograms. Figure 1A shows a frequency distribution of receptor states (most of which are inactive with respect to physiological signaling) of a given receptor population. The most prevalent spontaneously formed active state is given as a *filled bar*. If an agonist selectively bound to this spontaneous active state, then agonism would ensue (Fig. 1B). Also shown is another active state, the spontaneous production of which is extremely rare in a quiescent (no agonist present) receptor system. The apparently conflicting ideas of conformational selection and induction can then be reconciled as extremes of the same mechanism. In this

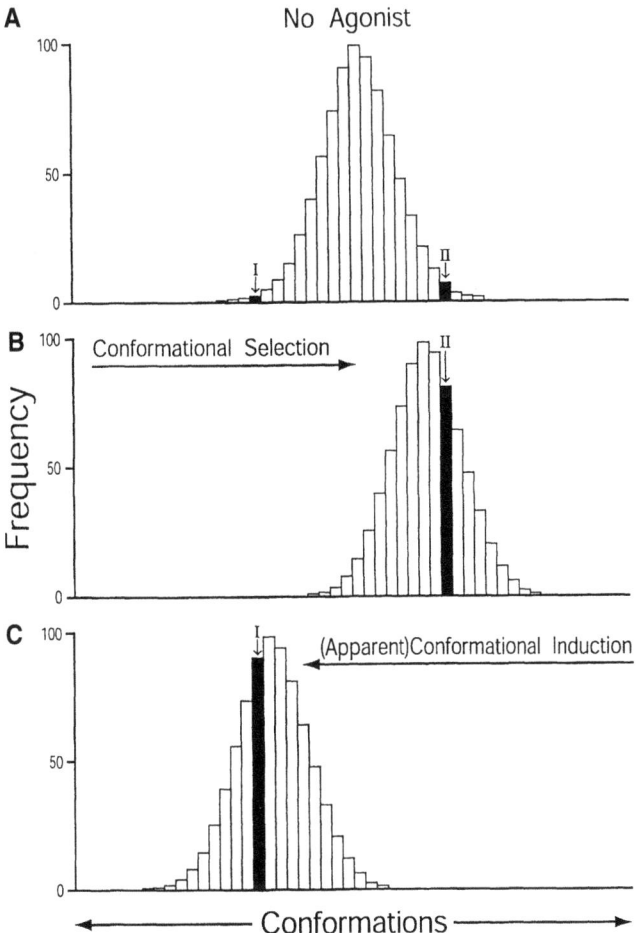

Fig. 1A–C. Histograms depicting the frequency of occurrence of different receptor conformations. **A** While most of the conformations do not activate G proteins, the states labeled I and II produce G protein activation. The spontaneous occurrence of state II is low but measurable. State I occurs so infrequently that it is insignificant. A large concentration of receptors allows constitutive activity to be observed by virtue of the spontaneous formation of state II. **B** Conformational selection. Agonist A produces a bias in the receptor conformation spectrum, thus enriching active state II; the agonist selects the naturally most prominent active state. **C** Conformational induction. Agonist B appears to "induce" a special active conformation by shifting the spectrum of conformations toward the extremely rare conformation I

scheme, the selection of an extremely rare spontaneously active conformation would appear to occur via conformational induction, i.e., the conformation would essentially not exist, in appreciable quantities, in the absence of the agonist (apparent conformational induction; Fig. 1C). This would reconcile the thermodynamics in that the agonist would not be required to produce a conformation that the receptor did not naturally assume (KENAKIN 1996a).

This concept is discussed in statistical terms by ONARAN and COSTA (1997). They describe a population of receptors existing in a very large number of interconvertible microscopic states. Thus, the conformational universe of that receptor could be described as a distribution in terms of the probability or fraction of receptors existing in those states. When a ligand is introduced into the system, it binds according to its microscopic affinity for each of the states and thus changes the distribution of conformations. The binding of G protein to the native receptor confers yet another distribution of states, and the introduction of a ligand into the receptor–G protein milieu forms a collection of distributions of the receptor. From this concept, it can be seen that different collections of receptor distributions could result from the binding of different ligands to the system.

One scenario for the production of physiologic response would be that agonists, by selectively binding to receptor states, present the cell with arrays of active-state receptors, which then go on to interact with G proteins. There are numerous examples of receptor–G protein pleiotropy (KENAKIN 1996b). It is also known that different regions of receptor cytosolic loops activate different G proteins. For example, point mutation studies on α_2-adrenergic receptors have shown that sequences in the second cytosolic loop are essential for activation of G_i protein, whereas sequences in the third cytosolic loop active G_s protein (IKEZU et al. 1992). If it can be assumed that different receptor conformations alternately expose and conceal these different regions in the cytosol, then it is also logical to assume that different conformations of the receptor can result in differential activation of G proteins.

This opens the possibility of agonist-directed trafficking of receptor stimulus (KENAKIN 1995a; KRUMINS and BARBER 1997). By virtue of differential stabilization of receptor active states, different agonists could traffic receptor stimuli to different G proteins. Under these conditions, the creation of agonists for selective receptor states would represent a new level of agonist selectivity (KENAKIN 1998; Fig. 2). If it is assumed that limiting the cytosolic cascades activated by an agonist in different cell types will lead to a more selective agonist profile, then it might also be assumed that receptor state-selective agonists should offer a more selective spectrum of agonism.

C. Positive and Negative Efficacy

Before the advent of constitutively active receptor systems, agonists and antagonists were studied in quiescent systems, i.e., there was little physiological activity present in the absence of an agonist. The production of agonist-independent activation of opioid receptor by the removal of sodium ions (COSTA and HERZ 1989) heralded the introduction of concepts relating the natural ability of G protein-coupled receptors to signal in the absence of ligands. Thus, the allosteric constant (Eq. 1) L could be altered chemically to induce an amount of R_a sufficient to produce a measurable response. The

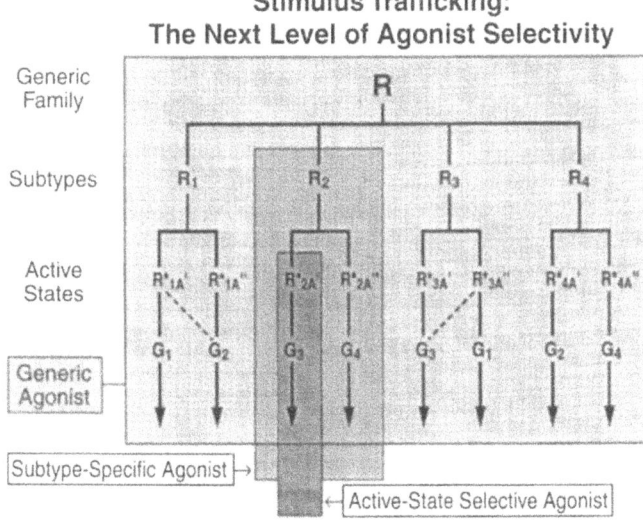

Fig. 2. A wide range of tissue responses is produced by the endogenous agonist for the general family of receptors R. A decrease in the range of responses can be achieved by producing an agonist for a receptor subtype of the general family. If agonists produce different active receptor conformations, then a further degree of selectivity can be achieved by utilizing agonists that produce a single active conformation (KENAKIN 1998)

reversal of this constitutive receptor activity by ligands introduced a new type of efficacy into pharmacological awareness. For example, the opioid ligand ICI-174864 produced a concentration-dependent blockade of the constitutive opioid receptor activity produced by removal of sodium ions in NIH-3T3 cells (COSTA and HERZ 1989). The term "inverse agonist" was coined for these ligands.

Subsequent work in many receptor systems has confirmed both the existence of constitutive receptor activity and inverse agonists. Even without ionic involvement, the allosteric constants of some receptors allow the production of measurable constitutive receptor activity from simple stoichiometry. The increased expression of receptors (SAMAMA et al. 1993) or G proteins (SENOGLES et al. 1990) in recombinant systems leads to the production of constitutive, receptor-mediated, physiological, basal response. Thus, the set point of the receptor systems determines whether a ligand will be classified as a neutral antagonist or an inverse agonist.

The reversal of constitutive receptor activity most probably emanates from the very same mechanisms as those mediating positive agonist responses: specifically, a selective affinity for the inactive state of the receptor. Under these circumstances, efficacy cannot be thought of only in terms of positive activation of cellular pathways. An alternate definition would suggest that *efficacy is that property of a ligand that changes the behavior of the receptor towards its host* (KENAKIN 1996a) Within this definition, inverse agonists possess negative efficacy, and agonists that activate cellular-response pathways possess positive efficacy.

D. The Operational Measurement of Relative Efficacy

At present, there is no absolute scale of efficacy, and it can only be measured as a ratio of proportionality factors relating the ability of agonists to occupy the receptor and produce a biological response. There are considerable advantages to measuring the efficacy of an agonist, since it is a molecular property that transcends the measuring system. Under ideal circumstances, the relative efficacies of agonists can be measured and used to predict agonism in every tissue thereafter, including human tissue in the therapeutic arena.

The location parameter of a dose–response curve for an agonist (potency) depends upon the affinity of the agonist for the receptor, and its efficacy. For this reason, it is not possible to judge the relative efficacy of two agonists if both produce the tissue-maximal response. Therefore, a dissimulation can occur if the agonist is discovered and initially tested in a system where it produces a maximal response and is then utilized in subsequent testing systems where the receptor density and/or efficiency of coupling is lower. In these latter systems, efficacy primarily determines the observed agonism and, if a low-efficacy agonist has been chosen from the initial test system, no response may be observed in these latter systems. By quantifying the relative dependence of agonist potency on efficacy (versus affinity), predictions can be made about the sensitivity of the agonism in a range of different organs containing the receptor (*vide infra*). As a preface to the description of a new method of doing this, the various operational methods of measuring the relative efficacies of agonists will be discussed in terms of two experimental approaches: binding experiments and functional experiments.

I. Binding Studies

1. Guanosine Triphosphate γS Shift

In keeping with the notion that the more efficacious an agonist is, the more it will promote the interactions of receptors and G proteins, the difference between the observed affinities of agonists for receptors that can undergo interaction with G proteins and their affinities when this process is canceled has been used as a measure of efficacy. If receptor–G protein interaction is not allowed to produce a steady-state accumulation of ternary complex (agonist–receptor–G protein), then the observed affinity of the agonist for the receptor will reflect the equilibrium dissociation constant of only the agonist–receptor complex (denoted K_A where $K_A = 1/K_a$; Eq. 2). However, if G protein complexation is possible, then the observed affinity of the agonist will be controlled by the avidity of this secondary complexation. Schematically, the following interactions between receptors ([R]), G proteins ([G]) and ligands ([A]) are relevant:

$$
\begin{array}{ccc}
\text{G} & & \text{G} \\
+ & K_a & + \\
\text{A} + \text{R} & \rightleftharpoons & \text{AR} \\
K_g \updownarrow & & \updownarrow \gamma K_g \\
\text{A} + \text{RG} & \rightleftharpoons & \text{ARG} \\
& \gamma K_a &
\end{array}
\tag{3}
$$

where K_a and K_g are equilibrium association constants of the ligand–receptor and receptor–G protein complexes, respectively, and γ is a multiplicative factor denoting the effects of ligand binding on the subsequent interaction of the receptor and G protein. Thus, an agonist ligand with positive efficacy would have a value of γ greater than 1. A value of γ equal to 10 indicates that the production of ARG is tenfold more likely after the initial interaction of agonist with receptor, i,e. there is high degree of isomerization produced by the agonist (it has high efficacy). It can be seen that, from a thermodynamic point of view, the overall reaction from the mixing of [A] and [R] to final [ARG] depends on both K_A and the second reaction. Since the [AR] complex formed by initial interaction of [A] and [R] is removed from the initial equilibrium by complexation with [G], more [A] is formed to take its place. Thus, the presence of the second reaction drives the initial reaction forward beyond the original constraints of K_A.

The fraction of receptors in the ternary complex form ([ARG]) can be derived from the following equilibrium equations:

$$[R] = \frac{[ARG]}{[A]\gamma K_a [G] K_g} \tag{4}$$

$$[RG] = \frac{[ARG]}{[A]\gamma K_a} \tag{5}$$

$$[AR] = \frac{[ARG]}{\gamma [G] K_g} \tag{6}$$

$$[G] = \frac{[ARG]}{[A]\gamma [R] K_a K_g} \tag{7}$$

The fraction of receptor forming the ternary complex species ARG can be calculated for use in the guanosine triphosphate (GTP)-shift approach. Under these circumstances,

$$\rho = \frac{[ARG]}{[R] + [RG] + [AR] + [ARG]} \tag{8}$$

From Eq. 8 and the equilibrium equations:

$$\rho = \frac{\gamma[A]/K_A}{[A]/K_A(\gamma[G]/K_G+1)+1+[G]/K_G} \qquad (9)$$

where K_A and K_G are now equilibrium dissociation constants for the agonist–receptor and G protein–receptor complexes, respectively. From Eq. 9, the observed affinity of binding in the presence of an operative G protein, complexation is given by:

$$K_H = K_A \frac{([G]/K_G+1)}{(\gamma[G]/K_G+1)} \qquad (10)$$

It can be seen from Eq. 10 that, if the ligand promotes G protein complexation (i.e., $\gamma > 1$), then $K_H < K_A$ and the observed affinity of the agonist will be greater than the affinity of the ligand for the bare receptor. This will be referred to as the high-affinity state. In the absence of G protein complexation, the observed affinity will be equal to K_A (the low-affinity state). Therefore, the ratio of affinities for an agonist ligand in the absence and presence of G protein coupling is given by:

$$\text{RATIO} = \frac{K_A}{K_H} = \frac{(\gamma[G]/K_G+1)}{([G]/K_G+1)} \qquad (11)$$

The ratio described by Eq. 11 has been termed the "GTP shift" experimentally since, in practical terms, it is the ratio of the affinity of agonist ligand obtained in the presence of G protein complexation and to that obtained after cancellation of that complexation by facilitation of guanosine diphosphate (GDP)–GTP exchange by excess GTP (or stable GTP analogue, such as GTPγS). An example of this method is shown in Fig. 3. Thus, the magnitude of the ratio is defined as the GTP shift for a given agonist. The relative GTP shift for two agonists is given as:

$$\text{AFFINITY RATIO} = \frac{\text{Agonist A}}{\text{Agonist B}} = \frac{(\gamma_A[G]/K_G+1)}{(\gamma_B[G]/K_G+1)} \qquad (12)$$

which reduces to an approximation of γ_A/γ_B at appreciable ratios of $[G]/K_G$. Thus, the GTP shift can be used to estimate the relative efficacy of two agonist ligands.

The strength of the GTP shift method is that there is a theoretical and mechanistic rationale for the estimate of relative efficacy. In practical terms, however, there are systems where the kinetics of GDP–GTP exchange are not sufficiently rapid to cancel accumulation of the [ARG] complex. Under these circumstances, the cancellation of ternary complex formation is incomplete, and erroneous estimates of relative efficacy may result.

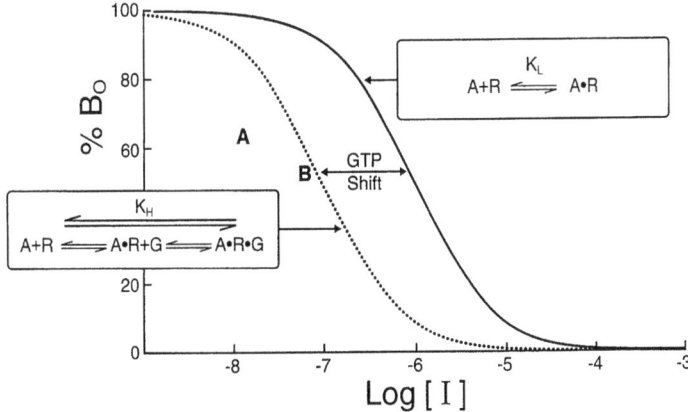

Fig. 3. Efficacy reflected as the magnitude of the guanosine triphosphate (GTP) shift. A displacement curve for radioligand displacement by an agonist in a G protein-rich environment reflects the affinity of agonist for the receptor and its ability to produce a ternary complex (efficacy). In the presence of GTP, this secondary receptor coupling is canceled, and the displacement curve reflects only agonist affinity. The difference between the two curves is a measure of efficacy

2. High-Affinity Selection Binding

An alternative to the GTP-shift approach for the estimation of efficacy in biochemical binding experiments takes advantage of the stoichiometry between receptors and G proteins. This approach works best in receptor systems where the stoichiometric relationship between receptors and G proteins is such that there is an overabundance of receptors. Under these circumstances, agonists will form a relatively small amount of ternary complex on binding, while antagonists will stochastically sample the complete receptor population. With appropriate radioligands, the receptor population can be "selected", and the ensuing difference in observed affinity can be used as an index of efficacy.

In receptor-overexpressed systems, agonists will only bind with high affinity to receptors that can couple to G proteins. If the agonist is a radiolabel, then the population of radioactive receptor species will be a "selected" population of ternary complexes. In contrast, an antagonist radioligand will bind to a random sample of receptors, with no regard for G protein coupling (CHEN et al. 1997; KENAKIN 1997a, 1997b).

When a non-radioactive agonist is used to displace an antagonist radioligand, it will induce G protein coupling (high-affinity binding) and then, at higher concentrations, compete for uncomplexed receptors. This would result in the well-known biphasic displacement curves seen with agonists but, since the G protein population is so much smaller than the receptor population, the high-affinity binding will be inconsequential (or result in a small biphasic component of the displacement curve). When a non-radioactive agonist is used to displace a radioactive agonist in a system where the only radioactive species are ternary complexes, then the production of ternary complex by the non-radioactive ligand will necessarily displace the radioactive agonist ternary complexes. This results in high-affinity displacement, in contrast to what is observed with a radioactive antagonist. Thus, there will be a difference in the observed potency of agonists in displacing radioactive agonists and antagonists in receptor-overexpressed systems, and this difference will be proportional to the "power" of the ligand to induce ternary complex production (i.e., efficacy).

Figure 4A shows a series of saturation curves for a radioactive agonist under conditions of increasing G protein concentration. Low levels of G protein (10%) produce a small high-affinity component of binding followed by a large low-affinity component. The distinction becomes less evident as the stoichiometry between receptor and G protein approaches unity and excessive amounts of G protein. Figure 4B panels I–VI show the disappearance of the difference in observed affinity of an agonist radioligand as G protein levels increase (KENAKIN 1998b). For example, this phenomenon has been noted in recombinant systems, such as cells transfected with receptor complementary DNA for human calcitonin receptor type 2 (hCTR2; Fig. 5). Theoretically, recombinant systems are ideal for the utilization of high-affinity selection if receptor expression levels are high or the G protein coupling is inefficient (i.e., baculovirus expression in *Ti ni* cells).

High-affinity selection may offer an additional method to detect and measure ligand efficacy in binding studies. Two disadvantages of this technique are the need for both agonist and antagonist radioligands and the dependence of the absolute differential in affinity on the relative quantity of G protein (usually an unknown). However, null techniques can be used effectively in the appropriate system to determine the *relative* efficacy of agonists.

II. Function

As a preface to the discussion of functional methods to estimate the relative efficacies of agonists, it is useful to describe agonism in terms of receptor occupancy theory. The ability of any given agonist to produce response depends upon its concentration in the receptor compartment (the fractional receptor occupancy, as defined by mass action and K_A, the equilibrium dissociation constant of the agonist–receptor complex), its efficacy (a dimensionless proportionality constant) and how well the organs having the receptor can process receptor signaling. This last ability depends upon the receptor density and the efficiency of coupling of the cellular stimulus–response mechanisms. A mathematical description of this stimulus-processing ability is the location parameter of a general logistic function linking receptor occupancy and tissue response (KENAKIN and BEEK 1980). This function can accurately model the stimulus–response characteristics of any tissue for any receptor. Thus, the response to an agonist is given by:

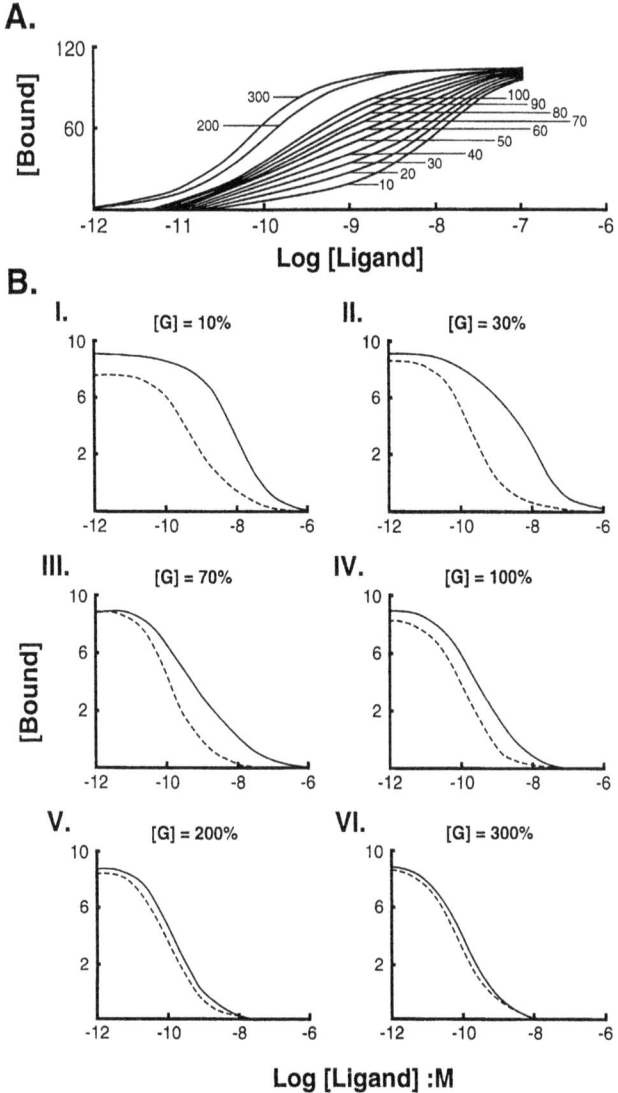

Fig. 4A,B. High-affinity selection. **A** Saturation curves for a radioactive agonist in the presence of increasing amounts of G protein. It can be seen that, when G protein is limiting (i.e., 10%), a small ridge of high-affinity binding is observed, followed by a much larger secondary phase of binding to the bare receptor. As the G protein content of the system is increased, this ridge of high-affinity binding increases until, in a G protein-rich environment (i.e., a natural system), only the high-affinity state is seen. **B** Displacement of a radioactive agonist (*dotted line*) and antagonist (*solid line*) by an agonist in a G protein-deprived (10%) system (panel I). Note the disparity in the observed potencies. *Panels II–VI*: same curves as for *panel I*, with increasing amounts of G protein. The curves converge as the G protein content increases

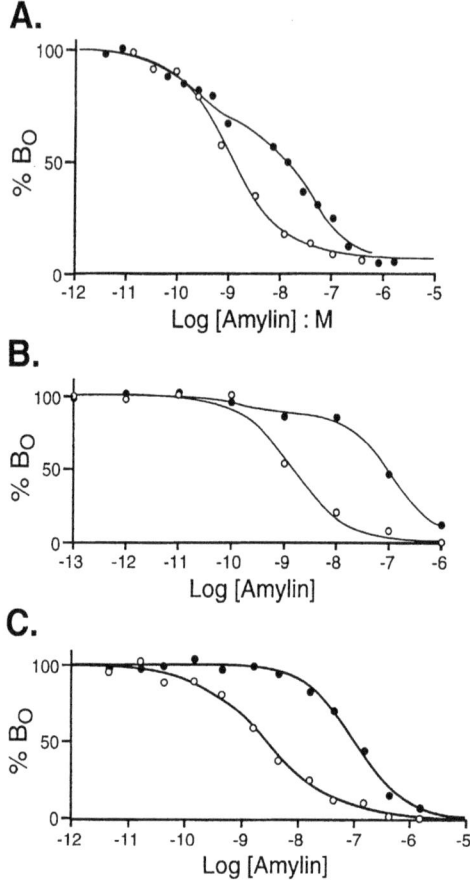

Fig. 5. The displacement of the amylin receptor agonist radioligand [^{125}I]-rat amylin (*open circles*) and peptide amylin receptor antagonist [^{125}I]-AC512 (*filled circles*) by non-radioactive amylin in (**A**) human MCF-7 cells, (**B**) transfected COS cells and (**C**) transfected HEK-293 cells. Note (1) the difference in the displacing potency of non-radioactive amylin (the relationships between each radioligand and its respective K_d for binding were very similar) for displacement of agonist versus antagonist radioligand and (2) the monophasic nature of the displacement curves in the recombinant systems. The total number of bound radioactive counts was adjusted to be equal for the displacement experiment (KENAKIN 1997b)

$$\text{RESPONSE} = \frac{\rho\varepsilon[R_t]}{\rho\varepsilon[R_t]+\beta} \tag{13}$$

where ε is the intrinsic efficacy of the agonist, $[R_t]$ is the receptor concentration, ρ is the fractional receptor occupancy and β is a fitting parameter denoting the efficiency of stimulus–response coupling. This parameter will be referred to as the receptor coupling constant (β) and is a property of the tissue. Redefining ρ by the Langmuir adsorption isotherm for receptor occupancy shows response to be the following function of agonist concentration:

$$\text{RESPONSE} = \frac{[A]\varepsilon[R_t]}{[A](\varepsilon[R_t]+\beta)+\beta K_A} \tag{14}$$

where K_A is the equilibrium dissociation constant of the agonist–receptor complex (the reciprocal of affinity) and $[A]$ is the agonist concentration. Equation 14 is a general equation describing agonism in terms of receptor occupancy theory.

1. The Method of Furchgott

In functional studies, the most common method used to measure the relative efficacies of agonists was derived by FURCHGOTT (1966). Comparing equivalent responses to another agonist B yields the equality:

$$\frac{\rho_A \varepsilon_A [R_t]}{\rho_A \varepsilon_A [R_t]+\beta} = \frac{\rho_B \varepsilon_A [R_t]}{\rho_B \varepsilon_B [R_t]+\beta} \tag{15}$$

From this relationship, it can be shown that the ratio of the relative receptor occupancies of the two agonists at an equivalent response yields:

$$\text{RELATIVE EFFICACY} = \frac{\varepsilon_A}{\varepsilon_B} = \frac{\rho_B}{\rho_A} \tag{16}$$

This method compares the ability of a compound to produce response with its concomitant ability to occupy receptors and yields the "power" of an agonist to produce response as a unit of response per receptor. For example, an agonist that produces 50% response while occupying 20% of the receptors is less efficacious than one that produces 50% response while occupying 2% of the receptors. In this method, the relative efficacy of agonists is obtained by plotting the response as a logarithmic function of the receptor occupancy. When this is done for two agonists in the same system (null procedures cancel tissue effects), a measure of the relative efficacy of the two agonists results (Fig. 6).

The method breaks down at low efficacies, because the curves become non-parallel, and all curves approach a limit of 100% occupancy. Even for agonists of high efficacy, there is a serious practical problem with this and other functional methods, namely the requirement for an unbiased estimate of the affinity of the agonist. This is because agonists produce isomerization of the receptor, a process that causes two-stage binding to G proteins. Under these circumstances, the observed affinity results from the two-stage binding process, and how well this process occurs depends on how well the ligand promotes G protein activation (i.e., the magnitude of the efficacy). Thus, the observed affinity is a system property, not a receptor property dependent on the efficacy and availability of cofactors, such as G proteins. BLACK and SHANKLEY (1990) have termed this effect "receptor distribution" to denote the ability of agonists to redistribute the receptor population into various species bound and not bound to G proteins. Thus, for agonists of high efficacy, there is a possibility

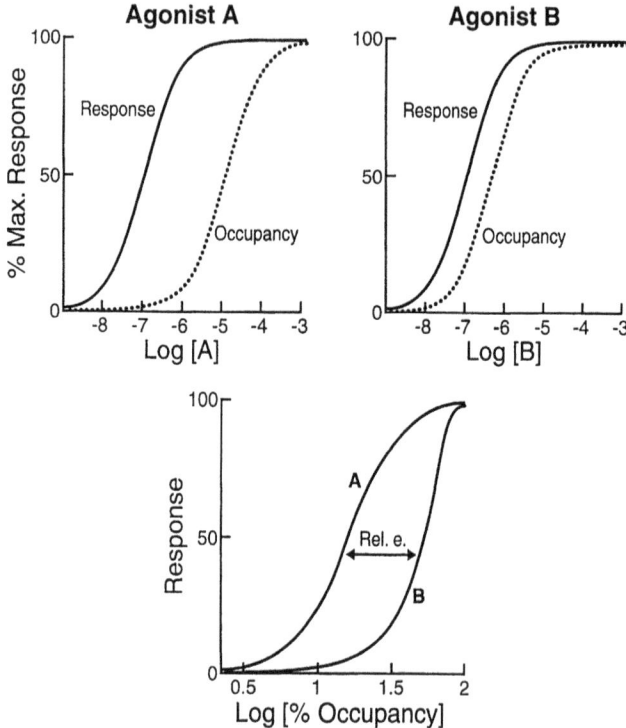

Fig. 6. The Furchgott method for the estimation of relative efficacy. Given curves for the response (*solid line*) and receptor occupancy of an agonist (*dotted line*), the efficacy is estimated as the multiplicative factor between them. For two agonists, the relative efficacy can be obtained by expressing the response curve as a function of receptor occupancy. The relative location parameters of these occupancy–response curves yields a measure of efficacy

of a higher observed affinity and a subsequent underestimation of efficacy in the Furchgott method.

2. The Method of Stephenson

Another, less widely used method in functional studies is the method of STEPHENSON (1956). This approach simply compares the dose–response curves of two agonists of equiactive concentrations in a double-reciprocal plot to yield a factor denoting the relative efficacy of the agonists (Fig. 7). However, if the agonists are of comparable efficacy, the intercept of this plot approaches zero, leading to enormous error in the efficacy ratio. This inordinate sensitivity to the relative efficacy of the agonists makes the Stephenson method limited.

Method of Stephenson

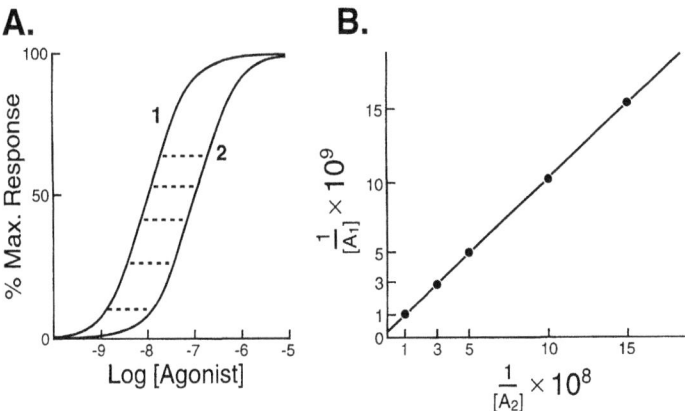

Fig. 7A,B. The Stephenson method for the estimation of relative efficacy. The reciprocals of equiactive doses of two agonists are compared in a double-reciprocal plot. The intercept of this plot yields the value $(\varepsilon_1/\varepsilon_2 K_{A1})(1-(\varepsilon_2/\varepsilon_1))$, where ε_1 and ε_2 refer to the intrinsic efficacies of agonists 1 and 2, and K_{A1} refers to the equilibrium dissociation constant of the agonist–receptor complex for agonist 1. Even though K_{A1} may not be known, the sign of the intercept can provide an estimate of the rank order of efficacy of the two agonists

3. Comparison of Relative Maximal Responses

Perhaps the most robust method to estimate the relative efficacy of agonists is to test them under null conditions, in receptor systems so compromised that the agonists cannot produce the system maximal response. This is because, when the maximal response to that agonist is submaximal (with respect to the tissue maximum), its magnitude is dependent upon efficacy and is not affected by differences in affinity (provided the concentration of agonist producing maximal response is saturating). For example, Fig. 8A shows dose–response curves for a number of α-adrenergic receptor agonists in rat aorta (data from RUFFOLO et al. 1979). It can be seen from this figure that seven of the agonists produce sub-maximal responses. It can be shown theoretically that the maximal response to an agonist under these conditions is dependent on efficacy and is completely independent of affinity (*vide infra*). Therefore, this system can definitively rank the relative order of the efficacies of the agonists as: phenylephrine > oxymetazoline > naphazoline > xylometazoline > clonidine > tenaphtoxaline > tolazoline > tetrahydrozoline. Measurement of the efficacies of these agonists by the method of Furchgott indicates that the maximal responses correlate well with efficacy (Fig. 8B), but it should be stressed that there are many conditions under which such an occurrence may not yield a good correlation (i.e., receptor distribution may cause the Furchgott method to be in error). The actual correspondence of the efficacy to the

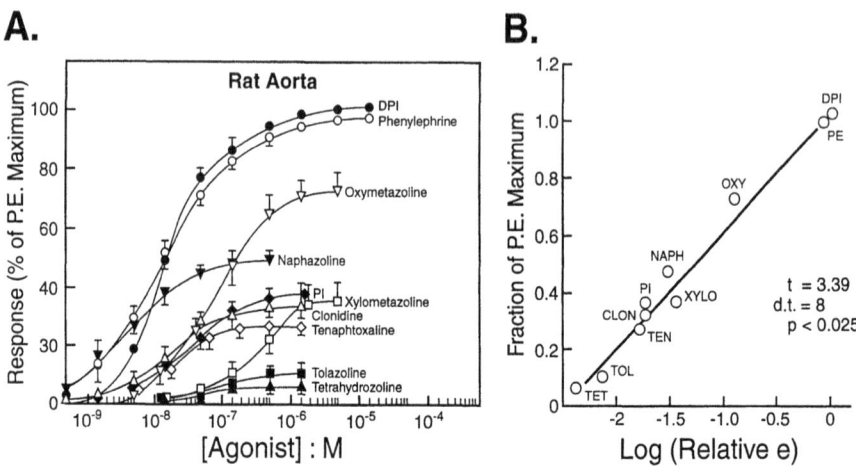

Fig. 8. A Dose–response curves for some α-adrenergic receptor agonists in rat aorta. **B** The relative maximal responses to these agonists correlated with their relative efficacies, as measured by the method of Furchgott (KENAKIN 1997a)

maximal responses may differ somewhat and depends upon the coupling set point of the system (*vide infra*). However, the rank order is definitive and can be used to classify agonists.

Historically, the usual method of compromising the efficiency of a receptor system is to chemically reduce the receptor density to a point where submaximal system stimulation is produced by activation of 100% of the available receptor pool. Traditionally, this has been accomplished by chemical alkylation of receptors by affinity labels that irreversibly bind to the receptor protein and prevent agonist activation (i.e., with β-haloalkylamines; NICKERSON and GUMP 1949). The effect of altering receptor density on agonist response can be observed by expressing response as a fraction of the total G protein. The total G protein is given by the conservation equation:

$$[G_{tot}] = [G] + [RG] + [ARG] \tag{17}$$

It is assumed that observed physiological response results from activated receptor and G protein interaction, i.e., the quantities [ARG] + [RG]. This is given by the ratio:

$$\rho = \frac{[RG] + [ARG]}{[G] + [RG] + [ARG]} \tag{18}$$

which can be rewritten as:

$$\rho = \frac{\gamma[A]/K_A + 1}{[A]/K_A (K_G/[R] + \gamma) + 1 + K_G/[R]} \tag{19}$$

Under these circumstances, the relative response of two agonists (denoted ξ) can be shown to be:

$$\xi = \frac{(\gamma_1[A]/K_A + 1)(\gamma_2[A]/K_A + K_G/[R] + 1)}{(\gamma_2[A]/K_A + 1)(\gamma_1[A]/K_A + K_G/[R] + 1)} \quad (20)$$

As the concentration of receptors is progressively decreased, the following equation for the ratio of maximal responses results:

$$\xi = \frac{(\gamma_1[A]/K_A + 1)}{(\gamma_2[A]/K_A + 1)} \quad (21)$$

An inspection of Eq. 21 shows that, as $[A]/K_A$ approaches infinity, the ratio of maximal responses equals γ_1/γ_2, the ratio of the abilities of the agonists to induce the active state of the receptor (efficacy).

Another approach, when alkylating agents for the receptor are not available, is to compromise the system such that the activated receptor cannot produce the maximal response. This can be done either by reducing the amount of G protein (i.e., pertussis toxin treatment for G_i) or by increasing the equilibrium dissociation constant of the ternary complex (increasing the dissociation constant K_G). For this calculation, the response is written as a function of the occupancy of the total receptor population. The conservation equation for receptors is:

$$[R_{tot}] = [R] + [RG] + [AR] + [ARG] \quad (22)$$

The response is assumed to emanate from the species [RG] and [ARG] and is given by the ratio:

$$\rho = \frac{[RG] + [ARG]}{[R] + [RG] + [AR] + [ARG]} \quad (23)$$

which can be rewritten:

$$\rho = \frac{\gamma[A]/K_A + 1}{[A]/K_A (K_G/[G] + \gamma) + 1 + K_G/[G]} \quad (24)$$

The maximal response to an agonist can be estimated by letting $[A]/K_A$ approach infinity. Under these circumstances, R_{max} is given by:

$$R_{max} = \frac{\gamma}{\gamma + K_G/[G]} \quad (25)$$

The relative maximal response to two agonists [A1] and [A2] (denoted ξ) is then given by:

$$\xi = \frac{\gamma_1(\gamma_2 + K_G/[G])}{\gamma_2(\gamma_1 + K_G/[G])} \quad (26)$$

It can be seen from Eq. 26 that two conditions could make the relative maximal response depend upon the relative efficacy of the two agonists (i.e., γ_1/γ_2). This will occur when either the coupling of receptor to the G protein is severely compromised (K_G becomes very large) or the relative quantity of the

G protein is diminished ([G]↔0). Under either of these circumstances, ξ reduces to γ_1/γ_2.

Figure 9 shows dose–response curves for two agonists, one with efficacy $\gamma = 100$ and one with $\gamma = 30$ in a system where the receptor coupling is progressively diminished (i.e., increasing values of $K_G/[G]$). Pairing of the curves for the two agonists at different levels of $K_G/[G]$ shows that, as the receptor coupling becomes increasingly compromised, the relative maximal responses of the two agonists approach the asymptote of the relative efficacy for these agonists ($\gamma_1/\gamma_2 = 100/30 = 3.3$). This raises the question of the error involved in this method as the maximal responses become depressed; i.e., how quickly does ξ approach γ_1/γ_2 as receptor coupling is diminished? This can be done by defining a metric, ψ, as $\gamma_2[G]/K_G$, which allows the relative coupling efficiency of the receptor system to be linked to the efficacy of one of the agonists. Under these circumstances, the relative efficacy of the two agonists (defined as γ_1/γ_2) is related to the observed maximal responses of the two agonists (ξ) by the following equation (derived from Eq. 26):

$$\xi = \frac{\gamma_1/\gamma_2(\psi+1)}{(\gamma_1/\gamma_2)\psi+1} \tag{27}$$

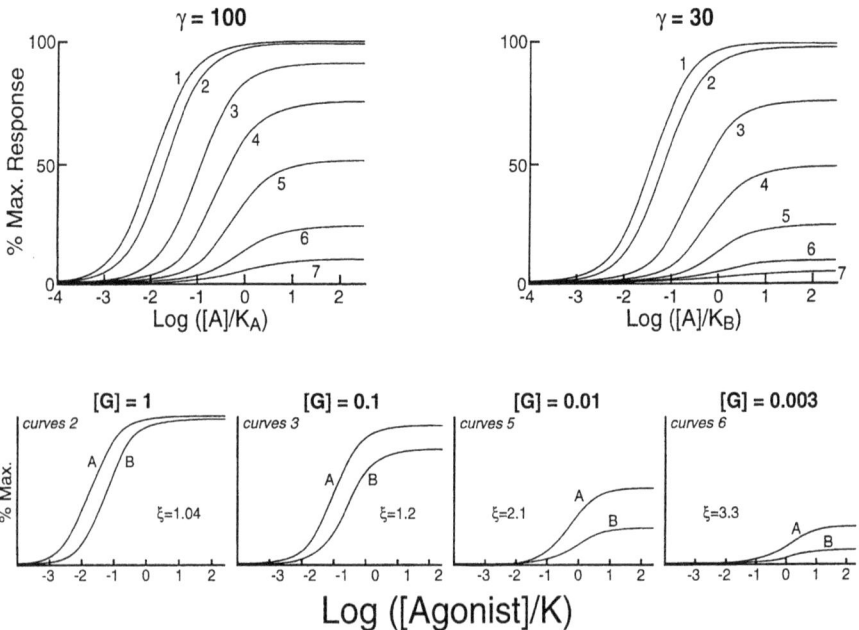

Fig. 9. The effect of diminishing the receptor coupling (and, therefore, response-producing ability) of a receptor system on the responses to two agonists of differing efficacy. Pairing of the curves examines the effects of a given change in receptor coupling on the relative responses to the two agonists. The true relative efficacy of the agonists is 100/30 = 3.3

Table 1. Strengths and weaknesses of methods of estimating agonist efficacy

Method	Advantages	Disadvantages
Furchgott	Simple	Requires unbiased estimate of affinity; inaccurate at low efficacies
Stephenson	Simple	Errors at comparable efficacies
GTP-shift	Theoretical rationale	Some GTP-insensitive systems
High-affinity selection	Theoretical rationale; robust	Requires radioactive agonist and antagonist; dependent on [R]/[G]
Relative maximal response	Sensitive to low efficacies; theoretical rationale	Inability to compromise maximal response in some systems; errors at widely divergent efficacies

GTP, guanosine triphosphate.

Values of relative maximal response that yield accurate estimates of the relative efficacy indicate that the greater the difference in efficacy of the two agonists, the more the receptor coupling must be reduced to produce an accurate estimate.

The various strengths and weaknesses of the operational methods available to estimate the relative efficacies of agonists are given in Table 1. It can be seen from this table that some methods are more suitable for lower-efficacy agonists rather than higher-efficacy agonists.

E. Limitations of Agonist Potency Ratios

The most common method of comparing agonists in pharmacologic systems is with potency ratios. It can be shown that the potency ratio of two agonists, if measured in the same receptor system under the same conditions, yields a value dependent only upon the drug-specific factors of affinity and efficacy. This null procedure cancels the translating tissue effects of receptor number and stimulus–response mechanisms. Thus, it can be shown that the relative equiactive potency ratio of two agonists A_1 and A_2 is given by:

$$\frac{[A_2]}{[A_1]} = \frac{[A_2](\varepsilon_1 - \varepsilon_2)}{K_{A2}\varepsilon_2} + \frac{K_{A2}\varepsilon_1}{K_{A1}\varepsilon_2} \tag{28}$$

where the relative equilibrium dissociation constants of the agonist–receptor complexes are denoted K_A and the efficacy is denoted ε. If the agonists are of comparable efficacy, the potency ratio can be estimated with a simple ratio of affinity and efficacy:

$$\frac{[A_2]}{[A_1]} = \frac{K_{A2}\varepsilon_1}{K_{A1}\varepsilon_2} \tag{29}$$

The use of potency ratios to characterize agonist activity has been extensively used in pharmacology for numerous years. While the ratio can be immutable under a number of experimental circumstances, it should be noted that it is a ratio of two completely independent properties. Thus, an agonist could have a given potency because of high efficacy and relatively low affinity, and another could be equiactive because of lower efficacy but higher affinity. This constitutes a major shortcoming in the use of isolated potency ratios.

Clearly, if the receptor-coupling efficiency of the test system is sufficient to allow both agonists to produce maximal response (or near-maximal response), then the potency ratios will transcend the specific coupling efficiencies of particular systems, and the estimate will not depend upon the type of test system used to measure agonism. Problems arise if the receptor test system is of low sensitivity. It should be stressed that low sensitivity is a relative term with respect to the strength of the agonists being tested; i.e., a natural hormone may produce a powerful response, but the system will still be inadequate to yield correct potency ratio values if the test agonists are of such low efficacy that they do not produce maximal response. Figure 10 shows the potency ratios of two agonists A and B differing in intrinsic efficacy by a factor of 100. It can be seen that the potency ratio is relatively stable in

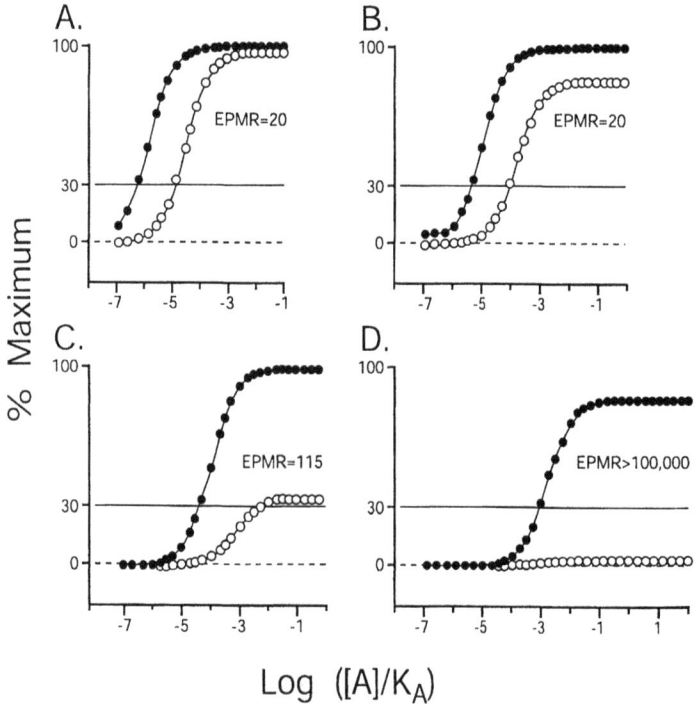

Fig. 10A–D. Estimates of potency ratios (defined as an equipotent molar ratio and denoted as EPMR) in systems of various sensitivities. System **A** is well coupled ($\beta = 0.01$), as is system **B** ($\beta = 0.1$). However, as receptor coupling becomes less efficient, the EPMR becomes ambiguous (system **C**) and meaningless (system **D**)

systems I and II but that completely errant values are obtained in systems C and D.

Another shortcoming of single relative-potency measurements relates to how the "snapshot" of relative potency in one test system can lead to dissimulation in other systems with differing signal processing abilities. Two agonists could be equiactive in a given test system for different reasons, i.e., one agonist could derive its potency from high efficacy and low affinity and another from high affinity but low efficacy. This could be likened to a snapshot of two horses jumping over a fence (Fig. 11A). More informative would be to see the complete film of the horses as they jump over the fence; this could be likened to observing the agonism in a range of tissues of differing receptor-coupling abilities (Fig. 11B). In essence, this is what occurs in agonist drug discovery programs. Lead compounds are chosen on the basis of a single relative-potency

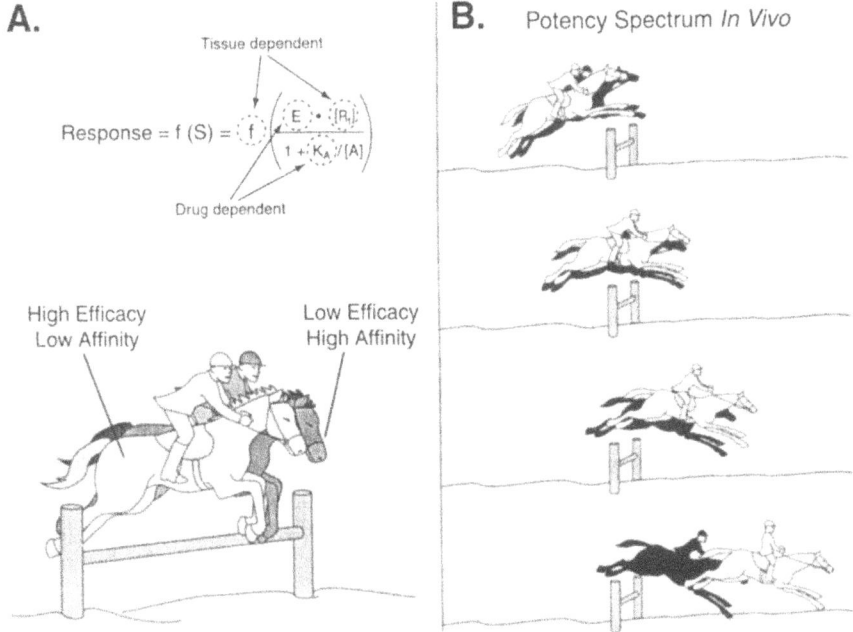

Fig. 11A,B. Potency ratios as snapshots of a continual process (depicted as horses jumping over a fence). **A** Since the observed potency of full agonists is an amalgam of a ratio of intrinsic efficacy and affinity, two agonists could be equipotent, one by virtue of high affinity and low intrinsic efficacy and the other by low affinity and high intrinsic efficacy. A single estimate of potency in one system (i.e., one receptor density and receptor coupling efficiency) cannot predict what will be observed under different tissue conditions. **B** The complete film of the horses jumping over the fence (i.e., the relative positions of the horses depict the relative potencies of the agonists under different tissue conditions) is observed when the agonists are tested in vivo, where they encounter numerous tissues under different conditions

estimate (the snapshot) and are then tested in vivo, where they encounter a wide range of tissues (the complete film). For these reasons, it is desirable to have a measure of the relative potencies of agonists that does not depend upon the coupling efficiency of the receptor system. There are examples of how such snapshots can be misleading and can create potential hazards of relying on estimates of relative agonist potency. Specifically, it can be shown that differences in receptor coupling can greatly affect the magnitude of agonism observed. This underscores the limitation of single agonist potency ratios.

The location parameter of a dose-response curve depends on the drug factors affinity and efficacy and the tissue factors receptor density and efficiency of receptor coupling. When two agonists are compared in the same test system, this null procedure cancels the tissue effects and allows estimation of only drug effects. However, if both agonists produce system-maximal responses, then it is not possible to distinguish the relative contributions affinity and efficacy make to agonist potency. This is important, since the contributions of these drug factors differ considerably when receptor coupling is less efficient. Figure 12A shows the relative activity of the α-adrenoceptor agonists norepinephrine and oxymetazoline in an efficiently receptor coupled tissue (rat anococcygeus muscle). It can be seen that oxymetazoline is twofold more potent than norepinephrine. In contrast, a less well-coupled tissue (rat vas deferens) bearing the same receptors shows norepinephrine to be the more active agonist. This is because oxymetazoline is a high-affinity but low-efficacy agonist, while norepinephrine is a low-affinity but high-efficacy agonist (KENAKIN 1984). These data show that the relative agonist profiles for oxymetazoline and norepinephrine differ considerably in two different test systems.

Compromise of system signal processing ability also can demonstrate differences. Fig. 12B shows the effects of progressive diminution of α-adrenoceptor number by treatment with phenoxybenzamine in the rat anococcygeus muscle. It can be seen that the relative profile of the agonists changes (toward that seen in vas deferentia) with decreases in receptor number. The same type of effect can be seen with the muscarinic agonists oxotremorine (low efficacy/high affinity) and carbachol (high efficacy/low affinity) in guinea pig ileum (KENAKIN 1997).

F. Why Measure the Relative Efficacies of Agonists?

The estimation of the relative efficacy of agonists, while not a trivial procedure, can be a useful method of removing system-dependence from relative estimates of agonism. For example, Table 2 shows a wide range of potencies for the β-adrenoceptor agonist prenalterol (EC_{50} values) and maximal responses (intrinsic activities); these are system-dependent measures of receptor activity. In contrast, the molecular parameters of affinity (equilibrium dissociation constant of the agonist-receptor complex) and relative efficacy for prenalterol

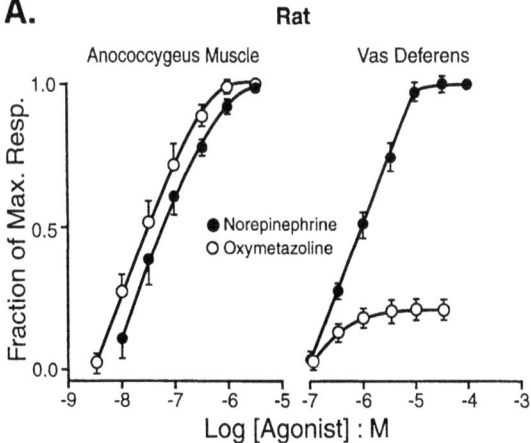

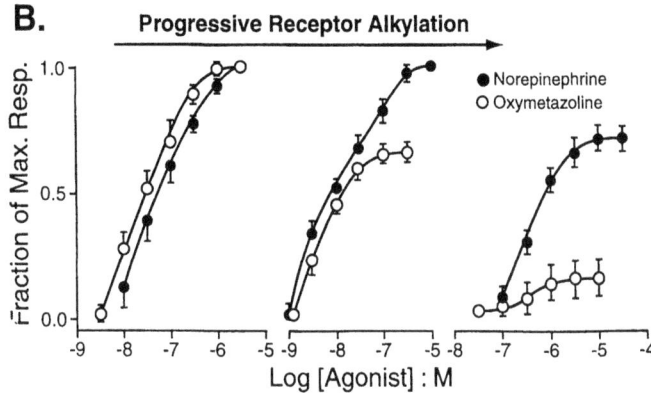

Fig. 12A,B. Differences in potency ratios with different receptor densities and/or coupling. **A** The relative potency of the α-adrenergic agonists norepinephrine and oxymetazoline in the rat anococcygeus muscle and vas deferens. Whereas oxymetazoline is the more powerful agonist in the anococcygeus muscle, it is less active in vasa deferentia. Data from KENAKIN (1984). **B** The effect of α-adrenergic receptor diminution (by alkylation with phenoxybenzamine) on the relative agonism produced by norepinephrine and oxymetazoline in rat anococcygeus muscle. The response to the lower-efficacy agonist (oxymetazoline) is more sensitive to decreases in receptor number (KENAKIN 1997)

(versus isoproterenol) vary little from tissue to tissue. Thus, the estimation of relative efficacy allows the prediction of agonism across receptor systems and, thus, is a good parameter for therapeutically directed medicinal chemistry.

Another reason for estimating relative efficacy is to determine whether the observed potency of a given agonist results from high affinity (affinity-driven agonist) or high efficacy (efficacy-driven agonist). As seen in Fig. 12B, efficacy-driven agonists are less sensitive to compromise of stimulus–response abilities of tissues and, therefore, desensitizing effects may be overcome with increased dosing (unlike affinity-driven, lower-efficacy agonists).

Table 2. System-dependent and -independent properties of prenalterol as a β-adrenergic receptor agonist. Data from KENAKIN and BEEK (1980)

Tissue	Observed potency[a] (nm)	Maximal agonist response[b] (%)	Affinity[c] (nm)	Relative intrinsic efficacy[d]
Guinea pig trachea	0.2	79	31.6	0.005
Cat left atrium	0.37	64	32	0.005
Rat left atrium	0.6	65	32	0.004
Cat papillary	1.4	40	46	0.0045
Guinea pig left atrium	3.2	24	32	0.005
Guinea pig EDL	NA[e]	0	39.8	NA[e]

EDL, extensor digitorum longus muscle; NA, not applicable.
[a] Molar concentration producing 50% maximal response.
[b] Fraction of the maximal tissue response, as measured by the response to the full β-adrenoceptor agonist isoproterenol.
[c] Molar concentration of prenalterol that occupies 50% of the receptor population at equilibrium (a chemical term).
[d] The relative power of prenalterol to produce response, relative to that of the full agonist isoproterenol. This is a measurement of agonism that transcends the receptor preparation.
[e] Since no agonist response was observed in the EDL, no potency as an agonist or no estimate of relative efficacy could be obtained.

G. A Simple Algorithm for the Prediction of Agonist Side Effects Using Efficacy and Affinity Estimates

The algorithm described below is designed to extend the information obtained in a single assay depicting relative potencies of agonists into a general pattern of behavior applicable to all agonist situations in vivo. The calculations are based on a term that characterizes the efficiency of signal transduction in a tissue; this term will be termed the receptor coupling constant. As discussed previously, various tissues process and amplify receptor stimuli by application of a succession of saturable biochemical cytosolic reactions. Each can be approximated by a hyperbolic function, and the sum total can also be generalized by a hyperbolic function (KENAKIN and BEEK 1980). Thus, a hyperbola with a fitting parameter β can be used to mathematically model stimulus–response functions (Eq. 13). It is known that different tissues have different amplification properties, i.e., some are more efficient at processing receptor stimuli than others. For example, Fig. 13 shows the receptor occupancy–response curves for isoproterenol in a range of β-adrenoceptor-containing tissues. It can be seen that, while nearly 90% response is achieved in guinea pig trachea with an isoproterenol receptor occupancy of 2%, this same occupancy produces only a 14% response in guinea pig extensor digitorum longus muscle (Fig. 13). A unique fitting parameter β can be assigned for each tissue/receptor combination. This value will be referred to as the *receptor cou-*

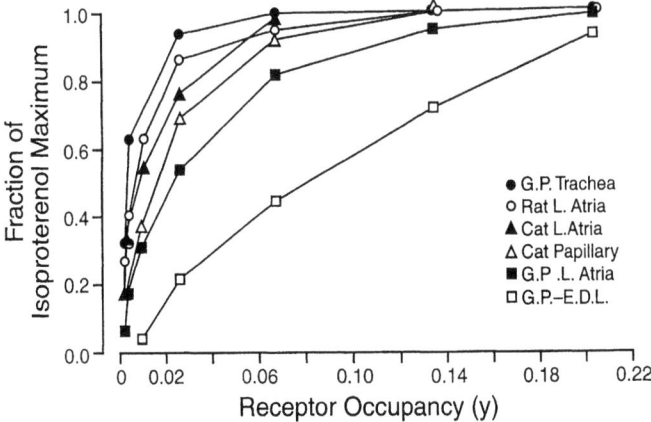

Fig. 13. Receptor occupancy–response curves for different isolated tissues: isoproterenol responses mediated by β-adrenoceptors. *Ordinates*: fraction of the maximal response to isoproterenol. *Abscissae*: fractional receptor occupancy by the concentrations of isoproterenol producing the responses. Data from KENAKIN and BEEK (1980) for guinea pig trachea (*filled circles*), rat left atria (*open circles*), cat left atria (*filled triangles*), cat papillary muscle (*open triangles*), guinea pig left atria (*filled squares*) and guinea pig extensor digitorum longus muscle (*open squares*) (KENAKIN 1997)

pling constant in the following calculations. The efficiency of receptor coupling is inversely proportional to the magnitude of the receptor coupling constant β. Figure 14 shows the response to a given agonist in a range of tissues possessing different efficiencies of stimulus–response coupling. As can be seen from this figure, the agonist produces a maximal response at low concentrations in highly coupled tissues (low value of β) and partial agonism in less efficiently coupled tissues (high β).

It can be seen from the previous discussion that the observed agonist response for a high-efficacy agonist will be more resistant to differences in receptor density and/or coupling efficiency in various organs than the response for a low-efficacy agonist. Assuming that a wider range of receptor coupling efficiencies implies a wider range of organs in which the agonist can produce response, it would be predicted that the high-efficacy agonist will produce agonism in a much larger number of target organs than the low-efficacy agonist. This concept can be used to predict the relative propensity of two agonists to produce responses in vivo.

One way to quantify the respective subsets of organs that will generate response to an agonist is to define the limiting value of the coupling constant β, where agonism disappears. Thus, one agonist may be powerful and have a limiting value of β of 1000 (i.e., a high-efficacy agonist will produce agonism in even poorly coupled organs; response will be seen in all organs with $\beta <$ 1000) while a weaker agonist may have a limiting value of β of 10 (response

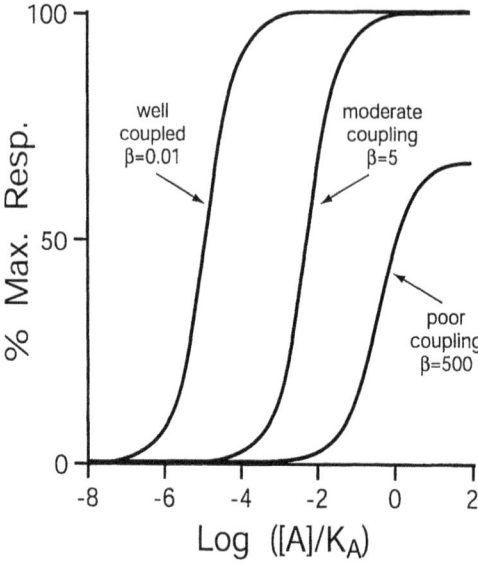

Fig. 14. The effects of differing values of the receptor coupling constant (Eq. 13) on the location and maximal asymptote of dose–response curves for a given agonist

will be seen in all organs with $\beta < 10$). This would be a smaller subset of organs, and this would be quantified by the ratio of limiting values of β.

I. Therapeutic Versus Secondary Agonism: Side Effect Versus Coupling Constant Profiles

Using Eq. 14 for an arbitrary level of coupling efficiency (for this simulation, $\beta = 0.25$) and measured values for the affinity and intrinsic efficacy of an agonist for the primary receptor (the receptor mediating the therapeutic effect), the dose–response curve for therapeutic agonism can be calculated. An example is shown in Fig. 15A (*double line*). With knowledge of the intrinsic efficacy and affinity of the same agonist at a secondary receptor (mediating the unwanted side effects), a range of dose–response curves depicting secondary agonism can be calculated for a range of coupling efficiencies of the secondary receptor; these are shown as *dotted lines* in Fig. 15A. These curves represent the dose–response curves possible for secondary effects in vivo. The magnitudes of the secondary effects at various values of β can be related to an ED_{50} concentration of agonist for the therapeutic effect. These are shown by the intersection of a vertical line at the concentration producing 50% therapeutic effect and the various dose–response curves for the secondary effect (Fig. 15A). Thus, the magnitude of the secondary effect can be related to a given efficiency of coupling of the secondary receptor; this defines an inverse sigmoidal function of secondary agonism versus the logarithm of the efficiency of coupling [$\log(\beta)$; Fig. 15B]. Not surprisingly, the lower the value of β (i.e.,

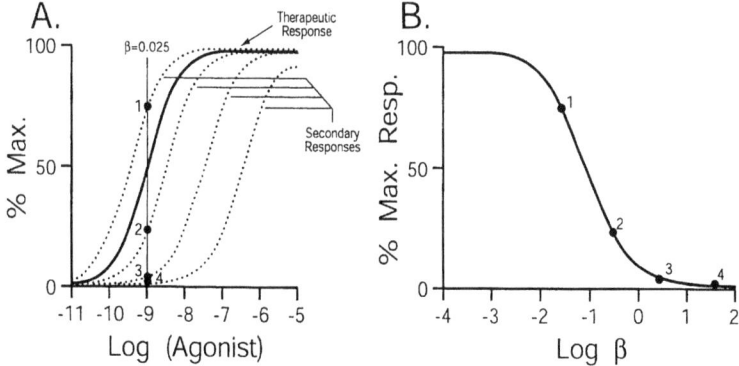

Fig. 15. A The relationship between secondary responses at various receptor coupling efficiencies for the secondary response (*dotted lines*) and the therapeutic response shown as a double-line dose–response curve (calculated for an arbitrary coupling constant for the receptors mediating the therapeutic response). At the ED_{50} concentration for therapeutic effect, the magnitudes of the secondary responses for different values of β for the secondary response are denoted by the numbers 1 to 4. **B** These secondary responses can be displayed as a function of the various coupling constants for the secondary response. This defines a side effect versus coupling constant profile for the agonist

the more efficiently coupled the secondary receptor), the greater the magnitude of the secondary effect. It is useful, at this point, to define a characteristic parameter of the agonist and the in vivo system, namely the β_{50}. This is the value of the receptor coupling constant where the secondary response to a concentration of agonist producing 50% therapeutic effect is also 50% of the maximum of the secondary response. The curve depicting this behavior will be referred to as the side effect versus coupling constant profile (Fig. 15B).

It can be seen that the magnitude of the β_{50} is dependent on the coupling efficiencies of the receptors mediating the therapeutic effect. This latter parameter is not known in vivo. However, at any given coupling efficiency for the therapeutic effect, the relative β_{50} values for different agonists can give a measure of the relative propensity of the agonists to produce side effects. Since the β value for the therapeutic effect will be constant for all of the agonists tested in the same system, the relative β_{50} values for the secondary responses will be related to each other and will reflect the *relative* tendencies of the agonists to produce side effects in the same system. Also, it will be seen that this algorithm can be used to calculate the relative β_{50} values for a collection of agonists and that it does not depend on the coupling efficiency of the therapeutic response.

An example of how the β_{50} can predict the relative ability to produce agonism at the secondary receptor (side effects) is shown in Fig. 16. This graph shows the side effect versus coupling constant profiles for two agonists, denoted A and B. It can be seen from this figure that the selectivity of agonist A is greater than that of agonist B. The relative β_{50} value for this example is

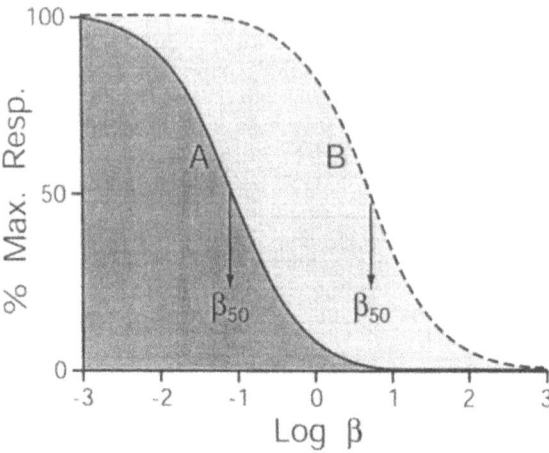

Fig. 16. Side effect versus coupling constant profiles for two hypothetical agonists. The midpoint of this curve is the β_{50}, defined as the receptor coupling constant at which the therapeutic concentration of the agonist (producing 50% therapeutic effect) also produces 50% maximal side effect. The *shading* refers to the spectrum of tissues (i.e., spectrum of coupling efficiencies) where the side effect to the particular agonist would be expected. The *darker shading* (smaller area) is for agonist A and *lighter* is for agonist B. It can be seen that a smaller subset of organs would be expected to respond to agonist A; thus, a higher degree of selectivity with this agonist would be expected

60, and this translates to a situation in which the secondary responses to agonist B will be observed in a greater number of tissues (a wider range of β values) than secondary responses to agonist A.

The relevance of the ratio of β_{50} values to the choice of agonists for use in vivo is the relative nature of the measurement. The coupling constant at which the side effects diminish to 50% maximal value is calculated from experimentally derived measures of efficacy and affinity and depends on the chemical features of the agonists. While any particular β_{50} value depends on the specific coupling constant of the therapeutic receptor system (which is not known), in a clinical setting, the agonist will be given until a therapeutic effect is observed, and this will automatically set the operative coupling constant for the therapeutic system. This level will be constant for all agonists tested in vivo; thus, the relative β_{50} values generalize to all systems (*vide infra*).

II. Algorithm for Calculation of Relative β_{50}

Prerequisites for calculation of relative β_{50} values are the affinities (as equilibrium dissociation constants obtained in binding studies) and relative efficacies for the therapeutic receptor and secondary receptor. The relative efficacy can be estimated by the method of Furchgott or can be approximated by the relative ratio of potency in binding and response in a functional assay (previous section). For this calculation, the subscript T refers to the therapeutic receptor, and subscript S refers to the secondary receptor.

From Eq. 14, the location parameter for the function–dose–response curve is:

$$ED_{50} = \frac{\beta K_A}{(\varepsilon[R_t] + \beta)} \qquad (30)$$

From Eq. 30, an expression for the concentration producing half-maximal therapeutic response (ED_{50T}) is given as:

$$ED_{50T} = \frac{\beta_T K_T}{(\varepsilon_T[R_t] + \beta_T)} \qquad (31)$$

This fixes the concentration being tested in the in vivo system (it is assumed the dose of agonist is adjusted to produce a 50% maximal therapeutic response). For Eq. 31, ε_T refers to the efficacy of the agonist at the therapeutic receptor, $[R_T]$ is the receptor density in the tissue mediating the therapeutic response, and β_T the efficiency of coupling of the therapeutic receptor.

This concentration then is substituted into the equation for response of the secondary receptor (denoted $RESP_s$ and obtained from Eq. 14) to yield:

$$RESP_s = \frac{\varepsilon_S K_T}{\beta_S(K_T + K_S) + (((\beta_S/\beta_T)K_S\varepsilon_T) + \varepsilon_S K_T)} \qquad (32)$$

which can be rewritten as an inverse sigmoidal curve of the form $c_1/(c_2 + x)$:

$$RESP_s = \frac{\varepsilon_S K_T}{\beta_S(K_T + K_S + K_S\varepsilon_T/\beta_T) + \varepsilon_S K_T} \qquad (33)$$

The location parameter of the curve described by Eq. 33 is defined as the β_{50}, the value of receptor coupling efficiency where this particular concentration of agonist (the concentration set by observation of the therapeutic effect) produces a 50% maximal response at the secondary receptor. For an equation of the form $c_1/(c_2 + x)$, this is given as:

$$\beta_{50} = \frac{\varepsilon_S K_T}{(K_T + K_S + (K_S\varepsilon_T/\beta_T))} \qquad (34)$$

which can be rewritten as:

$$\beta_{50} = \frac{\varepsilon_S K_T}{K_T + K_S + ((\varepsilon_T/\beta_T) + 1)} \qquad (35)$$

The lower this value is, the smaller the subset of tissues where this concentration of agonist will produce side effects. It can be seen from an inspection of Eq. 33 that, under normal circumstances, $(\varepsilon_T/\beta_T) \gg 1$ and that $K_S\varepsilon_T/\beta_T \gg K_T$. Thus, Eq. 35 can be approximated by:

$$\beta_{50} = \frac{\varepsilon_S K_T \beta_T}{\varepsilon_T K_S} \tag{36}$$

From Eq. 36, it can be seen that the value of the coupling constant producing 50% secondary response is linked to the sensitivity of the therapeutic agonism. This is a reasonable relationship, since a lower concentration of agonist will be needed to produce therapeutic effect if the target organ is highly coupled (low β), and this lower concentration will produce a correspondingly lower incidence of side effects. This is represented by a low value for β_{50}, corresponding to a smaller subset of secondary organs able to respond to the agonist to produce side effects.

Finally, the ratio of β_{50} values can eliminate the dependence on the coupling of the therapeutic organ(s); this eliminates the dependence on β_T. Thus, for agonists A and B,

$$\frac{\beta_{50A}}{\beta_{50B}} = \frac{\varepsilon_{SA} K_{TA} \varepsilon_{TB} K_{SB}}{\varepsilon_{TB} K_{SA} \varepsilon_{SB} K_{TB}} \tag{37}$$

Experiments yield relative efficacies; thus, the relative efficacy for the therapeutic effect is defined as $\varepsilon_{TA}/\varepsilon_{TB}$ and, for the secondary effects, is defined as $\varepsilon_{SA}/\varepsilon_{SB}$. Similarly, the relative affinities for the therapeutic and secondary effects (as equilibrium dissociation constants) are given as rel $K_{T(A/B)} = K_{TA}/K_{TB}$ and rel $K_{S(A/B)} = K_{SA}/K_{SB}$, respectively. Under these circumstances, Eq. 37 reduces to:

$$\beta_{50(A/B)} = \frac{\text{rel } \varepsilon_{S(A/B)} \times \text{rel } K_{T(A/B)}}{\text{rel } \varepsilon_{T(A/B)} \times \text{rel } K_{S(A/B)}} \tag{38}$$

where ε refers to efficacy and K is the equilibrium dissociation constant for binding (the reciprocal of affinity). Thus, rel $\varepsilon_{T(A/B)}$ and rel $\varepsilon_{S(A/B)}$ refer to the relative efficacies of agonists A and B on the receptors mediating the therapeutic and secondary receptors, respectively. Similarly, rel $K_{T(A/B)}$ and rel $K_{S(A/B)}$ refer to the relative affinities (binding pK_i values) for the therapeutic and secondary receptors, respectively.

III. Application of the Algorithm to β_3-Adrenoceptor Agonists

An example of how this method could be used to quantify agonist selectivity is shown below. Table 3 shows the relative intrinsic efficacies and affinities for three agonists on β_3-adrenoceptors (therapeutic agonism is defined, in this case, as increased rate of metabolism), two potential secondary receptors mediating side effects (β_2-adrenergic receptor activation leading to digital tremor) and β_1-adrenergic receptors (mediating tachycardia). From these data, side effect versus coupling constant profiles were calculated for β_2-adrenoceptors (Fig. 17A) and β_1-adrenoceptors (Fig. 17B) for two different coupling efficiencies of the therapeutic (β_3-adrenoceptor) response. These

Table 3. Relative affinities and efficacies for β-adrenoceptor-mediated effects

Compound	pK_i (binding)[a]	Relative ε[b]
β_2-adrenoceptors[c]		
Isoproterenol	5.0	1.0
BRL-37344	6.4	0.003
CL-31643	4.6	0.025
β_2-adrenoceptors[d]		
Isoproterenol	7.5	1.0
BRL-37344	7.2	0.016
CL-31643	5.3	0.03
β_1-adrenoceptors[e]		
Isoproterenol	6.9	1.0
BRL-37344	5.8	0.015
CL-31643	4.0	0.1

[a] Logarithm of the equilibrium dissociation constant of the compound–receptor complex. This corresponds to the molar concentration of compound producing 50% maximal receptor occupancy.
[b] Calculated according to the method of Furchgott from functional (adenylate cyclase assay; Tim True, Glaxo Wellcome Dept. of Receptor Biochemistry) and binding data (Tim True and Conrad Cowan, Glaxo Wellcome Dept. of Receptor Biochemistry).
[c] Chinese hamster ovary cell line expressing human β_3-adrenoceptor clone 6.
[d] Chinese hamster ovary cell line expressing human β_2-adrenoceptor clone 10.
[e] Chinese hamster ovary cell line expressing human β_1-adrenoceptor clone 1.

curves show the relative immutability of the order of potency to the coupling efficiency of the therapeutic event (slight differences occur at very inefficient receptor coupling efficiencies and require the complete form of Eq. 38, namely Eq. 35). Under normal circumstances, the differences are small, and the general application of Eq. 38 to data is recommended. The relative β_{50} values can be obtained graphically from Figs. 17A and B or can be estimated with Eq. 36; relative β_{50} values are shown in Table 4.

From the data in Table 4, it can be seen that BRL-37344 and CL31643 are 11 and 50 times more selective (respectively) for β_3-adrenoceptors than β_2-adrenoceptors compared with isoproterenol. These data suggest that BRL-37344 and CL31643 should be much less likely than isoproterenol to produce tremor in a range of β_2-adrenoceptor-containing tissues. The data in Table 3 also show that BRL-37344 and CL31643 have respective selectivity ratios of 75 and 83 for β_3-adrenoceptors over β_1-adrenoceptors (compared with isoproterenol). This, in turn, also suggests that BRL-37344 and CL31643 should

Metabolic Receptor Coupling

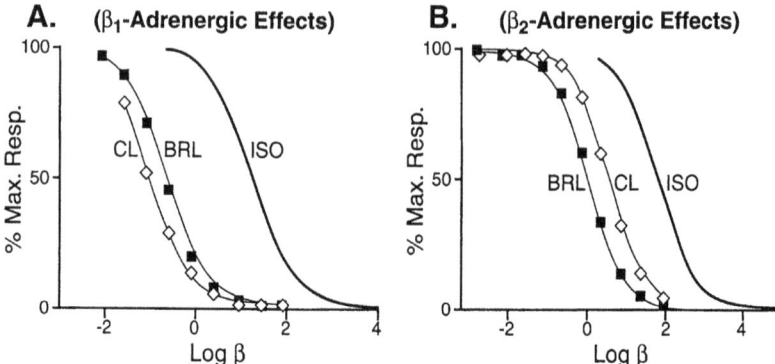

Fig. 17A,B. Side effect versus coupling constant profiles for β-adrenoceptor agonists at various efficiencies of receptor coupling for the therapeutic effect. **A** Data calculated for β_1-adrenoceptors (*I*, highly efficient signal processing of the therapeutic effect; *II*, low-efficiency coupling). **B** Data calculated for β_2-adrenoceptor (*I*, highly efficient signal processing of the therapeutic response; *II*, low-efficiency coupling)

Table 4. Relative β_{50} values for β_3-adrenoceptor agonists relative to isoproterenol. Reciprocals are given so that relative selectivity may be displayed as a multiple

Compound	$1/\beta_{50}$
β_2-adrenergic receptors (digital tremor)	
CL-31643	50
BRL-37344	11
Isoproterenol	1
β_1 (tachycardia)	
CL-31643	83
BRL-37344	75
Isoproterenol	1

be much less likely than isoproterenol to produce tachycardia in a range of β_1-adrenoceptor-containing tissues.

IV. Limitations of the Algorithm

It should be stressed that this algorithm predicts relative effects between agonists but that secondary responses in vivo may be so efficiently coupled that even extremely selective agonists may still produce the secondary effect. Secondly, if the receptors mediating the therapeutic effect are very poorly coupled (i.e., there is a low receptor density on the target organ), then the algorithm estimated by Eq. 28 should be substituted for the more complete equation (Eq.

35). However, under these circumstances, some estimated values for β_T must be used to calculate the relative β_{50}. Finally, it should be stressed that the algorithm depends on an accurate estimate of relative efficacy which, in turn, depends on an accurate estimate of affinity. It is well known that the affinities of agonists can be modified by G protein coupling and that, by this mechanism, the efficacy of the compound can artificially elevate the apparent affinity. This effect can be canceled, in some cases, by GTPγS, but it is not clear that this obviates the problem in all cases. Therefore, it is possible that the very property that is being estimated also affects the measurement of a vital parameter needed for its calculation. This can lead to circular reasoning. Under most circumstances, the null method of measuring relative efficacies for a range of agonists in the same test system tends to reduce the error produced by this effect. However, it is possible that the sensitivity of affinity measurements may not be linearly related to efficacy in some regions of the scale of efficacy. Under these circumstances, the affinities of some agonists may be more prone to G protein-coupling error than others, resulting in corresponding errors in the estimation of relative efficacies.

H. Conclusions

Research aimed at the discovery of agonists can consider both the quality and quantity of efficacy possessed by ligands. The quality of efficacy relates to the mechanism by which different agonists produce changes in the behavior of receptors towards their hosts. One particular arena where there may be practical applications of this approach is the differentiation of agonists that produce different populations of receptor active states, and may thus target stimulus toward different G proteins.

The quantity of efficacy possessed by an agonist can be estimated operationally by a number of means utilizing both binding and functional assays. There are examples where the quantification of efficacy has advantages over the simple measurement of relative agonist potency in the prediction of agonism in other receptor systems. In particular, an algorithm to predict the complete spectrum of agonism in a range of receptor systems is presented here, with application to the prediction of digital tremor and tachycardia for β_3-adrenoceptor agonists designed for metabolic thermogenesis.

References

Black JW, Shankley NP (1990) Interpretation of agonist affinity estimations: the question of distributed receptor states. Proc R Soc Lond [Biol] 240:503–518

Burgen ASV (1966) Conformational changes and drug action. Fed Proc 40:2723–2728

Chen W-J, Armour S, Way J, Chen G, Watson C, Irving P, Cobb J, Kadwell S, Beaumont K, Rimele T, Kenakin TP (1997) Expression cloning and receptor pharmacology of human calcitonin receptors from MCF-7 cells and their relationship to amylin receptors. Mol Pharmacol 52:1164–1175

Costa T, Herz A (1989) Antagonists with negative intrinsic activity at δ-opioid receptors coupled to GTP-binding proteins. Proc Natl Acad Sci USA 86:7321–7325
Fraunfelder H, Parak F, Young RD (1988) Conformational substates in proteins. Annu Rev Biophys Biophys Chem 17:451–479
Fraunfelder H, Sligar SG, Wolynes PG (1991) The energy landscapes and motions of proteins. Science 254:1598–1603
Furchgott R (1966) The use of β-haloalkylamines in the differentiation of receptors and in the determination of dissociation constants of receptor–agonist complexes. In: Harper J, Simmonds AB (eds) Advances in drug research, vol. 3. Academic Press, New York. pp 21–55
Ikezu T, Okamoto T, Ogata E, Nishimoto I (1992) Amino acids 356–372 constitute a G_i-activator sequence of the α_2-adrenergic receptor and have a Phe substitute in the G protein-activator sequence motif. FEBS Lett 311:29–32
Kenakin TP (1984) The relative contribution of affinity and efficacy to agonist activity: organ selectivity of noradrenaline and oxymetazoline with reference to the classification of drug receptors. Br J Pharmacol 81:131–143
Kenakin TP (1995) The nature of agonist-receptor efficacy. II. Agonist trafficking of receptor signals. Trends Pharmacol Sci 16:232–238
Kenakin TP (1996a) The classification of seven transmembrane receptors in cellular expression systems. Pharmacol Rev 48:413–463
Kenakin TP (1996b) Receptor conformational induction versus selection: all part of the same energy landscape. Trends Pharmacol Sci 17:190–191
Kenakin TP (1997a) In: Pharmacologic analysis of drug–receptor interaction. Lippincott-Raven, New York
Kenakin TP (1997b) Differences between natural and recombinant G-protein-coupled receptor systems with varying receptor/G-protein stoichiometry. Trends Pharmacol Sci 18:456–464
Kenakin TP (1998) Agonist selective receptor active states: the next level of selectivity. Pharmaceutical News 5:20–25
Kenakin TP, Beek D (1980) Is prenalterol (H133/80) really a selective β_1-adrenoceptor agonist? Tissue selectivity resulting from differences in stimulus-response relationships. J Pharmacol Exp Ther 213:406–412
Koshland DE (1960) The active site of enzyme action. Adv Enzymol 22:45–97
Krumins AM, Barber R (1997) The stability of the agonist β2-adrenergic receptor-G_s complex: evidence for agonist specific states. Mol Pharmacol 52:144–154
Monod J, Wyman J, Changeux JP (1965) On the nature of allosteric transitions. J Biol Chem 12:88–118
Nickerson M, Gump W (1949) The chemical basis for adrenergic blocking activity in compounds related to dibenamine. J Pharmacol 97:25–42
Onaran HO, Costa T (1997) Agonist efficacy and allosteric models of receptor action. Ann NY Acad Sci 812:98–115
Ruffolo RR, Rosing EL, Waddell JE (1979) Receptor interactions of imidazolines. I. Affinity and efficacy for α-adrenergic receptors in rat aorta. J Pharmacol Exp Ther 209:429–436
Samama P, Cotechhia S, Costa T, Lefkowitz RJ (1993) A mutation-induced activated state of the β_2-adrenergic receptor: extending the ternary complex model. J Biol Chem 268:4625–4636
Senogles SE, Spiegel AM, Pardrell E, Iyengar R, Caron M (1990) Specificity of receptor-G protein interactions. J Biol Chem 265:4507–4514
Stephenson RP (1956) A modification of receptor theory. Br J Pharmacol 11:379–393

CHAPTER 8
A Look at Receptor Efficacy. From the Signalling Network of the Cell to the Intramolecular Motion of the Receptor

H.O. ONARAN, A. SCHEER, S. COTECCHIA, and T. COSTA

A. Introduction

In this article, we examine the multiple connotations of the idea of ligand efficacy, from the macroscopic complexity of the signalling network in the living cell to the microscopic complexity of the single-protein macromolecule. Our analysis consists of two parts.

In the first (Sects. B, C), we give a discussion-oriented overview of the concept. Rather than be systematic or comprehensive, we try to identify the logical threads that link the definitions of efficacy based on stimulus–response relationships of classical pharmacology to the current developments of cellular and molecular pharmacology.

In the second (Sects. D, E), we focus more closely on molecular definitions. We illustrate the similarity of efficacy to the ideas of allosteric linkage and free-energy coupling and propose a paradigm of microscopic generalisation of these concepts. Finally, we present a stochastic model of molecular efficacy. We use such tools to examine the complex relationship between physics and function in a macromolecule and the influence of this relationship on the macroscopic observables that we call affinity and efficacy.

B. Biological Receptors and the Dualism of Affinity and Efficacy

I. Signal Transfer and Conformational Change in Membrane Receptor

Membrane receptors are proteins committed to the vectorial transfer of external signals to the intracellular environment. The input message is usually a freely diffusible extracellular mediator, such as hormone, neurotransmitter, or any mimicking drug. It is detected when such a molecule binds to a specific recognition site on the receptor. The output signal is generated in response to the binding event if and only if the activity state of the ligand-receptor complex is different from that of the unbound receptor. The result is a cascade of biological reactions. The outcome depends on (1) the intracellular network of biochemical transducers and effector proteins to which the receptor is connected and (2) the physiological role of the cells in the organism.

Analogies in molecular structure, biochemical properties and signalling mechanisms let us group receptors into major families, such as ion channels, various types of protein phosphokinases, guanylyl cyclases and so on (MORGAN 1989; IISMAA et al. 1995). The largest of such families, the G protein-coupled receptors, lacks any known enzymatic or channel activity. It consists of monomeric polypeptides folded to form seven hydrophobic membrane-spanning helices (STOECKENIUS and BOGOMOLNI 1982; DOHLMAN et al. 1987; HARGRAVE and MCDOWELL 1992; BALDWIN 1994; STRADER et al. 1994; IISMAA et al. 1995; STROSEBERG 1996; WESS 1997). Signal transfer through these proteins relies entirely on the ability of guanosine triphosphate (GTP)-binding regulatory proteins to "read out" a conformational change that ligand binding can apparently induce in the receptor (RODBELL 1980; GILMAN 1987; BOURNE 1997; SPRANG 1997; HAMM 1998). The idea of "ligand-induced perturbation" actually applies to all sorts of receptors. The leading step of signal transduction is always an intramolecular change intimately linked to the ligand-binding process. Thus, the ability of proteins to adapt their structures in response to ligand binding is a fundamental feature of all biological signalling strategies.

II. The Distinction Between Affinity and Efficacy

The conversion of receptors into active forms, however, does not necessarily result from binding. When many chemical analogues of a ligand for the same recognition site are available, only some of them can trigger a response. Among the effective ligands, the intensities of the triggered responses can be vastly different (ARIËNS et al. 1964). To describe empirically this phenomenon, we may say that not all conformational changes induced by occupation of the binding site result in equivalent functional changes of the receptor. We thus make an implicit distinction in underlying molecular mechanisms between the recognition of the ligand as a binding partner and the conversion of the bound receptor into a functionally relevant form.

This divergence was noted early in pharmacology and medicinal chemistry. Clark first realised that the binding of a drug to its receptor is a necessary (but not sufficient) condition for the production of a biological response (CLARK 1933, 1937). Virtually any theoretical model of receptor action considers ligand binding and ligand-induced change in receptor activity as distinct and distinguishable properties. The first property is analysed using the concept of *affinity*, which comes from chemical thermodynamics and the law of mass action (DENBIGH 1968). It indicates the tendency of a given ligand and receptor to exist in a complexed form. The second property is conceived as *efficacy*, to specify the ability of the receptor-bound ligand to excite or produce a stimulus. As explained in more detail later, its analysis largely depends on the nature and sampling mode of the observable called "biological response". Ligands that have efficacy are named agonists. The extent of response that they produce is related to the degree of intrinsic efficacy. Ligands lacking efficacy are called antagonists since their ability to compete for agonist binding results in impairment of the response elicited by an agonist.

III. Generality of the Concept of Efficacy in Functional Proteins

The distinction between affinity and efficacy is often felt as a mundane puzzle distilled from the imprecise jargon. However, the problem lies at the root of the relationship between intermolecular recognition and intramolecular structural change in proteins. It is crucial to understand not only receptors but any protein in which ligand binding is linked to a specific functional task (WYMAN and GILL 1990; WEBER 1992). For instance, the purpose of an enzyme is to bind to the substrate and accelerate a chemical change of the molecule, which would probably not take place otherwise (KOSHLAND 1968). Transporters or carriers bind ligands, which physically move across different compartments or are released in response to a proper shift of the environment (KOTYK and JANACEK 1970). Ion channels form solvated pores on phospholipid membranes, allowing ions to flow through selectively (MILLER 1987). For membrane receptors, the goal is to detect selected chemicals outside and transmit a recognisable signal inside the cell. In all such molecules, the change in functional condition (catalytic rate, translocation efficiency, ion conductance, signalling state) follows a structural perturbation that the binding of "efficacious" ligands can induce. In all such cases, ligands that do not produce the "effective kind" of perturbation on binding are inhibitors. In this widest sense, ligand efficacy is the endeavour to separate the binding of ligand (which holds two molecules together against the law of diffusion) from the resulting *perturbation of intramolecular energy distribution* (which makes a protein catalyst, transporter, ion filter or signal transducer despite the tendency to minimise energy).

IV. Asking the Questions

This raises a number of questions. One, obviously, concerns the validity of the objective. Does a distinction between affinity and efficacy indeed exist? What is the theoretical justification for such a distinction? The physics of proteins indicates that such a distinction cannot be made. The forces involved in the interaction of a protein with an external ligand are of the same type and magnitude as those holding together its many residues within the folded macromolecule (WEBER 1992). Hence, it is not possible to find a principle-based explanation that can assign these two properties (i.e. affinity and efficacy) to diverse molecular mechanisms of the protein macromolecule.

We think that the question to raise is phenomenological, not theoretical. Experiments show that affinity and efficacy are distinct kinds of information, both encoded within the ligand molecule. What simplifying metaphor of the protein physico-chemical behaviour can we use to understand such phenomena? Does the metaphor help build valuable models of receptor action? Most importantly, can the metaphor provide a code to translate molecular structure and dynamics into elements of biological information? These questions are the centre of this chapter.

The concept of allosteric modulation is an example of the metaphor most often used to understand the relationship between binding and function in proteins (WYMAN and GILL 1990). The underlying idea, as discussed later in this article, is that diverse functional forms of the protein directly map onto correspondingly different conformations. This lets us draw simple correlations between structural attributes and matching functional behaviour. For instance, if we know that an agonist can make a receptor active, all we must do is solve the structure of the agonist-bound receptor to identify the functionally active receptor form. Similarly, a binding-induced allosteric transition between high- and low-conductance conformations allows us to explain how a ligand can regulate ion flow through a channel. Metaphors are useful in science when considering complexity but become deceptive if mistaken for reality. The allosteric idea is a powerful tool of insight and experimental design. However, allosteric conformations of a protein do not represent its true physical states, nor do allosteric transitions say anything at all about the true dynamics of molecular motion underlying function. Therefore, we can use this concept to study the functional behaviour of the receptor, but not its physics.

V. Two Flavours in the Definition of Efficacy: Biological and Molecular

We mentioned the generality of the concept of ligand efficacy among proteins. The signalling purpose of the functional task, however, distinguishes the receptor from any other sort of "molecular machine". This makes it difficult to define a set of variables that can gauge the functional efficiency of the protein at the molecular level. In contrast, the efficiency of an enzyme on the purified molecule can easily be measured through a suitable combination of equilibrium and rate constants governing the reaction that is catalysed (RADZICKA and WOLFENDEN 1995). Similar estimates can be made for ion channels or carrier molecules when studied in isolation. However, the signalling efficiency of the receptor – especially the G protein-coupled kind – can only be appraised in the presence of at least one other partner of the signal-transducing network in which it operates.

The implication is that there is a fundamental difference among models of affinity and efficacy, depending on whether the receptor is described as a gregarious or an isolated molecule. In the first case, the biochemical network of signal-transducing molecules in the cell hosting the receptor sets the stage of the phenomenology that should be predicted. The molecular properties of the receptor itself need no explicit address or can conveniently be hidden in a conceptual "black box". In the second case, the transduction network is only an accessory reading tool, while the dominant problem is the search for "information" within the intramolecular universe of the receptor macromolecule. To acknowledge this difference, we speak of *"biological"* and *"molecular"* definitions of efficacy for the two cases illustrated above.

C. Biological Definitions of Efficacy

I. The Nature of Signal Strength

The aim of a theoretical model of receptor action is to predict all possible relationships between the strength of the input signal and the extent of the response. This requires a precise definition of quantities. What is the strength of the signal, and what are its components?

Experiments show that (1) receptors establish high-affinity but reversible bonds with signalling chemicals and (2) biological responses increase uniformly with the number of molecules of the ligand (VAN ROSSUM 1963; ARIËNS et al. 1964; RANG 1973). Thus, one component of signal strength is the number of ligand-bound receptors, which is related to the ligand concentration by the equilibrium affinity that governs their interaction. A second component appears when we analyse a collection of ligands all capable of binding to the same site on the receptor. If compared at saturating concentrations, the range of output they produce can be very broad – from null to full, with all possible gradations in between. This means that amplitudes of the signal triggered by bound ligands differs from one ligand to another.

Thus, the *number* and *amplitudes* of signals are two experimentally observable components of signal strength, and both are required for its definition. Experiments also show that these two components look like independent quantities, for chemical modification of the ligand can change its ability to trigger a response without altering the binding affinity and vice versa (KENAKIN 1984).

Even from this intuitive description, it is clear that the components of signal strength pose a conceptual problem. They appear to be set on quite different scales. The first (number of signals) has probabilistic connotations. What makes the cell respond in gradual fashion to the concentration of signalling chemicals is the presence of an ensemble of receptors on its surface. It is the bound fraction of such an ensemble that tells the cell how much ligand is present. However, at the molecular level, each receptor can be either bound or not so. As an individual molecule, the receptor is a binary source of signs (bound/not bound) that can transfer at most one bit of information. Conversely, the amplitude of the signal encrypted in the ligand is a graded quantity that each single molecule of receptors can decode. Its boundaries enclose all possible molecular structures that "fit" the ligand-binding site of the receptor. The range can be very large and, for many practical purposes, may not be discernible from infinite. In this case, each receptor acts like a continuous signalling source. However, we do not measure output from single receptors but from the (whole) population existing in a membrane.

Clearly, a normalisation for the two components of signal strength is needed before any relationship to cell response can be established. In receptor theory, this was achieved simply and cleverly, with the introduction of the quantity called stimulus (S; STEPHENSON 1956). S indicates the "sum of stimulations" produced by all receptors that are occupied by the ligand on the cell surface. The best definition of S was given in FURCHGOTT (1966) as the product of three terms: $[R_t]$, O_f, and ε. $[R_t]$ is the molar concentration of the total number of receptors. O_f is the fractional occupancy (i.e. the ligand-bound fraction of the total receptor population), computed by the mass-action law: $O_f = K[H]/(K[H] + 1)$, where [H] is the concentration of free ligand and K is the equilibrium affinity constant. The term ε, the *intrinsic efficacy* of the ligand, indicates the "unit stimulus" produced by each ligand–receptor complex. It is dimensionless (although FURCHGOTT himself suggests that its dimension is M^{-1}) and covers the range of positive real numbers and zero. Thus defined, efficacy looks like a "weight" given to the bound receptor over the unbound form. However, it may also be interpreted as a measure of the "extra information" gained by each receptor upon ligand binding.

Of the four parameters defining S, affinity and efficacy (K and ε) are constants, determined only by the characteristics of the ligand and the receptor, and independent of the cellular system and the experimental settings in which they are measured. Although both are molecular attributes, they show an important difference. Affinity is – to a crude approximation, at least – a true physicochemical quantity: the standard free energy change for the reaction of association between ligand and receptor. It can be expressed in fundamental energy units and measured in the test tube with isolated receptor molecules, if necessary. Efficacy, however, is a molecular property that cannot be formulated using molecular dimensions, for it needs the signal transduction equipment of a cell to be revealed. Since it lacks physical or chemical units, it must be expressed as a relative measure whose sole reference frame is the collection of ligands that are available for the receptor under study. The term "biological", as used here, denotes that efficacy thus defined is only gauged through the biochemical effectors of response.

II. Stimulus–Response Relationship

The signal-transduction network of the cell forges the relationship between stimulus and response. In the final and most general theory of STEPHENSON and FURCHGOTT, response r was defined as an unspecified but monotone function of the stimulus: $r = f(S)$. The exact shape of this function depends on the cell (or tissue) where the receptor is located and must be found by experiments. ARIËNS had previously adopted the far more restrictive assumption that f is a linear function (ARIËNS 1954), which prompted enough experimental work to show that such an assumption is seldom true in practice.

Efficacy, of course, can always be measured whatever the nature of f is, but the effort required differs dramatically in the two cases. If f is strictly linear, as in the limiting case of ARIËNS, relative efficacy (i.e. *intrinsic* activity in his words) can be deduced simply (from concentration–response curves) as the ratio of maximal effects induced by different agonists. For non-linear f, efficacy affects all the parameters of the concentration–response curves. Thus, we must derive the stimulus–response relationship numerically to calculate the relative efficacy of ligands (as, for example, in the null method, which relies on the researcher's ability to manipulate the number of membrane receptors; FURCHGOTT 1966).

At the time the essentials of receptor theory were developed, biological response was an end-point measurement. Isolated tissue preparations and their contractions were the sole instrument used to quantify the effects of drugs and hormones, while second messengers and their biosynthesis were the objectives of discovery rather than of routine determination. Despite such limitations – or perhaps because of them – the models of receptor action derived from that work are so robust and versatile that they are still applicable today (KENAKIN 1992).

Receptor-mediated responses can now be assessed at multiple levels: intact cells, permeabilised-cell preparations, isolated plasma membranes and even synthetic vesicles, where purified receptors are reconstituted with selected proteins of diverse signal-transduction pathways. Preparations in which signal amplification is low, such as isolated membranes or reconstituted systems, are more likely to exhibit linear stimulus–response relationships. For instance, this is often the case for receptor-dependent stimulation of GTPase or adenylate-cyclase activity (KAUMANN and BIRNBAUMER 1974; LEVY et al. 1993). Conversely, responses measured in intact cell systems, such as secretory activity or gene expression, are likely examples of strongly hyperbolic relationships between stimulus and response.

No general a priori assumption can be made, however, and an experimental verification of the stimulus–response function is always needed to compute efficacy. To evaluate the nature of the stimulus–response relationship, we can measure affinity and receptor number by radioligand binding assay. Linearity between observed response and receptor occupancy also indicates linearity of $f(S)$. To avoid artifacts, however, the variable time should be eliminated from both determinations (i.e. occupancy must be measured at equilibrium, and response must be measured under steady-state conditions), which is not always a trivial task. Another interesting strategy is the possibility of generating a large variation in receptor number by engineering cells expressing receptor complementary DNAs under the control of inducible promoters (GOSSEN et al. 1995; HOWE et al. 1995; THEROUX et al. 1996) or selecting transfected cells exhibiting a suitable range of receptor densities (CHIDIAC et al. 1996). In such cells, stimulus–response relationships for agonists can be derived at several levels of signal transduction, and efficacy can be precisely measured.

III. The Scale of Agonism and Antagonism

Affinity and efficacy are unique molecular markers of both the ligand and receptor and, thus, provide a major means to their classification (KENAKIN

1984). Since the reference frame of receptor efficacy is determined by its ligands, receptors lacking a sufficient number of ligands of varying efficacy cannot be studied. An orphan receptor, even if well characterised with respect to molecular structure and signalling pathway, is basically an unknown receptor, and it will remain so until ligands of varying efficacy become available.

The numerical values of efficacy can be arbitrarily ranged between zero and one. If this is done, ligands are antagonists when $\varepsilon = 0$, full agonists when $\varepsilon = 1$ and partial agonists when $0 < \varepsilon < 1$. The extent of efficacy dictates the "virtual" effect of a ligand, i.e. the theoretical stimulus that it can trigger from the receptor. However, the actual biological effect depends on the stimulus–response function. The nature of this function, therefore, sets the biological significance of the scale of efficacy.

For example, a weak partial agonist ($\varepsilon \approx 0$) would produce a barely detectable biological effect if we measure a linearly coupled response (such as stimulation of adenylate-cyclase activity in isolated plasma membranes) so that it may be considered as an antagonist in practice. However, the same ligand may produce as much stimulation as a full agonist in that cell if we measure a response with strong hyperbolic dependence on the stimulus, such as cyclic-adenosine-monophosphate-dependent secretory activity or gene induction.

The divergence between efficacy and effect is not only a trivial cause of discrepant literature regarding whether a certain ligand is an agonist or antagonist. It is also the source of a more serious misconception: the idea that ligand efficacy may vary among cell types. It is not the efficacy that is changing, of course, but the relationship between the stimulus and response. Similar misconceptions are generated by the common notion of "receptor reserve" or "spare receptors" that describe non-linear stimulus–response relationships in cells. Since a hyperbolic $f(S)$ generates a discrepancy between agonist occupancy and response, the number of receptors appears to be "in excess" with respect to the fractional effect in this case, from which the idea of "spareness" comes. However, this notion causes only confusion, because it fosters two prejudices: (1) that the existence of spare receptors depends on the receptor only, rather than *both* the ligand and receptor, and (2) that the non-linearity of stimulus–response relationships is the result of the existing stoichiometry between receptors and effectors. The dullness of such conceptions has little to do with the refined subtlety of the theory they attempt to simplify.

Originally, the efficacy of antagonists was set to zero. The underlying assumption was that receptors not bound to ligand are basically "silent" signalling sources. There is plenty of experimental evidence now for a different scenario. Under proper experimental conditions, receptors induce basal or "constitutive" signalling activity in the absence of ligands (COSTA and HERTZ 1989, 1990; LEUNG et al. 1990; HILF and JAKOBS 1992; MEWES et al. 1993; ADIE and MILLIGAN 1994; TIBERI and CARON 1994; COHEN et al. 1997; JIN et al. 1997). This constitutive activity can be enhanced dramatically by engineered (KJESLBERG et al. 1992; ROBINSON et al. 1992; REN et al. 1993; SPALDING et al.

1995; Cho et al. 1996; Parent et al. 1996; Hjorth et al. 1998) or spontaneous (Parma et al. 1993; Robbins et al. 1993; Shenker et al. 1993; Rao et al. 1994; Schipani et al. 1995; Ferris et al. 1997) mutations of their sequences. For some such mutants, any possible amino-acidic replacement (except for the original one) invariably results in some basal response (Kjeslberg et al. 1992; Scheer et al. 1997). Thus, the emerging scenario is that receptors are intrinsically active molecules by default and that their inactive states are maintained by one or more crucial structural constraints. Removal of such constraints, as may be reversibly achieved by agonist binding or irreversibly caused by some mutations, would then be the key mechanism of activation (Kjeslberg et al. 1992; Lefkowitz et al. 1993).

The notion that receptors have ligand-independent activity extends the scale of efficacy for antagonists. Along with those with $\varepsilon = 0$, we can have antagonists with negative efficacy ($\varepsilon < 0$), as has been experimentally shown (Costa et al. 1989, 1992; Barker et al. 1994; Samama et al. 1994; Smit et al. 1996). As far as the classical theory, there is no problem, of course, in accounting for negative efficacy. We simply let the range of ε include negative numbers. However, this is only a mathematical gimmick. How can efficacy be negative? What is a negative stimulus, and what is its reference frame? The strict logical abstraction that defines efficacy in the Stephenson–Furchgott model cannot provide answers to such questions.

IV. Steps of Signal Transduction and the Indeterminacy of Stimulus–Response Relationships

Suppose we wish to draw a simple scheme of a receptor-operated membrane signalling process. The variety of molecules and biochemical mechanisms is so large, and the fraction of those known in sufficient biochemical detail is so small (Alberts et al. 1994), that any attempt to generalise seems pointless. We can roughly identify, however, two separate phases of signal transfer in every membrane receptor system. The first includes detection and transmission of the signal, and the second includes modulation and processing. A precise attribution of these functions to diverse molecular entities is hard to make in general, but the family of G protein-coupled receptors offers one special example where such subdivision appears more evident.

Here, the first phase involves the receptor and G protein exclusively, and it is confined to the membrane. The receptor (i.e. the receiver) detects the message and decodes its information by promoting subunit dissociation of the G protein heterotrimer (Gilman 1987; Bourne 1997; Hamm 1998). G protein subunits (i.e. the transducer) broadcast the message by stimulating or inhibiting a number of membrane-bound proteins having enzymatic or channel activity (Neer and Clapham 1988; Birnbaumer 1990). Pre-amplification of the signal also happens at this stage (Ross 1989).

The second phase involves the membrane and the whole cells via the intervention of multiple-enzyme systems and gene expression. Modulation (i.e.

control of the sensitivity of the detection system) is achieved by a variety of mechanisms. Phosphorylation operated by specific cytosolic kinases segregates the receptor from the G protein, e.g. by promoting binding to arrestins (LEFKOWITZ and CARON 1986; HAUSDORFF et al. 1990; LOHSE et al. 1990). Acceleration of receptor recycling and repression of messenger-RNA expression leads to reduction of the steady-state levels of receptor units (KIRSHHAUSEN et al. 1997; FERGUSON et al. 1998). Processing accomplishes the conversion of the signal into cellular output. Both high-gain amplification and integration of the signal occur at this stage. Amplification is due to the inherent power of the sequence of hyperbolic enzymatic reactions that link, in a non-linear fashion, the initial membrane stimulus to the final cell response. Integration results from the convergence of multiple signals onto pre-programmed circuits of response and the mutual exclusion or synergy of opposite or parallel signalling pathways.

It is obvious from this layout that the relationship between ligand concentration and final output continuously changes with the step of signal transduction that is the object of observation. The signal decoded at the receptor gains emergent properties as it travels through the network of biochemical interactions sketched above. These properties, in turn, influence our perception of the initial stimulus generated at the receptor level. According to such perspective, we may define the relationship between initial stimulus and final cellular response as a "macro" function whose domain includes the subsets of all stimulus–response functions related to the range of signals that are propagated within the cell. Even if we were able to dissect and analyse in detail each of those subfunctions, it would be impossible to define precisely how extensive or minimal changes in their parameters affect the global function in which they are all enclosed. Therefore, the undetermined nature of the function $f(S)$ is the result of the inherent complexity of the signalling network, not of insufficient knowledge about its components.

One implication of these considerations is that for efficacy to be an exclusive constant of ligand and receptor, it must be stripped of any dimension that may link it to the biochemical interactions that reveal its existence. Biologically defined efficacy, in other words, must be dimensionless to retain invariance. As we add explicit biochemical attributes to its definition, we also add indeterminacy. The operational model of receptor action introduced by BLACK and LEFF illustrates this situation (BLACK and LEFF 1983). They proposed that the stimulus–response relationship may be empirically defined as a hyperbolic function describing the operational interaction of the agonist–receptor complex with elements of the transduction process. The dissociation constant for such interaction (normalised to receptor concentration or its reciprocal, named τ by the authors of the model) yields a measure of ligand efficacy. However, efficacy now does not depend on ligand and receptor only but also on the characteristics of the effector system for which the interaction is specified.

D. Molecular Definitions of Efficacy

I. A Molecular Link Between Affinity and Efficacy

Can ligand efficacy be defined using molecular dimensions independently of the biological signalling network serving the receptor? The reason for seeking an answer to such a question is experimental evidence suggesting that affinity and efficacy are not independent at the molecular level. The idea of their independence comes from years of studies of the structure–activity relationships on congeneric analogues of receptor ligands. That work, taken collectively, indicates that affinity and efficacy can be manipulated independently upon modification of ligand structure. Thus, K and ε are truly independent constants in receptor theory, except for the obvious condition that makes the second impossible in the absence of the first.

However, a different view emerges from biochemical investigations of agonist binding to G protein-coupled receptors. Two essential facts sum up a large number of diverse experimental observations: (1) the apparent affinity of agonists is diminished between one and two orders of magnitude by preventing the interaction of the receptor with the G protein and (2) the extent of such diminution is related to the magnitude of ligand efficacy (DE LEAN et al. 1980). The fundamental implication of this phenomenon is that a component of the free-energy change of agonist binding is somehow created by the molecular perturbation that the ligand induces when triggering a signal from the receptor. Therefore, efficacy can, in principle, be sought within the same interface of interaction that gives rise to affinity and, like affinity, efficacy can be expressed in physico-chemical terms.

II. Allosteric Equilibrium, Free-Energy Coupling, and Thermodynamic Definitions of Efficacy

A very general approach to analysis of how the binding of ligands modifies the function of proteins is the concept of allosteric equilibrium (WYMAN 1967; GILL et al. 1985; WYMAN and GILL 1990). The core idea is that a protein can be thought of as existing in equilibrium among several tautomeric conformations, each of which corresponds to a different biologically functional state (DEL CASTILLO and KATZ 1957). If the protein has one (or more) ligand-binding site(s), the introduction of the ligand to the system will shift the tautomeric equilibrium towards the conformation(s) having the highest affinity for that ligand and, consequently, will turn the average activity of the protein into one primarily determined by that favoured conformation(s) (MONOD et al. 1963, 1965).

A simple example illustrates how the allosteric idea leads to a definition of efficacy. Let us imagine an ion-channel-receptor existing roughly in two functionally distinct states: closed (non conducting, R) and open (conducting, R*; NEHER and SAKMANN 1976; HAMILL et al. 1981). In the absence of any ligand, the equilibrium between R and R* is given by $J = [R^*]/[R]$, which can be interpreted either as the ratio of the number of channels in the two states (for an ensemble of channels) or the ratio of average fractional times the channel spends in each state (for a single channel; COLQUHOUN and HAWKES 1983). In the presence of the ligand H, the equilibrium becomes Scheme 1, where K is the affinity of formation of ligand–receptor complex for the "closed" state, and β is the measure of the change in relative abundance of R* induced by the bound ligand or, equivalently, the degree of selectivity of the ligand for the two channel states. Thus, β directly gauges the ligand's power to change the activity of the channel protein. For $\beta > 1$, the binding of the ligand shifts the equilibrium toward the active state (J versus βJ) or, equally, the ligand binds to the active state better than it does to the inactive one (K versus βK). The opposite happens for $\beta < 1$, while $\beta = 1$ means that the ligand does not perturb channel activity at all.

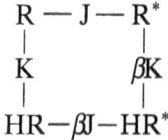

Let us assume that the measured biological function is the mean current carried by the population of channels in the cell (Fig. 1a). The functional change imparted by ligand binding (expressed as the increase in mean conductance) is $\beta(J + 1)/(\beta J + 1)$ (Fig. 1b). The dependence of this quantity on β (Fig. 1b) has a plateau given by $(1 + J)^{-1}$, i.e. the inverse of the fraction of channels existing in open state before the introduction of ligand. The smaller J is (smaller basal activity), the larger the maximum possible change in conductance that can be induced by a ligand, which implies that there is an upper limit for production of a unit stimulus in a single channel by a ligand.

The parallelism between β and the efficacy defined by classical theory (ε) is obvious. Necessarily, β greater than, equal to or less than one implies ε greater than, equal to or less than zero, respectively, and classifies ligands as agonists, neutral or negative antagonists, respectively. Unlike ε, however, β has both physical meaning and energy units. It is the standard free energy of the coupling between the first-order transition in the protein (R→R*) and the second-order ligand-binding process (H + R→HR). While, in receptor theory, affinity is independent of efficacy, in the allosteric model, they are linked (COLQUHOUN 1987). In fact, the measurable binding affinity, given by $K_{obs} = K(\beta J + 1)/(J + 1)$, includes both efficacy and the equilibrium of the channel among its possible states. The pure second-order constant K can only be calculated if J and β are known but never directly observed (except if we conceptually "freeze" the R→R* interconversion prior to or during ligand saturation – an interesting exercise, perhaps, but unquestionably useless).

Let us now examine a different kind of receptor, one for which function can be described by an additional ligand-binding process – for example, a

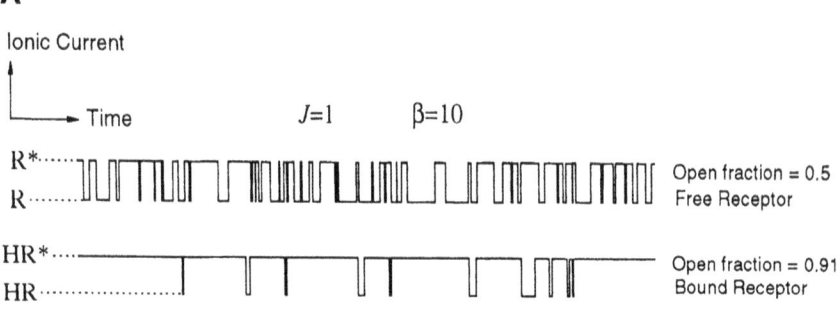

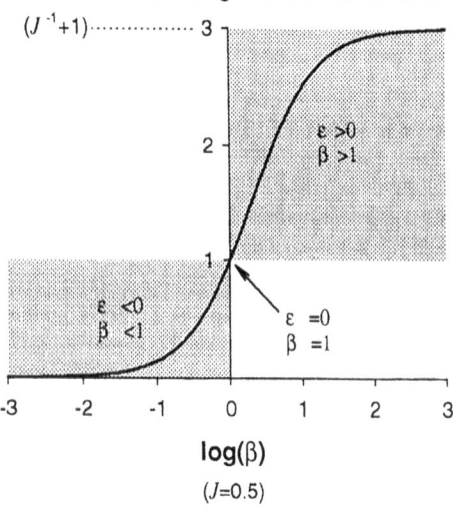

Fig. 1A,B. An ion-channel example illustrates allosteric equilibrium. **A** Computer-simulated sample recording of a single channel in free (*upper trace*) or ligand-bound forms (*lower trace*). In either form, the channel randomly shuttles between a conducting (R* or HR*) and non-conducting (R or HR) state. The ionic current carried by the open state (equivalent to its unit conductance, u.c.) is assumed constant and independent of the state of ligation of the molecule. However, in free form, the channel spends equal average times in the two states [fractional open time (f.o.t.) is 0.5] while, in ligated form, the open state is visited more frequently (f.o.t. = 0.91). Thus, the mean conductance (i.e. f.o.t. × u.c.) differs between the free and bound forms. According to the allosteric model, the equilibrium between open and closed states in unligated form ([R*]/[R] = 0.5/0.5) is J = 1, and, in ligated form ([HR*]/[HR] = 0.91/0.09), becomes $\beta J \approx 10$. Thus, $\beta \approx 10$ is the efficacy (unit stimulus) of the ligand. **B** Ligand-induced change in biological activity as a function of $\log(\beta)$ with J = 0.5. Biological activity (i.e. the change in mean conductance upon ligand binding) is measured as the ratio of f.o.t. in bound versus free forms (y-axis). According to the allosteric equilibrium of Scheme 1, it is given as [HR*]([R*] + [R])/[R*]([HR*] + [HR]) = (J + 1)/(βJ + 1). The *shaded rectangles* show the regions where β lowers (*left*) or enhances (*right*) the biological activity of the channel. They correspond to regions of negative and positive efficacies, respectively. At the junction is null efficacy, equivalent to $\beta = 1$ (i.e. no change in channel conductance upon ligation). To translate from the single-channel situation shown here to an ensemble of channels, fractional times can be replaced with the fractional number of channels existing in the two possible states at any given instance

member of the G protein-coupled family, where biological effect depends on the interaction of the G protein transducer at a binding site different from that of the ligand. We may imagine that this receptor is in equilibrium among various isomeric forms R, R*, R**, and so forth. Each of the two ligands H and G would induce or stabilise a particular receptor state – R* when H is bound, and R** when G is bound. What state would be predominant when both H and G are bound? No matter how many pre-existing states we postulate in the conformational space of the free receptor, it is clear that the state of the ternary complexed receptor must be neither entirely different nor exactly identical to those bound to either G or H.

We encounter here the major limitation of the allosteric metaphor: the idea that a protein can be described by a restricted number – however large – of equilibrium conformations. This assumption cannot account for the quasi-continuous nature of the state transitions generated by the stochastic motion of the macromolecule.

To solve this problem, WEBER introduced the concept of free-energy transfer between simultaneous binding processes on distinct sites of a protein (WEBER 1972, 1973, 1992). Accordingly, the interactions of H, G and R can be depicted using the equilibrium Scheme 2, where K and M are the unconditional affinities for the formation of ligand-receptor and G protein-receptor complexes, respectively, and α represents the reciprocal effect that ligand and G protein impart on the binding of each other when they are simultaneously bound into a ternary complex. It measures the standard free energy transferred from the binding of ligand to the binding of G protein (or vice versa).

$$\begin{array}{ccc} R & -K- & HR \\ | & & | \\ M & & \alpha M \\ | & & | \\ RG & -\alpha K- & HRG \end{array}$$

In the allosteric model discussed before (Scheme 1), the change in the functional state of the receptor is addressed explicitly by drawing an equilibrium between two postulated conformations having pre-existing activities. Here (Scheme 2), the change is assumed implicitly with the statement that the association of each of the two ligands (H and G) to their binding sites may be different depending on whether the second ligand is bound or not bound to the receptor (K versus αK and M versus αM). The equivalence between ligand efficacy and α is also obvious. α greater than, equal to or less than one implies that the ligand enhances, leaves unchanged or reduces the tendency of R to bind G, respectively; thus, it has positive, null or negative efficacy, respectively.

Also, in this case, ligand affinity ($K_{obs} = K(1 + \alpha M[G])/(1 + M[G])$) is a composite parameter and includes efficacy, M and [G]. Unlike K in Scheme 1, however, "pure" K here can be measured directly in the absence of G. Efficacy defined by α depends on H, R and G. It is trivial to extend Scheme 2 to describe the interaction of R with $G_1, G_2, \ldots G_n$, which yields $\alpha_1, \alpha_2, \ldots \alpha_n$.

This means that a given ligand H interacting with the same receptor R will have different efficacies when R binds to different types of G protein (KENAKIN 1988, 1995a, 1995b). Indirect experimental evidence for this theoretical prediction are available (SPENGLER et al. 1993; GURWITZ et al. 1994; ROBB et al. 1994; PEREZ et al. 1996; RIITANO et al. 1997).

As suggested by the similarity of Schemes 1 and 2, β and α have identical thermodynamic meaning: the conservation of free energy for multiple perturbations applied to the same protein. Therefore, they lead to a unique definition of efficacy based on fundamental principles and, thus, valid for any protein regardless of differences in function, detail and experimental background. However, the two approaches give rise to a potentially large series of apparently different models of receptor action (a small sample of which is compared in Table 1). The reason is that the property "biological function" appears as an intrinsic property in the receptor in the first case ($R \rightarrow R^*$) and an extrinsic one in the second ($R+G \rightarrow RG$). In reality, biological function, by definition, is not part of any of the two formulations. It is merely an attribution that we make in their interpretation.

III. Linkage Between Macroscopic Perturbations in the Receptor

The constants β and α are two equivalent ways to assert the existence of a macroscopic linkage between standard free-energy changes concurrently taking place on the same macromolecule. For instance, if we could experimentally alter the equilibrium J (Scheme 1), we would find a correlated change in the apparent affinity of the ligand H, the extent and direction of which is given by the value of β. Similarly, changes in M in the presence of G (Scheme 2) will produce correlated variations in ligand affinity, dictated by the value of α.

The term "macroscopic linkage" here has a dual meaning. First, it declares the level of theoretical formulation. What α or β really intend to express are the microscopic changes occurring within the intramolecular interactions of the individual receptor. However, thermodynamics, by definition, can only formulate those changes macroscopically as the linkage they impose on the interaction of the receptor with the reacting partners of the macroscopic system in which they are all enclosed (WYMAN and ALLEN 1951; WYMAN 1967, 1984). The second meaning of "macroscopic linkage" is a condition of validity. The exactness of the linkage holds only for macroscopic observables, i.e. for measurements that can be considered the result of time- or ensemble-averaged properties of receptor molecules (DENBIGH 1968).

To illustrate this point, the ion-channel example used before is still a very useful aid. Let us assume that the channel of Fig. 1 emits a spectroscopic signal strictly proportional to the number of molecules in the open state. If we measured how that is modified at saturating concentrations of several ligands, we would find good agreement between ligand-induced spectral changes and modification of conductance and would also obtain similar values of β. At com-

Table 1. Comparison of receptor models derived from biological and molecular definitions of efficacy

Type of definition Constant	Biological ε[a]	Molecular β[b]	α[c]	$\alpha\beta$[d]
Receptor model	Occupancy	Two-state or allosteric model	TCM	Allosteric TCM and "cubic"
Physical meaning	None	Free-energy coupling (isomerisation vs binding process)	Free-energy coupling (two binding processes)	A combination of the meanings for α and β
Units	Arbitrary (M^{-1}?)	kcal/mol [for $-\ln(\beta)/RT$]	kcal/mol [for $-\ln(\alpha)/RT$]	kcal/mol [for $-\ln(\alpha\beta)/RT$]
Independence of efficacy and affinity?	Yes	No	No	No
Measured affinity	K	$K \cdot \dfrac{\beta J+1}{J+1}$	$K \cdot \dfrac{\alpha M[G]+1}{M[G]+1}$	$K \cdot \dfrac{\beta J(1+\alpha M[G])+1}{J(1+M[G])+1}$
Resolution of efficacy from affinity (condition)	Yes (stimulus–response relation must be computed)	Yes (equilibrium J between receptor states must be measurable)	Yes (concentration of G must be experimentally varied over a wide range)	Yes (same conditions as for both α and β)
Efficacy depends on:	Ligand and receptor	Ligand and receptor	Ligand, receptor and G protein	Ligand, receptor and G protein

TCM, ternary-complex model.
[a] STEPHENSON 1956; FURCHGOTT 1966.
[b] KARLIN 1967; COLQUHOUN 1973, 1987; THRON 1973; LEFF 1995.
[c] DE LEAN et al. 1980; WREGGET and DE LEAN 1984; EHLERT 1985; COSTA et al. 1992.
[d] SAMAMA et al. 1993, 1994; WEISS et al. 1996.

parable levels of macroscopicity, in other words, physical and functional determinations appear to be well linked and lead to a singular value of ligand efficacy.

However, let us now imagine that, at some stage of the study, the resolving power of our experiments improves dramatically. Detailed analysis of single-channel statistics reveals the existence of a large number of discrete states that are physically distinct, perhaps, but functionally very similar (several closed and open states with diverging time statistics but comparable conductances, for example). Also, suppose that time-resolved half-life distributions of the spectral signal lead us to discover that every ligand can induce a different pattern of many distinguishable physical states on binding. We now face two problems. First, we can no longer match physical and functional states with a simple correlation. Second, we cannot unequivocally quantify efficacy as the "conformational consequence" of ligand binding, because every ligand appears to perturb the state composition of the receptor regardless of its activity on the channel. In Fig. 2, we show a very schematic example of the disparity between ligand-induced physical states and macroscopic functional forms of the receptor.

IV. Functional and Physical States in Proteins

If proteins could be equated to conventional solids, there would be straightforward correspondence between a well-defined biological function and a given conformation of the atomic structure of the receptor. Efficacy would be unambiguously identified and measured by the conformational displacement that the agonist-bound receptor undergoes with respect to the empty form. However, proteins posses all the characteristics of what is known in physics as a complex system (FRAUENFELDER 1995). The energy of a protein as a function of its conformational coordinates draws a rugged hypersurface. Any cross-section through this surface traces an "energy landscape" where distributed minima indicate that a macromolecule in a given conformation can exist in a large number of slightly different substates. States and substates are organised in a branched hierarchical arrangement whose knottiness is progressively unveiled as the energy level decreases (FRAUENFELDER et al. 1988, 1991; BRYNGELSON et al. 1995). This means that each receptor state that we may infer from a macroscopically observed biological function or measure from an averaging physical determination, such as X-ray crystallography or nuclear magnetic resonance, corresponds to a large number of conformationally similar substates of approximately equal energy. Different substates may perform the same function with different kinetics, perhaps, or may even be totally inactive. This "graininess" of protein conformations implies that any functional state must always be considered as a macroscopic or heterogeneous entity according to some more resolving physical criterion. In essence, we may say that there is a stochastic relationship between functional and physical states in proteins.

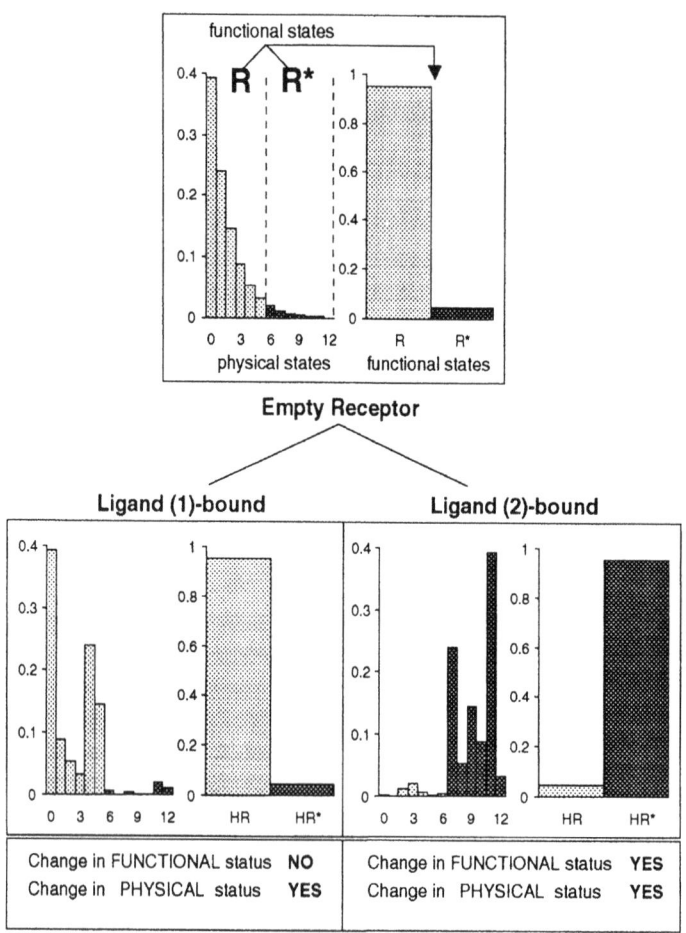

Fig. 2. Schematic representation of the relationship between functional and physical states in a receptor. A putative receptor that carries a recordable biological activity exists in 13 physically distinct, interconvertible microstates (as in the reaction scheme of Fig. 3). Some substates (6–12) contribute to function, others (0–5) do not. From biological observation, the unbound form of the receptor (upper panel) appears to be in equilibrium between two "functional" conformations (R and R*), within which the "hidden" microstates are grouped according to their biological activity. The arbitrary fractional abundance of physical states is shown on the left-hand histograms and is accumulated in the plot of the functional forms on the right-hand side. Two effects of ligand binding are envisioned (*lower panels*). In case 1, the ligand changes the stability of all microstates, but the relative abundance between the active and inactive groups of states is not changed. Therefore, there is no visible effect on the two macroscopic functional forms (*left-hand panel*), and the ligand is inactive (antagonism). Conversely, in case 2, the stabilities of all microstates are affected so that the relative abundances of the active states are enhanced. The result is a ligand-induced shift of the macroscopic equilibrium towards the R* form (*right-hand panel*; agonism). Thus, both ligands are "conformationally active", but only the agonist (ligand 2) is biologically active

V. Microscopic Interpretation of Allosteric Equilibrium

The microscopic nature of efficacy becomes explicit if we define allosteric equilibrium microscopically over the entire conformational space of the receptor (ONARAN and COSTA 1997). Let us define the receptor as existing in equilibrium among all its possible conformational states. By "possible states" we mean all those that can be predicted from the primary sequence of the protein, regardless of the actual energy level. For example, given that each residue in a protein can assume, on average, two to three conformations of approximately equal energy, a typical G protein-coupled receptor of 450 amino acids has a total number of possible states of $(2-3)^{450} \geq 3.3 \times 10^{135}$. This is equivalent to an approximately continuous scale of interconvertible states. Each transition from state (i–1) to i (i > 0) is given by a corresponding stability constant. Microscopic reversibility (DENBIGH 1951) lets us relate all the transitions to an arbitrarily chosen state s_0, i.e. $j_i = [s_i]/[s_0]$ (Fig. 3). Thus, the mass conservation of a receptor existing in n possible receptor states is given by:

$$[R_t] = [s_0]\left(1 + \sum_{i=1}^{n} j_i\right)$$

If the receptor is bound to a ligand H, the stabilities of the entire set of receptor states are potentially altered. Depending on the chemical characteristics of the ligand, some states will become more and some less stable, and others will not be affected. Thus, the state transitions in the bound receptor can be given as $b_i j_i = [s_i H]/[s_0 H]$, where b_i is a numerical multiplier of j_i; it tells how the stability of state i is altered when that ligand is bound. The macroscopic equilibrium affinity of the ligand is:

$$K_{macro} = \frac{\sum[\text{bound states}]}{[H]\sum[\text{free states}]} = k_0 \frac{\left(1 + \sum_{i=1}^{n} b_i j_i\right)}{\left(1 + \sum_{i=1}^{n} j_i\right)} \quad (1)$$

where $k_0 = [Hs_0]/([H][s_0])$ and represents the second-order association constant that we would measure if all receptors were simultaneously "immobilised" in that single reference state s_0. This equation asserts that the Gibbs free-energy change for the association of every ligand to a macromolecule should always be regarded as the result of two contributions. One is the change due to the intermolecular association between the reactants (the "virtual" constant k_0). The other is the displacement of intramolecular energy that the protein endures in the bound state (the fractional term of Eq. 1). Therefore, every ligand is expected to perturb the receptor conformation regardless of the biological effect. Where, then, is the microscopic equivalent of efficacy?

We can define the collection of biologically active states of the receptor as a subset drawn within the universal set of all its possible states. No physical criterion can identify such a subset, because a precise physical attribution would require that each active microstate be studied in isolation. We may call

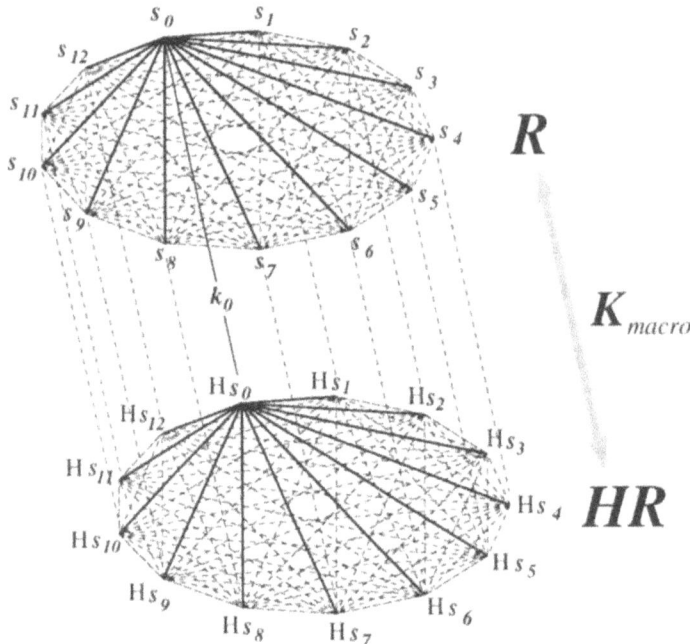

Fig. 3. Allosteric equilibria among microscopic states of the receptor (R). The receptor is schematically shown as consisting of 13 microscopic states (s, *indexed from 0 to 12*) in equilibrium with each other. The *upper and lower layers* represent the free and ligand-bound forms of the protein, respectively. All possible paths of state transitions within layers (first-order constants) and between layers (second-order constants) are drawn as *thin dotted lines*. The transitions that are sufficient to fully describe the entire system in thermodynamic equilibrium are shown as *thick solid lines*. Within layers, they connect the arbitrarily chosen reference state s_0 (or Hs_0 in the ligand-bound form) to all the others. The corresponding parameters are j_i for the free layer, and b_ij_i for the bound layer ($i = 1-12$). Between layers, a single transition (governed by the virtual affinity k_0) is enough to connect the two receptor forms. The parameters of such transitions collectively define the macroscopic affinity, K_{macro}, describing the formation of the HR complex as explained in the text

this the "informational" set I, for its existence can only be guessed from the signalling activity of the receptor. Let us also consider physically "predominant" subsets of states that may appear in either the free or bound form of the receptor. A predominant subset P encloses all states whose stability exceeds a conveniently chosen critical value (all substates in a given energy minimum). Obviously, there will be a different P for each type of ligand that can bind to the receptor. We can now define ligand efficacy with a series of Boolean rules that specify the degree of intersection between the subset I and all possible subsets P in the conformational space (Fig. 4). The intersection is small in the empty receptor and determines the extent of constitutive receptor activity. If a full agonist is bound, the intersection is maximal, so I is a

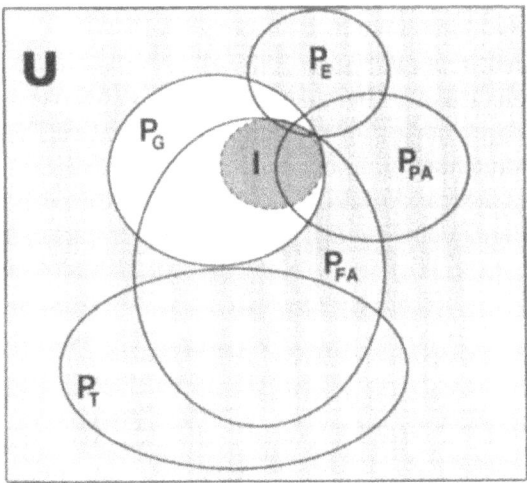

Fig. 4. Venn diagram of ligand efficacy, according to the microscopic allosteric model. The conformational space of all possible states of the receptor is represented as the universal set, U. The subset I contains biologically relevant states. Physically favoured subsets P of states (see text) are traced either for the unbound form (subscript E), or for the forms bound to G protein (G), full agonist (FA), partial agonist (PA) or antagonist (T), respectively. The intersection between I and all the others defines microscopic efficacy

proper subset of P (including the "perfect" and unlikely case of I = P). Intersections anywhere between the previous two extremes identify partial agonists, while antagonists are defined by an intersection equal to or smaller than that in the empty receptor or by the case where I and P are disjoint sets.

The relevant implication of this microscopic definition of efficacy is that even the most sophisticated tool of comparison of the structures of the bound or empty form of the receptor cannot warrant positive identification of the biologically relevant conformation. For example, it is often thought that an analysis of the difference in X-ray-resolved average structures of the receptor crystalised in either full-agonist or antagonist-bound forms would lead to identification of the biologically active configuration. However, the identification of active states among all those that are not shared between the two ligand-bound conformations may still be a virtually unsolvable problem. The more ligands we include in the study, the greater the chance that the biologically relevant states may be identified, but that still depends on largely serendipitous factors, such as choice of available ligands, their chemical properties and the range of their efficacies.

G protein-coupled receptors, however, constitute a special case. Here, the predominant set of the G protein-bound receptor (P_G in Fig. 4) provides a "physical domain" where biological states must be present. Therefore, the examination of the similarities in state composition between the agonist-bound and G protein-bound receptor forms may be far more efficient than the analysis of the differences between agonist- and antagonist-bound receptors. A conceptually related strategy is the study of the conformations of receptors constitutively activated by mutagenesis. Since it is reasonable to assume that such mutations make the receptor active in a way similar to agonist binding, the predominant states of constitutively active receptors must include the active states, which allows one to search them even in the absence of precise information about the ligand-bound receptor forms. The usefulness of this strategy has been documented by a combination of molecular-dynamics simulations and site-directed mutagenesis (SCHEER et al. 1996, 1997; SHEER and COTECCHIA 1997).

E. A Stochastic Model of Molecular Efficacy

I. Protein Motion and Fluctuations in Its Conformational Space

Allosteric states – even if microscopic – lead to a fundamentally static view of protein conformation. However, protein function depends on both structure and dynamics (BROOKS et al. 1987), and the second is conceivably much more important than the first. In a folded protein, the sharp inequality in the distribution of forces (strong covalent bonds along the backbone versus much weaker bonds across all residues) is not fixed but fluctuates as weak interactions are constantly broken and formed by thermal oscillations (KARPLUS and PETSKO 1990; GERSTEIN et al. 1994; HALTIA and FREIRE 1995). This generates large-scale, plastic conformational movements that seamlessly coexist with the vibrational motions of all the atoms (AMADEI et al. 1993). Through such complex dynamics, each macromolecule at room temperature stochastically explores its energy landscape. Small energy fluctuations around the equilibrium promote transitions among all substates in a local minimum. Larger, less frequent perturbations may allow jumps over energy barriers and movement towards other minima of the hypersurface. As a consequence, certain states and substates will emerge more frequently than others, depending on the intrinsic nature of the energy landscape and the interactions of the receptor with its external partners. Functions may be the intrinsic properties of clusters of substates that are visited more or less often but, just as likely, they may be the collective results of the way the protein moves across its landscape (KARPLUS and PETSKO 1990). Thus, the internal equilibrium of the protein should be regarded as a probability distribution of occurrences whose domain is the entire conformational space of the macromolecule, rather than as a collection of states given by a fixed matrix of stability constants.

II. Probability Distribution of Microscopic States and Derivation of Macroscopic Constants

Let us consider again the space of all theoretically possible states of a protein and assume that each member (s_i) of this space is visited by the protein with a certain probability (p_i) in the unit time. Then $\sum_{i=0}^{n}[p_i]=1$ for the entire conformational space. In thermodynamic equilibrium, each p_i is expected to be time-independent, and their collection (p) defines a time-independent probability distribution over the entire conformational space of the receptor protein (R). In an ensemble of such receptors, these probabilities can be seen as the relative abundance of microstates (TOLMAN 1938; HILL 1956): $p_i = [s_i]/[R]$ (where brackets reads as the number or concentration of the corresponding species). It follows that: $\sum_{i=0}^{n}[s_i]=[R]$. Given the probability distribution p, we can further define a set of constants (j_i) that govern the transitions between an arbitrary reference state (s_0) and all the others (s_i, $i \neq 0$): $j_i = [s_i]/[s_0] = p_i/p_0$, with $j_0 = 1$. Since j also serves as a definition of free-energy difference between any state s_i and s_0, then $j_i = e^{-\Delta G_{i0}/kT}$. The sum of j_i over states for the empty receptor is

$$\Omega = 1 + \sum_{i=1}^{n} j_i \left[= (p_0)^{-1} \right] \qquad (2)$$

and represents, in a loose sense, a conditional partition function for the empty receptor, because it yields the standard result of $p_i = e^{-\Delta G_{i0}/kT}/\Omega$.

Since the binding of a ligand X can be expected to change the frequency of appearance of microstates in the macromolecule, the probability distribution of the X-bound receptor over the conformational space is given by: $^xp_i = [^xs_i]/[XR]$ and $^xj_i = [^xs_i]/[^xs_0] = {}^xp_i/{}^xp_0$, (with $^xj_0 = 1$), where left-hand superscripts indicate the presence of bound ligand. By defining the effect that ligand binding imparts on the equilibrium between s_0 and s_i with $^xb_i = {}^xj_i/j_i$, the conditional partition function of the ligand-bound receptor is:

$$\Omega_x = 1 + \sum_{i=1}^{n} {}^xb_i j_i$$

or

$$\Omega_x = 1 + \Omega \sum_{i=1}^{n} {}^xb_i p_i \qquad (3)$$

For a second ligand Y, we likewise have:

$$\Omega_y = 1 + \Omega \sum_{i=1}^{n} {}^yb_i p_i \qquad (4)$$

If both X and Y can be simultaneously bound,

$$\Omega_{xy} = 1 + \Omega \sum_{i=1}^{n} {}^xb^y b_i p_i \qquad (5)$$

Thus, the macroscopic free-energy coupling (α in Scheme 2) between the binding of X and Y is microscopically defined as follows:

$$\alpha = (\Omega \Omega_{xy}) \times (\Omega_x \Omega_y)^{-1} \qquad (6)$$

provided that the reference microstate (s_0) is the same for all partition functions (Eq. 2–5).

The virtual microscopic affinity, k_0, for the formation of the ligand–s_0 complex is $^xk_0 = [^xs_0]/([X][s_0])$ and $^yk_0 = [^ys_0]/([Y][s_0])$, for ligands X and Y, respectively. Thus, the macroscopic affinities for the two ligands (K_x and K_y), that govern the macroscopic-binding equilibria, $X + R \leftrightarrow XR$ and $Y + R \leftrightarrow YR$, are defined as:

$$K_X = {^xk_0}(\Omega_x)(\Omega)^{-1}; K_y = {^yk_0}(\Omega_y)(\Omega)^{-1} \qquad (7)$$

Equation 7 is the probabilistic equivalent of the macroscopic affinity derived before (Eq. 1) and shows again the separate contributions of protein motion (the terms Ω_x/Ω and Ω_y/Ω) and intermolecular forces ($^xk_0$ and $^yk_0$) to a macroscopic ligand-binding process. We can interpret k_0 as related to the sum of interaction energies between ligand and receptor (i.e. what might be computed from an average structural model of their complex, which does, in fact, constitute a "motionless" reference state of the receptor). It is evident from Eq. 7 that a mutation in the receptor may change the macroscopic affinity through either k_0 or Ω. Therefore, a map of the ligand "docking site" inferred from the effect of mutations on binding affinity cannot possibly coincide with that deduced from an X-ray structure, because the contribution of molecular motion will influence the first far more than the second.

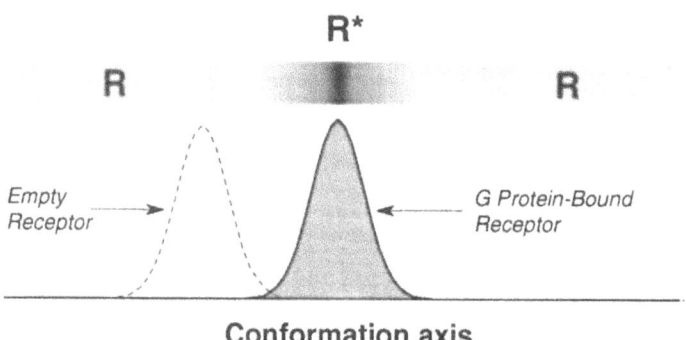

Fig. 5. Relationship between the probability of emergence of function and the macroscopic active conformation R* of the receptor. The figure is a schematic representation of the logical process that generates macroscopic conformations from the probability distributions of microstates. The conformational space of the receptor is schematically shown as a unidimensional axis. The state distributions over the conformational space in the empty (*dotted*) or G protein-bound (*shaded*) receptor forms are drawn as *bell-shaped lines*. On top, the same conformational space is re-drawn as a *gray-scaled string* (where density is proportional to the probability level) and represents the projection of this space onto the macroscopic observation of biological response. Here, the G protein-bound distribution (*dark smearing band*) shows as a functional target for agonism (the functional form). The receptor is seen as a bipartite macroscopic entity switching between active (R*) and inactive (R) forms

III. Probabilistic Interpretation of Ligand Efficacy

The model presented above provides a general microscopic formulation from which all macroscopic definitions of efficacy (β and α in Scheme 1 and 2) can be derived. In the case of a G protein-coupled receptor, the subset of states that have a relatively high probability of occurrence when the receptor is in the G protein-bound form identifies the target of biological relevance. Macroscopically, this is equivalent to generating a partition in the receptor universe between an active form (R*), which localises the probability of emergence of function, and an inactive form (R), which is the rest (Fig. 5). Efficacy depends on how ligand-induced distributions of states intersect the biological target distribution (Fig. 6 for a schematic example).

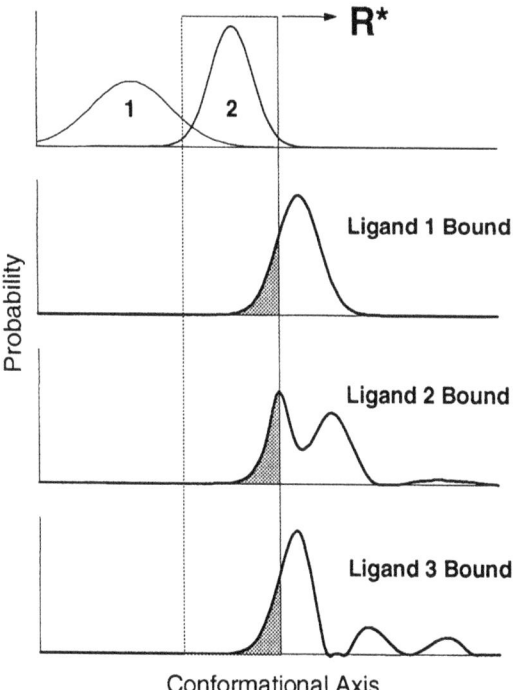

Fig. 6. Ligand-induced perturbations of the state distribution of the receptor and efficacy. The topmost panel shows schematic state distributions over the conformational space of a receptor in empty (1) or G protein-bound (2) form. The conformational space is drawn over a single dimension, as in Fig. 5. The next three panels show three possible state distributions on the same conformational axis for the receptor bound to three different ligands. The virtual subspace (R*) defined by the distribution of G protein-bound receptor is *framed*, and the intersecting areas of the ligand-induced distributions are *shaded*. The calculated α values of the three ligands (Eq. 6 in the text) range between 98 and 102. The shapes of the distributions are arbitrarily chosen for demonstration purposes. Note that the three ligands induce different physical distortions of the conformational space but result in equal efficacies

To illustrate the relationship between ligand efficacy and physical perturbation, we provide an example based on computer simulations of a simplified conformational space (Fig. 7). We assume a receptor consisting of 100 microstates (arbitrarily indexed as 0–99 on one dimension), the distribution of which (in the empty form) is exponential in state number (Fig. 7a). For the G protein-bound receptor, we set a bell-shaped distribution over the same conformational axis (a Gaussian in state number, with the expected value located at state index 25). We next assume a panel of ligands (consisting of 56 members) defined as follows. Each ligand, when bound to the receptor, results in a bell-shaped distribution having a different expected value on the conformational axis. This value is placed at state number 5 for the first ligand, and it is gradually shifted rightwards for succeeding ligands (i.e. 6 for the second, 7 for the third and so forth). The variances of the distributions are set identical for all 56 ligands (Fig. 7a). From all distributions, the macroscopic efficacies α of each ligand are computed using Eqs. 1–5. The relationship between these numerical values [$\log(\alpha)$] and the intersections among the probability distributions of microstates is shown in Fig. 7b.

It is evident from this simulation that efficacy cannot be related to the extent of physical perturbation that ligands produce in the receptor. As shown in the example (Fig. 7a), pairs of ligands producing vastly different "distortions" of the distributions of states of the empty receptor (zone 1 and zone 2 in Fig. 7b) result in perfectly identical extents (or lacks) of macroscopic efficacy. This means that the same level of efficacy can be produced by an endless number of diverse possible modes of receptor perturbation. This number, however, becomes progressively smaller as the ligand approaches full agonism (ligand 1 in Fig. 7b), because there are fewer ways to match the G protein-

Fig. 7A,B. Numerical simulation of the conformational space of a receptor bound to ligands of varying efficacies. **A** A monodimensional conformational space of 100 elements (as numbered in the x-axis) is defined as explained in the text. r, RG, and HR are the conditional distributions of microstates for empty, G protein-bound and ligand-bound receptors, respectively, over the same conformational axis. The probabilities of states are given in the ordinates. The exponential (empty receptor) and Gaussian distributions (G protein- or ligand-bound receptor) with equal variances and different means were chosen arbitrarily for simulation purposes. Probability values less than 10^{-10} were not allowed, so all distributions become uniform when the probability falls below that value. For clarity, of the distributions for the 56 ligands (assigned as explained in the text), only the first and the last are shown. **B** *Upper panel*: $\log(\alpha)$ values for different ligands are calculated (Eq. 6) from the distributions in A and are plotted against the positions of the mean values of the corresponding Gaussians on the conformational axis. Five ligands further illustrated in the lower panel (the negative antagonists 1 and 5, the neutral antagonists 2 and 4, the full agonist 3) are marked. Ligands are divided into two zones, as indicated on the picture. *Lower panel*: the state distributions of the bound receptor (HR, *black, filled Gaussians*) related with the five ligands marked in the upper panel, are compared with those of empty (R exponential) and G protein-bound (RG *gray-shaded Gaussians*) receptor, respectively. Note that, in this particular simulation, full agonism [maximum $\log(\alpha)$ value with the ligand labelled 3)] results from a perfect match between G protein- and ligand-bound receptor distributions

A Look at Receptor Efficacy 243

A

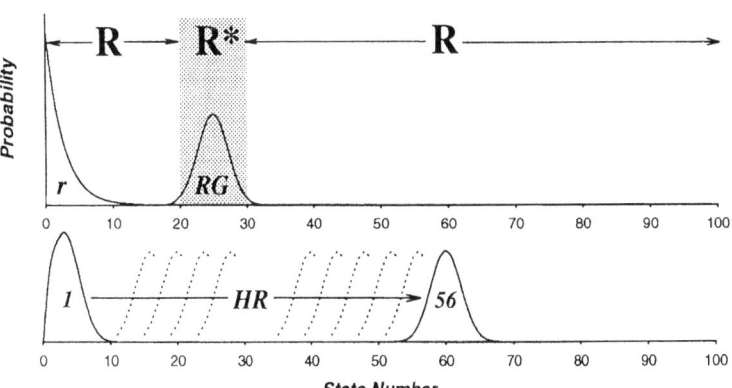

B

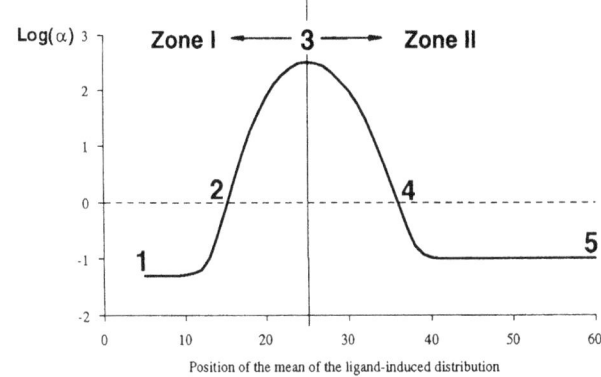

Position of the mean of the ligand-induced distribution

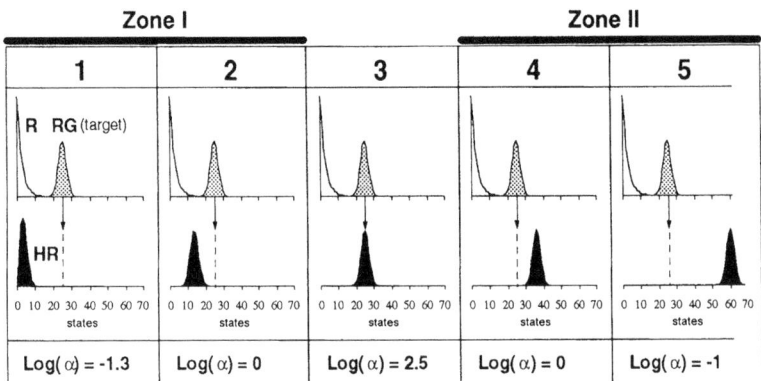

target perturbation than to miss it. This also explains why, at the macroscopic level, there is an upper limit to the range of efficacy (Fig. 1).

IV. Relationship Between Physical States and Biological Function

There are two fundamental differences of interpretation between the probabilistic model described here and the general microscopic definition of allosteric equilibria given in the previous section.

1. In the microscopic allosteric model, we are forced to view biological activity as a static property present in a hidden set of microstates (the informational set). This implies that we conceptually decompose function into microscopic quanta like the protein itself so that each microstate either is or is not microscopically active. In the probabilistic model, ligand activity can only be defined as the probability of occurrence of all the microstates of the receptor among concurrent perturbations. This intersection emerges and gains functional meaning en route from the microscopic to the macroscopic level. Therefore, function is seen as a collective property that results from the microscopic elements of the receptors and their motions but cannot be individually attributed to any of them.
2. Consequently, the problem of relating receptor physics to function is differently formulated in the two models. In the allosteric model, it is the trivial question of how individually active or inactive microstates are embodied in physically identified clusters. In the stochastic model, receptor function and physics move along opposite vectors. The first is evanescent or blurred microscopically but sharpens at the macroscopic level, and vice versa for the second. They may overlap, depending on the reference frame of observation, but do not converge.

This has crucial implications for the interpretation of the results of receptor-engineering studies. A single residue replacement in a receptor is a microscopic change within the space of the macromolecule. However, since proteins are complex systems, the perturbation that such change generates may either remain microscopic or have vast repercussions at the macroscopic level. To understand the results of mutations, the stochastic nature of the conformational space in which they are created cannot be overlooked. Here, by examining two experimental examples, we show that the stochastic model discussed above can be of help.

V. Relationship Between Efficacy and Fluorescence Changes in β_2-Adrenoceptors

For the rationale and experimental strategies of this elegant study, we refer to the original paper (GETHER et al. 1995). Only the essence of the results will be discussed here. Purified β_2-adrenoceptors were covalently labeled on cysteines with the environment-sensitive fluorophore 4-[(iodoacetoxy)ethylmethylamino]-7-nitro-2,1,3-benzoxadiazole (IANBD). Ligand-induced changes of fluorescence were determined at saturating concentrations of a panel of ligands with a wide range of efficacy. As shown in a replot of the data (Fig. 8b, left), agonists and antagonists produced opposite shifts in fluorescence, the magnitude of which was correlated to their intrinsic activity on adenylate cyclase. Although not obvious from the overall outstanding correlation between fluorescence and efficacy (Fig. 8b, left), these data reveal a systematic departure from the predictions of a two-state macroscopic, allosteric model.

To show why, we need to define how the change in fluorescence can be predicted from the model. We assume that the inactive (R) and the active (R*) receptor conformations have two different arbitrary weights (w_1 and w_2) on the observed total fluorescence. We thus define fluorescence per mole of receptor (R_{total}), in the absence (^0f) or presence of saturating concentrations of ligand (^hf) as follows:

$$^0f = (w_1[R]+w_2[R^*])/R_{total}; \quad ^hf = (w_1[HR]+w_2[HR^*])/R_{total} \tag{8}$$

The ligand-induced shift in fluorescence is $\Delta f = {^hf} - {^0f}$ and can be derived by combining Eq. 8 and the mass-conservation relationships for the equilibrium given in Scheme 1 (Sect. D.II):

$$\Delta f = F\Delta w(\beta - 1) \tag{9}$$

where $\Delta w = w_2 - w_1$, and F is the positive-valued function $F(J,\beta) = J/[(1 + J)(1 + \beta J)]$. Since ligands shift fluorescence (Fig. 8a, left), w_2 and w_1 must be different. If we choose $w_2 < w_1$, then $\Delta f < 0$ for agonists ($\beta > 1$), and $\Delta f > 0$ for negative antagonists ($\beta < 1$), as observed in the experimental data. However, $\Delta f = 0$ means $\beta = 1$ and implies that neutral antagonists are the only ligands that should result in no change of fluorescence. This is in conflict with the experiments, because the trend shown in Fig. 8a (left) predicts that zero shift occurs with some partial agonists. The implication is that R* is different – albeit slightly – with respect to function and fluorescence, for there are ligand-induced changes in the receptor that are visible to either the one or the other, but not to both. In fact, (1) IANBD senses a perturbation induced by neutral antagonists that has no effect on the cyclase response and (2) the molecular change induced by weak partial agonists is detected by the cyclase but not by the fluorophore.

At this point, there are two choices. One is to blame "unknown factors" that may warp (to some extent) either fluorescence or efficacy measurements and to be satisfied with the prominent correlation between the two criteria. Another is to search for the root of the "imperfection" within the microscopic nature of the receptor. As shown next, the second choice brings further insight.

For the simulation of fluorescence according to the probabilistic model, we use exactly the same example as that illustrated in Fig. 7. To calculate the macroscopic shift induced by each of the 56 ligands, we assume that each microstate of the conformational space contributes to the total fluorescence with a microscopic weight (w_i). The macroscopically observed fluorescences in empty or ligated receptors are $^0f = \sum_i w_i p_i$ and $^h f_m = \sum_i w_i^h p_{im}$, respectively, where the index i enumerates the states and m stands for different ligands. The ligand-induced change in fluorescence intensity can be calculated as $\Delta f_m = {}^h f_m - {}^0 f$ Fluorescence weights – arbitrarily given otherwise – were chosen

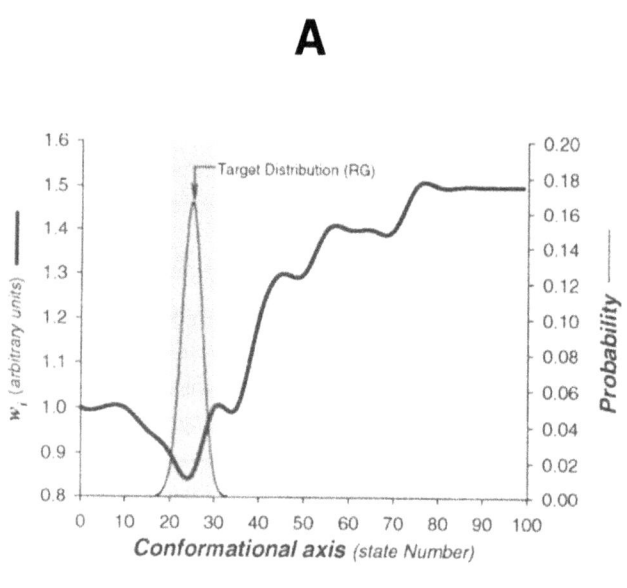

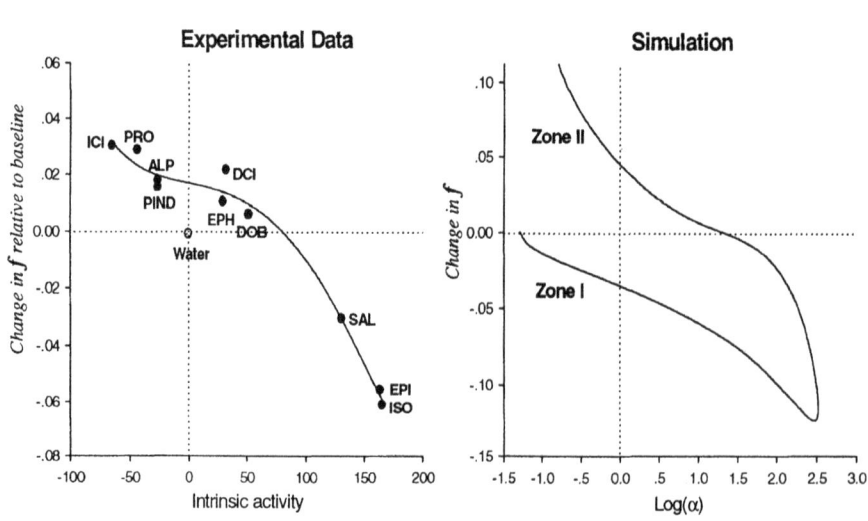

to have a global minimum on the conformational axis around the R* region. That is, they point to the macroscopic "active" form of the receptor (Fig. 8a).

Now the relationship between Δf_m and efficacy (Fig. 8b, right) draws a sharply bent line in which ligands in zone I and zone II (Fig. 7b) apparently trace two possible trends. In the second, simulated and experimental data are in good agreement. In fact: (1) the relationships are both monotone and equally shaped; (2) fluorescence shifts have opposite signs for negative antagonists and strong agonists but are positive for neutral antagonists; (3) there is a critical value of efficacy (partial agonist) at which the shift in fluorescence is zero.

As shown in Fig. 8b (right), there are two different relationships between fluorescence and efficacy which converge in the full-agonism "kink" (zones I and II). If we analyse the data with a slightly more realistic model (i.e. by computing a multi-dimensional rather than linear conformational space), the relationship between Δf and efficacy becomes a funnel-shaped, uneven hypersurface that encloses all possible sections of space where different shifts of fluorescence project onto equal regions of efficacy (data not shown). This reflects the variety of ligand-induced physical perturbations that can lead to equal levels of efficacy. For example, in the two-state analysis of the data, dichloroisoproterenol (DCI) seems to be an outlier, because it has as much efficacy as ephedrine (EPH) but has a fluorescence shift like that of the antagonist alprenolol (Fig. 8a, left). In the probabilistic model, however, this means that DCI and EPH produce two alternative physical perturbations of the receptor that result in quite similar intersections with the biological target.

◀───────────────────────────────

Fig.8. A Fluorescence weights of microstates in the receptor conformational space. Each member of the conformational space described in Fig. 7 is given an arbitrary fluorescence weight (w_i). The weights are plotted on the same conformational axis as the state distribution of G protein-bound receptor (indicated as RG). The virtual subset that defines the macroscopic, "active" conformation of the receptor (R*) is *shaded*. Fluorescence weights have a global minimum centred in the R* region but slightly off the peak of the target distribution (RG). **B** Relationships between ligand-induced changes in fluorescence and ligand efficacy. *Left panel*: experimentally recorded ligand-induced shifts in fluorescence intensity in 4-[(iodoacetoxy)ethylmethylamino]-7-nitro-2,1,3-benzoxadiazole-labelled human β_2-adrenoceptors were digitised from GETHER et al. (1995) and are re-plotted against the intrinsic activities of ligands. The shifts are given as relative to the background fluorescence of the receptor (indicated as water in the picture). The abbreviations used for the ligands are: ICI (ICI118551), PRO (propranolol), ALP (alprenolol), PIND (pindolol), DCI (dichloroisoproterenol), EPH (ephedrine), DOB (dobutamine), SAL (salbutamol), EPI (epinephrine), and (ISO) isoproterenol. Data points were interpolated using a fourth-order polynomial. *Right panel*: computer simulation of experimental data using the set of ligands described in Fig. 7 and fluorescence weights as in the *left panel*. The ligand-induced shift in receptor fluorescence (calculated as described in the text) is plotted against the logarithm of the molecular efficacy of the ligand (α). The Δf curve bends in the full agonist [$\log(\alpha) = 2.5$] region and traces two alternative relationships, reflecting the two ligand zones (*zones I and II*) described in Fig. 7b. Note that ligands aligned over *zone II* agree best with the experimental data

We can predict that the more ligands we include in the study, the more scattered will be the relationship from the perspective of the two-state model. If we reacted to such discrepancies by postulating additional macroscopic forms of the receptor (R*, R**, R***, etc.), we would certainly crowd our brain with a growing number of ill-defined parameters, but would gain little understanding. It is not the scarcity of states, in fact, but the deterministic nature of the macroscopic model that imposes one-to-one relationships between physical and functional changes, which is both impossible and unreasonable to expect in a protein.

VI. Correlated Macroscopic Changes in Constitutively Active Adrenoceptors

Mutagenesis of residues in critical areas of adrenoceptors produces constitutive activity (KJELSBERG et al. 1992; REN et al. 1993; SAMAMA et al. 1993; SCHEER and COTECCHIA 1997). The "activating" sites are generally located at the cytosolic side of the molecule, which is not accessible to ligands. However, agonist affinity is enhanced by these mutations in a manner that is related to both the extent of constitutive activity and the magnitude of ligand efficacy. Therefore, the effect on ligand affinity of such mutations is allosteric in nature. This linkage, as stated in Sect. D.III, could in principle be explained according to Scheme 2 (the ternary complex model, or TCM, in pharmacologists' jargon), assuming that G protein affinity M is enhanced by the mutations. However, experiments show that the presence of G protein is not indispensable for observation of the constitutively active phenotype (SAMAMA et al. 1993; GETHER et al. 1997). Instead, they suggest that spontaneous activity is the effect of an intrinsic shift towards the "active" conformation of the receptor itself (SCHEER and COTECCHIA 1997; GETHER et al. 1997).

A pragmatic way to model such a situation is to combine the two equilibria given in Schemes 1 and 2 into one to produce either a two-cycle (if only R* can bind to G; SAMAMA et al. 1993) or a "cubic", six-cycle (if R also binds to G; WEISS et al. 1996) thermodynamic reaction scheme. Either model can explain activating mutations as resulting from the increase of the equilibrium constant $J = [R^*]/[R]$. Mathematically, this is equivalent to stating that if there was an exclusive change of J (which is explicit in Scheme 1 and "hidden" in Scheme 2) in a receptor studied according to Scheme 2, then the three parameters K, M and α of the TCM must all be affected as follows:

$$\alpha_{mut} = \alpha\beta(1+J)/(1+\beta J); \quad K_{mut} = K(1+\beta J)/(1+J); \quad M_{mut} = MJ/(1+J) \quad (10)$$

Since J is present in all three equations, it affects (in a linked fashion) efficacy (α), ligand affinity (K) and G protein affinity (M) simultaneously. The correlation among all three parameters as a function of the variation of J is plotted in Fig. 9.

The existence of such correlations is the fundamental criterion for deciding whether a mutation exclusively affects the intramolecular equilibria of G

A Look at Receptor Efficacy 249

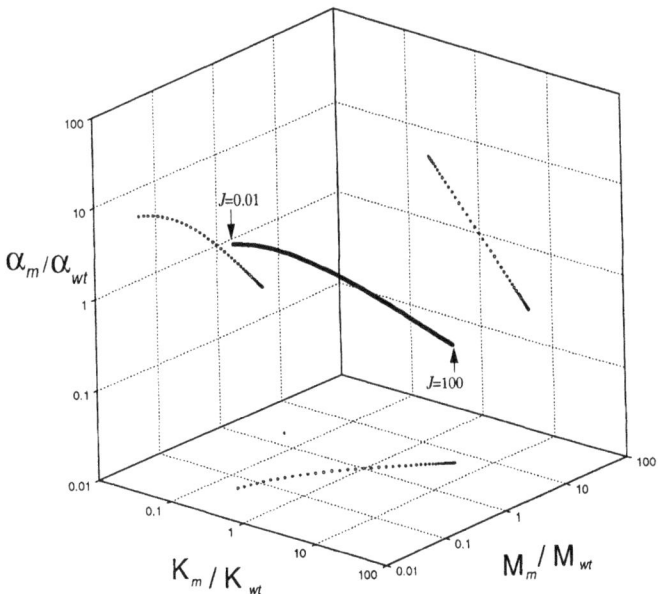

Fig. 9. Relationship between the variation of J and the macroscopic parameters of the ternary-complex cycle. The graph shows how a change in the tautomeric equilibrium of the receptor imposes correlated changes in the parameters of the ternary complex model (TCM). It is envisioned that, in a receptor studied according to Scheme 2 in the text (TCM), a perturbation applied through mutagenesis changes only J (which is hidden in Scheme 2 and is defined in Scheme 1). Data are computed according to Eq. 10. J is varied from 0.01 to 100, and β is maintained fixed at 30. The data are plotted in a three-dimensional Cartesian space where the values of the three parameters are scaled relative to J = 1, arbitrarily chosen as the "wild-type" receptor point (subscripts m and wt stand for mutant and wild type, respectively). The projections of the curve on the three surfaces show all the correlations between the three pairs of parameters. J > 1 and J < 1 correspond to activating and inactivating mutations, respectively

protein-coupled receptors, because direct diagnostic tools are lacking. A remarkable exception is the fluorescence example discussed above (GETHER et al. 1997). Important issues are how perfect such correlations should be and how to detect outliers.

For example, A293 in the third cytosolic loop (KJELSBERG et al. 1992) and the acidic residue of the conserved DRY sequence (SCHEER et al. 1996, 1997) of α_1-adrenoceptors, are crucial "activating" spots. All-residue substitutions at those sites produce uniform increases of basal activity and enhancement of agonist affinity. The changes in the two parameters are broadly correlated, but there are deviations. Also, replacement of arginine in DRY enhances affinity in the same way that replacement of the adjacent aspartate does, but does not affect constitutive activity. Shall we assume that the correlated effects are caused by a change of J, while the others are caused by a different parameter, such as M? If so, is it conceivable that in the same point of the sequence, one residue affects the equilibrium conformation of the receptor while another only changes the intermolecular forces that hold it bound to an external ligand? Shall we need to build a different model of receptor action for each different residue that is replaced? It is clear from such questions that the static nature of the R↔R* equilibrium eventually drives us into stunning confusion.

According to the stochastic model of the receptor conformational space, a mutation that alters the state distribution of the free receptor enhances constitutive activity if such distortion improves the intersection between the probability distributions of empty and G protein-bound receptors. However, any perturbation (activating or not) in the probability distribution of the free receptor must also affect efficacy and affinity, for both depend on the probability distribution of free receptor, as shown in Eqs 2–6.

To illustrate how this happens, let us assume a mutation that perturbs the distribution of states in the free receptor but alters neither the tendency of a ligand to distort such distribution (i.e. b_i in Eqs. 4–6) nor the intermolecular forces holding it bound to the receptor (k_0). Wild-type and mutated receptors are defined by two different probability distributions of states in the unligated form called $p1$ and $p2$. Let us examine the probability distribution of a hypothetical ligand X, which we may equally well regard, for the purpose of this argument, as either an agonist or the G protein. We must compare three state distributions in the ligand-bound receptor (Fig. 10a): (1) the distribution of $^x b$, which is invariant and represents the "projected" or potential effect of the ligand, (2) the calculated distribution $^x p1$, which indicates how that prospect is "realised" in the wild-type receptor, and (3) the distribution $^x p2$, which shows how $^x p1$ is changed in the mutant.

The discrepancy between b_i and $^x p1$ may be defined as resistance of the receptor to the distortion induced by the ligand upon binding. It is the difference between the actual distribution in ligand-bound form ($^x p1$) and what would result if the bound receptor perfectly complied with the distribution "intended" by a ligand with $^x b_i$. It is evident from Fig. 10a that the mutation diminishes such resistance ($^x p2$ is shifted towards $^x b_i$). Numerical calculations (data not shown) indicate that, if p is varied with $^x k_0$ and $^x b_i$ held constant, the macroscopic affinity of the ligand is inversely proportional to receptor resistance. This means that the increase in affinity induced by an activating mutation can be interpreted as diminished thermodynamic work done (against the resistance of the receptor) by agonists changing the average conformation. Similar conclusions have been proposed from molecular-mechanics arguments (GETHER et al. 1997). Analogous considerations apply to G protein binding. Therefore, the increase in G protein and agonist macroscopic affinities induced by the activating mutations is the result of the concurrent reduction of the receptor resistances to the tendencies of the two ligands to alter the intrinsic motion of the receptor when both ligands are simultaneously bound.

To illustrate numerically the correlations among macroscopic parameters, we again assumed a conformational axis of 100 elements. The distributions of b_i for agonist and G protein over the conformational axis were set similar on

A

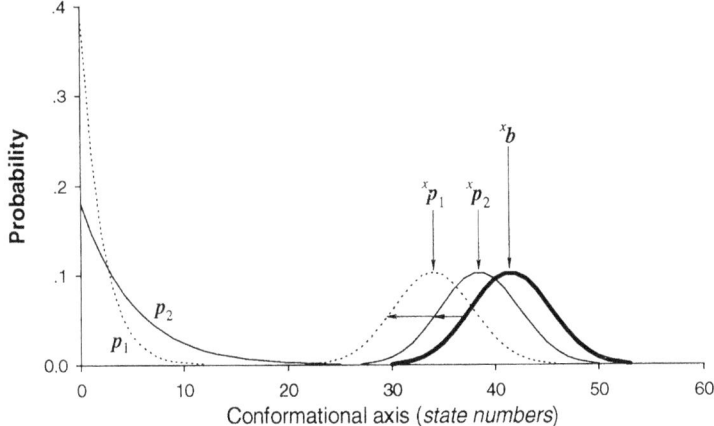

B

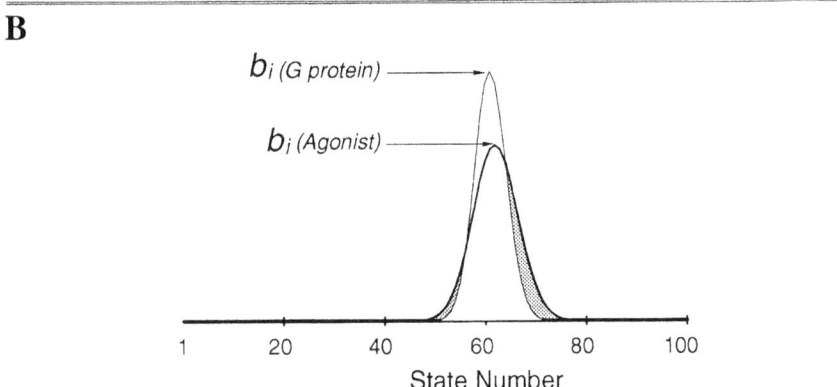

Fig. 10. A Effect of state distribution of empty receptors on the ligand-induced state distribution. The conformational space is given as a unidimensional axis consisting of 100 elements, as in Fig. 7. The curve $^x b$ represents an arbitrarily distributed set of allosteric factors for the ligand (see text). The values of $^x b$ are normalised with respect to the area under the entire curve of $^x b$ in order to show them on the same probability scale as the others. p_1 And p_2 are two exponential distributions representing the state distributions of two different empty receptors (wild type and mutant). The ligand-bound counterparts of the empty-receptor distributions, i.e. $^x p_1$ and $^x p_2$, are deduced from p_1, p_2 and $^x b$ using Eqs. 2 and 3. Note that $^x b$ is what ligands tend to produce upon binding to the receptor, but what is actually produced is $^x p_1$ or $^x p_2$ in wild type and mutant, respectively. The discrepancy between "intended" and "realised" distributions of ligand-bound receptors is larger for $^x p_1$ than $^x p_2$. The macroscopic affinity K_x (calculated with Eq. 7) was enhanced 4.2 orders of magnitude in the mutant (not shown). **B** Distributions of b values for agonist and G protein. The two distributions are scaled as in **A** on the same conformational axis. Their intersection is the "projected" efficacy of that agonist. The actual efficacy (i.e. the intersection between the probability distributions $^x p$ of agonist or G protein-bound receptor) depends on the state distribution of the empty receptor and changes if that receptor is modified by mutations. The slight offset in the two distributions reflects the idea that the conformation of the receptor bound to both ligands cannot be exactly identical to that bound to either ligand alone (Sect. D.II). These distributions were used for the simulations shown in Fig. 11

average but were slightly different in details (as in Fig. 10b). To simulate mutants, 5000 different random distributions of the free receptor were generated. The results are plotted (Fig. 11) as previously shown for the macroscopic two-state calculations (Fig. 9). The output shows how mutation-induced changes in the internal equilibrium of the receptor result in correlated changes of the macroscopic parameters of agonist–receptor–G protein interactions. On average, the macroscopic interdependence of the parameters is quite similar to that predicted from the two-state model TCM (compare Figs 9 and 11) but, in detail, individual mutations that violate such correlations are clearly evident.

To compare theoretical predictions and experimental data, the macroscopic quantities of Fig. 11 were converted into something close to experi-

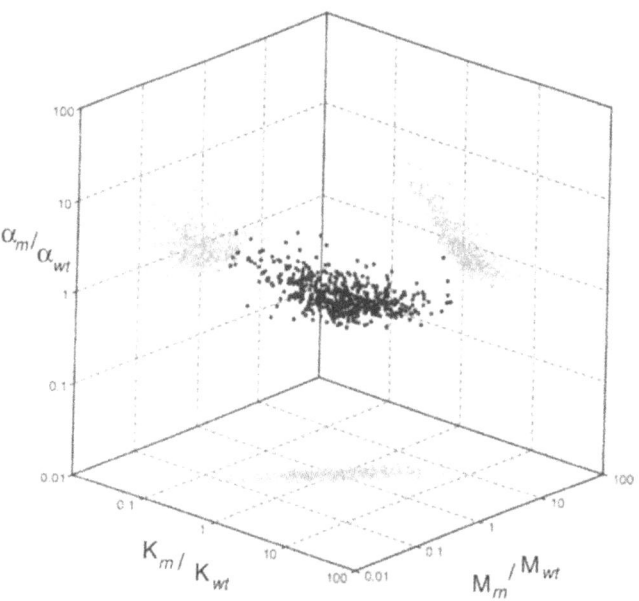

Fig. 11. Perturbations of internal equilibrium of unbound receptor and macroscopic parameters of the ternary complex cycle (K, M, a). Each *point* in the picture corresponds to a possible change of the state distributions of the empty receptor, upon which ligand and G protein act, with invariant sets of b values (Fig. 10b). A total of 5000 variations were computed to simulate random mutations. Each point was generated by a two-step procedure. First, probability values from a discrete exponential distribution $\left(\text{with } \sum_{i=0}^{99} p_i = 1\right)$ were randomly assigned to the elements of the conformational space (defined as in Figs. 7–9). The result is taken as the state distribution of the empty receptor. Next, a single point on the parameter space (K, M, α) is calculated using Eqs. 1–7 with the b values shown in Fig. 10b. *Each point* thus obtained corresponds to a mutation of the receptor that alters only the internal equilibrium of the empty receptor (see text for details). An arbitrary point in the macroscopic-parameter space is selected as the wild-type receptor, and all others are shown relative to that. Data are plotted and projected as in Fig. 9

mental measurables (legend in Fig. 12) and were compared with the results of "real" mutations. The experimental set included single- or double-residue replacements made in "activating sites" of the α_{1B}-adrenoceptor. Although prepared with a precise purpose in mind (SCHEER and COTECCHIA 1997), these mutants were treated as a "random" sample.

The experimental mutants fit within the region of variation predicted by the "theoretical" ones (Fig. 12) and may thus be consistent with a shift in the internal equilibria of the receptor. Analysed individually or as small selected groups (which would be the case if only a few mutations had been made in the study), we find examples that apparently call for more complex interpretations. In fact, examining the results – (1) enhancement of basal activity and

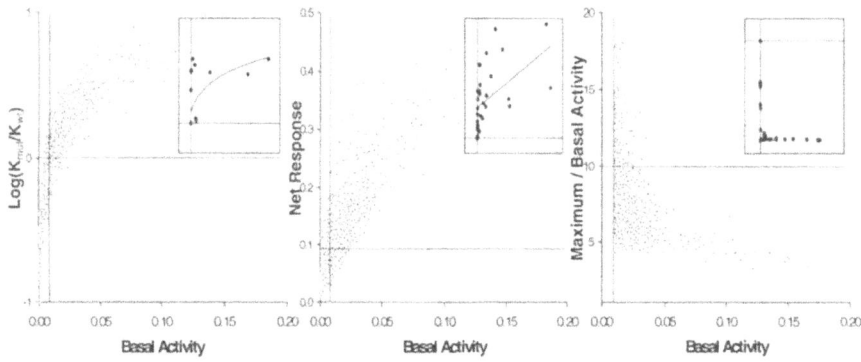

Fig. 12. Comparison of computer-simulated and experimental mutations. Each point from Fig. 11 is translated into experimental variables as follows: $\dfrac{[RG]}{R_{total}} \approx \dfrac{M[G]}{1+M[G]}$ for basal activity (receptor activity in the absence of ligand); $\dfrac{[HRG]}{R_{total}} \approx \dfrac{\alpha M[G]}{1+M[G]}$ for maximum response (at saturating ligand concentration); and $\Delta RG \approx \dfrac{[HRG]-[RG]}{R_{total}}$ for net response. Apparent agonist affinities are given in a logarithmic scale relative to that obtained in the wild-type receptor (K_{mut}/K_{wt}). We assumed $M[G] = 0.01$ for the wild-type receptor. The *crossings* between the *thin lines* of each graph indicate the position of the wild-type receptor. In the simulated data, both activating (increased basal activity) and inactivating (decreased basal activity) mutations are shown in the first two plots, but only the activating mutations are shown in the third to avoid scaling problems. *Insets* are the corresponding experimentally measured variables obtained from a series of α_{1b}-adrenoceptor mutants. Experimental details for the preparation of the mutants and their study are given elsewhere (SCHEER et al. 1996, 1997). The following groups of human α_{1B}-adrenergic-receptor mutants were plotted in the *insets*: (i) constitutively active mutants: N63A; D142 to N, E, F, R, G, K, L, S A; A293E; R288K/K280H/A293L (cam3); cam3/D142A; (ii) wild-type-like mutants: N63D; R160A; N344D; (iii) less- or non-coupled receptor mutants: N63 to V,K; D91 to E, N; N344A; Y348A; D142R/R143D; D142A/D91N; D91N/N63D; Cam3/R143 to A, N; Cam3/N344A; Cam3/D91N

(2) efficacy-dependent increase of ligand affinity – demonstrates that all three outcomes (i.e. 1, 2, or both) are found as possible results of the mutations. A deterministic interpretation would again force us to introduce multiple conformations and distribute the changes over new parameters defined in the hyperspace of complex reaction mechanisms. However, the validity of such models would only last until the next mutation was made.

We find, in this example, a situation complementary to the fluorescence case examined before. There, a single version of the receptor conformational space was confronted with many possible ligand-induced perturbations. Here, a single perturbation imposed by the same agonist is faced with many possible versions (mutations) of receptor space. The final message, however, is identical. The relationship between physical changes (measured in the first case or engineered in the second) and receptor function is likely to be evident on average but unclear in the details. These imperfections are not necessarily the outcome of experimental noise or poor resolution. They result from our insistence that the protein be viewed as a mechanical or deterministic object, which it is not.

Acknowledgements. The authors acknowledge support from the European Union BIOMED 2 programme "Inverse agonism. Implications for drug design".

References

Adie EJ, Milligan G (1994) Regulation of basal adenylyl-cyclase activity in NG10815 cells transfected to express the human β_2-adrenoceptor. Evidence for empty receptor stimulation of the adenylyl-cyclase cascade. Biochem J 303:803–808

Alberts B, Bray D, Lewis J, Raff M, Roberts K, Watson JD (1994) Molecular biology of the cell. Garland Publishing, Inc, New York, London

Amadei A, Linssen ABM, Berendsen HJC (1993) Essential dynamics of protein. Proteins 17:412–425

Ariëns EJ (1954) Affinity and intrinsic activity in the theory of competitive inhibition. Arch Int Pharmacodyn Ther 99:32–49

Ariëns EJ, Simonis AM, Van Rossum JM (1964) Drug receptor interaction: interaction of one or more drugs with one receptor system. In Ariëns EJ (ed) Molecular pharmacology: the mode of action of biologically active compounds. Academic Press, New York, pp 119–286

Baldwin JM (1994) Structure and function of receptors coupled to G proteins. Curr Opin Cell Biol 6:180–90

Barker EL, Westphal RS, Schmidt D, Sanders-Bush E (1994). Constitutively active 5-hydroxytryptamine 2C receptors reveal novel inverse agonist activity of receptor ligands. J Biol Chem 269:11687–11690

Birnbaumer L (1990) G proteins in signal transduction. Annu Rev Pharmacol Toxicol 30:675–705

Black JW, Leff P (1983) Operational models of pharmacological agonism. Proc R Soc Lond (Biol Sci) 220:141–162

Bourne HR (1997) How receptors talk to trimeric G proteins. Curr Opin Cell Biol 9:134–142

Brooks CL, Karplus M, Pettitt BM (1987) Proteins: a theoretical perspective of dynamics, structure and thermodynamics. In Prigogine I, Rice SA (series eds) Advances in Chemical Physics vol 71, John Wiley and Sons, New York

Bryngelson JD, Onuchic JN, Socci ND, Wolynes PG (1995) Funnels, pathways, and energy landscape of protein folding: a synthesis. Proteins: Structure, function and genetics 21:167–195

Chidiac P, Nouet S, Bouvier M (1996) Agonist-induced modulation of inverse agonist efficacy at the β_2-adrenergic receptor. Mol Pharmacol 50:662–669

Cho W, Taylor LP, Akil H (1996) Mutagenesis of residues adjacent to transmembrane prolines alters D1 dopamine receptor binding and signal transduction. Mol Pharmacol 50:1338–1345

Clark AJ (1933) The mode of action of drugs on cells. Edward Arnold, London

Clark AJ (1937) General pharmacology. In: Haffter's Handbuch der Exp. Pharmacol. Springer, Berlin

Cohen DP, Thaw CN, Varma A, Gershengorn MC, Nussenzveig DR (1997) Human calcitonin receptors exhibit agonist-independent (constitutive) signaling activity. Endocrinology 138:1400–1405

Colquhoun D (1973) The relationship between classical and cooperative drug action. In: Rang HP (ed.) Drug receptors. University Park Press, Baltimore, pp 149–182

Colquhoun D (1987) Affinity, efficacy, and receptor classification: is the classical theory still useful? In: Black JW, Jenkinson DH, Gerskowitch VP (eds) Perspectives on receptor classification. Alan R Liss, New York, pp 103–114

Colquhoun D, Hawkes AG (1983) The principles of stochastic interpretation of ion-channel mechanisms. In Sakmann B, Neher E (eds) Single-channel recording. Plenum Press, New York, pp 135–175

Costa T, Herz A (1989) Antagonists with negative intrinsic activity at delta opioid receptors coupled to GTP-binding proteins. Proc Natl Acad Sci USA 86:7321–7325

Costa T, Lang J, Gless C, Herz A (1990) Spontaneous association between opioid receptors and GTP-binding regulatory proteins in native membranes: specific regulation by antagonists and sodium ions. Mol Pharmacol 37:383–394

Costa T, Ogino Y, Munson PJ, Onaran HO, Rodbard D (1992) Drug efficacy at guanine nucleotide-binding regulatory protein-linked receptors: thermodynamic interpretation of negative antagonism and of receptor activity in the absence of ligand. Mol Pharmacol 41:549–560

De Lean A, Stadel JM, Lefkowitz RJ (1980) A ternary complex model explains the agonist-specific binding properties of the adenylate-cyclase-coupled β-adrenergic receptor. J Biol Chem 255:7108–7117

Del Castillo J, Katz B (1957) Interaction at end-plate receptors between different choline derivatives. Proc R Soc Lond (Biol Sci) 146:369–381

Denbigh K (1951) The thermodynamics of the steady-state. Methuen, London; Welry, New York

Denbigh K (1968) The principle of chemical equilibrium. Cambridge University Press, London, New York

Dohlman HG, Bouvier M, Benovic JL, Caron MG, Lefkowitz RJ (1987) The multiple membrane spanning topography of the β_2-adrenergic receptor. Localization of the sites of binding, glycosylation and regulatory phosphorylation by limited proteolysis. J Biol Chem 262:14282–14288

Ehlert FJ (1985) The relationship between muscarinic receptor occupancy and adenylate-cyclase inhibition in the rabbit myocardium. Mol Pharmacol 28:410–421

Ferguson SS, Zhang J, Barak LS, Caron MG (1998) Role of β-arrestins in the intracellular trafficking of G-protein-coupled receptors. Adv Pharmacol 42:420–424

Ferris HA, Carroll RE, Rasenick MM, Benya RV (1997) Constitutive activation of the gastrin-releasing peptide receptor expressed by the nonmalignant human colon epithelial cell line NCM460. Clin Invest 100:2530–2537

Frauenfelder H (1995) Proteins-paradigms of complex systems. Experientia 51:200–203

Frauenfelder H, Parak F, Young RD (1988) Conformational substates in proteins. Annu Rev Biophys Biophys Chem 17:451–479

Frauenfelder H, Sligar SG, Wolynes PG (1991) The energy landscapes and motions of proteins. Science 254:1598–1603

Furchgott RF (1966) The use of β-haloalkylamines in the differentiation of receptors and in the determination of dissociation constants of receptor-agonist complexes. In Harper NJ, Simmonds AB (eds) Advances in drug research vol 3. Academic Press, London New York, pp 21–55

Gerstein M, Lesk AM, Chothia C (1994) Structural mechanisms for domain movements in proteins. Biochemistry 33:6739–6749

Gether U, Lin S, Kobilka BK (1995) Fluorescent labeling of purified β2-adrenergic receptor. Evidence for ligand-specific conformational changes. J Biol Chem 270:28268–28275

Gether U, Ballesteros JA, Seifert R, Sanders-Bush E, Weinstein H, Kobilka BK (1997) Structural instability of a constitutively active G-protein-coupled receptor: agonist-independent activation due to conformational flexibility. J Biol Chem 272:2587–2590

Gill SJ, Richey B, Bishop G, Wyman J (1985) Generalized binding phenomena in an allosteric macromolecule. Biophys Chem 21:1–14

Gilman AG (1987) G Proteins: Transducers of G-protein generated signals. Ann Rev Biochem 56:615–649

Gossen M, Freundlieb S, Bender G, Muller G, Hillen W, Bujard H (1995) Transcriptional activation by tetracyclines in mammalian cells. Science 268:1766–1769

Gurwitz D, Haring R, Heldman E, Fraser CM, Manor D, Fisher A (1994) Discrete activation of transduction pathways associated with acetylcholine m1 receptor by several muscarinic ligands. Eur J Pharmacol 267:21–31

Haltia T, Freire E (1995) Forces and factors that contribute to the structural stability of membrane proteins. Biochim Biophys Acta 1228:1–27

Hamill OP, Marty A, Neher E, Sakmann B, Sigworth FJ (1981) Improved patch-clamp techniques for high-resolution current recording from cells and cell-free membrane patches. Pflugers Arch 391:85–100

Hamm HE (1998) The many faces of G-protein signalling J Biol Chem 273:669–672

Hausdorff WP, Caron MG, Lefkowitz RJ (1990) Turning off the signal: desensitization of β-adrenergic receptor function. FASEB J 4:2881–2889

Hilf G, Jakobs KH (1992) Agonist-independent inhibition of G-protein activation by muscarinic acetylcholine receptor antagonists in cardiac membranes. Eur J Pharmacol 225:245–252

Hill TL (1956) An introduction to statistical thermodynamics. McGraw-Hill, New York

Hjorth SA, Orskov C, Schwartz TW (1998) Constitutive activity of glucagon-receptor mutants. Mol Endocrinol 12:78–86

Howe JR, Skryabin BV, Belcher SM, Zerillo CA, Schmauss C (1995) The responsiveness of a tetracycline-sensitive expression system differs in different cell lines J Biol Chem 270:14168–74

Iismaa TP, Biden TJ, Shine J (1995) G-protein-coupled receptors. Springer, Berlin Heidelberg New York

Jin J, Mao GF, Ashby B (1997) Constitutive activity of human prostaglandin-E receptor EP3 isoforms. Br J Pharmacol 121:317–323

Karlin A (1967) On the application of "a plausible model" of allosteric proteins to the receptor for acetylcholine. J Theor Biol 16:306–320

Karplus M, Petsko GA (1990) Molecular dynamics simulations in biology. Nature 347:631–639

Kauman AJ, Birnbaumer L (1974) Studies on receptor-mediated activation of adenylyl cyclases. IV. Characteristics of the adrenergic receptor coupled to myocardial adenylyl cyclase: stereospecificity for ligands and determination of apparent affinity constants for β-blockers. J Biol Chem 249:7874–7885

Kenakin TP (1984) The classification of drugs and drug receptors in isolated tissues. Pharmacol Rev 36:165–222

Kenakin TP (1988) Are receptors promiscuous? Intrinsic efficacy as a transduction phenomenon. Life Sci 43:1095–1101

Kenakin TP (1992) Pharmacological analysis of drug receptor interaction. Raven Press, New York

Kenakin TP (1995a) Agonist-receptor efficacy. I: Mechanisms of efficacy and receptor promiscuity. Trends Pharmacol Sci 16:188–192
Kenakin TP (1995b) Agonist-receptor efficacy. II. Agonist trafficking of receptor signals. Trends Pharmacol Sci 16:232–238
Kirchhausen T, Bonifacino JS, Riezman H (1997) Linking cargo to vesicle formation: receptor tail interactions with coat proteins. Curr Opin Cell Biol 9:488–495
Kjelsberg MA, Cotecchia S, Ostrowski J, Caron MG, Lefkowitz JR (1992) Constitutive activation of the α1B-adrenergic receptor by all amino acid substitutions at a single site. Evidence for a region which constrains receptor activation. J Biol Chem 267:1430–1433
Koshland DE, Neet E (1968) The catalytic and regulatory properties of enzymes. Annu Rev Biochem 37:359–410
Kotyk A, Janacek K (1970) Cell membrane transport: principles and techniques. Plenum Press, New York London
Leff P (1995) The two-state model of receptor activation. Trends Pharmacol Sci 16:89–97
Lefkowitz RJ, Caron MG (1986) Regulation of adrenergic-receptor function by phosphorylation. Curr Top Cell Regul 28:209–231
Lefkowitz RJ, Cotecchia S, Samama P, Costa T (1993) Constitutive activity of receptors coupled to guanine nucleotide regulatory proteins. TIPS 14:303–7
Leung E., Jacobson KA, Green RD (1990) Analysis of agonist–antagonist interactions at A1 adenosine receptors. Mol. Pharmacol 38:72–83
Levy FO, Zhu X, Kaumann AJ, Birnbaumer L (1993) Efficacy of β_1-adrenergic receptors is lower than that of β_2-adrenergic receptors. Proc Natl Acad Sci USA 90: 10798–802
Lohse MJ, Benovic JL, Codina J, Caron MG, Lefkowitz RJ (1990) β-Arrestin: a protein that regulates β-adrenergic-receptor function. Science 248:1547–1550
Mewes T, Dutz S, Ravens U, Jakobs KH (1993) Activation of calcium currents in cardiac myocytes by empty β-adrenoceptors. Circulation 88:2916–2922
Miller C (1987) How ion channel proteins work. In Kaczmarek LK, Levitan IB (eds) Neuromodulation. The biochemical control of neuronal excitability. Oxford University Press, New York, Oxford, pp 39–63
Monod J, Changeux JP, Jacob F (1963) Allosteric proteins and cellular control systems. J Mol Biol 6:306–329
Monod J, Wyman J, Changeux JP (1965) On the nature of allosteric transitions: a plausible model. J Mol Biol 12:88–118
Morgan NG (1989) Cell signalling. Guilford Press, New York
Neer EJ, Clapham DE (1988) Role of G-protein subunits in transmembrane signalling. Nature 333:129–133
Neher E, Sakmann B (1976) Single channel currents recorded from membrane of denervated frog muscle fibers. Nature 260:799–802
Onaran HO, Costa T (1997) Agonist efficacy and allosteric models of receptor action. Ann NY Acad Sci 812:98–115
Parent JL, Le Gouill C, de Brum-Fernandes AJ, Rola-Pleszczynski M, Stankova J (1996) Mutations of two adjacent amino acids generate inactive and constitutively active forms of the human platelet-activating factor receptor. J Biol Chem 271:7949–7955
Parma J, Duprez L, Van Sande J, Cochaux P, Gervy C, Mockel J, Dumont J, Vassart G (1993) Somatic mutations in the thyrotropin receptor gene cause hyperfunctioning thyroid adenomas. Nature 365:649–651
Perez DM, Hwa J, Gaivin R, Mathur M, Brown F, Graham RM (1996) Constitutive activation of a single effector pathway: evidence for multiple activation states of a G-protein-coupled receptor. Mol Pharmacol 49:112–122
Radzicka A, Wolfenden R (1995) A proficient enzyme. Science 267:90–93
Rang HP (1973) Receptor Mechanisms. Br J Pharmacol 48:475–495
Rao VR, Cohen GB, Oprian DD (1994) Rhodopsin mutation G90D and a molecular mechanism for congenital night blindness. Nature 367:639–642

Ren Q, Kurose H, Lefkowitz RJ, Cotecchia S (1993) Constitutively active mutants of the α2-adrenergic receptor. J Biol Chem 268:16483–16487

Riitano D, Werge TM, Costa T (1997) A mutation changes ligand selectivity and transmembrane signaling preference of the neurokinin-1 receptor. 272:7646–7655

Robb S, Cheek TR, Hanan FL, Hall LM, Midgly JM, Eveans PD (1994) Agonist-specific coupling of a cloned *Drosophila* octopamine/tyramine receptor to multiple second-messenger systems. EMBO J 13:1325–1330

Robbins LS, Nadeau JH, Johnson KR, Kelly MA, Roselli-Rehfuss L, Baack E, Mountjoy KG, Cone RD (1993) Pigmentation phenotypes of variant extension locus alleles result from point mutations that alter MSH receptor function. Cell 72:827–834

Robinson PR, Cohen GB, Zhukovsky EA, Oprian DD (1992) Constitutively active mutants of rhodopsin. Neuron 9:719–725

Rodbell M (1980) The role of hormone receptors and GTP-regulatory proteins in membrane transduction. Nature 284:17–22

Ross EM (1989) Signal Sorting and Amplification through G-protein-coupled receptors. Neuron. 3:141–152

Samama P, Cotecchia S, Costa T, Lefkowitz RJ (1993) A mutation-induced activated state of the β2-adrenergic receptor. J Biol Chem 268:4625–4636

Samama P, Pei G, Costa T, Cotecchia S, Lefkowitz RJ (1994) Negative antagonists promote an inactive conformation of the β2-adrenergic receptor. Mol Pharmacol 45:390–394

Scheer A, Cotecchia S (1997) Constitutively active G-protein-coupled receptors: potential mechanisms of receptor activation. J Recept Signal Transduct Res 17:57–73

Scheer A, Fanelli F, Costa T, De Benedetti PG, Cotecchia S (1996) Constitutively active mutants of the α1B-adrenergic receptor: role of highly conserved polar amino acids in receptor activation. EMBO J 15:3566–3578

Scheer A, Fanelli F, Costa T, De Benedetti PG, Cotecchia S (1997) The activation process of the α_{1B}-adrenergic receptor: potential role of protonation and hydrophobicity of a highly conserved aspartate. Proc Natl Acad Sci USA 94:808–813

Schipani E, Kruse K, Juppner H (1995) A constitutively active mutant PTH-PTHrP receptor in Jansen-type metaphyseal chondrodysplasia. Science 268:98–100

Shenker A, Laue L, Kosugi S, Merendino JJ Jr, Minegishi T, Cutler GB Jr (1993) A constitutively activating mutation of the luteinizing hormone receptor in familial male precocious puberty. Nature 365:652–654

Smit MJ, Leurs R, Alewijnse AE, Blauw J, Van Nieuw Amerongen GP, Van DeVrede Y, Roovers E, Timmerman H (1996) Inverse agonism of histamine-H2 antagonist accounts for upregulation of spontaneously active histamine-H2 receptors. Proc Natl Acad Sci USA 93:6802–6807

Spalding TA, Burstein ES, Brauner-Osborne H, Hill-Eubanks D, Brann MR (1995) Pharmacology of a constitutively active muscarinic receptor generated by random mutagenesis. J Pharmacol Exp Ther 275:1274–1279

Spengler D, Waeber C, Pantolini C, Holsboer F, Bockaert J, Seeburg PH, Journot L (1993) Differential signal transduction by five splice variants of the PACAP receptor. Nature 365:170–175

Sprang SR (1997) G-protein mechanisms: insights from structural analysis. Annu Rev Biochem. 66:639–678

Stephenson RP (1956) A modification of receptor theory. Br J Pharmacol 11:379–393

Stoeckenius W, Bogomolni RA (1982) Bacteriorhodopsin and related pigments of halobacteria. Ann Rev Biochem 52:587–616

Strader CD, Fong TM, Tota MR, Underwood D, Dixon RA (1994) Structure and function of G-protein-coupled receptors. Annu Rev Biochem 63:101–132

Strosberg AD (1996) G-protein coupled R7G receptors. Cancer Surv 27:65–83

Theroux TL, Esbenshade TA, Peavy RD, Minneman KP (1996) Coupling efficiencies of human α1-adrenergic receptor subtypes: titration of receptor density and responsiveness with inducible and repressible expression vectors. Mol Pharmacol 50:1376–1387

Thron CD (1973) On the analysis of pharmacological experiments in terms of an allosteric receptor model. Mol Pharmacol 9:1–9

Tiberi M, Caron MG (1994) High agonist-independent activity is a distinguishing feature of the dopamine D1B receptor subtype J Biol Chem 269:27925–27931

Tolman RC (1938) The principles of statistical mechanics. Oxford University Press, New York

Van Rossum JM (1963) The relationship between chemical structure and biological activity. J Pharm Pharmacol 15:285–316

Weber G (1972) Ligand binding and internal equilibria in proteins. Biochemistry 11:864–878

Weber G (1973) Energetics of ligand binding to proteins. Adv Protein Chemistry 29:1–83

Weber G (1992) Protein interactions. Chapman and Hall, New York

Weiss JM, Morgan PH, Lutz MW, Kenakin TP (1996) The cubic ternary complex receptor-occupancy model I. Model description. J Theor Biol 178:151–167

Wess J (1997) G-protein-coupled receptors: molecular mechanisms involved in receptor activation and selectivity of G-protein recognition. FASEB J 11:346–354

Wregget KA, DeLean A (1984) The ternary complex model: Its properties and application to ligand interaction with the D2-dopamine receptor of the anterior pituitary gland. Mol Pharmacol 26:214–227

Wyman J (1967) Allosteric linkage. J American Chem Soc 89:2202–2218

Wyman J (1984) Linkage graphs: a study in the thermodynamics of macromolecules. Q Rev Biophys 17:453–488

Wyman J, Allen DW (1951) The problem of the heme interactions in hemoglobin and the basis of the Bohr effect. J Polymer Sci 7:499–518

Wyman J, Gill SJ (1990) Binding and linkage: Functional chemistry of biological macromolecules. University Science Books, Mill Valley, California

CHAPTER 9
Mechanisms of Non-Competitive Antagonism and Co-Agonism

D.G. TRIST and M. CORSI

A. Non-Competitive Antagonism

This chapter has the objective of furnishing the reader with some understanding of non-competitive antagonism. A simple version is first discussed; this will then be followed by a specific case applied to N-methyl-D-aspartate (NMDA) receptors. Since the latter is a particular system that is activated by the simultaneous presence of two agonists, it will be seen that competitive antagonists for one agonist-binding site appear to be non-competitive for the other agonist.

I. Definition

Antagonism is the term generally used to indicate the process of inhibition of agonist-induced responses. Antagonists can be classified on the basis of the molecular mechanism(s) by which they block agonist responses. A drug can chemically interact with the agonist and, therefore, block receptor activation (chemical antagonist). An antagonist may bind to the same site on the receptor as the agonist (competitive antagonist). The antagonist may bind to a site on the receptor different from the agonist-recognition site but, through an allosteric or other type of mechanism, may prevent the agonist activation of the receptor (non-competitive antagonist). Moreover, to inhibit the response, a functional antagonist does not interact with the receptor but with the process of activation.

II. Analysis of the Effect of Non-Competitive Antagonists

The hypothesis to be considered is that the antagonist binds to a site on the receptor different from the agonist-binding site without modifying the equilibrium dissociation constant of the agonist–receptor complex. In this way, the antagonist interferes with the process of receptor activation and produces a depression of the concentration–response curve of the agonist rather than dextral displacement with retention of the maximal asymptote. Considering the interaction between the agonist [A] and the receptor [R],

$$A + R \underset{K1}{\overset{K2}{\rightleftharpoons}} AR$$

at the equilibrium, this equation can be written:

$$\frac{[A]}{(KA+[A])} = \frac{[AR]}{[Rt]} \tag{1}$$

where Rt is the total receptor concentration and K_A is the dissociation constant of the agonist–receptor complex (K2/K1).

In the presence of a non-competitive antagonist, we can define the receptor fraction bound to the agonist as Y and the fraction bound to the antagonist as δ_B. At equilibrium, [R] is equal to $(1-Y-\delta_B)$.

$$A + R \underset{K1}{\overset{K2}{\rightleftharpoons}} AR$$
$$+$$
$$B$$
$$K1b \updownarrow K\text{-}1b$$
$$BR$$

Therefore,

$$[A] \cdot (1-Y-\delta_B) = KA \cdot Y \tag{2}$$

which rearranges to:

$$[A] \cdot (1-\delta_B) = Y \cdot (KA+[A]) \tag{3}$$

Substituting yields the equation:

$$\frac{[A]}{(KA+[A])} \cdot (1-\delta_B) = \frac{[AR]}{[R_t]} \tag{4}$$

Calculating δ_B by the mass-action law yields:

$$\delta B = \frac{[BR]}{R_t} = \frac{1}{1+\dfrac{KB}{[B]}} \tag{5}$$

where [BR] is the receptor concentration bound to the antagonist [B], and KB (K_{-1b}/K_{1b}) is the dissociation constant of the antagonist–receptor complex.

Therefore, an equation that links the receptor occupancy ([AR]/[Rt]) and potency of the non-competitive antagonist can be derived as follows:

$$\frac{[AR]}{[R_t]} = \frac{[A]}{(KA+[A])} \cdot \left(\frac{1}{1+\dfrac{[B]}{[KB]}} \right) \tag{6}$$

Equation 6 describes the concentration of a non-competitive antagonist that modulates the receptor occupancy. As depicted in Fig. 1, the removal of the receptors by increasing the amount of antagonist produces a progressive reduction of receptor occupancy.

Considering the production of response by the AR species gives the equation:

$$A + R \underset{K1}{\overset{K2}{\rightleftharpoons}} AR \longrightarrow \text{Effect}$$

Assuming a hyperbolic relationship between receptor stimuli and tissue response (E),

$$\frac{E}{Em} = \frac{[AR]}{(Ke + [AR])} \quad (7)$$

where Ke is the value of [AR] which elicits half the maximal tissue response (denoted Em). Ke is generally used to quantify relative agonist efficacy.

Combining Eq. 7 with Eq. 1 as described by BLACK and LEFF (1983), the following equation can be obtained:

$$E = \frac{Em \cdot \tau \cdot [A]}{(KA + [A]) + \tau \cdot [A]} \quad (8)$$

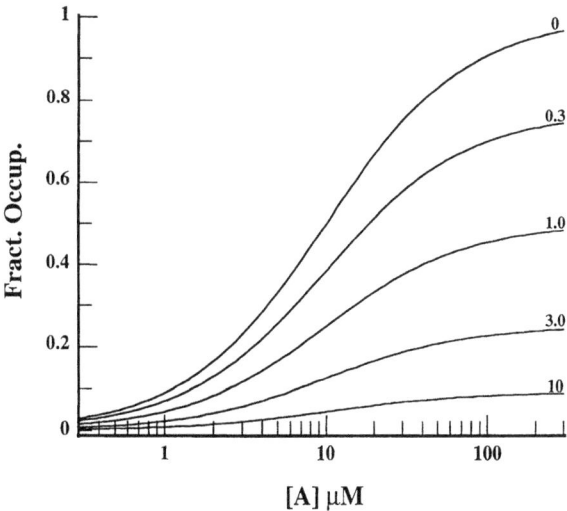

Fig. 1. Simulation of the effect of a non-competitive antagonist on receptor occupancy (AR/Rt). Curves have been simulated by setting the agonist K_A equal to $10\,\mu M$ and the antagonist K_B equal to $1\,\mu M$, and the antagonist concentration ([B]) was set to the values given above the curves

where τ is defined as [Rt]/Ke. As described by BLACK and LEFF (1983), Eq. 8 is the essential one in the operational model, as it defines the production of tissue response as a function of maximal tissue response, agonist concentration [A], agonist binding to the receptor K_A and the parameter τ. The latter is defined as a transducer constant and measures the efficacy of the agonist in that particular system.

In the presence of a non-competitive antagonist, Eq. 8 becomes:

$$E = \frac{Em \cdot \tau \cdot [A] \cdot (1-\delta_B)}{(KA+[A])+\tau \cdot [A] \cdot (1-\delta_B)} \tag{9}$$

Substituting $(1-\delta_B)$ for $(1/1 + [B]/KB)$, Eq. 9 becomes:

$$E = \frac{Em \cdot \tau \cdot [A] \cdot \left(\frac{1}{1+[B]/KB}\right)}{(KA+[A])+\tau \cdot [A] \cdot \left(\frac{1}{1+[B]/KB}\right)} \tag{10}$$

Equation 9 has been used to simulate the effect of a non-competitive antagonist on agonist response. In particular, Fig. 2 shows that a reduction of the agonist maximal response is produced by increasing the antagonist concentration.

The effect induced by a non-competitive antagonist on agonist response may differ considerably because of the efficacy (Ke) of the agonist. Figure 3

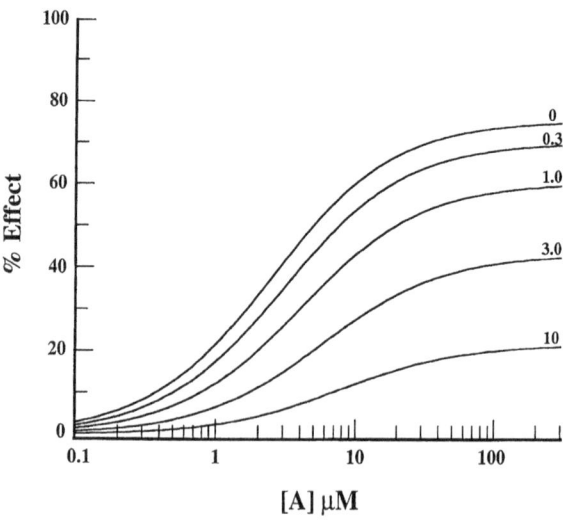

Fig. 2. Simulation of the effect of a non-competitive antagonist on agonist-induced effects by using Eq. 10. Parameters have been set as: agonist $K_A = 10\,\mu M$, antagonist $K_B = 1\,\mu M$, $\tau = 3$ and Em = 100. Antagonist concentrations ([B]) are given above the curves

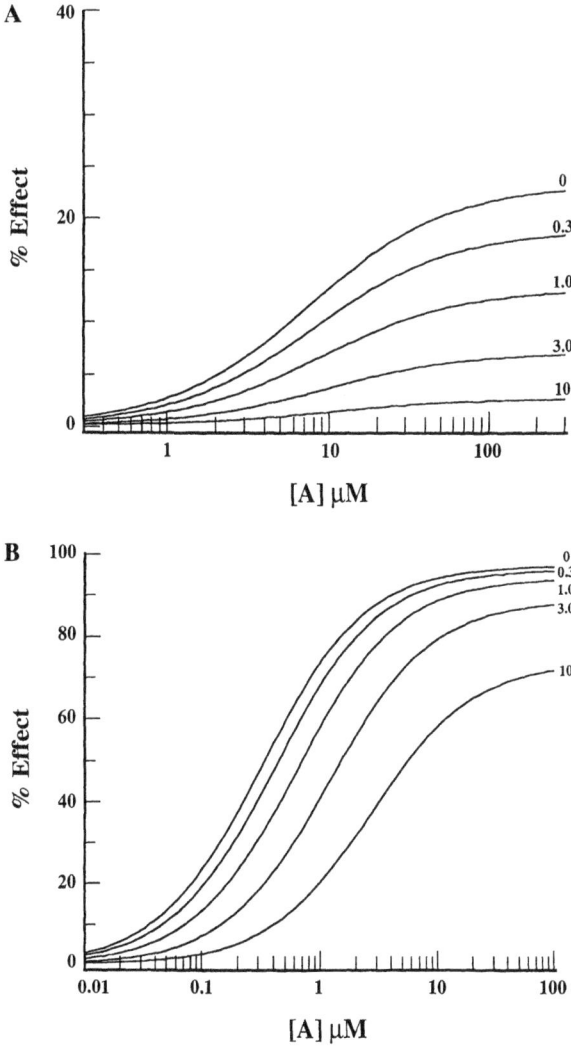

Fig. 3A,B. Simulation of the effect of a non-competitive antagonist on agonist-induced effects by using Eq. 10. Parameters have been set as: agonist $K_A = 10\,\mu M$, antagonist $K_B = 1\,\mu M$, Rt = 3 and Em = 100. In **A**, Ke was set equal to 10 whereas, in **B**, Ke was set equal to 0.1

shows the effect of a non-competitive antagonist on an agonist with a low receptor reserve (3A; low efficacy) and a large receptor reserve (3B; high efficacy). In the latter event, the lower concentrations of the antagonist seem to produce a competitive antagonism, with a right shift of the agonist curve with or without a minimal effect on the agonist maximum response. However, by increasing the concentration of the antagonist, a clear depression of the agonist-induced response can be observed.

A similar result can be obtained by testing a non-competitive antagonist in systems with different numbers of receptors (Rt). It appears to be more potent in a system with a low receptor number as opposed to one with a high receptor number, thus showing that the effect of a non-competitive antagonist is also influenced by tissue factors. Therefore, both tissue and agonist factors influence the activity of non-competitive antagonists, thus making their effects unpredictable, unlike the case for competitive antagonists.

The estimation of the equilibrium dissociation constant for non-competitive antagonist–receptor complexes has been derived, as described by KENAKIN (1993). In practice, agonist concentration–response curves are carried out in the absence and presence of the non-competitive antagonist. Generally, a depression of more of than 50% of the maximal response of the agonist concentration–response curve in the presence of the non-competitive antagonist gives accurate KB values.

The curves for the control and the antagonist-treated samples are related, based on the assumption that the fraction of the receptors occupied by the agonist in the presence of the antagonist is:

$$\frac{[A]}{(KA+[A])} \cdot \left(\frac{1}{1+\frac{[B]}{[KB]}} \right) = \frac{[AR]}{[Rt]} \tag{11}$$

In the absence of the antagonist, the curve is described by Eq. 1. Assuming that equal receptor-occupancy fractions lead to equal responses, equiactive concentrations of the agonist obtained in the absence and presence of the non-competitive antagonist are related by the following equation:

$$\frac{[A]}{(KA+[A])} = \frac{[A^1]}{(KA+[A^1])} \cdot \frac{1}{1+\frac{[B]}{[KB]}} = \frac{[A^1]}{(KA+[A^1])} \cdot (1-\delta_B) \tag{12}$$

which can be rearranged to yield:

$$\frac{1}{[A]} = \frac{1}{[A^1]} \cdot \frac{1}{1-\delta_B} + \frac{1}{1-\delta_B KA} \tag{13}$$

A regression of 1/A versus $1/A^1$ yields a straight line of slope $1/(1-\delta_B)$ and intercept $\delta_B/(1-\delta_B)KA$. The antagonist K_B can be obtained by the following equation:

$$K_B = B/(\text{slope} - 1) \tag{14}$$

Equation 11 can be rewritten as:

$$\frac{[A^1]}{[A]} = [A^1] \cdot \frac{1}{1-\delta_B KA} + \frac{1}{1-\delta_B} \tag{15}$$

This yields $K_B = B/(\text{intercept}-1)$.

B. Co-Agonism

Co-agonism is defined as the condition whereby two agonists must bind to the receptor for physiological response to occur.

I. Theory as Applied to the Glutamate NMDA Receptor

1. Evidence for Co-Agonism

Within the vertebrate central nervous system (CNS), excitatory synapses that are glutamatergic in nature mostly function through the glutamate NMDA receptor (NMDAR). This receptor belongs to the family of ligand-gated ion-channel receptors and is oligomeric in nature. Altogether, there are five gene products that potentially can contribute to the receptor/channel. These can be subdivided into two types on the basis of sequence homology: (1) NMDAR1 and (2) NMDAR2A, NMDAR2B, NMDAR2C and NMDAR2D. The exact composition of these receptors is unknown, and pentamers, tetramers and trimers have all been proposed. What is known, however, is that NMDAR1 is essential in functional receptors, and at least two NMDAR2 subunits also seem to be involved.

Like other ligand-gated ion channels, the NMDA receptor is modulated by a number of different molecules, including polyamines, Zn^{2+}, H^+ and Mg^{2+} ions and glycine, each binding to their own unique site. It was JOHNSON and ASCHER (1987) who were the first to show that glycine for NMDAR, but not for the other ionic glutamate receptors, was a co-agonist. That is, both glycine and NMDA (or glutamate) are needed simultaneously to activate the receptor. This is demonstrated in Fig. 4 where, in hippocampal neurons from embryonic rats, it can be seen that, electrophysiologically, glycine-concentration–effect curves are only generated in the presence of NMDA. Recently, point-mutation studies on both NMDAR1 (WILLIAMS et al. 1996) and NMDAR2B (LAUBE et al. 1997) subunits has suggested that the two agonists bind differentially; glycine binds to NMDAR1, and glutamate binds to NMDAR2.

Importantly, the binding of one agonist seems to facilitate binding of the other although, in theory, co-agonism can be positive, neutral or even negative. Radioligand-binding experiments have shown that both glycine and glutamate increase each other's affinity for the receptor (FADDA et al. 1988). Thus, the binding of one agonist effectively confers activity to the action of the other. This was pointed out by MARVIZÒN and BAUDRY (1993) when they proposed a model of co-agonism to account for the data obtained from the binding of [^{3}H]dizocilpine (also called MK-801) binding to NMDAR. This ligand binds within the channel of NMDAR and, thus, can be used to measure channel opening in response to glycine and glutamate. It is a classical non-competitive antagonist, and it will be used as a reference standard when the behaviour of competitive antagonists for each of the co-agonists is discussed. Thus, electrophysiological, molecular-biological and biochemical evidence has

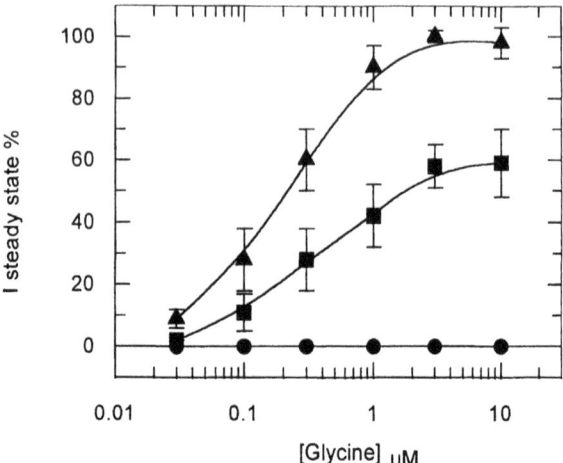

Fig. 4. The effect of N-methyl-D-aspartate (NMDA) on the response to glycine in hippocampal neurons from embryonic rats. Glycine-concentration–effect curves were constructed in the absence (●) or in the presence of 10 μM NMDA (■) or 100 μM NMDA (▲). The *ordinate* represents the steady-state current after each application, as measured by single-cell patch electrophysiology. Each *point* represents the mean of six replicates, with *standard error bars*

strongly supported the notion of co-agonism between glycine and glutamate at NMDAR.

2. The Theory of Co-Agonism

As mentioned above, MARVIZÒN and BAUDRY (1993) published a model which effectively described the interaction of two agonists and their binding sites on NMDAR. This model has been extended into an operational format and expanded to take account of the effect of antagonists (CORSI et al. 1996; FINA et al. 1997). This is represented in Fig. 5, which depicts the interaction of glycine, glutamate and a glycine antagonist. The model also takes into account the interaction of MK-801. The model is expressed in terms of steady states and, thus, does not take account of desensitisation, multiple inactive states, etc. However, the model does approximate the behaviour of the two agonists and their antagonists as seen in different functional systems.

Fixing the amount of glutamate (Glu), the fraction of receptors occupied following the interaction of glycine (Gly) with NMDAR (R) as shown in Fig. 5 can be described as follows:

$$[GluRGly] = \frac{[Gly] \cdot \dfrac{[Glu]}{\alpha \cdot K_{Gly} + [Glu]} \cdot [R_t]}{[Gly] + \dfrac{\alpha \cdot K_{Gly} \cdot (K_{Glu} + [Glu])}{\alpha \cdot K_{Glu} + [Glu]}} \quad (16)$$

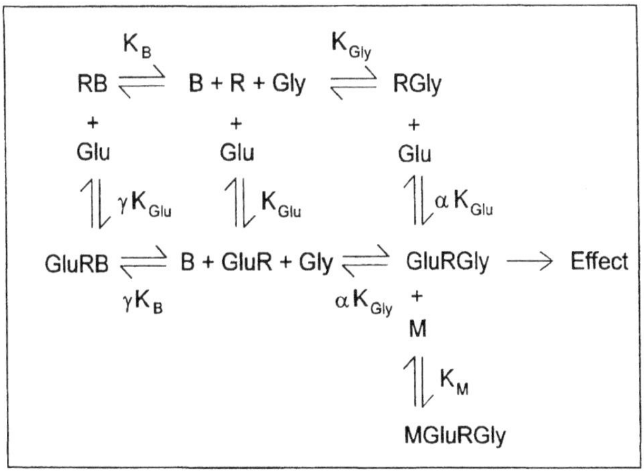

```
        K_B                    K_Gly
RB  ⇌  B + R + Gly  ⇌  RGly
+           +              +
Glu        Glu            Glu
⇵ γK_Glu   ⇵ K_Glu        ⇵ αK_Glu
GluRB  ⇌  B + GluR + Gly  ⇌  GluRGly  →  Effect
       γK_B              αK_Gly  +
                                  M
                                  ⇵ K_M
                                  MGluRGly
```

Fig. 5. A model of co-agonism for glutamate (*Glu*) and glycine (*Gly*) as activators of the *N*-methyl-D-aspartate receptor. The agonists Glu and Gly can bind to uncoupled receptor (*R*) or R pre-bound with the other agonist. Only when R is bound with both Glu and Gly can an effect (*E*) be elicited. Competitive glycine antagonists (*B*) can interact with either the uncoupled R or R bound with Glu. The dissociation constants (*K*) for each interaction are modified by the allosteric parameters α (for agonists) and γ (for antagonists) when R is already coupled to an agonist. MK-801 (abbreviated to *M* for simplicity) couples only with the activated form of the receptor when the channel is open

If this binding is transduced into an effect (E) by a hyperbolic function of the general form, the following equation is obtained:

$$E = \frac{Em \cdot [GluRGly]^n}{Ke^n + [GluRGly]^n} \tag{17}$$

By substituting Eq. 16 into Eq. 17, we get the following equation:

$$E = \frac{E_m \cdot [Gly]^n \cdot \left(\tau \cdot \frac{[Glu]}{\alpha \cdot K_{Glu} + [Glu]}\right)^n}{\left(K^*_{Gly} + [Gly]\right)^n + \left(\tau \cdot \frac{[Glu]}{\alpha \cdot K_{Glu} + [Glu]}\right)^n \cdot [Gly]^n} \tag{18}$$

where

$$K^*_{Gly} = K_{Gly} \cdot \left(\frac{\alpha \cdot K_{Glu} + \alpha \cdot [Glu]}{\alpha \cdot K_{Glu} + [Glu]}\right) \tag{19}$$

The meaning of the parameters α, K_{Gly}, and K_{Glu} is given in Fig. 5. The symbol τ is the transducer ratio and, as stated by BLACK and LEFF (1983), it links the number of receptors to the intrinsic efficacy of the agonist ($\tau = R_t/K_e$).

In this case, co-agonism means that each agonist alone has an efficacy close to zero when bound to the receptor. Thus, the system efficacy of both agonists bound together at the receptor needs to be taken into account, i.e. ($\tau = R_t/K_{e(Glu-Gly)}$). This is a fundamental difference from the operational model as applied to single-agonist/receptor interaction.

A major consequence of co-agonism is the effect it has upon the location and maximal response of the concentration–effect curve. Table 1 shows how these parameters have become modified. Thus, the location of the curve in functional experiments includes not only the dissociation constant (K_{Gly}), but also τ, K_{Glu}, [Glu] and α. Therefore, methods designed to estimate K_{Gly} or τ will be dependent on the value of the other parameters. For example,

$$\tau_{estimated} = \frac{\tau \cdot [Glu]}{\alpha \cdot K_{Glu} + [Glu]} \qquad (20)$$

By plotting $\tau_{estimated}$ against [Glu], a logistic-function curve whose midpoint is equal to $\alpha \times K_{Glu}$ is obtained. At high [Glu], $\tau_{estimated}$ approaches τ. Thus, experiments to determine τ need to be done in supra-maximal concentrations of this co-agonist. The location parameter of Gly will be shifted to the left when α is reduced or τ is increased. Thus, a highly potent agonist will have α less than one (positive allosteric effect) and a large τ value. Thus, positive allosterism contributes to potency but has little effect on the maximal response, which is more sensitive to τ. In theory, α can be less than, equal to or greater than unity. The latter system would exhibit negative allosterism between the two agonists.

Table 1. The location ([A_{50}]) and maximal effect (Max) for glycine-concentration–effect curves using radioligand binding and functional experiments with either single agonists or co-agonists

Description	[A_{50}]	Max
Single agonism, binding	K_{Gly}	B_{max}
Single agonism, functional	$\dfrac{K_{Gly}}{(2+\tau^n)^{1/n}-1}$	$E_m \dfrac{\tau^n}{\tau^n+1}$
Co-agonism, binding	$\dfrac{\alpha \cdot K_{Gly} \cdot (K_{Glu}+[Glu])}{\alpha \cdot K_{Glu}+[Glu]}$	$\dfrac{[Glu]}{\alpha \cdot K_{Glu}+[Glu]} \cdot [Rt]$
Co-agonism, functional	$\dfrac{K_{Gly} \cdot \left(\dfrac{\alpha \cdot K_{Glu}+\alpha \cdot [Glu]}{\alpha \cdot K_{Glu}+[Glu]}\right)}{\left(2+\left(\dfrac{\tau \cdot [Glu]}{\alpha \cdot K_{Glu}+[Glu]}\right)^n\right)^{1/n}-1}$	$\dfrac{E_m \cdot \left(\tau \cdot \dfrac{[Glu]}{\alpha \cdot K_{Glu}+[Glu]}\right)^n}{1+\left(\dfrac{\tau \cdot [Glu]}{\alpha \cdot K_{Glu}+[Glu]}\right)^n}$

3. The Effect of Competitive Antagonism on Co-Agonism

When a competitive antagonist (B) for glycine is introduced, the model must be modified to take into account the antagonist–receptor interaction (Fig. 4). In the same way that an agonist binding at one site might allosterically change the binding of the co-agonist, so also might a glycine-receptor antagonist modify the binding of glutamate. To account for this interaction, another parameter γ has been introduced into the model. Thus, the effect of B on the co-agonists causes Eq. 19 to be modified to:

$$E = \frac{Max \cdot [Gly]^n}{\left(Gly_{50} \cdot \left(1 + \frac{\frac{K_{Glu} + [Glu]}{\gamma \cdot K_{Glu} + \gamma \cdot [Glu]} \cdot [B]}{K_{Bapp}}\right)\right)^n + [B]^n} \quad (21)$$

where Max refers to the tissue maximal response and

$$K_{Bapp} = \frac{(\gamma \cdot K_{Glu} + \gamma \cdot [Glu]) \cdot K_B}{(\gamma \cdot K_{Glu} + [Glu])} \quad (22)$$

and Gly_{50} and max are as described in Table 1 for functional co-agonism.

As for agonists, the estimation of the antagonist dissociation constant (K_B) is complicated in that the apparent K_B is influenced by other parameters, namely γ, K_{Glu} and variable concentrations of Glu. The influence of γ on the estimation of the value of K_B can be observed in the simulations in Fig. 6. It can clearly be seen that, when γ is less than unity, the K_B is overestimated, and when it is smaller than one, K_B is underestimated.

The behaviour of the K_B under different values of [Glu] can be diagnostic as to whether γ is less than, equal to or greater than one. Figure 7 shows the influence of [Glu] on K_{Bapp} for values of γ of 0.1 and 10. In both cases, a logistic function describes the simulation. However, when $\gamma = 0.1$, K_{Bapp} decreases with increasing [Glu], whereas it increases when $\gamma = 10$. When $\gamma = 1$, $K_{Bapp} = K_B$ at all values of [Glu]. Experiments where K_B is estimated for a glycine antagonist with different [Glu] should predict the magnitude of γ even if the exact value of γ is unknown.

In general, differences in K_{Bapp} for a glycine-receptor antagonist when tested in different assays may be due to NMDAR heterogeneity, different γs and system effects, such as different [Glu]/K_{Glu} values.

Even if the dissociation constant of a competitive glycine antagonist is not accurately assessed, the behaviour of the antagonist against glycine is always insurmountable. More interesting is the behaviour of the glycine antagonist against the co-agonist glutamate. It has been suggested that glycine may normally be in supra-maximal concentrations (KEMP and LEESON 1993). Importantly, it seems that, in pathological conditions [such as focal ischaemia (stroke)], it is glutamate that increases during the first hours following artery

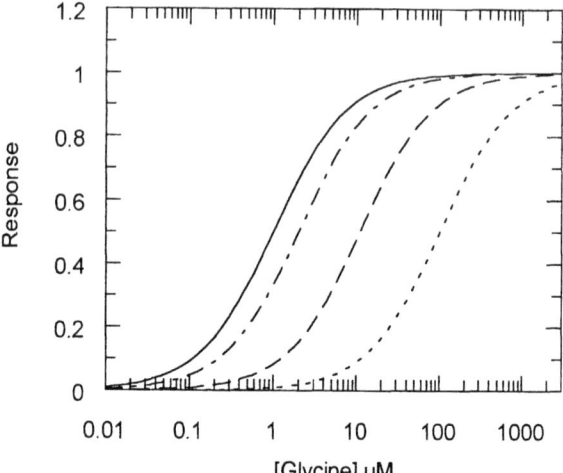

Fig. 6. Graphic simulation of the effect of the parameter γ on the K_B of a competitive glycine antagonist [as demonstrated by the shift of the glycine-concentration–effect curve in the presence of a fixed concentration of antagonist (B)], as proposed in the co-agonist model. The following values were assigned to the parameters: $K_{Glu} = 1$, [Glu] = 1, $K_B = 1$, $n = 1$, Max = 1, $Gly_{50} = 1$, [B] = 0 (control; –) or [B] = 10 with $\gamma = 0.1$ (....), 1 (–) or 10 (.-.-)

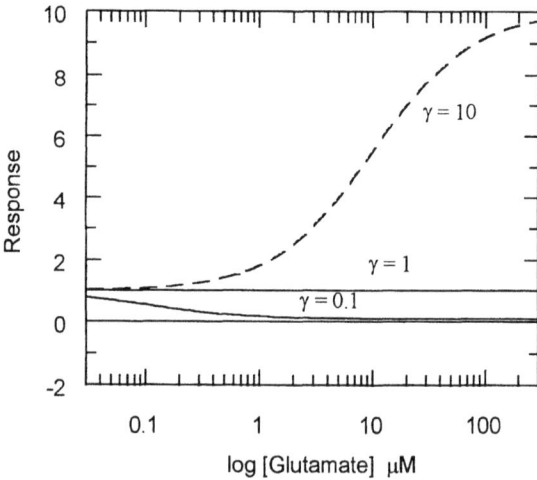

Fig. 7. Graphic simulations of the effect of increasing [Glu] on the estimated dissociation constant (K_B) of a competitive glycine antagonist in conditions where γ is less than, equal to or greater than one. Curves were simulated using Eq. 22, with the following parameters: $K_{Glu} = 1\,\mu M$, [Glu] = $1\,\mu M$ and $K_B = 1\,\mu M$

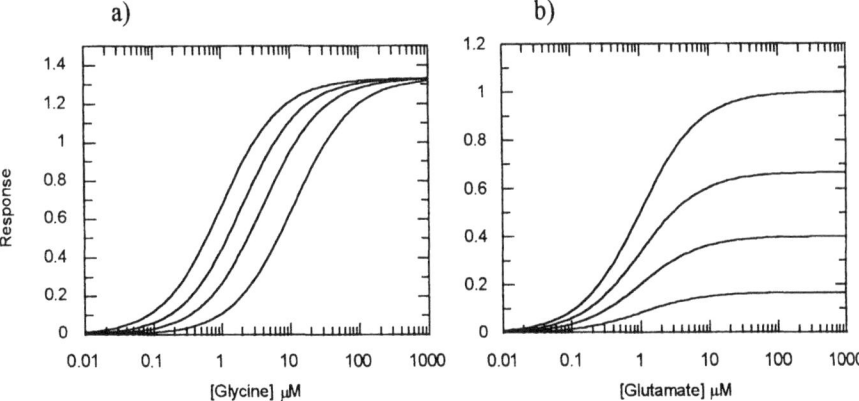

Fig. 8. Graphic simulations of antagonism by a competitive glycine antagonist of (**A**) a glycine-concentration–effect curve and (**B**) a glutamate-concentration–effect curve. Curves were simulated using Eq. 21. The following parameters were used: Max = 2, K_B = 1 μM, [B] = 0, 3, 10 and 30 μM, n = 1, Gly_{50} = 1 μM, Glu_{50} = 1 μM, α = 1 and γ = 1. In the case of **A**, increasing [B] caused a right shift of the glycine-concentration–effect curve, with no effect on the maximal response. However, in case **B**, increasing [B] reduced the maximal response in a concentration-dependent way, with no effect on the location of the curve

occlusion. In the case of glutamate, the antagonism can appear to be non-competitive. Figure 8 shows simulations of the antagonism of a glycine antagonist against glycine and against glutamate. Under the conditions chosen, the glycine antagonist, which shows classical competitive antagonism against glycine, reduces the maximal response of a glutamate-concentration–effect curve. However, the nature of this antagonism depends on the value of both α and of γ. Thus, a low value of γ or a high value of α predicts that the concentration effect collapses, with a left shift in the location of the curves. A high value of γ or a low value of α causes the antagonism to appear to be competitive at low concentrations of the antagonist and non-competitive at higher concentrations. This behaviour is different from that of a more traditional non-competitive antagonist and means that the nature of the antagonism can be tissue dependent, as α and γ can be considered to be properties of the receptor, which has been shown to contain different subunit compositions in different tissues (MONYER et al. 1992).

4. The Effect of a Non-Competitive Antagonist on the Response of Co-Agonists

In the presence of a non-competitive antagonist, such as dizocilpine, the model of co-agonism can be modified. In this case, the Gly_{50} and the maximal response become modified as shown below.

$$Gly_{50} = \frac{K_{Gly} \cdot \left(\dfrac{\alpha \cdot K_{Glu} + \alpha \cdot [Glu]}{\alpha \cdot K_{Glu} + [Glu] \cdot \left(1 + \dfrac{M}{K_m}\right)} \right)}{\left(2 + \left(\dfrac{\tau \cdot [Glu]}{\alpha \cdot K_{Glu} + [Glu]} \right)^n \right)^{1/n} - 1} \qquad (23)$$

where M is dizocilpine (MK-801, MK) and K_m is the equilibrium constant for dissociation of M from the receptor. Furthermore,

$$\text{Max} = \frac{E_m \cdot \left(\dfrac{[Glu]}{\alpha \cdot K_{Glu} + [Glu] \cdot \left(1 + \dfrac{M}{K_m}\right)} \right)^n}{1 + \left(\dfrac{\tau \cdot [Glu]}{\alpha \cdot K_{Glu} + [Glu] \cdot \left(1 + \dfrac{M}{K_m}\right)} \right)^n} \qquad (24)$$

Unlike the non-competitive antagonism of a glycine antagonist against glutamate, the affinity of MK is independent of α, γ and τ. In terms of behaviour of the antagonism, the concentration of MK required to give the same effect does not depend on α, but increasing τ does affect the sensitivity to MK. Thus, the larger the value of τ, the greater the amount of MK required for the same antagonism. Thus, in tissues where there is high efficacy co-agonism, a single concentration of MK will be less effective. This might well explain the observation that MK does not have the same effect on different combinations of subunits of NMDAR (SUCHER et al. 1996).

II. Experimental Data Supporting the Concept of Co-Agonism and Its Antagonism

1. Recombinant Experiments

Many experiments have been carried out on NMDAR subunits expressed in a number of different systems. These include *Xenopus* oocytes and human embryonic kidney cells (SUCHER et al. 1996). These have tended to be experiments where only two subunits have been expressed, namely NMDAR1 with each of the different NMDAR2 subunits. Electrophysiological studies on these recombinants have supported the idea that co-agonism exists, i.e. both glycine and glutamate or NMDA need to be present. Table 2 summarises some of the data in the literature for agonists. Experiments where glycine concentration–effect curves have been constructed in the presence of a fixed concentration of NMDA have shown that the Gly_{50}s differ for different combinations

Table 2. Selectivity of some agonists and antagonists for different combinations of N-methyl-D-aspartate (NMDA)-receptor-1/NMDA-receptor-2 (NMDAR1/NMDAR2) subunits (SUCHER et al. 1996)

Agonist/antagonist	Rank order of potency against different combinations of NMDA subunits
Glutamate	NMDAR1/2B>NMDAR1/2A>NMDAR1/2D>NMDAR1/2C
Glycine	NMDAR1/2C>NMDAR1/2B=NMDAR1/2D >NMDAR1/2A
NMDA	NMDAR1/2A=NMDAR1/2B=NMDAR1/2C=NMDAR1/2D
MK-801	NMDAR1/2A=NMDAR1/2B>NMDAR1/2C=NMDAR1/2D
ACEA-1021	NMDAR1/2A>NMDAR1/2C>NMDAR1/2B>NMDAR1/2D
CGP-39653	NMDAR1/2A>NMDAR1/2B>NMDAR1/2C=NMDAR1/2D

of NMDA subunits. Our data differs slightly from those reported in Table 2 in that we find the highest potency for the NMDAR1/NMDAR2D combination. This may well reflect the different recombinant systems used. Radioligand binding suggests that these differences might be affinity changes (LAURIE and SEEBURG 1994), but they are relatively small. Thus, selectivity might be explained in terms of changes in α or τ. The data also shows quite clearly that agonists may or may not show selectivity. Thus, NMDA appears to have the same [A_{50}] across the four combinations, whereas glutamate shows a higher preference for the NMDAR1/NMDAR2B receptor. Again, this could be due to system effects, such as variance in α, K_{Glu} or τ.

The finding that glycine and glutamate bind to distinct subunits, and in particular that glycine binds to the NMDAR1 subunit suggests that glycine antagonists might be less selective than glutamate antagonists. This, in fact, is not the case. It has been shown that the competitive glycine antagonist ACEA-1021 has a higher inhibition (IC_{50}) on NMDAR1/NMDAR2A than on NMDAR1/NMDAR2D (WOODWARD et al. 1995). This might indicate that glycine has a lower Gly_{50} (higher potency) on this latter combination and, therefore, that higher concentrations of antagonist are needed to generate the same inhibition. Experiments with other structurally different glycine antagonists confirm the data generated with ACEA-1021 (TRIST and CARIGNANI, unpublished). Thus, even though glycine may bind to only one type of subunit, some selectivity can still be seen with both agonists and antagonists, and this reflects differences in interaction (i.e. different values of α) between NMDAR1 and the different types of NMDAR2 subunits. Alternatively, the differences could also include differences in affinity (K_{Gly}) or in γ. Importantly, competitive glycine antagonists do appear to produce non-competitive antagonism of NMDA-concentration–effect curves. This supports a major tenet of the co-agonist hypothesis.

As postulated above, MK is not equally potent against each recombinant combination. Thus, it has been reported that the NMDAR1/NMDAR2A and NMDAR1/NMDAR2B combinations have a higher sensitivity to MK than the NMDAR1/NMDAR2C and NMDAR1/NMDAR2D associations. This is an important observation and argues against the assumption that MK is non-selective when it binds within the channel. A summary of the selectivities for the two antagonists discussed above for the different NMDAR1/NMDAR2 associations is shown in Table 2. These data indicate that, even if the exact receptor construction is not known, recombinant experiments can provide information that supports the concept of co-agonism.

2. Tissue Experiments

In theory, there are thousands of possible combinations of the NMDAR subunits, and many of these might be present not only in different parts of the CNS but also within individual synapses. Thus, responses are probably the mean of many subtypes, each with its own value of α and τ. The result of the potentially diverse receptors that might be present could be additive or subtractive, or they might all be the same. Thus, it was interesting to ask if the behaviour predicted by the theory of co-agonism, which is supported by recombinant experiments, is also seen at the level of the tissue. Thus, experiments were carried out to detect possible affinity changes of antagonists for one agonist in the presence of an antagonist of the co-agonist. In addition, functional experiments were undertaken to measure the effect of antagonists on the field-potential changes in brain slices activated by glycine and NMDA. A number of observations confirmed that the response predicted by co-agonism is also seen in CNS tissues. Responses to NMDA in both rat cortical slices and in rat hippocampal slices were blocked non-competitively by the two glycine antagonists 7-chlorokynurenic acid (7CK) and GV 150526A (GV; TRIST et al. 1997). However, the nature of the antagonism differed from tissue to tissue and between the two antagonists. In rat cortex, 7CK appeared to act as a low-α or high-γ antagonist (Fig. 9) in that the lower concentrations tested exhibited a competitive antagonism whereas the highest concentration showed a reduction in the maximal response. In hippocampal slices, the maximal response was reduced at all concentrations, but a right shift of the curve was maintained. Comparison of 7CK and GV in the same rat cortex tissue, where α and other tissue parameters should be constant, indicated that GV showed a small right shift of the NMDA curve at the lowest concentration (1 μM) followed by collapse of the curves at the higher concentrations tested (3 μM and 10 μM; Fig. 9). Thus, since GV did not act the same as 7CK, it can be assumed that the different behaviour between the two reflects different values of γ for the two antagonists. As the NMDA A_{50}s were not left-shifted by GV, it seems that GV should have a value of γ close to one, whereas that of 7CK should be much higher.

Experiments were carried out in which the response of a fixed concentration of glycine and NMDA was antagonised by increasing concentrations of two competitive glycine antagonists. By assuming that the concentration of NMDA is much greater than its dissociation constant (K_{NMDA}), the ratio of the concentrations of each antagonist producing half maximal inhibition (IC_{50}) can be written as:

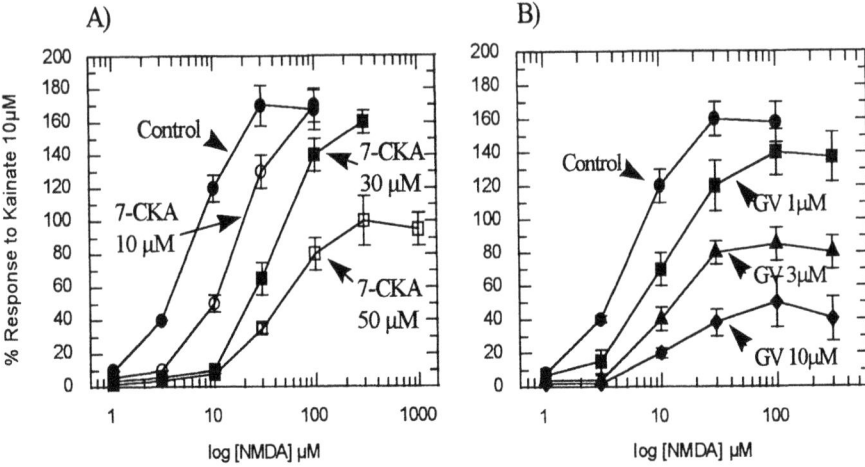

Fig. 9. The antagonism of N-methyl-D-aspartate-induced depolarisations in rat cortical slices by (**A**) 7-chlorokynurenic acid (7CK) and (**B**) GV 150526. It is assumed that the tissue parameters remained the same for both antagonists, as did the concentration of glycine. Therefore, since α and τ should be constant, the behaviour of the two antagonists reflects different values of γ. For GV, it would appear that γ is close to unity, as the curves collapse around the ED_{50} of the control curve. In contrast, for 7CK, γ should be large, as there is an appreciable right shift of the ED_{50} before the maximal effect is affected

$$\frac{IC_{50B1}}{IC_{50B2}} = \frac{\gamma_{B1} \cdot K_{BB1}}{\gamma_{B2} \cdot K_{BB2}} \tag{25}$$

Obviously, if γ is equal to unity, then the ratio of the IC_{50}s becomes the ratio of the K_Bs. Applying this concept to two tissues, namely the immature rat spinal cord (BRSC) and the adult rat cortical slice (ARCS), IC_{50}s were generated for ACEA-1021 and 7CK, and their ratios were calculated. These values are given in Table 3. It can be seen that the difference in ratios obtained between the two tissues are expected only if the receptors are heterogeneous or if the allosteric factors (γ) for the two glycine antagonists are different. As reported above, the effect of 7CK on NMDA-concentration–response curves suggests a high-γ antagonist and implies that either solution is possible.

Not only did glycine antagonists behave as expected in the whole tissue, MK also showed the non-competitive antagonism expected. In both slice preparations, MK reduced the maximal response of the NMDA curve and caused a small left shift of the $NMDA_{50}$, as predicted by the model (Fig. 5). However, the hippocampus was less sensitive to MK than the cortex (0.1 μM MK caused 49% inhibition in the cortex and 17% inhibition in the hippocampus), suggesting that τ might be higher in the hippocampus.

Radiolabelled-binding experiments have been carried out in washed rat cortical membranes in which the effects of co-agonists on the binding of a competitive NMDAR antagonist, [^{3}H]CGP-39653 (MUGNAINI et al. 1993), have

Table 3. pIC_{50}s (± standard error) calculated for ACEA-1021 and 7-chlorokynurenic acid (7CK) in immature-rat spinal cord (BRSC) and adult-rat cortical slice (ARCS) in the presence of 30 μM N-methyl-D-aspartate and 100 μM glycine (CORSI et al. 1997)

	pIC_{50} (ACEA)	pIC_{50} (7CK)	ACEA/7CK
BRSC	5.04 ± 0.06	5.17 ± 0.14	1.35
ARCS	5.67 ± 0.01	4.67 ± 0.01	0.10

been studied. Dissociation of the antagonist was enhanced in the presence of glycine, but not with NMDA. This negative allosterism of glycine on NMDA antagonists has been reported for other compounds (MUGNAINI et al. 1996). Interestingly, at least one antagonist, [^{3}H]-3-((R)-2-carboxypiperazin-4-yl)propyl-1-phosphonic acid (CPP), is not allosterically modified by glycine (MUGNAINI et al. 1993). In addition, similar observations were made in a quantitative autoradiography study using [^{3}H]CGP-39653 (MUGNAINI et al. 1996). Again, the antagonist binding was inhibited by glycine, but binding was not equal throughout the brain; lower inhibition was observed in the striatum compared with the cortex and hippocampus. These experiments support the concept that antagonists for one site differentially interact with the co-agonist site.

III. Implications of Co-Agonism

As already mentioned above, glycine may be the constant partner of the two agonists, and pathologies involving over-activation of the NMDA receptor are probably a manifestation of changing glutamate concentrations. Thus, the non-competitive nature of the glycine-antagonist–glutamate interaction is a considerable bonus in that the antagonism is insurmountable. Therefore, it can be predicted that a glycine antagonist should be effective in blocking NMDARs in vivo. This has been shown both with intracerebroventricularly (icv) administered NMDA itself and in cerebral ischaemia models (REGGIANI et al. 1995), where significant enhancements of glutamate have been reported (ANDINE et al. 1991).

Co-agonism predicts that the behaviour of a glycine antagonist can depend on the allosterism between the two agonists, and it allows for an allosteric interaction between the antagonist of one agonist with the co-agonist. Thus, it might explain why glycine-receptor antagonists have been shown in animals to lack the side effects seen with competitive NMDAR antagonists and with MK. In experiments where 7CK, CPP and MK were injected icv into mice and anti-convulsive activity against NMDAR antagonists was compared with cognitive function and ataxia, CPP was more potent in producing memory impairment than as an anti-convulsant or in the production of ataxia. MK showed a small separation between the desired effect and cognition and ataxia, while 7CK demonstrated good anti-convulsant activity and practically no undesired effects over the dose range tested (CHIAMULERA et al. 1990). Experiments with other glycine antagonists have confirmed that this approach to NMDAR inhibition does not exhibit the expected side-effect profile, as seen particularly with competitive NMDAR antagonists. Co-agonism can offer two possible explanations. The first is that, in pathological situations, the receptor subunits activated are those selective for glycine antagonists, namely NMDA2A-like subunits. This may also involve disease-modifying mechanisms at the level of the receptor, such as phosphorylation states. NMDAR can be phosphorylated. This phosphorylation has been shown to be combination specific. In general, the phosphorylated receptor is more active when NMDAR2A and NMDAR2B are involved and much less when NMDAR2C is in the combination (KUTSUWADA et al. 1992). Phosphorylated NMDAR2B persisted longer than phosphorylated NMDAR2A. The second explanation is that the behaviour of the glycine antagonist may be similar in different brain regions, but the level of glutamate is much lower for physiological functions than in pathological conditions. Thus, the nature of non-competitive antagonism means that the inhibition is more effective with higher concentrations and much less effective at the lower end of the dose–response curve.

C. General Observations

In general, pharmacologists have, on the whole, approached the problem of inhibiting receptors involved in physiological and pathological processes by designing reversible, competitive antagonists. They can be discovered logically by modifying the natural agonist structure, removing those attributes that cause agonism and, often, by increasing affinity. Since the starting point is the hormone or transmitter, it is not surprising that the antagonism should be surmountable with increasing agonist concentration. This has been a very successful approach to discovering medicines. Examples exist for a wide number of receptors. These include β-adrenoceptor antagonists used for angina and hypertension, histamine-H_2-receptor antagonists used for gastric ulcers, $5HT_3$ antagonists used to treat radiation- and chemotherapy-induced emesis, dopamine-D_2-receptor antagonists used for psychosis and many more examples.

The emphasis on competitive antagonists probably results from the fact that this type of antagonism is independent of tissue factors, such as receptor number and different transduction mechanisms. However, non-competitive antagonists can be advantageous when high concentrations of agonist are causing pathological conditions and therapeutic windows need to be maintained. In this condition, the concentration of competitive antagonist has to be increased to surmount the increase in agonist present and, as a consequence, the antagonist may well lose its selectivity, leading to unwanted side effects. An alternative approach that has not been overly exploited by industrial pharmacologists is to design insurmountable antagonists that are independent of agonist concentration and, in some cases (such as for co-agonism), can be selective, with fewer side-effects than conventional competitive antagonists. This article has attempted to describe non-competitive antagonists as potential medicines.

References

Andine P, Sandberg M, Bagenholm R, Lehmann A, Hagberg H (1991) Intra- and extracellular changes of amino acids in the cerebral cortex of the neonatal rat during hypoxic ischemia. Dev Brain Res 64:115–120

Black JW, Leff P (1983) Operational models of pharmacological agonism. Proc R Soc London Ser. B 220:141–162

Chiamulera C, Costa S, Reggiani A (1990) Effect of NMDA- and strychnine-insensitive glycine site antagonists on NMDA-mediated convulsions and learning. Psychopharmacology 102:551–552

Corsi M, Fina P, Trist DG (1996) Co-agonism in drug-receptor interaction: illustrated by the NMDA receptors. TiPS 17:220–222

Corsi M, Bettelini L, Fina P, Trist DG (1997) Estimation of the relative potency of two glycine antagonists on NMDA-induced depolarisations in BRSC and ARCS. In: Trist DG, Humphrey PPA, Leff P, Shankley N (eds) Receptor classification: the integration of operational, structural, and transductional information. Ann New York Acad Sciences 812:187–188

Fadda E, Danysz W, Wroblewski JT, Costa E (1988) Glycine and D-serine increase the affinity of N-methyl-D-aspartate-sensitive glutamate-binding sites in rat brain synaptic membranes. Neuropharmacology 27:1183–1185

Fina P, Corsi M, Carignani C, Trist DG (1997) Co-agonism between glutamate and glycine on the NMDA receptor: does it influence the estimation of antagonist dissociation constant? In: Trist DG, Humphrey PPA, Leff P, Shankley N (eds) Receptor classification: the integration of operational, structural, and transductional information. Ann New York Acad Sciences 812:184–186

Johnson JW, Ascher P (1987) Glycine potentiates the NMDA response in cultured mouse-brain neurons. Nature 325:529–531

Kemp JA, Leeson PD (1993) The glycine site of the NMDA receptor – five years on. TiPS 14:20–25

Kenakin T (1993) Allotopic, non-competitive and irreversible antagonism. In: Pharmacological analysis of drug-receptor interaction (2nd Edition). Raven Press, New York, pp 323–343

Kutsuwada T, Kashiwabuchi N, Mori H, Sakimura K, Kushiya E, Araki K, Meguro H, Masaki H, Kumanishi T, Arakawa M, Mishina M (1992) Molecular diversity of the NMDA receptor channel. Nature 358:36–41

Laube B, Hirai H, Sturgess M, Betz H, Kuhse J (1997) Molecular determinants of agonist discrimination by NMDA receptor subunits: analysis of the glutamate binding site on the NR2B subunit. Neuron 18:493–503

Laurie DJ, Seeburg PH (1994) Ligand affinities at recombinant N-methyl-D-aspartate receptors depend on subunit composition. Eur J Pharmacol 268:335–345

Marvizón J-C, Baudry M (1993) Receptor activation by two agonists: analysis by non-linear regression and application to N-methyl-D-aspartate receptors. Analytical Biochemistry 213:3–11

Monyer H, Sprengel R, Schoepfer R, Herb A, Higuchi M, Lomelli H, Burnashev N, Sakmann B, Seeburg PH (1992) Heteromeric NMDA receptors: Molecular and functional distinction of subtypes. Science 256:1217–1221

Mugnaini M, Giberti A, Ratti E, van Amsterdam FThM (1993) Allosteric modulation of [^{3}H]CGP 39653 binding by glycine in rat brain. J Neurochem 61:1492–1497

Mugnaini M, Van Amsterdam FThM, Ratti E, Trist DG, Bowery NG (1996) Regionally different N-methyl-D-aspartate receptors distinguished by ligand binding and quantitative autoradiography of [^{3}H]-CGP 39653 in rat brain. Br J Pharmacol 119:819–828

Reggiani A, Costa S, Pietra C, Ratti E, Trist D, Ziviani L, Gaviraghi G (1995) Neuroprotective activity of GV 150526A, a novel potent and selective glycine antagonist. Europ J Neurol (Suppl 2) 2:6–7

Sucheret NJ, Awobuluyi M, Choi Y-B, Lipton SA (1996) NMDA receptors: from genes to channels. TiPS 17:348–355

Trist DG, Marcon C, Fina P, Reggiani A, Corsi M (1997) A dual agonism model explains the behaviour of GV 150526A, 7-chlorokynurenic acid and MK801 against NMDA receptor activation in cortical and hippocampal slices. The Pharmacologist 39:56

Williams K, Chao J, Kashiwagi K, Masuko T, Igarashi K (1996) Activation of N-methyl-D-aspartate receptors by glycine: role of an aspartate residue in the M3-M4 loop of the NR1 subunit. Mol Pharmacol 50:701–708

Woodward RM, Huettner JE, Guastella J, Keana JF, Weber E (1995) In vitro pharmacology of ACEA-1021 and ACEA-1031: systemically active quinoxalinediones with high affinity and selectivity for N-methyl-D-aspartate receptor glycine sites. Molecular Pharmacology 47:568–581

CHAPTER 10
Mechanisms of Receptor Activation and the Relationship to Receptor Structure

D.M. PEREZ and S.S. KARNIK

A. Introduction

Current receptor theories try to relate drug effects to the interaction of a drug molecule with its specific receptor. While many theories have evolved and are explained in great detail in other chapters of this book, most are derived from the widely accepted and broadly based occupation theory by CLARK (1937) and the laws of mass action. The most important corollary to this theory is that the magnitude of the drug effect is directly proportional to the number of receptors occupied by the drug. The maximum response occurs when all of the receptors are occupied, and it is assumed that one drug molecule interacts with one molecule of receptor. Clark recognized that the ability of a drug to cause an effect depends on the drug "fixing" or binding to the receptor and the ability of the drug to produce its action on the receptor after binding. Although subsequent theories to explain spare receptors, efficacy and other pharmacological effects appeared, the basic tenets of Clark's model and his insights into drug action and recognition have held. With the advent of molecular biology and the ability to clone, purify, and mutagenize receptors, the principles of how drugs bind and subsequently activate receptors at the molecular level are becoming apparent. This chapter will review these activational paradigms in the G protein-coupled receptor (GPCR) field and how this field relates to the receptor's structure, and will try to bring together a general mechanistic view of drug action.

B. Common GPCR Structure/Function

Common features of the primary and secondary structures of GPCRs and how they relate to functional mechanisms will be briefly reviewed here. The GPCR superfamily includes several hundred distinct but related proteins. They are found in a wide range of organisms and are involved in the transmission of signals across membranes. Over 80% of all hormones signal through these types of receptors, which makes their interest to the pharmaceutical industry and pharmacology apparent. They are composed of single polypeptide chains, each containing seven stretches of 20–28 hydrophobic amino acids that represent transmembrane (TM) domains (Fig. 1). The TM segments are believed to be α-helices oriented approximately perpendicular to the membrane, as shown in rhodopsin (SCHERTLER et al. 1993), although helices such as TM5 might have distinct differences or may be not be fully α-helical (JAVITCH et al. 1995). The N-terminus of each receptor is extracellular and contains several glycosylation sites (APPLEBURY and HARGRAVE 1986). The C-terminus is located on the intracellular side and contains sites for phosphorylation, which are used in the regulation of the receptor in desensitization and sequestration. The TM domains are linked by three intracellular and three extracellular loops. There is also a highly conserved disulfide bond between cysteines in the second and third extracellular loops. This bond is needed to maintain proper folding of the protein and the attainment of the high-affinity site in binding (KARNIK and KHORANA 1990).

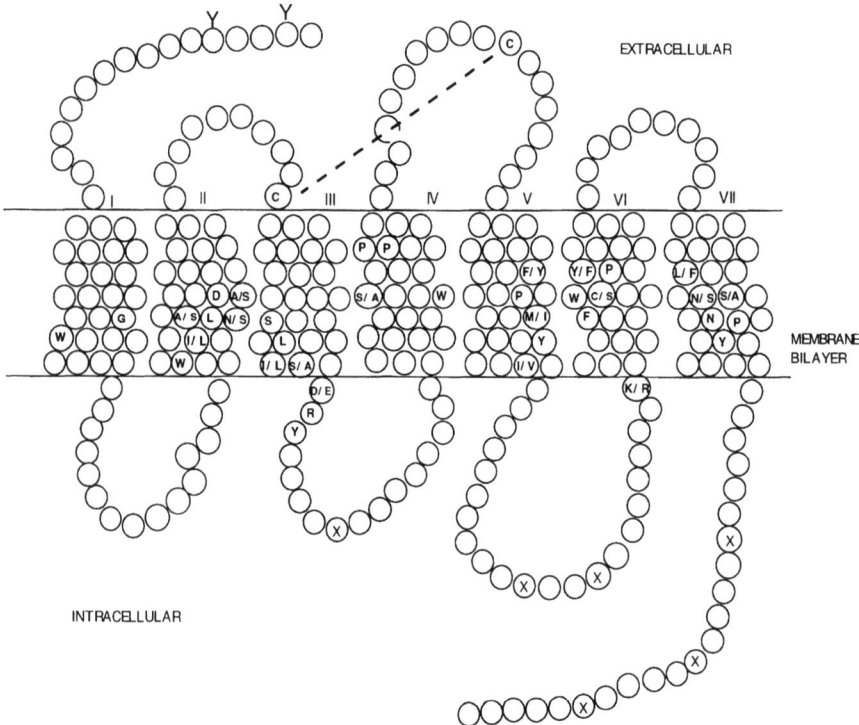

Fig. 1. General topography and conserved amino acids and motifs in G-protein-coupled receptors. The model is generated from data in BALDWIN (1993). Designated amino acids shown are conserved in at least 70% of all G-protein-coupled receptors. The *dashed line* indicates a disulfide bond that is also conserved in (>70%) of all G-protein-coupled receptors. *X* represents potential sites for phosphorylation in the regulation of receptor function. *Roman numerals* designate transmembrane helices or domains 1–7. *Y* in the N-terminus indicates sites for potential glycosylation

The receptors bind a signaling molecule on the extracellular side and then, following activation by the drug, cause a conformational change which causes the intracellular side to bind and activate the heterotrimeric guanine nucleotide-binding protein (G protein). The activated G protein then dissociates from the receptor and subsequently initiates a second messenger response by activating various effector molecules, such as phospholipases and channels. The exact mechanism of the receptor–G protein coupling is still unclear, since there is no direct structural information. However, the intracellular loops and C-terminus are implicated, with particular emphasis on intracellular loop 3, where the amino acids near the lipid bilayer have been found to impart specificity of coupling to a particular G protein (WONG et al. 1990; WONG and Ross 1994; BLIN et al. 1995).

Historically, the cloning of the β-adrenoceptor (β-AR) and its homology to rhodopsin (DIXON et al. 1986) led researchers to recognize that all GPCRs are encoded by genes with similar features. There is conservation of particular sequence(s) and spacing between key functional amino acids, especially in the TM domains and where the G proteins are predicted to bind and activate. This has suggested that the entire GPCR family arose from a single ancestral gene. A comparison of a large number of GPCRs by BALDWIN (1993) has highlighted the conserved motifs in the TM domains (Fig. 1). Of special consideration here is the conserved DRY at the end of helix 3, which has been shown to be involved in the binding and activation of G proteins. Specifically, the Arg is involved in the activation process, since mutations at this region still allow the G protein to bind but are unable to cause the exchange of guanosine diphosphate for guanosine triphosphate, a necessary step in the activation of the G protein (ACHARYA and KARNIK 1996). The Asp or Glu, in some cases, may be involved in some kind of proton shuttle, as suggested by the mutagenesis work of Cotecchia (SCHEER et al. 1997), since a number of substitutions at this region caused constitutive activation, with the protonation/deprotonation of this residue speculated to determine the transition of the inactive receptor, R, to the active receptor, R*. Also involved in the binding and activation process are conserved amphiphilic helices located in the intracellular loop regions. It is believed that these positive helices bind to oppositely charged helices on the G protein, as suggested by mutagenesis (KUNKEL and PERALTA 1993) and work on the peptide toxin mastoparan (Ross and HIGASHIJIMA 1994), in which this highly positively charged α-helix can effectively block receptor–G protein activation.

Although the tilt and orientation of the helices in the membrane may be different, all members of the GPCR family are believed to have basically the same structure in a membrane environment and, thus, may share common ligand-binding and G protein-activating paradigms. It has been assumed that the GPCRs have the same structure as bacteriorhodopsin (bR), an integral membrane protein from *Halobacterium halobium* which contains the same distinctive pattern of seven TM domains and has had its structure determined to 2.5 Å (PEBAY-PEYROULA et al. 1997). Many molecular models of GPCRs have based their α-carbon coordinates on bR. Mechanistically, bR also shares with rhodopsin its response to light through the isomerization of a retinal chromophore, but bR is involved in proton pumping and is not coupled to G proteins, and its sequence has none of the conserved motifs associated with other GPCRs. The projection map of rhodopsin also suggests that the tilt and orientation of the helices are distinct from bR. Nonetheless, the activational paradigms are analogous to rhodopsin and other GPCRs and will be included in this chapter. The lessons from bR demonstrate the concluding observations of this chapter, in which paradigms of binding and activation may be conserved even though there is a wide variety of differences in receptor primary sequence and type of ligand.

To understand the molecular mechanisms involved in receptor activation, a brief review of our current understanding of receptor theory is needed. Our current model of receptor theory is based upon a key paper in which mutations in the C-terminal region of the third intracellular loop of the β_2-AR resulted in its constitutive activation (SAMAMA et al. 1993). The mutant exhibited dramatic increases in affinity for agonists (even in the absence of G protein) but not for antagonists, with the extent of affinity increase being correlated with the intrinsic activity of the ligand. In addition, the constitutively active mutant (called CAM) exhibits an increased potency of agonists for stimulation of second messengers and an increased intrinsic activity of partial agonists. This prompted the speculation that this mutant receptor might have an increased tendency to adopt an active conformation, which could be responsible for the observed agonist-binding behavior and the spontaneous-activation properties. From a structural viewpoint, these mutants can be envisioned as being impaired in their key constraining function, thus spontaneously "relaxing" into their active conformation. This mutation lead to the revision of the old ternary-complex model (DELEAN et al. 1980), which postulated that receptor activation required the agonist-promoted formation of an active, "ternary" complex of agonist, receptor, and G protein. The revised and extended model (called the two-state model) includes an explicit isomerization (allosteric) of the receptor to an active state (R*) before it can couple to the G protein (SAMAMA et al. 1993). According to this model, constitutive activation is explained as a disturbance of the normal equilibrium between the inactive state (R) and the active state (R*), leading to a higher proportion of receptor molecules in the active R* state. Inverse agonists, formally called negative antagonists, such as ICI-118551, have a higher affinity for the inactive state, R. Therefore, inverse agonists can reverse a constitutively active phenotype of higher basal activity by shifting the equilibrium of the constitutively active receptor (in the R* state) back to the inactive state, R. Neutral antagonists, by definition, bind with equal affinity to both R and R*. Therefore, neutral antagonists cannot shift equilibrium and have no effect on the basal activity of constitutively active receptors. A number of inverse agonists, including ICI-118551, and neutral antagonists have been described and verified for β_2-AR (CHIDIAC et al. 1994, 1996; BOND et al. 1995).

C. Rhodopsin and BR Activation: Light as the Ligand

The first GPCR to be cloned and studied was rhodopsin, the photoreceptor molecule of the retinal rod cell. Because it is highly expressed in cell lines and can be purified in large amounts for spectrophotometric analysis, structure–function studies of this receptor are extensive and have served as a paradigm for the structure–function relationship in other GPCRs. At first glance, this highly specialized protein of the photoreceptor signaling cascade seems unrelated to other G protein receptors, but the structure–function relationship that is now known in other receptor systems can always draw parallels to rhodopsin. The photoreactive chromophore of rhodopsin is 11-*cis*-retinal, which is covalently bound in the interior of the protein as a protonated Schiff base. Rhodopsin initiates a cascade of visual transduction when a photon induces isomerization of its 11-*cis*-retinal chromophore to the all-*trans*-retinal chromophore (HUBBARD and KROPF 1958). This leads to an increase in the overall volume of the protein, which induces formation of a new binding site for the G protein on its cytoplasmic surface. Therefore, photoisomerization of 11-*cis*-retinal to all-*trans*-retinal causes receptor activation. Direct evidence for the arrangement of seven α-helices was obtained from cryo-electron microscopy projection maps of bovine and frog rhodopsins at 7.5-Å resolution (SCHERTLER et al. 1993; UNGER et al. 1997). The structural features delineated so far support much biochemical evidence that describes rhodopsin functions. Hence, the structural features of rhodopsin have been described in some detail.

I. Direct Structural Information

The projection map from two-dimensional crystals allows both the assignment of sequence segments of rhodopsins to the density peaks and the calculation of tilt angles for the seven TM helices in the membrane plane, but does not permit clear assignment of connections (loops) between the helices. Helices 4, 6, and 7 are the least tilted. Helices 1, 2, 3, and 5 are tilted and may also be curved. Helix 1 is more exposed to lipid and forms a short connection, with helix 2 in the cytoplasmic side. Helix 2 interacts with helices 1, 3, and 7 at the extracellular side and also with helix 4 at the intracellular side. Helix 3 is the most tilted of helices, with its cytoplasmic end closer to helix 5 and the extracellular end away from helix 5. This helix would then form the floor of the retinal-binding pocket, closing it off from access from the cytoplasmic side. Helix 4 is the shortest and least tilted in the entire density map. Helix 5 is most ambiguous in the map, since it is exposed to lipid throughout its length. Helix 6 is vertical and may be bent towards helix 5 on the extracellular side to limit lipid access to the interior of the protein. Helix 7 is the most perpendicular to the membrane and is in interaction with helices 1, 2, and 6 along its entire length.

The helices are somewhat loosely packed towards the extracellular side, with a retinal binding cavity formed by helices 3, 4, 5, 6, and 7. The helix arrangement close to the cytoplasmic surface of rhodopsin is significantly more compact (~25% smaller cross-sectional area) than at the extracellular surface. This is consistent with light-induced activation of rhodopsin, which has now been shown to involve movement of the helices to increase the cytoplasmic surface area and formation of the binding site for the G protein.

The electron-density map of the extracellular domain in bovine and frog rhodopsins suggests that the TM helices do not extend considerably beyond the membrane border. The projected amount of density represents part of the glycosylated N-terminus and the connecting loops. An important part of this extracellular domain structure is the loop connecting TM helices 4 and 5, which forms the roof of the retinal-binding pocket. In contrast to the extracellular domain, limited electron-density regions are visible outside of the lipid bilayer in the intracellular domain. A small segment connecting helices 1 and 2 is visible. Near this 1–2 loop, density corresponding to an ordered part of the C-terminal tail is also visible. TM helix 6 appears to extend farthest into the intracellular region. Thus, the cytoplasmic loops are less packed than the extracellular loops. Portions of cytoplasmic loop connecting TM3 and TM4 in addition to the loop connecting TM5 and TM6 have been functionally implicated in interactions with G protein, receptor kinase and arrestin. However, large segments of these two loops can be deleted without significantly hampering the TM-domain function. Therefore, the low-resolution structure of the cytoplasmic loops suggests that rhodopsin lacks a constitutive binding site for the G protein in the ground state. Light activation would lead to formation of a complete binding site through the rearrangement of TM helices and, presumably, by induction of an ordered structure in the cytoplasmic loops. The molecular mechanism underlying this rearrangement of TM helices is reviewed in some detail below.

II. The Activation Mechanism

1. How Retinal Binds

In rhodopsin, the basal, inactive pigment contains 11-*cis*-retinal covalently attached to Lys296 in TM7 through a protonated Schiff-base linkage. Formation of this pigment suppresses a small but measurable basal activity of the apoprotein, especially at low pH (i.e. inverse agonist). Preference for the bent geometry of 11-*cis*-retinal is conserved among all visual receptors. Linear analogs, such as 13-*cis*-retinal and all-*trans*-retinal, are not efficient at inhibiting the basal activity or quantitatively generating rhodopsin from opsin. The key residues in retinal binding and subsequent activation by retinal are expected to be conserved in the entire opsin family. Constraints from mutagenesis combined with spectroscopic studies suggest that the retinal-binding pocket is largely defined by conserved residues on TM helices 3, 5, 6 and 7. Gly121 in TM3 and Phe261 in TM6 form one of the possible contact sites along the chromophore (HAN et al. 1996a, 1996b). Cross-linking studies are consistent with the idea that the C3 carbon of the ionone ring is near Trp265 and Leu266 in TM6 (NAKAYAMA and KHORANA 1990). These likely favorable steric interactions are necessary to form a stable pigment whose binding pocket can discriminate between the inactive (11-*cis*-retinal) and active (all-*trans*-retinal) states of the chromophore.

2. Salt-Bridge Constraining Factor: Movement of TM3 and TM6

The strong positive charge of the protonated retinal-protein Schiff-base linkage in the middle of TM7 is compensated by the counter-ion, Glu113 in TM3. Mutation of either Lys296 or Glu113 results in constitutive activation of the apoprotein, opsin. The molecular mechanism of activation of opsin has been shown to be the same as that involved in the light activation of rhodopsin. The model for activation involves disruption of the salt-bridge interaction between the protonated retinal-Lys296 Schiff base and Glu113, followed by deprotonation of the Schiff base and net uptake of a proton by Glu113 which essentially generates a neutral pocket (COHEN at al. 1992). The developed strain in retinal that ultimately leads to disruption of the salt bridge is formed from steric interactions between the chromophore and apoprotein at two different sites, Trp265 and Tyr268 in TM6, which interact with the β-ionone ring, as suggested by ligand cross-linking, mutagenesis and retinal-analog reconstitution studies (NAKAYAMA and KHORANA 1990, 1991; BHATTACHARYA et al. 1992; RIDGE et al. 1992). Substitution of Trp265 and Tyr268 uncouples the photocycle from the biochemical activation. This uncoupling occurs since metarhodopsin II (i.e. the active transition-state intermediate) can be generated in these mutants, but less than 10% of the G protein can be activated. An identical phenotype results when opsin is reconstituted with an analog of 11-*cis*-retinal that is constrained from converting to the all-*trans* form. These studies suggest that Trp265 and Tyr268 in opsin undergo steric perturbation during the light-induced *cis*-to-*trans* isomerization of retinal. A Trp residue is highly conserved in TM6 of nearly all members of the GPCR super family, and occurrence of an aromatic residue three or four residues away from the conserved Trp is very frequent. Therefore, this region in the GPCR structure is likely to be important for a conserved function in other GPCRs.

Gly121, in the middle of TM3, is another crucial residue in the activation process; it has been shown to make direct contact with the C9 methyl group of retinal. The progressive loss of G protein activation and the stability of the pigment in the basal state result from regeneration of Gly121 mutants with 11-*cis*-retinal analogs bearing larger ethyl and propyl groups at the C9 position. Furthermore, Gly121 mutants regenerated with 11-*cis*-retinal show partial G protein activation without light activation, but the opsin form of the Gly121 mutants are not constitutively active (HAN et al. 1996a). This residue is conserved in opsins (but not in other GPCRs), but mechanisms analogous to this apparent van der Waals contact with the agonist and TM3 appear to be conserved in several GPCRs. Linking Gly121 in TM3 and Phe261 in TM6 in the activation mechanism are studies in which the Gly121 mutants can be partially rescued by mutation of Phe261 (HAN et al. 1996b), a residue highly conserved in all GPCRs. In addition, Glu122 in TM3 and His211 in TM5 are thought to interact with 11-*cis*-retinal and may also play a role in activation of rhodopsin, but they are not conserved in the entire family and are dispensable in bovine rhodopsin (SAKMAR et al. 1989; WEITZ and NATHANS 1992). Although more than 22 residues located on six out of seven TM helices of opsin stabilize interaction between retinal and the apoprotein, only a small number of interactions directly influence the functional activation. Thus, in rhodopsin, contacts between 11-*cis*-retinal and the Gly121, Trp265, Tyr268 residues of the apoprotein play a key role in controlling the rate of protein conformational changes initiated ultimately by breaking the critical salt-bridge interaction. Therefore, a reasonable goal for elucidating structure–function relationships of GPCRs without an easy experimental system should be the identification of interactions that critically influence the function. We believe the locations of such interactions are well conserved within the putative structure of the receptors.

From a different experimental system, spin-label studies indicate that photoexcitation involves a rigid-body movement of TM3 relative to the other helical bundles (FARAHBAKHSH et al. 1995). The net effect is similar to movement of a cylinder, with the extracellular top half of the helix tilting toward the binding pocket while the intracellular bottom half tilts outward. Other spin-label studies have also shown that TM6 movement is involved in rhodopsin activation (ALTENBACH et al. 1996). The sequences studied included the TM6 and TM7 intracellular regions, a transducin interaction site. Changes in the characteristics of the spin labeled rhodopsin upon photoactivation indicate that chromophore isomerization results in patterns of structural changes that can be interpreted in terms of movement of the TM6 helix that extends into the aqueous loop regions of the intracellular region.

Recent biochemical data have also linked the TM3 and TM6 helical movements together (SHEIKH et al. 1996). Metal-ion-binding sites between TM3 and TM6 were engineered by substituting His residues for the natural amino acids at the cytoplasmic ends of these helices. The resulting mutant proteins were able to activate the visual G protein transducin in the absence (but not the presence) of metal ions. This was due to the constraint formed by the chelation of the metal ions with the substituted His. These results indicate that TM helices 3 and 6 are in close proximity and suggests that movements of these helices relative to each other are required for transducin activation. Thus, both the spin-label and chelation studies confirm implications from the diffraction studies, in which light activation leads to an opening of the cytoplasmic surface, allowing the interaction site(s) with the G protein transducin to be exposed. Thus, a change in the orientation of TM3 and TM6 is likely to be a key element in the mechanism for the coupling of ligands to the activation of other GPCRs.

Indeed, several of the details involved in rhodopsin activation have also been shown to be involved in the photochemical coupling reactions that form the basis of the proton-pumping mechanism in bR (KHORANA 1988), halide-ion pumping by halorhodopsin (HAUPTS et al. 1997), and chemotaxis by sensory rhodopsin (HAUPTS et al. 1997). These proteins share with rhodopsin the retinylidene Schiff base deprotonation reaction as the primary light-triggered reaction that leads to protein conformational transitions. The photocycle of bR can be compared with that of rhodopsin, in which proton transport from the cytoplasm to the extracellular surface is initiated by the light-induced isomerization of retinal from all-*trans*-retinal (inactive) to 13-*cis*-retinal (active). The apoprotein structure contains a seven-TM-helical domain with a Lys residue in the TM7 also attached to the retinal chromophore via Schiff-base linkage. The third TM helix harbors the putative counter-ion (Asp85 in bR) for the protonated Schiff base, which directly extracts the proton from the Schiff base. In both charge and position, both of these residues are analogous to Lys296 and Glu113 in rhodopsin. A Leu residue (Leu93 in bR) in TM3 interacts with the C13 methyl group of retinal; this interaction controls the rate of protein conformational change associated with retinal isomerization (DELANEY et al. 1995). TM6 contains a Trp residue which, when mutated, slows down the protein-isomerization rate. Both of these residues are analogous to Trp265 and Gly121 in rhodopsin. Thus, the basic steps that trigger conformational transitions upon retinal isomerization are conserved in other retinal-based signal transducers, though they are not coupled to G proteins.

D. AR Activation: Small Organics as Ligands

After rhodopsin, the most widely studied GPCRs are the ARs. ARs represent a large class of receptors in which the endogenous ligand is a small organic, non-peptide molecule. It is also representative of a smaller family of biogenic amine receptors in which the endogenous ligand contains a conserved protonated amine that is essential for bioactivity. Members of the biogenic amine-receptor family are the AR amines, serotonins, dopamines, muscarinics, and histamine. The AR family (α_{1a}, α_{1b}, α_{1d}, α_{2a}, α_{2b}, α_{2c}, β_1, β_2, β_3) mediates the effects of the sympathetic nervous system through the actions of the catecholamines, (–)-epinephrine (adrenaline) and (–)-norepinephrine (noradrenaline). All of the above subtypes are cloned (BYLUND et al. 1994). Classification of ARs (α_1, α_2, and β-) is based on the primary amino acid sequences and potencies of selective AR ligands (STROSBERG 1990). All ARs contain highly conserved residues within the TM domains where the agonist binds. However, the ligand-binding pocket is distinct among the receptor subtypes, as they can discriminate among a wide variety of synthetic agonists and antagonists (HWA et al. 1995). The observed physiological response is dependent upon the type of AR expressed in the tissue being innervated.

I. Important Binding Contacts of the Endogenous Ligands

Norepinephrine is composed chemically of an ethylamine moiety attached to a di-hydroxyl (catechol) benzene ring. For positively charged ligands, such as the biogenic amines (and specifically norepinephrine), negatively charged counter-ions in the binding pocket are crucial to ligand binding and subsequent activation. Earlier work has shown that removing the aspartic acid in TM3 of the receptor (conserved to Glu113 of rhodopsin) reduces the binding of agonists and antagonists in the β_2-AR by 10,000-fold (STRADER et al. 1988), with similar results in the α_2-AR (WANG et al. 1991) and M_1 muscarinic (FRASER et al. 1989). However, the same mutation in α_1-AR still maintains high-affinity binding for antagonists and reduced (but not abolished) affinity for agonists (PORTER et al. 1996). A similar result for α_1-AR was also obtained in a serotonin-receptor subtype (Ho et al. 1992). The apparent contradiction may be due to the type of substitution, which also affects the correct folding of the receptor protein. Alternately, the strength of the ionic bond may be different among the adrenergic subtypes. Nonetheless, the aspartic acid residue in the receptor is a very important binding contact for the endogenous ligand and is analogous in both charge and position to Glu113 of rhodopsin.

Another important binding contact in ARs is how the catechol hydroxyls of epinephrine interact with the receptor. The catechol hydroxyls are orientated on the benzene ring *para* and *meta* to the ethylamine moiety. There are several Ser residues in TM5 that are highly conserved among receptors (generally three total) that bind catecholamines, but not in other GPCRs. In β_2-AR, two Ser residues in TM5, Ser204 and Ser207, have been shown to be involved in hydrogen-bond interactions with the hydroxyl groups on the catechol ring (STRADER et al. 1989). Both residues are required for high-affinity binding of agonists (either Ser204Ala or Ser207Ala result in a 100-fold reduction in binding affinity) and for the full activation of the receptor, with each Ser contributing 50% to the activation process. Using catechols that lack either a *m*- or *p*-hydroxyl, together with the two mutant Ser receptors, the authors proposed a model in which Ser204 hydrogen bonds with the *m*-hydroxyl group of the catechol ring while Ser207 interacts with the *p*-hydroxyl group. The validity of extrapolation of this model to α_2-AR is unclear. The equivalent serines in the α_2-AR (Ser200 and Ser204) both show a tenfold decreased affinity for (–)-epinephrine when mutated to Ala (suggesting both Ser residues and receptor hydroxyls are required) but no change in affinity with an optically racemic mixture of synephrine that lacks a *m*-hydroxyl group (suggesting the *p*-hydroxyl is unimportant; WANG et al. 1991). The Ser204 mutant significantly attenuated the functional activity (65% active), but only with synephrine, implicating *p*-hydroxyl interaction. No changes in activity were seen with the natural hormones with either Ser mutant.

With the recent cloning of the α_{1a} subtype and the receptor containing only two of the three conserved serines in TM5, the role of the serines in α_1-AR was explored and compared with the β_2-AR paradigm (HWA and PEREZ 1996). Replacement of either Ser188 or Ser192 in TM5 of α_{1a}-AR with an Ala did not significantly reduce the binding affinity for any of the agonists compared with the binding in wild-type (WT) protein. In fact, the binding affinity for phenylephrine was significantly increased (sevenfold) in the Ser192A mutant. These results are quite distinct from the β_2-paradigm, where either Ser mutation was able to reduce agonist-binding affinity. To confirm a hydrogen-bond interaction, the double mutant Ser188/192Ala was created and found to decrease the binding affinity by 25–100-fold for various agonists, consistent with a decrease in binding energy of $\Delta G = +2–3\,\text{kcal/mol}$ and equivalent to a disruption of a single hydrogen bond. Since either Ser residue is sufficient in itself in maintaining the WT binding affinity but the free energy values indicate only one hydrogen bond is formed, the data suggests that both serines contribute a weak hydrogen bond to the agonist. When one Ser is eliminated by mutagenesis, there are competition interactions occurring between the two hydroxyls on the agonist and the remaining Ser on the receptor in which the agonist moves to optimize its docking. This results in no net decrease in affinity when one Ser is mutated but decreases by 25–100-fold when both serines are mutated. However, because of the sevenfold increase in affinity for phenylephrine (the *m*-hydroxyl agonist) with the Ser192Ala mutant, it seems that the *m*-hydroxyl interaction with Ser188 is the stronger of the two interactions, and only the Ser188–*m*-hydroxyl contributes a major role in receptor activation, providing 70–90% of the WT response in inositol phosphate (IP) release. However, the effect of Ser192 on receptor activation was minimal.

To account for a stronger *m*-hydroxyl interaction and a weaker *p*-hydroxyl effect, the catecholamine would be modeled in a planar orientation (relative to the extracellular surface) in the ligand-binding pocket as opposed to a skewed (tilted) orientation in β-AR. Since there are three residues between the two serines in the TM5 helix in α_1-ARs, as opposed to two residues between the β-AR serines, the placement of the serines on the helix is different in these two receptor subtypes, which results in differences in the orientation of the catechol ring in the binding pocket. The *m*-hydroxyl–Ser188 interaction in α_1-AR is closest to TM4, while the *m*-hydroxyl–Ser204 interaction in β-AR is closer to TM6, resulting in rotation of the catechol ring about 120° in α_1-AR. Hence, it appears that the catechol docks in a unique manner in α_1-ARs as compared with the β-AR paradigm, and this may implicate activational differences.

There is also a report of a phenylalanine residue in TM6 of the β_2-AR that is involved in an aromatic/hydrophobic contact with the benzene ring of epinephrine (DIXON et al. 1988). Although not confirmed in the other ARs, it is likely to be analogously involved in binding, since this residue is strictly conserved (unlike the serines in TM5) in all AR subtypes. It is also analogous to Phe261 in rhodopsin in identity and position.

II. Insights on How Epinephrine Activates

1. Release of Constraining Factors

A "constraining factor" has been previously postulated for the α_{1b}-AR subtype, holding the receptor in a basal or inactive configuration until bound by a receptor agonist (KJELSBERG at al. 1992). This hypothesis is based on the characterization of mutant α_{1b}-ARs where the normal Ala at position 293 in the third cytosolic loop was changed to all possible amino acid combinations. KJELSBERG et al. observed that any amino acid substituted at position 293 caused the α_{1b}-AR to become constitutively active. This constitutive activity was characterized by higher binding-affinity values for AR agonists but not antagonists. Also, a higher AR-agonist potency for generating intracellular signals was observed and, more importantly, there was increased second-messenger production in the absence of AR agonist; this all reflects the active state, R*. In addition, KJELSBERG et al. suggested that any amino acid change at position 293 of the α_{1b}-AR releases a physical restraint that allows the isomerization of the receptor protein to an active configuration. This has also been proposed for the same mutation in the β_2-AR but analyzed structurally using fluorescent anisotropy (GETHER et al. 1997). Here, the mutation produced more pronounced structural changes with both agonists and antagonists than the WT receptor but was structurally unstable. It was proposed that these CAMs remove some stabilizing conformational-structural constraints, allowing CAM to more readily undergo transitions (i.e., have greater flexibility) between the inactive and active states and making the receptor more susceptible to denaturation. This was consistent with the CAM being a "high-energy" or "active-state" intermediate in the activation process. This notion of release of constraining factors may be an important conserved paradigm in the activation processes of all GPCRs in which the resulting release caused by agonist occupation is the direct cause of helical movements. This unification hypothesis of agonist-induced conformational changes would explain drug action at the molecular level and the rationale of conserved key pharmacophores in agonist drug classes. The implications of such a theory would make possible rationally designed pharmaceuticals with specific targets and lesser side effects.

2. Evidence for a Salt-Bridge as a Constraint: Movement of TM3 and TM7

As reviewed in Sect. C, previous work has identified an inter-helical salt bridge holding the rhodopsin receptor in a basal or inactive conformational structure until activated by light. These charged amino acid pairs of the rhodopsin-receptor salt bridge are conserved in other GPCRs that are activated by biogenic amines. The negatively charged Asp in TM3 involved in epinephrine binding is highly conserved among all aminergic receptors and is analogous in both position and charge to the Glu113 found in TM3 of rhodopsin. The positive counter-ion is also preserved between some aminergic and rhodopsin receptors. In non-conserved cases, a different basic amino acid may be substituted in TM7, comparable to what is observed for the dopamine receptor, or the Asp may be an Asn, forming a hydrogen-bond constraint. Similarities between the position and type of charged amino acid pairs found in rhodopsin versus the α_{1b}-AR suggest that a constraining salt bridge could also be formed. Analogous to rhodopsin, this potential salt bridge between Lys331 in TM7 and an Asp125 in TM3 could restrict the α_{1b}-AR to a basal conformation until bound by an AR agonist. Abolishing this ionic bond would allow the α_{1b}-AR to adopt an active conformation that would have properties of a constitutively active receptor, as in the rhodopsin receptor. To test this hypothesis, site-directed mutagenesis eliminated this charged amino acid pair in α_{1b}-AR by mutating the Lys331 to an Ala or a Glu and Asp125 to an Ala or a Lys (PORTER et al. 1996). In summary, both sets of mutations produced constitutively active receptors, strongly suggesting the existence of an Asp125–Lys331 salt bridge in the WT α_{1b}-AR that constrains the receptor in the inactive state. Docking of epinephrine initiates a competition between the protonated amine of the ligand and the positive charge of Lys331 for the negative Asp counter-ion. It is speculated that competition for the Asp125 disrupts the salt bridge, allowing a translational movement of TM3 towards the protonated amine of the ligand, a helical movement conserved in the rhodopsin paradigm.

To further test the hypothesis that a salt-bridge constraint was involved in the activation mechanism, triethylamine (TEA) was used to mimic the basic amine portion of the endogenous agonist to break this ionic constraint, leading to agonism (PORTER et al. 1998). TEA was able to generate concentration-dependent increases in IP release in transiently transfected COS-1 cells of the α_{1b} subtype and in stably transfected fibroblast cells with the α_{1a} subtype, though it was not a full agonist. TEA was able to synergistically potentiate the IP response of weak partial agonists, an ability which was fully inhibited by prazosin, a specific α_1 antagonist. However, this potentiation was not observed for full agonists; it shifted the dose-response to the right, indicating competitive antagonism and was consistent with the efficacy of TEA as a weak agonist. TEA can bind to α_1-ARs with a K_i of 28mM, which was consistent with its EC_{50}. The site of TEA binding was found to be at Asp125 in TM3, which is part of the constraining salt bridge. These results are consistent with a direct interaction of TEA in the binding pocket, which leads to a disruption of the salt bridge. It was postulated that full agonists break the salt bridge and, therefore, TEA can now only compete at the same binding site as the full agonist, thus displaying competitive antagonism. Weak partial agonists, because they cannot bind optimally, cannot break the salt bridge but instead weaken it, thus allowing TEA to bind to Asp125 and break the salt bridge, leading to potentiation. Thus, the use of a basic amine salt has supported the salt-bridge-breakage hypothesis and it is at least part of the activational mechanism in α_1-ARs.

This hypothesis of salt-bridge breakage as a step in the activation process also explains previous work on β_2-AR, in which substitution of the Asp125 with a longer Glu125 imparted to some antagonists the ability to display partial agonism (STRADER et al. 1989). Many antagonists in the ARs contain a protonated nitrogen analogous to adrenergic agonists; it is located, however, at a longer bond-length away from the aromatic moiety. This may impair the ability of antagonists to break the salt bridge because of sub-optimal positioning of the basic amine and accounts for why a Glu substitution partially rescues its potential agonism.

3. Evidence for Multiple Activation States or Mechanisms

Recently other constitutively active ARs have been characterized, and insights have been contributed to the activation process. A Cys-to-Phe mutation in TM3 of α_{1b}-AR a helix turn below the critical Asp125 involved in binding constitutively activates the receptor, resulting in G protein coupling in the absence of agonist and selective constitutive activation of a single-effector pathway (phospholipase C, not phospholipase A_2; PEREZ et al. 1996). It was shown previously that these two pathways in COS-1 cells are coupled to two different G proteins. It was found that phenethylamine agonists, such as epinephrine, were able to recognize this "selective active state" by binding and potency changes. However, a structurally distinct imidazoline agonist, such as oxymetazoline, could not. Since Cys was strictly conserved in β_2-AR, this mutation was created in β_2-AR (ZUSCIK et al. 1998) and gave analogous phenotypes. The β_2-AR C116F mutant can selectively constitutively activate the Na/H exchanger NHE-1 through the putative G protein $G_{\alpha 13}$ without constitutively activating the $G_{\alpha s}$/adenylate-cyclase pathway. Both studies indicate that a single receptor subtype forms multiple conformations for G protein interactions (i.e. different activation states) that are specific for a particular G protein/effector pathway and that multiple binding sites that promote or induce these specific interactions exist for different classes of agonists.

This notion that a receptor conformation is important in recognizing a G protein activated state is also supported by the observation in α_{2a}-AR in which a point mutation in TM2 uncoupled the receptor from activating potassium currents but not calcium currents (SUPRENANT et al. 1992). Since this mutation and the TM3 mutant are located in the TM domains and not in the intracellular loops, which are thought to interact directly with the G proteins, the receptor conformation must have changed to allow this differential coupling. In support, it has also been shown in β_2-AR that agonists and antagonists induce distinct conformational states of the receptor (GETHER et al. 1995). Ligand-dependent structural changes, as measured by fluorescent anisotropy, showed that agonists and antagonists have opposite effects on baseline fluorescence.

4. Evidence for Additional Constraining Factors: Movement of TM5 and TM6

Since the catechol hydroxyls bind to serines in TM5 and the benzene ring of epinephrine interacts with a phenylalanine in TM6, it might also be expected that residues in these helices are also involved in the activation process. Indeed, the salt-bridge mutants are not fully active, suggesting such a possibility. How can different mutations/residues located in diverse areas of the receptor all lead to the same, R* or active state of the receptor? It is possible that many of these mutations alter helical packing of important TM domains that are involved in the activation process. These mutations can be direct, such as the salt-bridge mutants, or indirect, such as what was found in β_2-AR, in which extracellular-loop mutants caused constitutive activity (ZHAO et al. 1998). Here, it was postulated that TM domains 6 and 7, attached to the extracellular loop mutants, were indirectly altered in their packing to allow a greater ability of the intracellular loops to bind G protein. In a similar conclusion (HWA et al. 1996), CAMs in TM5 and TM6 of the α_{1a} or α_{1b} subtypes were created by substituting a particular residue in either helix with the corresponding amino acid in the other subtype (TM5: Ala204Val, α_{1b} to α_{1a}; TM6: Met292Leu, α_{1a} to α_{1b}). The constitutive activity could be silenced or reversed when the complementary amino acid in the adjacent helix was substituted with the corresponding same-subtype residue. A simple interpretation of these findings is that the entire structure of the receptor has evolved to constrain the receptor in the inactive state such that even helix packing has important ramifications in the activation process. Whether these helices have specific constraining factors that are broken indirectly by the helix-packing differences or the implicit deduction that the helices move in bulk due to steric constraint from agonist binding is still unknown. The aromatic ring conserved in all ARs is positioned near TM5/6. Movement of these helices (TM5/6) to accommodate steric strain induced by agonist occupation would be analogous to that in rhodopsin, where the ionone (aromatic) ring of retinal flips toward TM6 during *cis*-to-*trans* isomerization. It is this resulting accommodation of steric bulk that is speculated to move TM6 in the activation process.

Recent evidence using the substituted Cys accessibility method on a constitutively active β_2-AR also suggests movement of TM6 in the activation process (JAVITCH et al. 1997). Using the CAM β_2-AR previously characterized (SAMAMA et al. 1993), JAVITCH has showed that Cys285 in TM6 is the sole Cys responsible for the increased susceptibility of the CAM for the polar sulfhydryl-specific reagent methane thiosulfonate ethylammonium (MTSEA), compared with the WT receptor. The results are consistent with a rotation and/or tilting of the sixth membrane-spanning segment associated with activation of the receptor and similar to the rhodopsin paradigm (ALTENBACH et al. 1996). This rearrangement of TM6 would bring Cys285 to the margin of the binding crevice, where it becomes accessible to MTSEA.

Further evidence of additional constraining factors comes from studies combining individual CAMs. Three CAMs [C128F in TM3 (PEREZ et al. 1996), A204V in TM5 (HWA et al. 1996) and A293E in the 5–6 intracellular loop (KJELSBERG et al. 1992), which moved TM6 above] in α_{1b}-AR displayed similar manifestations of constitutive activity. It was hypothesized that the individual mutations, because of their critical locations, alter the conformation or packing of the TM helices so that mimicry that partially conforms to the activated state, R*, occurs. To explore whether these potential conformations are independent, these three mutations were combined in all possible permutations (HWA et al. 1997). Each mutation contributed independently and synergistically to both receptor–agonist binding and constitutive activation. There was also a direct correlation between epinephrine's binding affinity and the degree of constitutive activity. The binding curves became more complex (reflected by multiple binding sites that were independent of G protein modulation, as assessed by the addition of guanylyl imidodiphosphate) as the degree of constitutive activity increased. This was consistent with the idea that these mutants allow a greater degree of flexibility (or a lowering of activational energy) between the R and R* states. Indeed, from a structural viewpoint, these mutants can be envisioned as being impaired in their key constraining functions, thus spontaneously "relaxing" into their active conformation. Because the mutations affect different TM domains, these results are consistent with a mechanism in which helical movement acts in a concerted fashion in agonist-induced activation, a synergism predicted if multiple helix movement (between TM3, 5, and 6) is involved in receptor activation. Multiple-helix movement implicates additional constraining factors that are released upon agonist binding, allowing this greater flexibility between the R and R* states.

E. Angiotensin-Receptor Activation: Peptides as Ligands

I. Peptide-Hormone GPCRs: How the Peptide Binds

The endogenous ligands for many GPCRs are peptide hormones with sizes varying from tripeptides to large glycoproteins. The primary-sequence identity between the peptide hormone GPCRs and opsins or biogenic amine receptors is less than 20%. Hydropathy analysis, the alignment of sequence motifs that define GPCR structure, suggest that the putative structure of peptide-hormone GPCRs is likely to be similar to that of rhodopsin and adrenergic receptors. However, the functional pharmacophore mapped on the surfaces of many peptide hormones suggests that the binding site on GPCRs might have a substantially larger surface area than that determined from the binding of retinal and biogenic amines. Consistent with this expectation, in several peptide-hormone GPCRs, the agonist contact residues identified are located in the putative extracellular domain. The first and second extracellular segments are critical for binding of different neurokinins to their receptors (HUANG et al. 1995). The high-affinity binding of thyrotropin and lutropin requires the N-terminal region of the respective receptors (LAAKONEN et al. 1996). In GPCRs for angiotensin II (AngII), formyl peptide, and interleukin-8, evidence for involvement of the extracellular loop has been obtained (PEREZ et al. 1994; SERVANT et al. 1997). Thus, in contrast to the retinal and biogenic amine receptors, the extracellular segments play a role in agonist interaction in peptide-hormone GPCRs. In light of this observation, the well-ordered structure of the extracellular segments in the rhodopsin electron-density map may be very pertinent for this group of receptors.

II. The Ang-II Receptor

The existence of at least two specific subtypes of the AngII receptor has been confirmed by molecular cloning studies (MURPHY et al. 1991; SASAKI et al. 1991; KAMBAYASHI et al. 1993; MUKOYAMA et al. 1993). The type-1 (AT_1) receptor is highly selective for the biphenylimidazole non-peptide antagonist DUP753 or losartan while displaying low affinity for pentapeptide analogs, such as CGP42112A. The other subtype, AT_2, binds a Park-Davis antagonist, PD123377, preferentially and essentially has affinities that are the opposite of those of the AT_1 receptor. AngII-dependent activation of the AT_1 receptor causes intracellular IP production through the activation of a G protein that is pertussis-toxin insensitive (MURPHY et al. 1991). Most of the physiology for AngII has been assigned to the AT_1 receptor (BUMPUS and KHOSLA 1977; DUDLEY et al. 1990), with the signal-transduction pathways and G protein coupling of AT_2 still unknown.

III. How AngII Peptides and Non-Peptides Bind

Despite significant advancement in defining the ligand pocket, the molecular mechanism of activation of the peptide hormone GPCRs is unclear at this time. An important insight for the striking similarities between the activation processes of rhodopsin, biogenic amine receptors, and the peptide-hormone GPCRs through the involvement of the TM domains comes from the mapping of the binding sites for the small-molecule, non-peptide agonists and antagonists. Recently, the complete binding site of the biphenyl imidazole antagonist and agonist compounds that are selective for the AngII receptor have been mapped (JI et al. 1995; NODA et al. 1995; SCHAMBYE et al. 1995; MONNOT et al. 1996; PERLMAN et al. 1997). The site is formed by residues located in TM3–7, buried in the middle of the plane of the lipid bilayer, similar to the biogenic amine receptors. Pharmacophore overlay suggests that the biphenyl analogs of the non-peptide ligands mimic the C-terminal tetrapeptide region of AngII, which carries the functional determinants. The binding epitopes for the neurokinin-receptor antagonists are also localized on TM2, 5, and 6 (FONG et al. 1993; GETHER et al. 1993; HUANG et al. 1995). The particular residues involved are located at positions equivalent to residues in adrenergic receptors that bind the catechol ring. Similarities in non-peptide binding suggest the existence of a site within the TM domain of the peptide-hormone GPCRs; this site must be accessed by the native peptide hormones to initiate the receptor-activation process. Whether this is achieved by direct contact or indirectly through extracellular loops is unclear.

The observation that TM3 and TM6 in the AT_1 receptor directly interact with the agonism-specifying functional groups of AngII suggests that the mechanism of activation of this peptide hormone GPCR conforms to a more general molecular mechanism of GPCR activation. The octapeptide hormone AngII (DRVYIHPF) plays an important role in regulating hydromineral balance and arterial blood pressure in species as diverse as fish and humans. The C-terminal tetrapeptide region of AngII has been proven important for receptor activation (BUMPUS and KHOSLA 1977). Substitution of the Phe8 side chain with an aliphatic group produces an agonist-to-antagonist transition without a change in binding affinity. Hydrophobic aliphatic substitutions at position 4 weaken agonist activity and reduce binding affinity for the receptor (BUMPUS and KHOSLA 1977). The remaining hormone residues are not considered to be crucial for agonist activity. Several predicted electrostatic/hydrogen-bonding interactions between AngII and its receptor have recently been assigned using group-specific modifications of AngII in combination with AT_1-receptor mutagenesis. The contact between the N-terminus of AngII and the extracellular domain involves two salt-bridge/hydrogen-bond interactions between His183 of the AT_1 receptor and Asp1 of AngII and between Asp 281 of the AT_1 receptor and Arg2 of AngII. His183 is located in the extracellular loop connecting TM4 and TM5, and Asp281 is in the extracellular loop connecting TM6 and TM7. These two points of contact are not utilized by the non-peptide agonists and antagonists in binding to AT_1 receptor, which is consistent with the lower affinity of the non-peptide analogs. A direct interaction between the backbone α-COO-group of Phe8 in AngII and Lys199 of the receptor was also demonstrated (NODA et al. 1995). Lys199 is located in TM5 of the AT_1 receptor. An interaction between the aromatic side group of Phe8 of AngII and His256 in TM6 of the AT_1 receptor is also necessary for AngII-dependent receptor activation (NODA et al. 1995). NODA et al. (1995) confirmed that the α-COO-group of AngII docked to Lys199 and demonstrated that Lys199, in combination with His256 in TM6, constitutes an important site where all classes of AT_1-selective ligands bind to exert agonist or antagonist effects.

IV. Insights into AngII Activation

1. Role of His256 in TM6

Replacement of the Phe8 side chain in AngII with aliphatic side chains (such as Ile, Ala, and Thr) and Gly produces poor agonists without substantial change of affinity. Likewise, the replacement of His256 of the AT_1 receptor also produced a functionally defective receptor with no change in affinity for AngII. The results obtained with the His256 mutants are consistent with van der Waals contact between His256 and the angiotensin position-8 side chain, based on small differences in the affinity of the analogs, implying that a direct contact of His256 with the Phe8 side chain (i.e., amino aromatic or cation π) is responsible for "transmitting" the agonist occupancy of the ligand pocket as a signal for receptor activation (NODA et al. 1995). The most significant conclusion from the present results, therefore, is that His256 is a point of contact between agonists and the AT_1 receptor, where the process of receptor activation is initiated.

2. Release of Constraining Factors: Role of TM3

Molecular modeling of AngII docked to the AT_1 receptor predicts interaction between Tyr4 of AngII and TM3, specifically with residues Ser107, Asn111, and Leu112 (NODA et al. 1996). Mutagenesis studies indicate a role for Asn111 in AngII binding but have excluded a significant role for Ser107, Leu112, and

Ser115, all located in TM3 (NODA et al. 1996). The binding affinity of AngII is specifically affected by the size of the residue substituted at position 111 of the AT_1 receptor, implying an interaction between them. The most surprising observation, however, was the constitutively active phenotype of Asn111Ala and Asn111Gly mutants of the AT_1 receptor. AT_1-receptor agonists and partial agonists show an increase in binding affinity for these mutants compared with WT receptor, whereas AT_1-receptor antagonists exhibited a markedly lower binding affinity. These mutations also induce a conformational change in the AngII-binding pocket, which leads to an increase in binding affinity for agonists in the absence of G protein coupling, suggesting that the mutation induced an "active-state" conformation of the receptor. This, in turn, promotes ligand-independent activation of intracellular signal transduction (SAMAMA et al. 1993). Basal IP production in the Gly111 and Ala111 receptor mutants was substantially elevated (NODA et al. 1996). Therefore, it appears that the Asn111Gly mutation induces a conformational change in the AngII-binding pocket, which causes misalignment of the residues required for losartan binding but favors the binding of agonists and partial-agonist analogs of AngII.

Because Asn111 is a receptor residue that is also involved in docking Tyr4, the Tyr4/Asn111 interaction is likely to be the switch that controls a conformation-dependent reconfiguration of the ground-state binding pocket, which concomitantly stabilizes AngII binding and induces the active state of the receptor (R*). In addition, experimental evidence suggests that His256 is not required when the Asn111 side chain is mutated and that the Phe8/His256 interaction enables the Tyr4/Asn111 switch to engage the WT AT_1 receptors in the activation process (NODA et al. 1996). This is seen experimentally in the Asn111Gly mutant, which is able to become fully activated with AngII analogs that substitute the aromaticity of the Tyr4 and Phe8 residues, but the WT receptor is unable to respond. The simplest way to explain this data is that the transition of a WT receptor into a fully activated receptor requires a determinative step (a true transitional intermediate or pre-active state) that occurs after AngII binds but does not occur when AngII analogs lacking Tyr4 and Phe8 groups bind to the WT receptor. The initial rate-limiting step in activation is achieved through thermodynamically linked interactions between AngII-[Tyr4] and Asn 111 of the AT_1 receptor, and between AngII-[Phe8] and His256 of the AT_1 receptor, resulting in helical changes in TM3 and 6. When the ligand-binding pocket is not occupied, the receptor is stabilized by a complex set of intramolecular interactions that chiefly constrain Asn111. Release of Asn111 from this interaction produces a conformational change in the ligand-binding pocket such that this pocket now favors agonists and partial agonists of the AT_1 receptor but not antagonists. Whether the paradigm of peptide-hormone GPCR activation through contact of agonism-specifying hormone groups with critical receptor residues in the TM domain is unique to AT_1 receptors or is a more general phenomenon among the peptide-hormone receptors remains to be established. However, the involvement of TM3 and 6 in the activation process is conserved among the rhodopsin receptor, AR, and angiotensin receptors.

A separate report has also postulated a role for Asn111 as a constraining factor, but differs from the previous report in that a hydrogen-bond constraint with Asn295 in TM7 is proposed (BALMFORTH et al. 1997). This would be analogous to the salt-bridge constraining factor between TM3 and 7 in rhodopsin and the α_1-AR. Mutation of Asn295 produced a constitutive active phenotype similar to that produced by substitution of Asn111. They also proposed that substitution of these two residues causes the loss of an interaction between TM3 and 7, which allows the receptor to "relax" into its active conformation. However, a molecular model of the rat AT_1 receptor (JOSEPH et al. 1995) has suggested not Asn295 but Tyr292 as the molecular constraining partner for Asn111. However, this particular residue has been mutated (MARIE et al. 1994) to a Phe and did not yield the predicted constitutive phenotype. This would imply that Asn111–Asn295 is a more likely interaction.

F. The Ties that Bind: Concluding Remarks

I. Conservation of Critical Binding Contacts and Resulting Helical Movements

Since all GPCRs are thought to have evolved from a single ancestral gene, there is key conservation of the structural motifs and the coupling regions in the intracellular loops of the G protein, it is likely that the binding determinants, the agonist-induced conformational changes, and the resulting helical movements that occur in rhodopsin will be conserved in other GPCRs. In support of this notion, evidence from both the AR and angiotensin receptors indicate a critical interaction in the binding of the endogenous agonists in TM3, 5, and 6. Indeed, the involvement of these helices in the binding of the endogenous agonist in many other GPCRs not reviewed in this chapter supports this consensus. This and the evidence of the resulting movements in TM3 and TM5/6, analogous to the rhodopsin paradigm, suggest these helices in the binding and subsequent activation mechanism in all GPCRs. As in rhodopsin, the conformational result in other GPCRs, regardless of the class of ligand, would be a widening of the cytoplasmic surface via the intracellular loops, allowing for higher affinity interactions with the G protein(s).

II. Conservation of Switches that Control Attainment of the Active State(s)

An interesting conservation in TM3 appears to be a possible "switch residue" or region that can control key aspects of the activation state(s). As shown in Fig. 2, alignment of the third TM domain has Gly121 in rhodopsin, which is conserved in all visual pigments; in bR, it is replaced by Leu93. In analogous positions in the ARs are Cys116 in the β_2-AR and Cys128 in the α_{1b}-AR, while

Transmembrane helix 3

```
         111
AT₁    N  L  Y  A  S  V  F  L  L  T  C  L  S  I  D  R  Y

         128
α1b    C  C  T  A  S  I  L  S  L  C  A  I  S  I  D  R  Y

         116
β2     C  V  T  A  S  I  E  T  L  C  V  I  A  V  D  R  Y

         121
Rho    G  G  E  I  A  L  W  S  L  V  V  L  A  I  E  R  Y

         93
Br     L  L  L  L  D  L  A  L  L  V  D  A  D  Q  G  T  I
```

Fig. 2. Alignment of residues in the third transmembrane helix of the angiotensin type-1 receptor (AT₁), the α_{1b} adrenoceptor (α_{1b}), the β_2 adrenoceptor (β_2), rhodopsin (Rho) and bacteriorhodopsin (Br). Using the highly conserved DRY domain located at the cytoplasmic surface of the helix as a reference (*bold*), Asn111 in AT₁, Cys128 in α_{1b}, Cys116 in β_2 and Gly121 in rhodopsin are in or near the same relative position in the helix. Bacteriorhodopsin is aligned according to the rhodopsin sequence (i.e., at the counter-ion for the Schiff base, not shown), which places Leu93 at the same position as Gly121 in rhodopsin. Bacteriorhodopsin does not have the DRY domain, because it is not G-protein coupled. All five amino acids cause aspects of constitutive activity when mutated and are implicated as switch positions in the α-helix that control receptor isomerization to selective and/or distinct activation states

the angiotension receptor contains Asn111. Substitution at all of these positions gives some aspects of constitutive activity. In rhodopsin, substitution of Gly121 causes 11-*cis*-retinal to become a pharmacological partial agonist (HAN et al. 1997), allowing the mutant rhodopsin to activate transducin in the dark. Replacement of Gly121 with residues of increasing size results in increased transducin activation in the presence of the agonist all-*trans*-retinal. Replacement of Leu93 in bR results in a 250-fold increase in the time needed to complete the photocycle with the continued presence of the 13-*cis*-retinal intermediate (DELANEY et al. 1995). Since bR's photocycle is opposite that of rhodopsin (proton transport is initiated by the light-induced isomerization from the all-*trans* to the 13-*cis* configuration), the 13-*cis*-retinal buildup represents an increase in the active-state intermediate. In the both the AR and angiotensin receptors, the constitutive activity from these substitutions was obvious and explained in their sections. However, since it is possible for these receptors to couple to more than one G protein, the data indicate that the constitutive activity represents a particular activation state. In the ARs, this is represented as preferential coupling to one particular G protein, suggesting that each G protein/receptor complex represents distinct activational states. The angiotensin receptor was analogous, with the mutant giving a distinct intermediate state, R', that was different from R*. All of these residues are predicted to face the water-accessible binding pocket and, in rhodopsin, the phenotype can be "rescued" by an appropriate substitution of Phe261 in TM6. Given the tilt of TM3 towards TM5 at the intracellular end of rhodopsin, it is possible for Gly121 to face the pocket towards TM6. This has not been confirmed in the ARs or angiotensin receptors, and the tilting of TM3 may be different in these molecules, which may account for the apparent one-base difference in the alignments.

The mechanistic reasons behind this switch position in TM3 may be the result in changes in Van der Waals contact, as was proposed for Asn111 (NODA et al. 1996) and Gly121 (HAN et al. 1996). The inactive-state partner for the switch residue may then be Phe261 in TM6 (as in rhodopsin) or the analogous Phe residue conserved in both the AR and angiotensin receptors. It is interesting to speculate that the substitution of a bulkier side chain at the Gly121 position may somehow mimic the structural interactions and movements that occur between TM3 and 6. Similar larger substitutions in the α_{1b}-AR at Cys128 (such as Tyr or Trp) also lead to greater levels of constitutive activity (PORTER et al. 1996). This Van der Waals interaction would then be a predicted "constraining factor" that is broken upon agonist occupation. The different degrees of flexibility or distinct conformational states that are imparted to the receptor may be directly dependent upon the changes in van der Waals contact resulting from the class of agonist occupying in the pocket. Alternately, this switch position may be invoked in the constraining process between TM3 and 7 (as in the ionic salt-bridge in rhodopsin, bR, and the α_1-AR) or in the postulated hydrogen-bond constraint (as in the AT_1). The substitution of the natural switch residue could interfere in this constraint. It is possible that the release of one constraint in the activation process allows competent coupling to one G protein, while another constraint imparts specificity to another G protein. Release of all constraints, as done with the endogenous agonist, leads to multiple couplings. If this is true, the goal is to design agonists that only remove a specific constraint to achieve true signal fidelity, which would lead to rationally designed drugs that impart fewer side effects.

Since it is unlikely that rhodopsin couples to more than one G protein (transducin), because of the high fidelity (on/off) needed in the visual system, the residue in TM3 can only alter the kinetics of the inactive–active-state isomerization in rhodopsin. The same is true for bR, in which the sole "signal-transduction" pathway is proton pumping. However, we speculate that, in other GPCRs in which multiple G proteins can couple (and, thus, multiple signaling pathways are activated), this residue acts as a "switch" in controlling agonist trafficking or the ability of the agonist to direct different activational states that are G protein dependent.

References

Acharya S, Karnik SS (1996) Modulation of GDP release from transducin by the conserved Glu 134–Arg 135 sequence in rhodopsin. J Biol Chem 271:25406–25411

Altenbach C, Farrens DL, Khorana HG, Hubbell WL (1996) Structural features and light dependent changes in the cytoplasmic interhelical E–F loop region of rhodopsin: a site-directed spin labeling study. Biochemistry 35:12470–12478
Applebury ML, Hargrave PA (1986) Molecular biology of the visual pigments. Vision Res 26:1881–1895
Baldwin JM (1993) The probable arrangement of the helices in G protein-coupled receptors. The EMBO Journal 12:1693–1703
Balmforth AJ, Lee AJ, Warburton P, Donnelly D, Ball SG (1997) The conformational change responsible for AT_1 receptor activation is dependent upon two juxtaposed asparagine residues on transmembrane helices III and VII. J Biol Chem 272:4245–4251
Bhattacharya S, Ridge KD, Knox BE, Khorana HG (1992) Light-stable rhodopsin: a rhodopsin analog reconstituted with a non-isomerizable 11-*cis*-retinal derivative. J Biol Chem 267:6763–6769
Blin N, Yun J, Wess J (1995) Mapping of single amino acid residues required for selective activation of $G_{q/11}$ by the m3 muscarinic acetylcholine receptor. J Biol Chem 270:17741–17748
Bond RA, Leff P, Johnson TD, Milano CA, Rockman HA, McMinn TR, Apparsundaram S, HyekMF, Kenakin TP, Allen LF, Lefkowitz RJ (1995) Physiological effects of inverse agonists in transgenic mice with myocardial overexpression of the β_2-adrenoceptor. Nature (Lond) 374:272–276
Bumpus FM, Khosla MC (1977) In: Hypertension: physiology and treatment. McGraw-Hill Publishing, New York, pp 183–201
Bylund DB, Eikenberg DC, Hieble JP, Langer SZ, Lefkowitz RJ, Minneman KP, Molinoff PB, Ruffolo RR Jr, Trendelenburg U (1994) International Union of Pharmacology nomenclature of adrenoceptors. Pharmacol Rev 46:121–136
Chidiac P, Hebert TE, Valiquette M, Dennis M, Bouvier M (1994) Inverse agonist activity of β-adrenergic antagonists. Mol Pharmacol 45:490–499
Chidiac P, Nouet S, Bouvier M (1996) Agonist-Induced modulation of inverse agonist efficacy at the β_2-adrenergic receptor. Mol Pharmacol 50:662–669
Clark AJ (1937) General pharmacology. In: Heffner's Handbuch der exp. Pharmacol. Suppl. vol. 4. Springer, Berlin
Cohen GB, Oprian DD, Robinson PR (1992) Mechanism of activation and inactivation of opsin: pole of Glu 113 and Lys 296. Biochemistry 31:12592–12601
Delaney JK, Schweiger U, Subramaniam S (1995) Molecular mechanism of protein–retinal coupling in bacteriorhodopsin. Proc Natl Acad Sci USA 92:11120–11124
DeLean A, Stadel JM, Lefkowitz RJ (1980) A ternary complex model explains the agonist-specific binding properties of the adenylate-cyclase-coupled β-adrenergic receptor. J Biol Chem 255:7108–7117
Dixon RA, Kobilka BK, Strader DJ, Benovic JL, Dohlman HG, Frielle T, Bolanowski MA, Bennett CD, Rands E, Diehl RE, Mumford RA, Slater EE, Sigal IS, Caron MG, Lefkowitz RJ, Strader CD (1986) Cloning of the gene and cDNA for mammalian β-adrenergic receptor and homology with rhodopsin. Nature 321:75–79
Dixon RA, Sigal IS, Strader CD (1988) Structure-function analysis of the β-adrenergic receptor. Cold Spring Harb Symp Quant Biol 53 Pt 1:487–497
Dudley DT, Panek RL, Major TC, Lu GH, Bruns RF, Klinkefus BA, Hodges JC, Weishaar RE (1990) Subclasses of angiotensin-II-binding sites and their functional significance. Mol Pharmacol 38:370–377
Farahbakhsh ZT, Ridge KD, Khorana HG, Hubbell WL (1995) Mapping light-dependent structural changes in the cytoplasmic loop connecting helices C and D in rhodopsin: a site-directed spin labeling study. Biochemistry 34:8812–8819
Fong T, Cascieri MA, Yu H, Bansal A, Swain C, Strader CD (1993) Amino-aromatic interaction between histidine 197 of the neurokinin-1 receptor and CP 96345. Nature 362:350–353

Fraser CM, Wang CD, Robinson RA, Gocayne GD, Venter JC (1989) Site-directed mutagenesis of m1 muscarinic acetylcholine receptors: conserved aspartic acids play important roles in receptor function. Mol Pharmacol 36:840–847

Gether U, Johansen TE, Snider RM, Lowe III JA, Nakanishi S, Schwartz TW (1993) Different binding epitopes on the NK1 receptor for substance P and a non-peptide antagonist. Nature 362:345–348

Gether U, Lin S, Kobilka BK (1995) Fluorescent labeling of purified β_2-adrenergic receptor: evidence for ligand-specific conformational changes. J Biol Chem 270:28268–28275

Gether U, Ballesteros JA, Seifert R, Sanders-Bush E, Weinstein H, Kobilka BK (1997) Structural instability of a constitutively active G protein-coupled receptor. J Biol Chem 272:2587–2590

Han M, Lin SW, Smith SO, Sakmar TP (1996a) The effects of amino acid replacements of glycine 121 on transmembrane helix 3 of rhodopsin. J Biol Chem 271:32330–32336

Han M, Lin SW, Minkova M, Smith SO, Sakmar TP (1996b) Functional interaction of transmembrane helices 3 and 6 in rhodopsin. J Biol Chem 271:32337–32342

Han M, Lou J, Nakanishi K, Sakmar TP, Smith SO (1997) Partial agonist activity of 11-cis-retinal in rhodopsin mutants. J Biol Chem 272:23081–23085

Haupts U, Tittor J, Bamberg E, Oesterhelt D (1997) General concept for ion translocation by halobacterial retinal proteins: The isomerization/switch/transfer (IST) model. Biochemistry 36:2–7

Ho BY, Karschin A, Branchek T, Davidson N, Lester HA (1992) The role of conserved aspartate and serine residues in ligand binding and function of the 5-HT_{1a} receptor: a site-directed mutation study. FEBS 312:259–26221

Huang R-RC, Vicario PP, Strader CD, Fong TM (1995) Identification of residues involved in ligand binding to the neurokinin-2 receptor. Biochemistry 34:10048–10055

Hubbard R and Kropf A (1958) The action of light on rhodopsin. Proc Natl Acad Sci USA 44:130–139

Hwa J and Perez DM (1996) The unique nature of the serine residues involved in α_1-adrenergic receptor binding and activation. J Biol Chem 271:6322–6327

Hwa J, Graham RM, Perez DM (1995) Identification of critical determinants of α_1-adrenergic-receptor subtype-selective agonist binding. J Biol Chem 270:23189–23195

Hwa J, Graham RM, Perez DM (1996) Chimeras of α_1-adrenergic receptor subtypes identify critical residues that modulate active-state isomerization J Biol Chem 271:7956–7964

Hwa J, Gaivin R, Porter J, Perez DM (1997) Synergism of constitutive activity in α_1-adrenergic receptor activation. Biochemistry 36:633–639

Javitch JA, Fu D, Chen J (1995) Residues in the fifth membrane-spanning segment of the dopamine D2 receptor exposed in the binding-site crevice. Biochemistry 34:16433–16439

Javitch JA, Fu D, Liapakis G, Chen J (1997) Constitutive activation of the β_2-adrenergic receptor alters the orientation of its sixth membrane-spanning segment. J Biol Chem 272:18546–18549

Ji H, Zheng W, Zhang Y, Catt KJ, Sandberg K (1995) Genetic transfer of a non-peptide antagonist binding site to a previously unresponsive angiotensin receptor. Proc Natl Acad Sci USA 92:9240–9244

Joseph M-P, Maigret BM, Bonnafous JC, Marie J, Scheraga HA (1995) A computer modeling postulated mechanism for angiotensin-II-receptor activation. J Protein Chem 14:381–398

Kambayashi Y, Bardhan S, Takahashi K, Tsuzuki S, Inui H, Hamakubo T, Inagami T (1993) Molecular cloning of a novel angiotensin-II-receptor isoform involved in phosphotyrosine phosphatase inhibition. J Biol Chem 268:24543–24546

Karnik SS, Khorana HG (1990) Assembly of functional rhodopsin requires a disulfide bond between cysteine residues 110 and 117. J Biol Chem 265:17520–17524

Khorana HG (1988) Bacteriorhodopsin, a membrane protein that uses light to translocate protons. J Biol Chem 263:7439–7442

Kjelsberg MA, Cotecchia S, Ostrowski J, Caron MG, Lefkowitz RJ (1992) Constitutive activation of the α_{1b}-adrenergic receptor by all amino acid substitutions at a single site. J Biol Chem 267:1430–1433

Kunkel MT, Peralta EG (1993) Charged amino acids required for signal transduction by the m3 muscarinic acetylcholine receptor. EMBO J 12:3809–3815

Laakkonen LJ, Guarnieri F, Perlman JH, Gershengorn MC, Osman R (1996) A refined model of he thyrotropin-releasing hormone (TRH) receptor binding pocket. Novel mixed mode Monte Carlo/stochastic dynamics simulations of the complex between TRH and TRH receptor. Biochemistry 35:7651–7663

Marie J, Maigret B, Joseph MP, Larguier R, Nouet S, Lombard C, Bonnafous JC (1994) Tyr292 in the seventh transmembrane domain of the AT_{1a} angiotensin-II receptor is essential for its coupling to phospholipase C. J Biol Chem 269:20815–20818

Monnot C, Bihoreau C, Conchon S, Curnow KM, Corvol P, Clauser E (1996) Polar residues in the transmembrane domains of the type-1 angiotensin-II receptor are required for binding and coupling. J Biol Chem 271:1507–1513

Mukoyama M, Nakajima M, Horiuchi M, Sasamura H, Pratt RE, Dzau VJ (1993) Expression cloning of type-2 angiotensin-II receptor reveals a unique class of 7-transmembrane receptors. J Biol Chem 268:24539–24542

Murphy TJ, Alexander RW, Griendling KK, Runge MS, Berstein KE (1991) Isolation of a cDNA encoding the vascular type-1 angiotensin-II receptor. Nature 351:233–236

Nakayama TA, Khorana HG (1990) Orientation of retinal in bovine rhodopsin determined by cross-linking using a photoactivatable analog of 11-*cis*-retinal. J Biol Chem 265:15762–15769

Nakayama TA, Khorana HG (1991) Mapping the amino acids in membrane-embedded helices that interact with the retinal chromophore in bovine rhodopsin J Biol Chem 266:4269–4275

Noda K, Saad Y, Karnik SS (1995a) Interaction of Phe[8] of angiotensin II with Lys199 and His256 of AT_1 receptor in agonist activation. J Biol Chem 270:1–5

Noda K, Saad Y, Kinoshita A, Boyle TP, Graham RM, Husain A, Karnik SS (1995b) Tetrazole and carboxylate groups of the angiotensin receptor antagonists bind to the same subsite by different mechanisms. J Biol Chem 270:2284–2289

Noda K, Feng Y-H, Liu X-P, Saad Y, Husain A, Karnik SS (1996) The active state of the AT_1 angiotensin receptor is generated by angiotensin-II induction. Biochemistry 35:16435–16442

Pebay-Peyroula E, Rummel G, Rosenbusch JP, Landau EM (1997) X-ray structure of bacteriorhodopsin at 2.5 Å from microcrystals grown in lipidic cubic phases. Science 277:1676–1681

Perez HD, Vilander L, Andrews WH, Holmes R (1994) Human formyl peptide receptor ligand binding domain(s) J Biol Chem 269:22485–22487

Perez DM, Hwa J, Gaivin R, Mathur M, Brown F, Graham RM (1996) Constitutive activation of a single effector pathway: evidence for multiple activation states of a G protein-coupled receptor. Mol Pharmacol 49:112–122

Perlman S, Costa-Neto CM, Miyakawa AA, Schambye HT, Hjorth SA, Paiva ACM, Rivero RA, Greenlee WJ, Schwartz TW (1997) Dual agonistic and antagonistic property of non-peptide angiotensin AT_1 ligands: susceptibility to receptor mutations. Mol Pharmacol 51:301–311

Porter J, Hwa J, Perez DM (1996) Activation of the α_{1b}-adrenergic receptor is initiated by disruption of an inter-helical salt-bridge constraint. J Biol Chem 271:28318–28323

Porter J, Edlemann S, Piascik MT, Perez DM (1998) The agonism and synergistic potentiation of partial agonists by triethylamine in a1-adrenergic receptor activation: evidence for a salt bridge as the initiating process. Mol Pharmacol 53:766–771

Ridge KD, Bhattacharya S, Nakayama TA, Khorana HG (1992) Light-stable rhodopsin: An opsin mutant (Trp265Phe) and a retinal analog with a nonisomerizable 11-*cis* configuration form a photostable chromophore. J Biol Chem 267:6770–6775

Ross EM, Higashijima T (1994) Regulation of G-protein activation by mastoparans and other cationic peptides. Methods Enzymol 237:26–37

Sakmar TP, Franke RR, Khorana HG (1989) Glutamic acid-113 serves as the retinylidene Schiff base counterion in bovine rhodopsin. Proc Natl Acad Sci USA 86:8309–8313

Samma P, Cotecchia S, Costa T, Lefkowitz RJ (1993) A mutation-induced activated state of the β_2-adrenergic receptor. J Biol Chem 268:4625–4636

Sasaki K, Yamano Y, Bardhan S, Iwai N, Murray JJ, Hasegawa M, Matsuda Y, Inagami T (1991) Cloning and expression of a complementary DNA encoding a bovine adrenal angiotensin-II type-1 receptor. Nature (Lond) 351:230–233

Schambye T, Hjorth SA, Weinstock J, Schwartz TW (1995) Interaction between the non-peptide angiotensin antagonist SKF-108,566 and histidine 256 of the angiotensin-type-1 receptor. Mol Pharmacol 47:425–431

Scheer A, Fanelli F, Costa T, De Benedetti PG, Cotecchia S (1997) The activation process of the α_{1b}-adrenergic receptor: potential role of protonation and hydrophobicity of a highly conserved aspartate. Proc Natl Acad Sci USA 94:808–813

Schertler GFX, Villa C, Henderson R (1993) Projection structure of rhodopsin. Nature (London) 362:770–772

Schertler GFX, Hargrave PA (1995) Projection structure of frog rhodopsin in two crystal forms. Pro Natl Acad Sci USA 92:11578–11582

Servant G, Laporte SA, Leduc R, Escher E, Guillemette G (1997) Identification of angiotensin-II-binding domains in the rat AT_2 receptor with photolabile angiotensin analogs J Biol Chem 272:8653–8659

Sheikh SP, Zvyaga TA, Lichtarge O, Sakmar TP, Bourne HR (1996) Rhodopsin activation blocked by metal-ion-binding sites linking transmembrane helices C and F. Nature 383:347–350

Strader CD, Sigal IS, Candelore MR, Rands E, Hill WS, Dixon RA (1988) Conserved aspartic acid residues 79 and 113 of the β-adrenergic receptor have different roles in receptor function. J Biol Chem 263:10267–10271

Strader CD, Candelore MR, Hill WS, Sigal IS, Dixon RA (1989a) Identification of two serines residues involved in agonist activation of the β-adrenergic receptor. J Biol Chem 264:13572–13578

Strader CD, Candelore MR, Hill WS, Dixon RA, Sigal IS (1989b) A single amino acid substitution in the β-adrenergic receptor promotes partial agonist activity from antagonists. J Biol Chem 264:16470–16477

Strosberg AD (1993) Structure, function, regulation of adrenergic receptors. Protein Science 2:1198–1209

Suprenant A, Horstman DA, Akbarali H, Limbird LE (1992) A point mutation of the α_2-adrenoceptor that blocks coupling to potassium but not calcium currents. Science 257:977–980

Unger VM, Hargrave PA, Baldwin JM, Schertler GFX (1997) Arrangement of rhodopsin transmembrane α-helices. Nature (London) 389:203–206

Wang CD, Buck MA, Fraser CM (1991) Site-directed mutagenesis of α_{2a}-adrenergic receptors: identification of amino acids involved in ligand binding and receptor activation by agonists. Mol Pharmacol 40:168–179

Weitz CJ, Nathans J (1992) Histidine residues regulate the transition of photoexcited rhodopsin to its active conformation, metarhodopsin II. Neuron 8:465–472

Wong SKF, Ross EM (1994) Chimeric muscarinic cholinergic: β-adrenergic receptors that are functionally promiscuous among G proteins. J Biol Chem 269:18968–18976

Wong SKF, Parker EM, Ross EM (1990) Chimeric muscarinic cholinergic: β-adrenergic receptors that activate G_s in response to muscarinic agonists. J Biol Chem 265: 6219–6224

Zhao M-M, Gaivin RJ, Perez DM (1998) The third extracellular loop of the β2-adrenergic receptor can modulate receptor/G protein affinity. Mol Pharmacol 53:524–529

Zuscik M, Porter J, Gaivin R, Perez DM (1998) Identification of a conserved switch residue responsible for selective constitutive activation of the β_2-adrenergic receptor. J Biol Chem 273:3401–3407

Section III
New Technologies for the Study of Drug Receptor Interaction

CHAPTER 11
The Assembly of Recombinant Signaling Systems and Their Use in Investigating Signaling Dynamics

S.M. LANIER

A. Introduction

Recombinant signaling systems are a basic component of research efforts concerned with the structure and function of members of the superfamily of membrane receptors coupled to heterotrimeric G proteins. The widespread use of such systems in pharmacology began in the mid 1980s following isolation of the complementary DNAs (cDNAs) encoding β-adrenergic receptors (ARs) and muscarinic receptors. The next decade witnessed the determination of the primary structure of several protein families involved in signal propagation, including receptors, G proteins and effectors. In the midst of this activity, transient and stable expression systems were used to determine the ligand-recognition properties and functional domains of the receptor protein in addition to the signal-transduction events initiated by agonist-activated receptors. Generally, this approach involved ectopic expression of the receptor in a mammalian cell line. With recent advances in gene technology, recombinant systems can also be generated in *Saccharomyces cerevisae*, aplysia, *Dictyostelium discoideum*, *Caenorhabditis elegans*, *Drosophila melanogaster*, transgenic rodents and mini-swine.

Recombinant signaling systems using G protein-coupled receptors have provided insight into several basic issues of cell signaling, including signaling efficiency/specificity, the mechanistic basis for partial/inverse agonists and multiple receptor conformations that may be drug or signaling-pathway specific. The use of stably transfected cell lines together with cells expressing the endogenous gene of interest provides a very powerful investigative tool when appropriate controls are in place. In almost every report, the signaling mechanisms (i.e. G proteins and effectors involved) associated with receptors ectopically expressed in mammalian cells at reasonable receptor densities are observed in non-transfected cells expressing the endogenous receptor gene. In some transfected systems, a receptor unexpectedly couples to signal-transduction pathways not normally observed to be activated by agonists in tissues or cell lines expressing the endogenous receptor gene. Such observations may be interpreted as artifacts of the transfection system; usually, however, these results provide key insights into the signaling pathways a receptor is capable of activating at specific stages of cell development or when the signaling system goes awry in certain diseases. This chapter focuses on the generation and use of recombinant signaling systems in mammalian cell lines to gain insight into the pharmacology and cell biology of G protein-coupled receptors.

B. Assembly of Recombinant Signaling Systems

Heterologous expression of a gene product in a mammalian cell line can be achieved by stable or transient transfection. Stable transfection essentially means that the gene/cDNA becomes incorporated into the host genome and is propagated with each mitotic event such that, theoretically, each cell expresses the same amount of the gene product. Most transfection studies use the cDNA clone rather than genomic clones for optimal expression. Stable transfections require ~2–3 months for complete characterization. Transient transfection refers to systems that express high levels of the gene product of interest over a defined period of time shortly following introduction of the cDNA (usually within 2–4 days). These two systems for gene expression in mammalian cells provide the basis for the great majority of recombinant signaling systems. The recombinant signaling system may be used to determine the ligand-recognition and structural properties of specific receptors, various aspects of receptor regulation or to determine the signal-transduction pathways regulated by a receptor. In addition, systems may be constructed in varying degrees of complexity to address broader issues relative to mechanisms of signal transfer and control of the signaling process.

I. Stable Transfection

A large number of cell types have been used for stable expression of G protein-coupled receptors and subsequent analysis of receptor function (ALBERT 1994; GEISS et al. 1996; TATE and GRISSHAMMER 1996; GERHARDT et al. 1997; KOLLER et al. 1997). Fibroblast cell lines are commonly used to generate recombinant signaling systems, because the cells grow well and are easily transfected. Such cells do not express a wide panel of G protein-coupled receptors; thus, in the great majority of cases, fibroblast cell lines would not normally express the endogenous counterpart of the transfected protein. Fibroblast cell lines stably transfected with a G protein-coupled receptor are routinely used to screen compounds for receptor interaction using radioligand-binding assays and for analysis of receptor–effector coupling. However, one must consider that the receptor's microenvironment in fibroblasts may be quite different from that in which the receptor normally functions in vivo. Indeed, it is quite clear that the signal-transduction pathways activated by a G protein-coupled receptor are dependent upon the cell type in which the receptor is functioning. Thus, for functional analysis of signaling events in recombinant systems, one should consider the use of additional, non-fibroblast cell lines related to the cell type expressing the protein of interest in vivo.

Plasmid vectors are used as vehicles to introduce the cDNA of interest (i.e. receptor, G protein subunit, effector) into the cell. The cDNA is incorporated into the vector in the sense direction downstream of viral promoter elements and upstream of a polyadenylation signal. If the gene product is toxic to the cells or if specific issues of receptor coupling are being addressed, one may select an expression vector that uses an inducible promoter (GOSSEN et al. 1995). These and other expression vectors are illustrated and described in detail elsewhere (SAMBROOK et al. 1989; KAUFMAN 1990). Generally, it is best to insert only the protein-coding region of the cDNA into the expression vector. Elimination of non-coding sequences 5' to the translational start site will remove any regulatory regions that may compromise gene expression. The 3' untranslated region (utr) is usually less problematic but can also present problems in various expression systems (YANG et al. 1997). The expression of a foreign gene can often be enhanced by inserting a Kozak's consensus sequence for translational initiation at the translational start site or by modifying the 3' utr to stabilize the mRNA. A drug-resistance cassette is required to isolate stable transfectants. Such a marker generally confers resistance of the transfected cell to a drug, such as G418 or hygromycin. The drug-resistance cassette may be included in the same vector containing the cDNA of interest, or cells may be co-transfected with an expression vector containing the gene of interest and another vector containing the drug-resistance cassette. Cell lines stably transfected using one selectable marker can also be further manipulated by transfection using a second drug-resistance cassette.

For stable transfection of cells as monolayers, we routinely use calcium phosphate/DNA co-precipitation, maintaining a fixed concentration of plasmid DNA (KEOWN et al. 1990; COUPRY et al. 1992; DUZIC et al. 1992). Our experience indicates essentially similar results using linearized or supercoiled plasmids. Cell lines resistant to transfection by the calcium phosphate/DNA co-precipitation technique may be transfected by electroporation or lipid-mediated gene-transfer reagents. Compared with monolayer cultures of cells, cell lines that do not attach to the tissue-culture dish are more difficult to transfect. For primary cultures and cell lines that are particularly resistant to transfection by standard procedures, one may explore the use of a viral-based packaging system. The viral-based systems are more involved than standard plasmid-based transfection systems in terms of both vector construction and transfection optimization.

Monolayer cultures of mammalian cell lines are generally transfected at 60% confluency, and a control transfection using the plasmid vector without insert should be processed in parallel. Cells are allowed to recover for 2–3 days, permitting expression of the drug-resistance cassette. At this stage, the cells are usually confluent, and G418 or hygromycin is added to the medium. Drug sensitivity will vary among cell lines, and a concentration–response curve for the drug must be determined in each cell line prior to transfection. Generally, the non-transfected cells will die within the first week, and the transfected cells will start to appear as isolated colonies 2–4 weeks later. Such colonies are easily removed from the plate using cloning cylinders for subsequent propagation as individual cell lines. Clonality may be further ensured by serial dilution.

Once sufficient numbers of clonal cells are propagated, the different cell lines are evaluated for expression of the transfected cDNA by radioligand binding, immunoblotting or functional readouts. Although each of the isolated transfectant cell lines are drug resistant, the expression of transfected cDNA may be variable. The drug-resistant cell lines can also be screened for expression of the transfected cDNA by RNA-blot analysis. Only transformants expressing the appropriate size of messenger RNA (mRNA) for the transfected cDNA are utilized for further studies. The expected size of mRNA is calculated from the expression-vector construct, and this analysis serves as a strong indicator that the coding region and regulatory regions are intact. Appropriate processing of the transfected cDNA may also be ascertained by characterizing the expressed protein (i.e. photoaffinity labeling or immunoblotting). One would expect an M_r for the expressed cDNA similar to that observed for the protein in tissue homogenates. The effects of receptor activation on any of the typical signal transduction pathways in a cell are determined by standard procedures. In some cases, the systems are constructed to provide an easily detectable readout as a reporter for receptor activation facilitating rapid screening for agonists (WELSH and KAY 1997).

In stable transfection systems, the receptor is generally expressed at densities of 50–7000 fmol/mg membrane protein. In contrast, receptor densities in tissue homogenates range from 20 fmol/mg to 1200 fmol/mg. The higher densities are observed in tissues where one cell type predominates (i.e., α_2-AR in basolateral membranes of renal proximal tubules or in enriched human platelet membranes or α_1-AR in liver membranes). The lower receptor densities in tissue homogenates are observed in tissues expressing multiple cell types and, thus, are an underestimate of receptor density in that percentage of cells actually expressing the protein of interest. Although the receptor protein is generally considered to be "overexpressed" in stable or transient transfection systems, one may select clonal cell lines expressing receptor densities similar to those observed in vivo. For studies using cell lines stably transfected with G protein-coupled receptors or other entities involved in signal transduction, the functional properties of the expressed protein should be studied in multiple clonal cell lines expressing a range of receptor densities to avoid clonal artifacts in data interpretation.

II. Transient Expression Systems

COS cells and HEK-293 cells are widely used for transient expression of G protein-coupled receptors and the various signaling proteins downstream of the receptor. COS cells are actually transformants of the CV-1 cell line derived from African green-monkey kidneys engineered to express the SV40 large T antigen. Expression in the COS system requires plasmids with an SV40 origin of replication to ensure high copy numbers. Otherwise, the various expression vectors are generally interchangeable between transient and stable expression systems. The expression vector pMT2 is particularly efficient in the COS system (KAUFMAN 1990). The systems used to introduce the expression vector into the cell are also interchangeable between stable and transient systems. Both the calcium phosphate/DNA co-precipitation and diethylaminoethyl-(DEAE-) dextran methods are widely used due to their simplicity, reproducibility and effectiveness in diverse cell types. The DEAE-dextran method described by Kaufmann is highly reproducible and efficient for protein expression in COS cells (KAUFMAN 1990; LANIER et al. 1991). Lipid-mediated gene-transfer reagents may result in a higher percentage of transfectants. It is fairly easy to titrate protein expression in the transient expression system by using increasing amounts of plasmid DNA during transfection, although a plateau is quickly reached. Such a strategy is quite useful when cells are transfected with multiple plasmid vectors containing cDNAs encoding different signaling molecules.

Transient expression of G protein-coupled receptors allows rapid analysis of their ligand-recognition properties and functionalities. The transient systems are particularly useful for analysis of signaling events in a single cell by confocal imaging and also for simultaneous transfection of multiple cDNAs. However, this approach is less suitable for detailed analysis of the cellular effects of receptor activation, since only a small percentage (10%–20%) of cells express the transfected cDNA. These cells often express the transfected cDNA far in excess of other components of the signaling pathway compared with the stoichiometry observed under normal conditions. In transient systems, quantitative measurements of the expressed protein (radioligand binding, immunoblots) underestimate the relative levels in the expressing cell population by five- to tenfold. The expression of the gene in a subpopulation of cells in the transient system complicates functional characterization of receptors that elicit inhibitory responses (i.e., inhibition of adenylyl cyclase). If only 10% of the cells express the desired inhibitory receptor, changes in the basal level of adenylyl cyclase in the transfected subpopulation of cells are difficult to detect and the effects of receptor activation on hormone- or forskolin-stimulated cyclase are compromised by the stimulation of adenylyl cyclase in the entire cell population. Receptors coupled in stimulatory fashion to effectors are more readily assayed in such a system. Several approaches are available to overcome these limitations. The transfected cells can be selected by fluorescence-activated cell sorters by co-transfection with an appropriate marker, providing an enriched population of transfectants. Alternatively, the G protein-coupled receptor of interest can be co-transfected with another cDNA (i.e., effector, regulatory molecule, different receptor) that would allow the functional readout to be restricted to the subpopulation of transfected cells. Co-transfection of multiple cDNAs is quite useful for transient systems as, generally, a high percentage (90%) of transfected cells take up and express each of the presented cDNAs. Nevertheless, even following selection, there will likely be individual variation in terms of protein expression among cells.

C. Drug–Receptor Interactions in Recombinant Signaling Systems

I. Cell-Type-Specific Signaling Events

Although the generation of various recombinant signaling systems and analysis of various signal-transduction events are fairly straightforward, there are several factors that influence data interpretation. Perhaps the most important factor is the realization that the responses mediated by the great majority of G protein-coupled receptors are cell specific (Fig. 1). The realization that a single receptor molecule functions in a cell-specific manner is a simple point but has broad implications. First, these observations indicate that the signaling system is dynamic and, thus, likely developmentally regulated and responsive to physiological and non-physiological challenges. Second, the action of an agonist/antagonist at a receptor in one cell may be different from that observed for the same receptor in another cell type, and the action of an

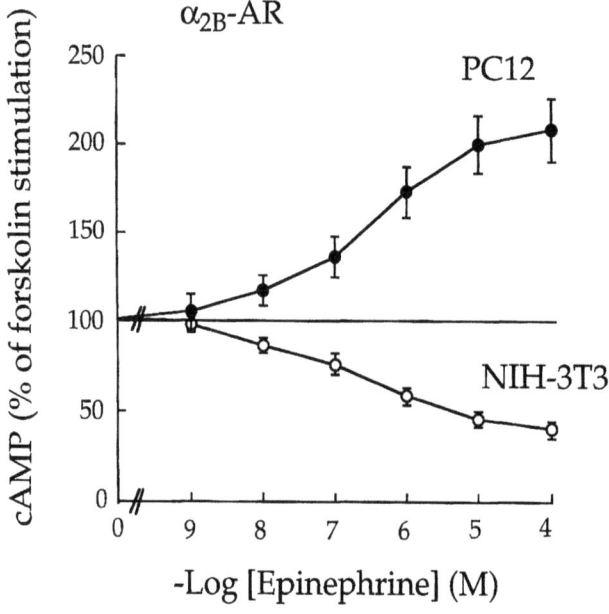

Fig. 1. Cell-type-specific regulation of adenylyl cyclase by the α_{2B}-adrenergic receptor (AR). NIH-3T3 fibroblasts or the pheochromocytoma cell line PC-12 were stably transfected with α_{2B}-AR. Receptor densities: ~1 pmol/mg membrane protein for NIH-3T3; ~1.8 pmol/mg membrane protein for PC-12 (Duzic and Lanier 1992)

agonist/antagonist in a normal cell can be quite different from its action in the same cell following initiation of a disease process.

The cell-specific responses observed for a specific receptor reflect the microenvironment in which the receptor is operating. Downstream signaling components and proteins involved in receptor regulation/trafficking are also expressed in a cell-type-specific manner. In addition, the architecture of each cell is unique, presenting the possibility of segregated signaling events in cellular microdomains where a receptor may have access to a subpopulation of the G proteins/effectors expressed in the cell. Thus, one is faced with issues related to stoichiometry of the signaling components and unknown cell-specific factors that regulate signal propagation. A role for the stoichiometry of the receptor/G protein and effector in signaling efficiency/specificity is indicated by altered G protein levels associated with signaling malfunction in various disease states and by receptor-inactivation studies indicating pathway-specific receptor reserves (Chap. 3, Sect. C.I.3).

These caveats are important considerations when one designs a recombinant system and in interpreting the data generated with such systems. These issues are illustrated by several examples in which one can manipulate receptor-mediated responses. For example, formation of the ternary complex can be manipulated by altering the type of G protein expressed in the cell (Fig. 2; COUPRY et al. 1992). Agonist-competition curves at the RG10 α_{2C}-AR expressed in NIH-3T3 fibroblasts were monophasic in the absence of the G protein Goα1 but became biphasic when cells were co-transfected with recep-

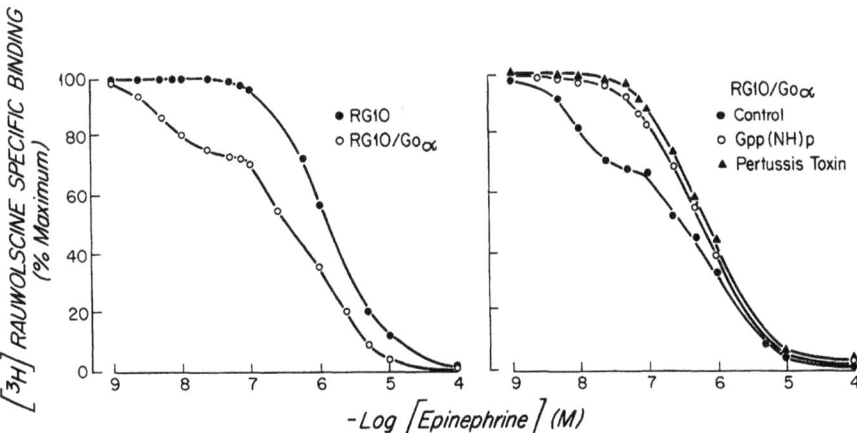

Fig. 2. Expression of Goα facilitates the formation of receptor conformations exhibiting high affinity for agonist. NIH-3T3 fibroblasts, which do not normally express Go, were stably transfected with the RG-10 α_{2C}-adrenergic receptor (AR) or co-transfected with the α_{2C}-AR and Goα1. The transfectants were characterized by radioligand binding and immunoblotting to verify the expression of receptor and G protein. Formation of the high-affinity state for agonists was determined in agonist-competition studies. Receptor density: 1.2–1.5 pmol/mg membrane protein (COUPRY et al. 1992)

tor and Goα1, indicating effective interaction of the receptor with G_o. The hormone-induced response mediated by a specific receptor is also dependent on the types of effector molecules expressed in the cell (Fig. 3; MARJAMAKI et al. 1997). In DDT_1-MF2 cells transfected with $α_{2A/D}$-AR and different adenylyl cyclases not normally expressed in the cell, receptor activation inhibited

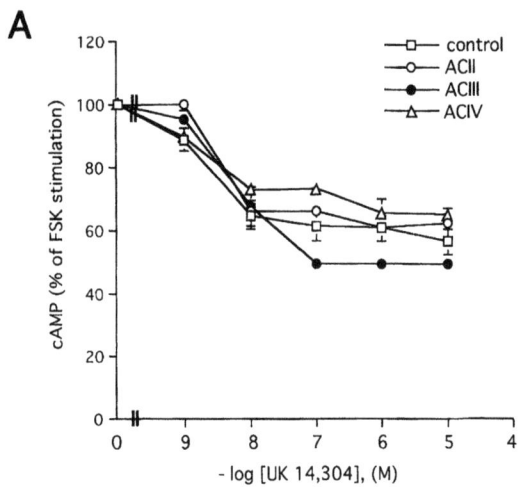

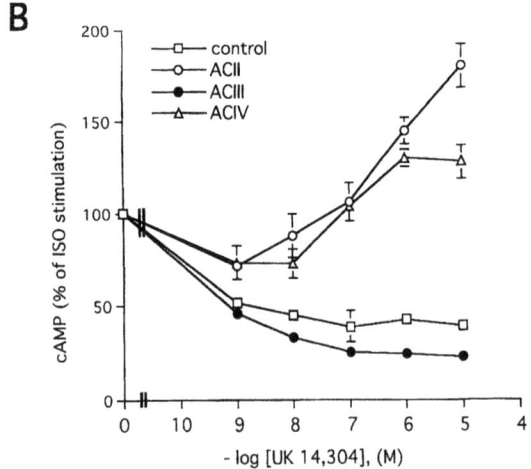

Fig. 3A,B. Effect of $α_{2A/D}$-adrenergic receptor (AR) activation on forskolin- and isoproterenol-stimulated adenylyl cyclase (AC) in DDT_1-MF2 cell lines expressing ACII, ACIII or ACIV. DDT1-MF2 cells stably transfected with $α_{2A/D}$-AR and selected by G418 resistance were transfected again with ACII, ACIII or ACIV complementary DNA using hygromycin selection. Cell membranes from control and AC transfectants were incubated with **A** forskolin (*FSK*) or **B** isoproterenol (*ISO*) and increasing concentrations of the $α_{2A/D}$-AR agonist UK-14,304 and enzyme activity measured by standard procedures (MARJAMAKI et al. 1997). Receptor densities: ~3–4 pmol/mg membrane protein

forskolin-induced activation of adenylyl cyclase in each cyclase transfectant. However, the effects of $\alpha_{2A/D}$-AR activation on enzyme activity stimulated by the β-AR agonist isoproterenol was adenylyl-cyclase-type specific. These examples of manipulation indicate the importance of cell-specific expression of the signaling molecules in determining the cell response to extracellular stimuli.

The importance of cell-type and stoichiometric considerations are also illustrated by the use of recombinant signaling systems in the analysis of partial and inverse agonists (Chap. 2, Sect. B.2) (Table 1). Partial agonists are defined

Table 1. Analysis of partial agonists in recombinant signaling systems

Receptor system	Cell line/readout	Results	Citation
M2	CHO/cAMP/PI	Partial agonists converted to full agonists at higher receptor densities	VOGEL et al. (1995)
5-HT$_{1A}$	CHO/GTPγ^{35}S binding	Partial agonists converted to full agonists at higher receptor densities	NEWMAN-TANCREDI et al. (1997)
α_{2C}-AR	CHO/cAMP	Partial agonists converted to full agonists at higher receptor densities	POHJANOKSA et al. (1997)
5-HT$_{1Db}$	C6-glial CHO/cAMP	Intrinsic activities for selected ligands are cell specific	PAUWELS and COLPAERT (1995)
v-Opioid	GTPγ^{35}S binding	Intrinsic activities for selected ligands are cell specific	SELLEY et al. (1997)
5-HT$_{1B}$	Y-1/cAMP	Receptor reserve masks partial agonist activity	ADHAM et al. (1992)
M1, M3, M5	NIH-3T3/proliferation	G$_q$ overexpression increases the efficacy of partial agonists	BURSTEIN et al. (1997)
5-HT$_{1A}$	HeLa/intracellular calcium	Antagonist converted to agonist at higher receptor densities	BODDEKE et al. (1992)
5-HT$_{1A}$	NIH-3T3/cAMP	Efficacy of full and partial agonists increased at higher receptor densities	VARRAULT et al. (1992)
$\alpha_{2A/D}$-AR	NIH-3T3/GTPγ^{35}S binding	Partial agonist converted to full agonist by co-transfection of receptor and G$o\alpha$1	YANG and LANIER (1999)

cAMP, cyclic adenosine monophosphate; *CHO*, Chinese-hamster ovary; *GTP*, guanosine triphosphate.

by their inability to produce the same maximum response observed for the endogenous receptor agonist (i.e., full agonist). The explanation for full versus partial agonism remains elusive. Possible explanations for this behavior must incorporate the concept of pre-coupling of receptor and G protein, conformational selection/induction by agonists, the "energy landscape" generated by multiple conformations of the receptor and the cell-specific manifestation of partial agonism (KENAKIN 1989; EASON et al. 1994; PEREZ et al. 1996; WEISS et al. 1996; Chap. 2, Sect. G). For receptors that couple to multiple G proteins, the relationships between partial and full agonists may be influenced by the type/amount of G protein expressed in the cell (GETTYS et al. 1994), the receptor density or accessory proteins that regulate the transfer of signal from receptor to G protein and to various effectors further downstream (Table 1). Figure 4 presents an example of how a drug may be partial agonist in one cell and a full agonist in another. In NIH-3T3 $\alpha_{2A/D}$-AR transfectants, the α_2-AR agonist clonidine was essentially converted from a partial to a full agonist by co-expression of Goα1, a pertussis-toxin-sensitive G protein not normally expressed in these cells (YANG and LANIER 1999). These observations have significance relative to the concepts of conformational induction vs conformational selection in terms of agonist efficacy and the development of cell-specific drug efficacy. The classification of a partial versus a full agonist in such recombinant systems and in vivo also depends on whether the readout is proximal (i.e., guanosine triphosphate (GTP) γ^{35}S binding) or distal (i.e., adenylyl cyclase activity, contraction/relaxation of smooth muscle) within the

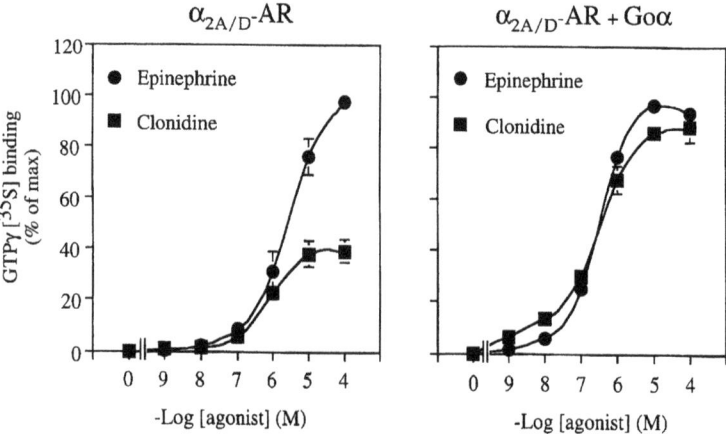

Fig. 4. Expression of Goα1 increases agonist efficacy. NIH-3T3 fibroblasts, which do not normally express Go, were stably transfected with $\alpha_{2A/D}$-AR or co-transfected with $\alpha_{2A/D}$-AR and Goα1. The transfectants were characterized by radioligand binding and immunoblotting to verify the expression of the receptor and G protein. G protein activation was determined by measuring agonist-induced increases in [^{35}S]-guanosine triphosphate-γS binding in membrane preparations (SATO et al. 1995). The receptor transfectants expressed 4–6 pmol of receptor per milligram of membrane protein. G protein activation was determined in the presence and absence of increasing concentrations of epinephrine or clonidine (24°C for 30 min)

signal-transduction cascade. The influence of such factors on the relationship between full and partial agonism indicates that the effects of a particular drug could be maximized in tissue where it behaves as a full agonist but would be diminished in tissues where it behaves as a partial agonist. The same thoughts may apply to inverse agonists.

II. Influence of Accessory Proteins

Based on results in cells expressing the endogenous receptor gene or in cells stably transfected with receptor subtypes, a specific receptor can couple to multiple effectors. For example, the α_{2A}-AR, a typical G protein-coupled receptor, may regulate phospholipases A_2, C and D, calcium flux, Na^+/H^+ exchangers, p21-ras or adenylyl cyclases (references in MARJAMAKI et al. 1997). The specific regulation of any of these effector molecules often depends upon receptor subtype, receptor density and the environment in which the receptor is operating. The variable efficiency/specificity of coupling observed for a specific receptor in different cells actually suggests that there are additional, unidentified, cell-specific proteins/lipids involved in the signaling process (SATO et al. 1995).

One working hypothesis that encompasses several recent observations relative to signaling events is that signaling efficiency/specificity is determined, in part, by proteins found in the receptor's microenvironment which, together with receptor, G protein and effector, contribute to the formation of a signal-transduction complex at the cytoplasmic face of the receptor. I refer to these proteins as accessory proteins, distinct from receptor, G protein and classical effectors (Table 2). This hypothesis is consistent with data suggesting the existence of multimeric G protein-subunit complexes, the isolation of receptor or G protein subunits together with some effectors (NAKAMURA and RODBELL 1990; VAILLANCOURT et al. 1990; AUSIELLO et al. 1992; COULTER and RODBELL 1992; MAH et al. 1992) and the existence of proteins that influence the activation state of G (Table 2). The signal-transduction network for this system may parallel that used by receptors with a single-membrane span motif, where binding of an agonist initiates a series of protein interactions dependent on protein phosphorylation.

Within the theoretical framework of a signal-transduction complex, one might list accessory proteins, as indicated in Table 2. Group I accessory proteins are those involved in various aspects of receptor regulation, including receptor kinases/phosphatases and arrestins. Group II accessory proteins are suggested to influence events at the receptor–G protein or G protein–effector interfaces. Group III proteins have the potential to interact with receptor or G protein, based on various types of protein interaction assays. A fourth group of proteins directly influence the activation state of G proteins and includes proteins that may inhibit/stimulate signal propagation and/or determine the specific pathway that the signal travels. The precise functional roles of several of the group II, III and IV proteins are undefined at present.

Table 2. Accessory proteins for G protein coupled receptor systems

Protein	Comment	Citation
GROUP I: RECEPTOR REGULATION		
GRKs	Phosphorylate activated receptor	Ferguson et al. (1996); Hosey et al. (1996)
Arrestins	Interact with phosphorylated/activated receptors; involved in signal termination	Dolph et al. (1993); Sterne-Marr (1995)
Phosphatases	Dephosphorylated receptor	Krueger (1997)
Protein kinase A/C	Phosphorylate-selected receptors; involved in crosstalk between receptor systems and signal specificity	Oppermann (1996); Daaka (1997)
Recoverin	Inhibits rhodopsin kinase	Polans et al. (1996)
GROUP II: REGULATION OF SIGNAL PROPAGATION		
CRAC Pianissimo	Required for receptor coupling to adenylyl cylcase in *Dictyostelium discoideum*	Kim et al. (1996); Chen et al. (1997)
CGRP-RCP Coupling cofactor Serotonin-receptor-related cDNA	Required for effective coupling of CGRP receptor Influences stability of receptor–G protein complex Required for receptor activation	Luebke et al. (1996) Nanoff et al. (1995) Ohya et al. (1997)
?	Membrane component influencing the efficiency of signal transfer from receptor to G protein	Sato et al. (1995)
RAMPS	Receptor trafficking and ligand recognition	McLatchie et al. (1998)
GROUP III: INTERACTING PROTEINS		
Jak2	Tyrosine kinase that associates with the C-terminus of the AT1 receptor	Marrero et al. (1995); Ali et al. (1997)
Spinophilin	Binds to third intracellular loop of D2-dopamine receptor	Smith et al. (1999)
14-3-3 Z	Binds to third intracellular loop of α_2-adrenergic receptors	Prezeau (1999)
Calcyon	Binds to carboxy terminus of D1-dopamine receptors and influences calcium signaling	Lezcano (2000)

Grb2/Nck	Interacts with third intracellular loop of D4-dopamine receptor	OLDENHOF et al. (1998)
NHERF	Binds to carboxyl terminus of B_2-adrenergic receptor	HALL ET AL (1998)
Src	Complexed with B_2-adrenergic receptor via arresting	LUTTRELL et al. (1999)
Homer	PDZ protein that associates with C-termini of metabotropic glutamate receptors; possible involvement in receptor trafficking	BRAKEMAN et al. (1997)
EIF2bα	Translation factor that associates with C-termini of α_{2A}-, α_{2B}-, α_{2C}- and β_2-adrenergic receptors	KLEIN et al. (1997)
G$\beta\gamma$	Directly associates with receptor subdomains	TAYLOR et al. (1994); WU et al. (1998)
Mosaic protein	Associates with Giα2	MOCHIZUKI et al. (1996)
Nucleobindin	Associates with Giα2	MOCHIZUKI et al. (1995)

GROUP IV: G PROTEIN REGULATORS

RGS proteins	Family of proteins that regulate G protein signaling; some members act as GTPase-activating proteins for G_i/G_o or G_q	DOHLMAN and THORNER (1997)
AGS proteins	Receptor-independent activators of heterotrimeric G-proteins	TAKESONO et al. (1999)
Pcp2 (PcpL7)	Associates with and activates Go	LUO and DENKER (1999)
Rap1gap	Associates with Gi, Go, G_z	MOCHIZUKI et al. (1999); JORDAN et al. (1999); MENG et al. (1999)
Caveolins	Enriched in cell microdomains; influence G protein activity	LI et al. (1992); SCHERER (1996)
β-APP, Presenilin I	Associate with and activate Go	NISHIMOTO et al. (1993); OKAMOTO et al. (1995)
Phosducin and phosducin-like proteins	Bind G$\beta\gamma$ and impede formation of heterotrimer	GAUDET et al. (1996); BOEKHOFF et al. (1997); THIBAULT et al. (1997)
Tubulin	Influences nucleotide exchange	ROYCHOWDHURY (1993); POPOVA (1994)
Neuromodulin (GAP-43)	Associates with and activates Go	STRITTMATTER (1991, 1993); VITALE (1994)
NG108-15 G protein activator	Increases GTPγ^{35}S binding to Go and possibly G_i	SATO et al. (1996)

APP, amyloid precursor protein; *AT1*, angiotensin II type I; *cDNA*, complementary DNA; *CGRP*, calcitonin gene-related peptide; *CRAC*, cytosolic regulator of adenylyl cyclase; *EIF*, eukaryotic initiation factor; *GAP*, growth-associated protein; *GRK*, G protein-coupled receptor kinase; *GTP*, guanosine triphosphate; *RCP*, receptor-component protein; *RGS*, regulator of G protein signaling.

Binding of agonist to a G protein-coupled receptor is a trigger for guanine-nucleotide exchange, and the bound GTP is hydrolyzed by Gα subunits at varying rates. Several observations suggested that the exchange of guanosine diphosphate (GDP) for GTP and/or subsequent GTP hydrolysis are regulated by accessory proteins. First, the kinetics of signal termination for many physiological responses is faster than the rate of GTP hydrolysis of purified Gα subunits. A second observation relates to the transfer of signal from receptor to G protein in a reconstituted system. G activation occurs with homogeneous preparations of receptor and G protein following reconstitution in phospholipid vesicles, suggesting that initiation of the cellular response to a biological stimulus involves only receptor and G protein. However, the efficiency with which the receptor mediates the agonist-induced activation of G protein and subsequent effector regulation/signal termination is often dramatically lower when the purified entities are reconstituted in phospholipid vesicles than what is observed or expected to occur in the intact cell or membrane preparations. Such a discrepancy may reflect technical issues associated with reconstitution studies or the absence of other cellular factors required for driving these events. A third point relates to the ability of non-receptor proteins/peptides to associate with and regulate the activation state of specific G proteins (Table 2). Fourth, the efficiency of G protein activation by a specific receptor and the influence of guanine nucleotides on receptor–G protein interactions varies among cell types (Duzic and Lanier 1992; Nanoff et al. 1995; Sato et al. 1995 and references therein). Nanoff et al. have partially purified a protein ("coupling cofactor") from brain tissue; it is suggested to trap the ternary complex and impede further signal propagation (Nanoff et al. 1995). Using a "signal-reconstitution system", Sato et al. demonstrated that cell-specific differences in the transfer of signal from a specific receptor to G protein was independent of G protein type or amount (Sato et al. 1995). Overall, these observations indicate the existence of accessory proteins that influence events at the receptor–G protein or G protein–effector interface and regulate the key steps of nucleotide exchange and/or GTP hydrolysis. Recombinant signaling systems have provided a powerful tool to study the functionalities of such proteins.

Relative to the first point in the preceding paragraph, the effectors for G$_q$ and transducin were found to actually accelerate the GTPase activity of the Gα subunit, and these data provide a partial explanation for earlier discrepancies regarding the catalytic process of GTP hydrolysis. In addition to this function of certain effectors, a family of proteins, "regulators of G protein coupled receptor signaling" (RGS), was recently discovered; these proteins also act to stimulate hydrolysis of bound GTP. Several RGS proteins stimulate the GTPase activity of the Gα subunit, effectively turning the signal off (Berman et al. 1996; Koelle and Horvitz 1996; Druey et al. 1996; Hunt et al. 1996; Watson et al. 1996; Helper et al. 1997). These proteins serve a function somewhat analogous to the GTPase-activating proteins for low-molecular-weight monomeric G proteins, such as p21-ras. Although the RGS proteins characterized to date impede signal propagation by accelerating signal termination or by blocking effector interaction, other members of this protein family might actually use the RGS core motif as a handle for G protein association but act in the opposite manner to accelerate/enhance signal propagation (SAITOH et al. 1997). The recently characterized family of caveolins are also listed in group IV as potential regulators of G protein function (LI et al. 1995; SCHERER et al. 1996).

Other group IV proteins may influence the rate of nucleotide exchange analogous to the guanine-nucleotide-exchange inhibitors/stimulators for several low-molecular-weight G proteins. For heterotrimeric G proteins, the activated receptor serves as a guanine-nucleotide exchange stimulator and, in one sense, $G\beta\gamma$ serves as a guanine-nucleotide-exchange inhibitor. In addition to receptor, other proteins or chemicals directly activate G, likely by increasing the rate of GDP dissociation. Examples of such entities include neuromodulin (GAP-43), tubulin, β-amyloid precursor protein and the wasp venom mastoparan (HIGASHIJIMA et al. 1990; MOUSLI et al. 1990; STRITTMATTER et al. 1991, 1993; SUDO et al. 1992; MILES et al. 1993; NISHIMOTO et al. 1993; ROYCHOWDHURY et al. 1993; POPOVA et al. 1994; VITALE et al. 1994; LI et al. 1995; OKAMOTO et al. 1995). As part of a broader approach to define a signal transduction complex for G protein-coupled receptors (WU et al. 1997, 1998), we developed a solution-phase assay system to identify proteins that directly influence the activation state of G (SATO and LANIER 1996; SATO et al. 1996). This approach resulted in the partial purification and characterization of the NG10815 G protein activator (SATO et al. 1996). This protein activates both heterotrimeric brain G protein and free $G\alpha$ and exhibits mechanistic properties that are distinct from receptor-mediated activation of G protein. The NG108-15 G protein activator is distinct from neuromodulin, tubulin, β-amyloid-precursor protein and caveolins, based on immunoblot analysis of partially purified material and its biochemical properties.

The functional role of the G protein activator and other members of the group IV accessory proteins within the cell is open to speculation. One might imagine the following:

1. Group IV proteins may provide a cell-specific mechanism for signal amplification by acting in concert with G protein-coupled receptors;
2. Group IV proteins may influence the population of activated G protein within the cell, independent of receptor activation; and/or
3. Group IV proteins may be "effectors" subject to receptor regulation, providing attractive targets for cross-talk between diverse signaling systems

Whatever the case, such accessory proteins are extremely interesting, with potentially broad physiological and pharmacological significance relative to the cell biology and functional properties of G proteins themselves. Such proteins may selectively interact with specific G proteins to regulate specific signaling pathways. Influences on signaling specificity may also be achieved by the cell-specific and developmentally regulated expression of such proteins and their ability to influence signal intensity and/or duration. By contributing to the amplification of biological stimuli commonly observed with signaling events involving heterotrimeric G protein, group IV proteins may be of particular significance in tissues requiring rapid signal processing or under conditions of aberrant cell growth and development.

D. Perspective

The molecular cloning of the various members of the signal-transduction pathways associated with G protein-coupled receptors and their analysis in recombinant signaling systems has provided a wealth of information regarding both how these molecules actually elicit a cellular response and the flexibility of the signaling system. One of the amazing aspects of G protein-coupled-receptor signaling systems is the inherent specificity of the cell response to the myriad of external stimuli that are processed by such systems. The apparent flexibility of G protein-coupled-receptor signaling is an important aspect of the dynamic ability of the cell to learn, remember, and respond. This refers not only to events in the central nervous system but also to the ability of peripheral tissues to reset and adjust to their environments. These observations present a major challenge to investigators in this field. What are the factors involved in regulating the apparent plasticity of signal transduction? Can this information be "exploited" to provide a new level of therapeutic specificity? A more complete understanding of cellular plasticity with respect to signal transduction may lead to the development of therapeutic approaches that perhaps target the receptor–G protein or G protein–effector interface, as opposed to the receptor's hormone-binding site. Such advances will require novel approaches to chemistry and drug delivery and a precise understanding of the molecular events involved in signal processing. This understanding includes dissection of the signaling pathways and their regulatory mechanisms and fine-structural analysis of the signaling molecules and the mis-signaling events that occur in the disease process.

Acknowledgements. Work from the author's laboratory presented in this text was supported by grants from the National Institutes of Health. My personal gratitude to all the students, fellows and visiting scientists that have spent time in the laboratory, each of whom made special contributions to various aspects of the research effort. My thanks to Sybil Moore for continued secretarial assistance. I also thank Drs. Kenakin and Angus for their efforts as editors of this volume and for the opportunity to contribute. My appreciation to Drs. Jerry Webb and Steve Rosenzweig for valuable comments and suggestions.

References

Adham N, Ellerbrock B, Hartig P, Weinshank RL, Branchek T (1993) Receptor reserve masks partial agonist activity of drugs in a cloned rat 5-hydroxytryptamine$_{1B}$ receptor expression system. Mol Pharmacol 43:427–433

Albert PR (1994) Heterologous expression of G protein-linked receptors in pituitary and fibroblast cell lines. Vit Horm 48:59–109

Ali MS, Sayeski PP, Dirksen LB, Hayzer DJ, Marrero MB, Bernstein KE (1997) Dependence on the motif YIPP for the physical association of Jak2 kinase with the intracellular carboxyl tail of the angiotensin II AT1 receptor. J Biol Chem 272: 23382–233888

Arshavsky VY, Bownds MD (1992) Regulation of deactivation of photoreceptor G protein by its target enzyme and cGMP. Nature 357:416–417

Ausiello DA, Stow JL, Cantiello HF, de Almeida JB, Benos DJ (1992) Purified epithelial Na^+-channel complex contains the pertussis toxin-sensitive $G\alpha i$-3 protein. J Biol Chem 267:4759–4765

Berman DM, Wilkie TM, Gilman AG (1996) GAIP and RGS4 are GTPase-activating proteins for the G_i subfamily of G protein α subunits. Cell 86:445–452

Berstein G, Blank JL, Jhon D-Y, Exton JH, Rhee SG, Ross EM (1992) Phospholipase C-β1 is a GTPase activating protein for $G_q/11$, its physiologic regulator. Cell 70:411–418

Boekhoff I, Touhara K, Danner S, Inglese J, Lohse MJ, Breer H, Lefkowitz RJ (1997) Phosducin, potential role in modulation of olfactory signalling. J Biol Chem 272: 4606–4612

Brakeman PR, Lanahan AA, O'Brien R, Roche K, Barnes CA, Huganir RL, Worley PF (1997) Homer: a protein that selectively binds metabotropic glutamate receptors. Nature 386(6622):284–288

Brauner-Osborne H, Ebert B, Brann MR, Falch E, Krogsgaard-Larsen P (1996) Functional partial agonism at cloned human muscarinic acetylcholine. Eur J Pharmacol 313:145–150

Burstein ES, Spalding TA, Brann MR (1997) Pharmacology of muscarinic receptor subtypes constitutively activated by G proteins. Mol Pharmacol 51:312–319

Coulter S, Rodbell M (1992) Heterotrimeric G proteins in synaptoneurosome membranes are cross-linked by p-phenylenedimaleimide, yielding structures comparable in size to cross-linked tubulin and F-actin. Proc Natl Acad Sci USA 89: 5842–5846

Coupry I, Duzic E, Lanier SM (1992) Factors determining the specificity of signal transduction by G protein-coupled receptors. II. Preferential coupling of the α_{2C}-adrenergic receptor to the guanine nucleotide binding protein, G_o. J Biol Chem 267:9852–9857

Daaka Y, Luttrell LM, Lefkowitz RJ (1997) Switching of the coupling of the $\beta(2)$-adrenergic receptor to switching of the coupling of the $\beta(2)$-adrenergic receptor to different G proteins by protein kinase A. Nature 390:88–91

Danley DE, Chuang TH, Bokoch GM (1996) Defective Rho GTPase regulation by IL-1β-converting enzyme-mediated cleavage of D4 GDP dissociation inhibitor. J Immunol 157:500–503

Dohlman HG, Thorner J (1997) RGS proteins and signalling by heterotrimeric G-proteins. J Biol Chem 272:3871–3874

Dolph PJ, Ranganathan R, Colley NJ, Hardy RW, Socolich M, Zuker CS (1993) Arrestin function in inactivation of G protein-coupled receptor rhodopsin in vivo. Science 260:1910–1916

Druey KM, Blumer KJ, Kang VH, Kehrl JH (1996) Inhibition of G-protein-mediated MAP kinase activation by a new mammalian gene family. Nature 379:742–746

Duzic E, Lanier SM (1992) Factors determining the specificity of signal transduction by G protein-coupled receptors III. Coupling of α_2-adrenergic receptor subtypes in a cell-type specific manner. J Biol Chem 267:24045–24052

Duzic E, Coupry I, Downing S, Lanier SM (1992) Factors determining the speci-ficity of signal transduction by G protein-coupled receptors I. Coupling of α_2-adrenergic receptor subtypes to distinct G proteins. J Biol Chem 267:9844–9851

Eason MG, Jacinto MT, Liggett SB (1994) Contribution of ligand structure to activation of α2-adrenergic receptor subtype coupling to G_s. Mol Pharmacol 45:696–702

Ferguson SS, Barak LS, Zhang J, Caron MG (1996) G-protein-coupled receptor regulation: role of G-protein-coupled receptor kinases and arrestins. Can J Physiol Pharmacol 74:1095–1110

Gaudet R, Bohm A, Sigler PB (1996) Crystal structure at 2.4 angstroms resolution of the complex of transducin $\beta\gamma$ and its regulator, phosducin. Cell 87:577–588

Geiss S, Gram H, Keuser B, Kocher HP (1996) Eukaryotic expression systems – a comparison. Prot Exp Purif 8:271–282

Gerhardt CC, Vanheerikhuizen H (1997) Functional characteristics of heterologously expressed 5-HT receptors. Eur J Pharmacol 334:1–23

Gettys TW, Fields TA, Raymond JR (1994) Selective activation of inhibitory G-protein α-subunits by partial agonists of the human 5-HT1A receptor. Biochemistry 33:4283–4290, 11404

Gossen M, Greundlieb S, Bender G, Muller G, Hillen W, Bujard H (1995) Transcriptional activation by tetracyclines in mammalian cells. Science 268:1766–1769

Helper JR, Berman DM, Gilman AG, Kozasa T (1997) RGS4 and GAIP are GTPase-activating proteins for Gqα and block activation of phospholipase Cβ by γ-thio-GTP-Gqα. Proc Natl Acad Sci USA 94:428–432

Higashijima T, Burnier J, Ross EM (1990) Regulation of G_i and G_o by mastoparan, related amphiphilic peptides, and hydrophobic amines. J Biol Chem 265:14176–14186

Hosey MM, DebBurman SK, Pals-Rylaarsdam R, Richardson RM, Benovic JL (1996) The role of G-protein coupled receptor kinases in the regulation of muscarinic cholinergic receptors. Prog Brain Res 109:169–179

Hunt TW, Fields TA, Casey PJ, Peralta EG (1996) RGS10 is a selective activator of Gαi GTPase activity. Nature 383:175–177

Kaufman RJ (1990) Vectors used for expression in mammalian cells. Methods Enzymol 185:487–511

Kenakin TP, Morgan PH (1989) Theoretical effects of single and multiple transducer receptor coupling proteins on estimates of the relative potency of agonists. Mol Pharmacol 35:214–222

Keown WA, Campbell CR, Kucherlapati RS (1990) Methods for introducing DNA into mammalian cells. Methods Enzymol 185:527–537

Kim JY, Haastert PV, Devreotes PN (1996) Social senses: G-protein-coupled receptor-signalling pathways in *Dictyostelium discoideum*. Chem Biol 3:239–243

Klein U, Ramirez MT, Kobilka BK, Zastrow MV (1997) A novel interaction between adrenergic receptors and the α-subunit of eukaryotic initiation factor 2B. J Biol Chem 272:19099–19102

Koelle MR, Horvitz HR (1996) EGL-10 regulates G protein signalling in the *C. elegans* nervous system and shares a conserved domain with many mammalian proteins. Cell 84:115–125

Koller KJ, Whitehorn EA, Tate E, Ries T, Aguilar B, Chernov-Rogan T, Davis AM, Dobbs A, Yen M, Barrett RW (1997) A generic method for the production of cell lines expressing high levels of 7-transmembrane receptors. Analyt Biochem 250:51–60

Krueger KM, Daaka Y, Pitcher JA, Lefkowitz RJ (1997) The role of sequestration in G protein-coupled receptor resensitization. Regulation of β2-adrenergic receptor dephosphorylation by vesicular acidification. J Biol Chem 272:5–8

Lanier SM, Downing S, Duzic E, Homcy CJ (1991) Isolation of rat genomic clones encoding subtypes of the α_2-adrenergic receptor: identification of a unique receptor subtype. J Biol Chem 266:10470–10478

Li S, Okamoto T, Chun M, Sargiacomo M, Casanova JE, Hansen SH, Nishimoto I, Lisanti MP (1995) Evidence for a regulated interaction between heterotrimeric G proteins and caveolin. J Biol Chem 270:15693–15701

Luebke AE, Dahl GP, Roos BA, Dickerson IM (1996) Identification of a protein that confers calcitonin gene-related peptide responsiveness to oocytes by using a cystic fibrosis transmembrane conductance regulator assay. Proc Natl Acad Sci USA 93:3455–3460

Mah SJ, Ades AM, Mir T, Siemens IR, Williamson JR, Fluharty SJ (1992) Association of solubilized angiotensin II receptors with phospholipase C-alpha in murine neuroblastoma NIE-115 cells. Mol Pharmacol 42:217–226

Marjamaki A, Sato M, Bouet-Alard R, Yang Q, Limon-Boulez I, Legrand C, Lanier SM (1997) Factors determining the specificity of signal transduction by G-protein coupled receptors. V. Integration of stimulatory and inhibitory input to the effector adenylyl cyclase. J Biol Chem 272:16466–16473

Marrero MB, Schieffer B, Paxton WG, Heerdt L, Berk BC (1995) Delafontaine P. Bernstein KE. Direct stimulation of Jak/STAT pathway by the angiotensin-II AT1 receptor. Nature 375:247–250

Miles MF, Barhite S, Sganga M, Elliott M (1993) Phosducin-like protein: an ethanol-responsive potential modulator of guanine nucleotide-binding protein function. Proc Natl Acad Sci USA 90:10831–10835

Mochizuki N, Hibi M, Kanai Y, Insel PA (1995) Interaction of the protein nucleobindin with G α i2, as revealed by the yeast two-hybrid system. FEBS Lett 373(2):155–158

Mochizuki N, Cho G, Wen B, Insel PA (1996) Identification and cDNA cloning of a novel human mosaic protein, LGN, based on interaction with G alpha i2. Gene 181:39–43

Mousli M, Bronner C, Landry Y, Bockaert J, Rouot B (1990) Direct activation of GTP-binding regulatory proteins (G-proteins) by substance P and compound 48/80. FEBS Lett 259:260–262

Nakamura S-I, Rodbell M (1990) Octyl glucoside extracts GTP-binding regulatory proteins from rat brain "synaptoneurosomes" as large, polydisperse structures devoid of $\beta\gamma$ complexes and sensitive to disaggregation by guanine nucleotides. Proc Natl Acad Sci USA 87:6413–6417

Nanoff C, Mitterauer T, Roka F, Hohenegger M, Freissmuth M (1995) Species differences in A_1 adenosine receptor/G protein coupling: identification of a membrane protein that stabilizes the association of the receptor/G protein complex. Mol Pharmacol 48:806–817

Newman-Tancredi A, Conte C, Chaput C, Verriele L, Millan MJ (1997) Agonist and inverse agonist efficacy at human recombinant serotonin 5-HT1A receptors as a function of receptor-G-protein stoichiometry. Neuropharmacology 36:451–459

Nishimoto I, Okamoto T, Matsuura Y, Takahashi S, Okamoto T, Murajama Y, Ogata E (1993) Alzheimer amyloid protein precursor complexes with brain GTP-binding protein G_o. Nature 362:75–79

Ohya S, Takii T, Yamazaki HF, Matsumori M, Onozaki K, Watanabe M, Imaizumi Y (1997) Molecular cloning of a novel gene involved in serotonin receptor-mediated signal transduction in rat stomach. FEBS Lett 401:252–258

Okamoto T, Takeda S, Murayama Y, Ogata E, Nishimoto I (1995) Ligand-dependent G protein coupling function of amyloid transmembrane precursor. J Biol Chem 270:4205–4208

Oppermann M, Freedman NJ, Alexander RW, Lefkowitz RJ (1996) Phosphorylation of the type-1A angiotensin-II receptor by G-protein-coupled receptor kinases and protein kinase C. J Biol Chem 271:13266–13272

Pauwels PJ, Colpaert FC (1995) Differentiation between partial and silent 5-HT1D beta receptor antagonists using rat C6-glial and Chinese hamster ovary cell lines permanently transfected with a cloned human 5-HT1D beta receptor gene. Biochem Pharmacol 50:1651–1658

Perez DM, Hwa J, Gaivin R, Mathur M, Brown F, Graham RM (1996) Constitutive activation of a single effector pathway: evidence for multiple activation states of a G protein-coupled receptor. Mol Pharmacol 49:112–122

Pohjanoksa K, Jansson CC, Luomala K, Marjamaki A, Savola JM, Scheinin M (1997) α(2)-Adrenoceptor regulation of adenylyl cyclase in CHO cells: dependence on receptor density, receptor subtype and current activity of adenylyl cyclase. Eur J Pharmacol 335:53–63

Popova JS, Johnson GL, Rasenick MM (1994) Chimeric Gαs/Gαi2 proteins define domains on Gαs that interact with tubulin for β-adrenergic activation of adenylyl cyclase. J Biol Chem 269:21748–21754

Roychowdhury S, Wang N, Rasenick MM (1993) Tubulin-G protein association stabilizes GTP binding and activates GTPase: cytoskeletal participation in neuronal signal transduction. Biochemistry 32:4955–4961

Saito O, Kubo Y, Miyatani Y, Asano T, Nakata H (1997) RGS8 accelerates G-protein-mediated modulation of K^+ currents. Nature 390:525–529

Sambrook J, Fritsch EF, Maniatis T (1989) Expression of cloned genes in mammalian cells. In: Chris Nolan (ed) Molecular cloning – a laboratory manual. Cold Spring Harbor Laboratory Press

Sato M, Kataoka R, Dingus J, Wilcox M, Hildebrandt J, Lanier SM (1995) Factors determining the specificity of signal transduction by G-protein coupled receptors. IV. Regulation of signal transfer from receptor to G-protein. J Biol Chem 270: 15269–1527

Sato M, Ribas C, Hildebrandt JD, Lanier SM (1996) Characterization of a G-protein activator in the neuroblastoma glioma cell hybrid NG108–15. J Biol Chem 271:30052–30060

Sato M, Wu G, Lanier SM (1996) Regulation of the transfer of signal from receptor to G-protein. In: Lanier SM, Limbird LE (eds) $α_2$-Adrenergic receptors: structure, function and therapeutic implications. Proceedings. Gordan and Breach Publishers

Scherer PE, Okamoto T, Chun M, Nishimoto I, Lodish HF, Lisanti MP (1996) Identification, sequence, and expression of caveolin-2 defines a caveolin gene family. Proc Natl Acad Sci USA 93:131–135

Selley DE, Sim LJ, Xiao R, Liu Q, Childers SR (1997) μ-opioid receptor-stimulated guanosine-5′-O-(γ-thio)-triphosphate binding in rat thalamus and cultured cell lines: signal transduction mechanisms underlying agonist efficacy. Mol Pharmacol 51:87–96

Sterne-Marr R, Benovic JL (1995) Regulation of G protein-coupled receptors by receptor kinases and arrestins. Vitamin Horm 51:193–234

Strittmatter SM, Valenzuela D, Sudo Y, Linder ME, Fishman MC (1991) An intracellular guanine-nucleotide-release protein for G_o. J Biol Chem 266:22465–22471

Strittmatter SM, Cannon SC, Ross EM, Higashijima T, Fishman MC (1993) GAP-43 augments G-protein-coupled receptor transduction in *Xenopus laevis* oocytes. Proc Natl Acad Sci USA 90:5327–5331

Sudo Y, Valenzuela D, Beck-Sickinger AG, Fishman MC, Strittmatter SM (1992) Palmitoylation alters protein activity: blockade of G_o stimulation by GAP-43. EMBO 11:2095–2102

Tate CG, Grisshammer R (1996) Heterologous expression of G-protein-coupled receptors. Trends Biotechnol 14:426–430

Taylor JM, Jacob-Mosier GG, Lawton RG, VanDort M, Neubig RR (1996) Receptor and membrane interaction sites on Gβ. A receptor-derived peptide binds to the carboxyl terminus. J Biol Chem 271:3336–3339

Thibault C, Sganga MW, Miles MF (1997) Interaction of phosducin-like protein with G protein βγ subunits. J Biol Chem 272:12253–12256

Vaillancourt RR, Dhanasekaran N, Johnson GL, Ruoho AE (1990) 2-Azido-[^{32}P]NAD$^+$, a photoactivatable probe for G-protein structure: evidence for holotransducin oligomers in which the ADP-ribosylated carboxyl terminus of alpha interacts with both α and γ subunits. Proc Natl Acad Sci USA 87:3645–3649

Varrault A, Journot L, Audigier Y, Bockaert J (••) Transfection of human 5-hydroxytryptamine 1A receptors in NIH-3T3 fibroblasts: effects of increasing receptor density on the coupling of 5-hydroxytryptamine 1A receptors to adenylyl cyclase

Vitale N, Deloulme JC, Thierse D, Aunis D, Bader MF (1994) GAP-43 controls the availability of secretory chromaffin granules for regulated exocytosis by stimulating a granule-associated Go. J Biol Chem 269:30293–30298

Vogel WK, Mosser VA, Bulseco DA, Schimerlik MI (1995) Porcine m2 muscarinic acetylcholine receptor-effector coupling in Chinese hamster ovary cells. J Biol Chem 270:15485–15493

Watson N, Linder ME, Druey KM, Kehrl JH, Blumer KJ (1996) RGS family members: GTPase-activating proteins for heterotrimeric G-protein alpha-subunits. Nature 383:172–175

Weiss JM, Morgan PH, Lutz MW, Kenakin TP (1996) The cubic ternary complex receptor-occupancy model. J Theor Biol 181:381–397

Welsh S, Kay SA (1997) Reporter gene expression for monitoring gene transfer. Curr Opin Biotech 8:617–622

Wu G, Krupnick JG, Benovic JL, Lanier SM (1997) Interaction of arrestins with intracellular domains of receptors coupled to heterotrimeric G-proteins. J Biol Chem 17836–17842

Wu G, Benovic JL, Hildebrandt JD, Lanier SM (1998) Receptor docking sites for G-protein bg subunits: implications for signal regulation. J Biol Chem 273:7197–7200

Yang Q, McDermott PJ, Sherlock JD, Lanier SM (1997) The 3' untranslated region of the α_{2C}-adrenergic receptor mRNA impedes translational processing of the receptor message. J Biol Chem 272:15466–15473

Yang Q, Lanier SM (1999) Influence of G protein type on agonist efficacy. Mol Pharmacol 56:651-656

CHAPTER 12
Insect Cell Systems to Study the Communication of Mammalian Receptors and G Proteins

R.T. WINDH, A.J. BARR, and D.R. MANNING

A. Introduction

Cells contain large and often diverse complements of receptors that bind circulating hormones, paracrine factors, and neurotransmitters and, in so doing, initiate events that culminate in the regulation of cell function. The largest family of cell-surface receptors are those coupled to guanosine triphosphate (GTP)-binding regulatory proteins (G proteins). These receptors, or GPCRs (G protein-coupled receptors), exhibit a characteristic seven-transmembrane-domain motif. The binding of an agonist to a GPCR induces a change in structure which, if GTP is present, culminates in the activation of one or more G proteins. Activated G proteins, in turn, regulate enzymes and channels responsible for the control of intracellular second messengers.

G proteins are $\alpha\beta\gamma$ heterotrimers present at the inner surface of the plasma membrane. The identity of the G protein is usually linked to that of its α subunit. Currently recognized α subunits range in size from 40kDa to 46kDa and, by virtue of structural similarities, support the classification of G proteins into four major families (Fig. 1). The activity of G proteins and the consequent status of target regulation are tightly linked to the binding and hydrolysis of GTP (BOURNE et al. 1991; CONKLIN and BOURNE 1993). Upon binding to receptors at the cell surface, agonists promote the release of guanosine diphosphate (GDP) from G proteins and, thus, an exchange for GTP present in the cytoplasm. Correlates of the exchange are an altered conformation of the α subunit and its dissociation from $\beta\gamma$. Regulation of effector activity can be achieved by the α subunit alone, the $\beta\gamma$ heterodimer alone, or the α and $\beta\gamma$ subunits together. The GTP on the α subunit is eventually hydrolyzed to allow reversion of the subunit to an inactive, GDP-bound form that can re-associate with $\beta\gamma$. Factors that can accelerate the hydrolysis of GTP by the α subunit include certain effectors and a growing number of regulator of G protein-signaling (RGS) proteins.

I. Selectivity of Receptor–G Protein Interactions

The selectivity of receptor-G protein coupling plays a key role in determining the nature of a cell's response to a particular ligand. At least three regions

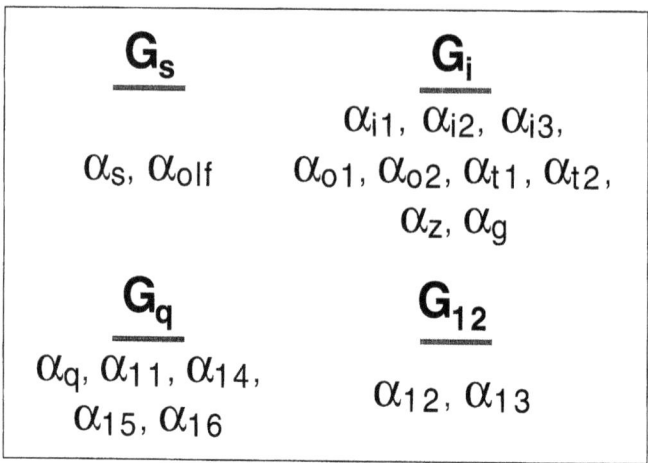

Fig. 1. The classification of α subunits among the four families of G proteins. The α subunits of heterotrimeric G proteins are grouped into four families according to primary structure. Subunits exhibiting greater than 50% homology are grouped into the same family (STRATHMANN and SIMON 1990; HEPLER and GILMAN 1992)

within α subunits support interactions with GPCRs – the N-terminus (probably by virtue of binding $\beta\gamma$, which helps to stabilize receptor–G protein complexation), a stretch of residues just prior to the C-terminus, and the C-terminal 4–10 residues (CONKLIN and BOURNE 1993). Recent work with chimeras developed from α_i, α_o, α_z, and α_q demonstrates that the C-terminus represents a point of discrimination for receptors among G proteins (CONKLIN et al. 1993). ADP-ribosylation catalyzed by pertussis toxin (PTX), which disrupts receptor–$G_{i/o}$ interaction, occurs at a cysteine four residues from the C-terminus. It is nevertheless clear that, just as a GPCR can recognize several structurally related ligands, so too can it interact with multiple G proteins. Communication of GPCRs with several G proteins in the same family is especially common, as α subunit family members often have identical or highly similar C-termini. Thus, in a variety of tissues, receptors that interact with G_i proper also couple with G_o and G_z. Interactions with G proteins from different families also occur quite commonly. A receptor for thrombin (protease-activated receptor 1), for example, couples with members of the G_i, G_q, and G_{12} families (HUNG et al. 1992; OFFERMANS et al. 1994). Going a step further, the receptor for human thyroid-stimulating hormone has been reported to interact with members of all four families of G proteins in the thyroid (LAUGWITZ et al. 1996).

II. Effector Modulation as a Measure of G Protein Activity

The G protein(s) with which a receptor interacts is often deduced from the effectors and/or second messengers regulated in response to the binding of an agonist. The accuracy of this kind of deduction depends on the specificity of the G protein for the effector of interest. In some cases, specificity is reasonably assumed. For example, the inhibition and stimulation of adenylyl cyclase most commonly result from activation of members of the G_i and G_s families, respectively (TANG and GILMAN 1992). The regulation of other effectors, however, can be achieved through several different G proteins. Both G_i and G_q stimulate the activity of phosphoinositide-specific phospholipase C-β (PLC-β; MORRIS and SCARLATA 1997), making it difficult to confidently assign the effect to one G protein or the other without resorting to PTX. Indeed, $\beta\gamma$ released from any G protein can theoretically activate PLC-β, modulate adenylyl cyclase activity (depending on the isotype), and stimulate phosphotidyl inositol (PI) 3-kinase (CLAPHAM and NEER 1997). The converse holds true; just as some effectors are regulated by several different G proteins, many G proteins interact with multiple effectors. Therefore, members of the G_i family inhibit adenylyl cyclase, stimulate phospholipase A_2 and PLC-β activities (STERNWEIS and SMRCKA 1992; AXELROD 1995; MORRIS and SCARLATA 1997), stimulate Na^+/H^+ exchange (GARNOVSKAYA et al. 1997), and regulate the activity of K^+ and Ca^{2+} channels (VAN DONGEN et al. 1988; YATANI et al. 1988; BIRNBAUMER et al. 1990). Adding to the confusion with respect to what kinds of deductions about coupling can be made is the fact that some G proteins exist (for example G_{12} and G_{13}) for which no effector has yet been conclusively defined. It is also important to point out that, while PTX ablates communication between receptor and most members of the G_i family, it does not permit discrimination among members within the family.

III. Direct Measures of G Protein Activity

More direct measurements of G protein activation markedly improve the resolution of receptor–G protein coupling. Assays developed for this purpose utilize one of two highly conserved aspects of G protein α-subunit function: GDP–GTP exchange and GTP hydrolysis. Assays of nucleotide exchange generally entail measurements of [α-^{32}P]GTP azidoanalide or [^{35}S] guanosine 5'-3-O-(thio)triphosphate (GTPγS) binding promoted by agonists. The basis of these assays is the fact that the rate-limiting step in exchange of GDP for GTP (or related analogues) on the α subunit is the release of GDP, which an agonist-activated receptor promotes. This kind of assay measures the initial activation event. Assays of [γ-^{32}P]GTP hydrolysis are based on the same principle, i.e. agonists promote an exchange of GDP for GTP but represent estimates of steady-state changes in G protein activity. By assaying the function of G protein α subunits themselves rather than their downstream effects, issues of differential amplification of signal and feedback or cross-regulation of effector function are avoided.

Whereas assays of G protein α subunit activity provide a more direct measure of receptor–G protein coupling, they do not necessarily enable the determination of which G proteins are activated. As normally performed in tissue or mammalian cell lines, the presence of many different GPCRs and G proteins precludes identification of receptor–G protein pairs, i.e., there is no easy way to determine which G protein α subunits account for the increases in [^{35}S]GTPγs binding or released ^{32}P$_i$. To some extent, protein-reconstitution systems increase the power of these assays by providing an environment in which the identity of all the components are known. However, purifying functional GPCRs and G proteins is time consuming and sometimes difficult. Furthermore, reconstitution requires disruption of the membrane environment, which can alter the fidelity of coupling.

B. Insect Cell Expression Systems

Insect cells provide an appealing alternative to mammalian cells or reconstitution of purified proteins for probing specific receptor–G protein interactions. Baculoviruses, such as the commonly used *Autographa californica* nuclear polyhedrosis virus (AcNPV), are a group of viruses that selectively infect insect cells. In the laboratory, cell lines, such as Sf9 and Sf21 derived from *Spodoptera frugiperda* (fall army worm), and High Five or TN-5 cells derived from *Trichoplusia ni*, are used as hosts for baculoviruses. The utility of these nuclear polyhedrosis viruses derives largely from the nature of their infection cycle (MILLER 1988; O'REILLY et al. 1994). In the first day following infection of cells with virus, newly formed virus particles are released from cells into the medium by an exocytotic (or budding) process. In later stages of infection, virions are packaged in polyhedra, crystalline matrices made primarily of the protein polyhedrin, which protect the virions from environmental conditions following cell lysis. Polyhedra are not essential for infection or viral replication, so replacing the polyhedrin gene (which can account for 30%–50% of the total insect protein at cell lysis) with foreign genes results in the expression of foreign protein 24–72 h following infection.

I. Expression of Receptors in Insect Cell Lines

Whereas the high levels of expression often attained using the baculovirus/insect cell system can be attributed to the characteristics of the virus, other advantages are derived from the insect host cells themselves. Using baculovirus infection, a wide variety of proteins involved in signal transduction, including hormones, receptors, G proteins, and effectors, have been expressed in insect cells, and these proteins are generally functionally equivalent to those expressed in mammalian cells or in tissue (KLAIBER et al. 1990; MILLS et al. 1993; SUTKOWSKI et al. 1994; TAUSSIG et al. 1994; DONG et al. 1995). In contrast to prokaryotic cells, insect cells support proper folding and disulfide bond formation, oligomerization, and extensive co- and post-translational modification of proteins. Fatty-acid acylation of most proteins (including G proteins and receptors) in Sf9 cells is essentially identical to that observed in mammalian cells (MILLER 1988; LABRECQUE et al. 1992; MOUILLAC et al. 1992; LINDER et al. 1993; NG et al. 1993; BUTKERAIT et al. 1995; GRÜNEWALD et al. 1996a;

LINDORFER et al. 1996). However, although N- and C-linked glycosylations occur in Sf9 cells, they are less well conserved between insect and mammalian cells. The extent of glycosylation is sometimes less than that in mammalian cells (GEORGE et al. 1989; REILANDER et al. 1991), and the oligosaccharides attached in insect cells tend to be simpler than those in mammalian cells (REILANDER et al. 1991; KUSUI et al. 1995; GRÜNEWALD et al. 1996a). Thus, it is not uncommon for proteins expressed in Sf9 cells to have lower apparent molecular weights than those expressed in mammalian cells. While differences in glycosylation can alter protein function (for example, the alternately glycoslyated follicle stimulating hormone purified from Sf9 cells activates different signaling pathways than that derived from mammalian cells; AREY et al. 1997), consequences for receptor function have not generally been reported.

The high yield and extensive post-translational modification of mammalian proteins expressed in insect cells using recombinant baculoviruses has made the system a popular choice for protein purification. A wide variety of mammalian GPCRs have been successfully expressed in insect cells (Table 1). Membrane densities of recombinant receptors are generally high; 1–30 pmol/mg protein is common (Table 2), though both the baculovirus construct and conditions of infection can influence the degree of expression.

Table 1. G protein-coupled receptors that have been expressed in insect cells. Below is a comprehensive tabulation of mammalian receptors that have been expressed in insect cells. The reader is referred to the original reports for further details

Receptor	References
Acetylcholine M1	WONG et al. 1990; PARKER et al. 1991; RICHARDSON and HOSEY 1992; RINKEN et al. 1994; NAKAMURA et al. 1995; KUKKONEN et al. 1996
M2	PARKER et al. 1991; RICHARDSON and HOSEY 1992; KAMEYAMA et al. 1994; RINKEN et al. 1994; NAKAMURA et al. 1995; HAYASHI and HAGA 1996; HEITZ et al. 1997
M3	PARKER et al. 1991; DEBBURMAN et al. 1995; KUKKONEN et al. 1996; VASUDEVAN et al. 1991; RINKEN et al. 1994
M4	PARKER et al. 1991; RINKEN et al. 1994
M5	HU et al. 1994; RINKEN et al. 1994; KUKKONEN et al. 1996
Adenosine	
A_1	FIGLER et al. 1996; YASUDA et al. 1996
A_{2A}	ROBEVA et al. 1996
Adrenergic	
α_{2A}	JANSSON et al. 1995; NASMAN et al. 1997
α_{2B}	PEI et al. 1994; JANSSON et al. 1995; NASMAN et al. 1997
β_1	KLEYMANN et al. 1993; BARR et al. 1997
β_2	GEORGE et al. 1989; PARKER et al. 1991; REILANDER et al. 1991; CHIDIAC et al. 1996
Chemokine	
IL-8	MOEPPS et al. 1997
LESTR (CXCR4)	MOEPPS et al. 1997

Table 1. (Continued)

Receptor	References
Dopamine	
D_1	NG et al. 1994a, 1995; SUGAMORI et al. 1995
D_2	JAVITCH et al. 1994; NG et al. 1994b; BOUNDY et al. 1996; GRÜNEWALD et al. 1996a, 1996b
D_3	BOUNDY et al. 1993; MACH et al. 1993; PREGENZER et al. 1997
D_4	MILLS et al. 1993; BERNARD et al. 1994
Histamine	
H_1	HARTENECK et al. 1995; KÜHN et al. 1996; LEOPOLDT et al. 1997
H_2	HARTENECK et al. 1995; KÜHN et al. 1996; LEOPOLDT et al. 1997
Neurokinin	
NK_1	MAZINA et al. 1996; BARR et al. 1997
NK_2	AHARONY et al. 1993; ALBLAS et al. 1995
Substance P	KWATRA et al. 1993
Odorant	
OR5	RAMING et al. 1993
OR12	RAMING et al. 1993
Opiate	
δ	OBERMEIER et al. 1996; WEHMEYER and SCHULZ 1997
κ	OBERMEIER et al. 1996
μ	OBERMEIER et al. 1996
Serotonin	
5-HT_{1A}	MULHERON et al. 1994; PARKER et al. 1994; BUTKERAIT et al. 1995; NEBIGIL et al. 1995; BARR et al. 1997; CLAWGES et al. 1997
5-HT_{1B}	NG et al. 1993; PARKER et al. 1994; CLAWGES et al. 1997
5-HT_{1D}	PARKER et al. 1994; CLAWGES et al. 1997
5-HT_{1E}	PARKER et al. 1994; CLAWGES et al. 1997
5-HT_{2C}	LABRECQUE et al. 1995; HARTMAN and NORTHUP 1996
5-HT_7	OBOSI et al. 1997
5-HT_{dro1}	OBOSI et al. 1996
5-HT_{dro2B}	OBOSI et al. 1996
Other	
FSH-R	LIU et al. 1994
$mGluR_{1\alpha}$	ROSS et al. 1994; PICKERING et al. 1995
GRPr	KUSUI et al. 1995
LH/CG-R	NARAYAN et al. 1996
N-formyl-peptide-receptor	QUEHENBERGER et al. 1992
Neuropeptide Y Y1	MUNOZ et al. 1995
Thrombin	HARTENECK et al. 1995; Barr et al. 1997
Thromboxane A_2	HARTENECK et al. 1995
Tyramine	VANDEN BROECK et al. 1995

5-HT, 5-hydroxytryptamine; *FSH-R*, follicle-stimulating-hormone receptor; *GRPr*, gastrin-releasing-peptide receptor; *IL*, interleukin; *LESTR*, leukocyte-derived seven-transmembrane-domain receptor; *LH/CG-R*, leutenizing-hormone/chorionic gonadotropin receptor; *mGluR*, metabotropic glutamate receptor.

Table 2. Parameters of agonist binding for mammalian receptors expressed alone in insect cells. Both the B_{max} and the proportion of receptors exhibiting high-affinity binding of agonist were determined by the original authors. High-affinity binding of agonist was typically determined by saturation binding, competition against radiolabeled antagonist, and guanine-nucleotide sensitivity

Receptor	B_{max} (pmol/mg protein)	Reference
No detectable high-affinity binding of agonist		
M1ACh	0.03–0.2	KUKKONEN et al. 1996
M2ACh	20–30	PARKER et al. 1991
	Not given	RICHARDSON and HOSEY 1992
β-adrenergic	5	PARKER et al. 1991
D_{2L} dopamine	4–12	BOUNDY et al. 1996
(40 h post-infection)		
D_{2S} dopamine	4–8	BOUNDY et al. 1996
(40 h post-infection)		
D_{2S} dopamine	1×10^6 sites/cell	GRÜNEWALD et al. 1996b
D_{2S} dopamine	6	GRÜNEWALD et al. 1996a
D_3 dopamine	5–15	BOUNDY et al. 1993
GRPr	4	KUSUI et al. 1995
(96 h post-infection)		
LH/CG-R	4500 sites/cell	NARAYAN et al. 1996
NFPr	0.1–0.8	QUEHENBERGER et al. 1992
5-HT_{1A}	5–34	BUTKERAIT et al. 1995
	1–2	BARR and MANNING 1997
5-HT_{dro1}	60	OBOSI et al. 1996
Detectable high-affinity binding of agonist, but less than 20% binding of receptor		
M1ACh	5	PARKER et al. 1991
M3ACh	0.1	KUKKONEN et al. 1996
M5ACh	0.01–0.2	KUKKONEN et al. 1996
A_1 adenosine	2–5	FIGLER et al. 1996
A_{2A} adenosine	18.7	ROBEVA et al. 1996
D_2 dopamine	3–5	JAVITCH et al. 1994
δ Opiate	1.4	WEHMEYER and SCHULZ 1997
5-HT_{1A}	3	PARKER et al. 1994
	16	CLAWGES et al. 1997
5-HT_{1B}	8	CLAWGES et al. 1997
5-$HT_{1D\beta}$	1.3	PARKER et al. 1994
5-HT_{dro2B}	2.1	OBOSI et al. 1996
Detectable high-affinity binding of agonist involving more than 20% of receptor		
D_1 dopamine	<33	NG et al. 1994a
(24 h post-infection)	7.3	NG et al. 1995
D_{2L} dopamine	2.6	NG et al. 1994b
GRPr	6	KUSUI et al. 1995
(24 h post-infection)		
NK_2	0.8	AHARONY et al. 1993
5-HT_{1A}	0.15	MULHERON et al. 1994
5-HT_{1B}	1–5	NG et al. 1993
Substance P	Not given	KWATRA et al. 1993

Table 2. (Continued)

Receptor	B_{max} (pmol/mg protein)	Reference
Guanine nucleotide-sensitive binding observed, but not quantified		
A_1 adenosine	4	YASUDA et al. 1996
D_{2L} dopamine (18h post-infection)	<1	BOUNDY et al. 1996
D_{2S} dopamine (18h post-infection)	0.06–0.6	BOUNDY et al. 1996
D_4 dopamine	5	MILLS et al. 1993

5-HT, 5-hydroxytryptamine; *GRPr*, gastrin-releasing-peptide receptor; *LH/CG-R*, leutenizing-hormone/chorionic gonadotropin receptor; *mACh*, muscarinic acetylcholine; *NFPr*, N-formyl peptide receptor; *NK*, neurokinin.

Though insect cell lines differ slightly in the level and time course of receptor expression following infection, comparable results are obtained in Sf9, Sf21 and TN-5 cells (JAVITCH et al. 1994; GRÜNEWALD et al. 1996a; HEITZ et al. 1997). An important advantage of insect cells for purification or characterization of mammalian GPCRs is that no mammalian GPCR has yet been detected in uninfected insect cells. Therefore, it is possible to purify and study an expressed receptor without interference from endogenous receptors.

II. Interaction of Receptors with Endogenous G Proteins and Effectors

Following infection of insect cells with baculoviruses encoding GPCRs, ligands for these receptors can modulate endogenous effectors in a manner similar to that observed in mammalian cells or tissue. Agonists activate potassium channels (VASUDEVAN et al. 1992), stimulate PI turnover (LABRECQUE et al. 1995; OBOSI et al. 1996), regulate $[Ca^{2+}]_i$ (HU et al. 1994; KUSUI et al. 1995; KUKKONEN et al. 1996), and both inhibit (PARKER et al. 1994; JANSSON et al. 1995; BOUNDY et al. 1996; NASMAN et al. 1997) and stimulate (PARKER et al. 1991; MOUILLAC et al. 1992; JANSSON et al. 1995; NG et al. 1995; NARAYAN et al. 1996) adenylyl cyclase in cells expressing recombinant receptors, but not in uninfected Sf9 cells. Downstream consequences of effector regulation, including receptor desensitization and phosphorylation following prolonged administration of agonists, further support compatibility of mammalian GPCRs and insect signal-transduction pathways (RICHARDSON and HOSEY 1992; NG et al. 1994a; NEBIGIL et al. 1995). That functional communication between mammalian GPCRs and insect effectors can be observed is not overly surprising, given the existence of insect G proteins and the conservation of genes for G protein subunits in evolution (WILKIE and YOKOYAMA 1994).

The range of effectors modulated by mammalian GPCRs through G proteins endogenous to insect cells suggests that these cells express G proteins from several families. This has been confirmed by several laboratories. Western blots, [^{35}S]GTPγS binding, immunoprecipitation, and PTX- and cholera toxin-mediated ADP-ribosylation have indicated that uninfected insect cells express homologues for a variety of α subunits. Whereas α subunits that interact with antibodies toward $α_q$ and $α_s$ are nearly uniformly observed (KLEYMANN et al. 1993; BUTKERAIT et al. 1995; OBOSI et al. 1996; LEOPOLDT et al. 1997), detection of $α_{i/o}$ subunits has been more controversial (QUEHENBERGER et al. 1992; OBOSI et al. 1996). The balance of both functional (VASUDEVAN et al. 1992; PARKER et al. 1994; JANSSON et al. 1995) and immunological data (NG et al. 1993; MULHERON et al. 1994; BUTKERAIT et al. 1995; LEOPOLDT et al. 1997) suggest that Sf9 cells probably express a protein resembling $α_o$ more than $α_i$. The presence of $α_{12}$ in membranes of Sf9 cells has recently been reported (LEOPOLDT et al. 1997). Antibodies for mammalian β subunits also react with endogenous Sf9 cell proteins of the corresponding molecular weight (KLEYMANN et al. 1993; OBOSI et al. 1996).

That recombinant GPCRs can interact with several families of endogenous insect G proteins is supported by direct measurement of G protein activation in insect cells. MULHERON et al. (1994) demonstrated that agonists for the 5-hydroxytryptamine (5-HT)$_{1A}$ receptor introduced by infection into Sf9 cells stimulated the incorporation of [$α$-^{32}P]GTP azidoanilide into a G_o-like protein. In Sf9 cells expressing D_3 (PREGENZER et al. 1997) or D_4 (CHABERT et al. 1994) dopamine receptors, or the muscarinic acetylcholine m2 receptor (RICHARDSON and HOSEY 1992), agonists stimulate [^{35}S]GTPγS binding to endogenous G proteins. Activation of endogenous G proteins by agonists for recombinant receptors is not uniformly observed, however (HEITZ et al. 1995; GRÜNEWALD et al. 1996b; LEOPOLDT et al. 1997). One possible explanation for the difference is that the expression of endogenous G proteins decreases in a time-dependent manner during infection with baculoviruses, probably due to a subversion of the machinery for protein synthesis for viral processes (LEOPOLDT et al. 1997). Agonist-induced incorporation of [$α$-^{32}P]GTP azidoanilide into $α_q$ in H_1 histamine receptor-containing membranes and into $α_q$ and $α_s$ in H_2 histamine receptor-containing membranes could be detected 28h and 48h (but not 72h) post-infection. However, although the same selectivity of activation of endogenous $α_q$ and $α_s$ by H_1 and H_2 histamine receptors was observed in High Five insect cells (KÜHN et al. 1996), expression of $α_q$ and $α_s$ did not decrease following infection. Therefore, expression of endogenous G proteins might be differentially regulated in different cell lines or by different baculovirus constructs.

III. Quantitation of Coupling Using Radioligand Binding

Although mammalian receptors expressed under the control of baculovirus promoters can regulate G proteins and effectors endogenous to insect cells, the precise degree of coupling between the receptors and G proteins can be ascertained only by other means. One method is to determine the fraction of receptor that exhibits a high affinity for agonists. The ternary-complex model and variants thereof (Chap. 2.2) posit the existence of receptors in at least two, interconvertible conformations: R and R*. R* represents a conformation of the receptor that is better able to interact with a G protein. A G protein therefore deforms the equilibrium between R and R* toward the latter. R* also represents a conformation that is better able to interact with an agonist. The greater the degree of coupling between a population of receptors and G proteins, therefore, the greater the percentage of receptor that binds agonist with a relatively high affinity. Antagonists exhibit an equal affinity for both R and R*, and the binding of an antagonist to a receptor is, therefore, unaffected by a G protein. Experimental support for the two states is extensive. In tissue and isolated cell systems, disruption of the interaction between receptor and G protein eliminates or greatly reduces high-affinity binding of agonist. Disruption can be achieved by GTP or non-hydrolyzable analogs of this nucleotide, which promote the dissociation of the G protein α subunit from $\beta\gamma$ and receptor, or, if the receptor couples through PTX-sensitive G proteins, disruption can be achieved by PTX.

Binding studies in insect cells made to express the typically large amounts of mammalian receptors suggest that, at most, only a small proportion of a given receptor is coupled to endogenous G proteins (Table 2). Both the binding of an agonist to receptors exhibiting an apparently uniform low affinity, as determined in saturation and competition binding assays, and the insensitivity of the binding to guanine nucleotides are commonly observed. Thus, the modulation of effector activity discussed above would appear to occur through an often undetectable population of coupled receptor, no doubt through signal amplification so that the effects on second messengers can be measured even in cases where no high-affinity binding is apparent (PARKER et al. 1991; JANSSON et al. 1995; OBOSI et al. 1996). Reconstitution of purified receptors and G proteins has indicated that a higher G protein:receptor ratio is necessary to observe high-affinity agonist binding than to detect agonist-induced activation of G proteins (HAGA et al. 1989). Therefore, the extent of coupling, as defined by high-affinity binding of agonist, is probably limited (in the case of highly expressed receptors) by the relatively low levels of endogenous G proteins.

Not surprisingly, levels of receptor expression can profoundly influence the relative degree of coupling. In some instances, a high level of expression of receptor cannot be attained under any circumstances and, in these instances, the percentage of receptors exhibiting a high affinity for agonist is easily distinguished (MULHERON et al. 1994). Harvesting cells at relatively early time points following infection also permits visualization of coupled receptor. BOUNDY et al. (1996), working with dopamine D_{2L} and D_{2S} receptors, for example, found that shortening the interval between infection and harvest from 40h to 18h, and therefore lowering amounts of receptor from approximately 8pmol/mg membrane protein to less than 1pmol/mg membrane protein, dramatically increased the degree of coupling as defined by high-affinity, guanine-nucleotide-sensitive binding. Yet, although the great majority of data suggest limited coupling of expressed receptors with endogenous G proteins, a high degree of coupling has been noted even at high receptor densities in a few cases. The reasons for these exceptions are unclear.

On balance, the baculovirus/insect cell expression system appears to provide a reasonably defined environment for studying the properties of (mostly) uncoupled mammalian receptors. Receptors can usually be expressed at high levels, and little if any interference is posed by endogenous receptors or G proteins at the level of radioligand binding. Of practical interest, mammalian receptors introduced into insect cells undergo the kinds of co- and post-translational modifications necessary for proper targeting and insertion into the membrane. That receptors expressed at routinely high levels exhibit an almost uniformly uncoupled phenotype provides the basis for yet another experimental manipulation, the reconstitution of receptors with G protein subunits.

C. Reconstitution of Mammalian Receptors and G Proteins: Reconstituted Properties of Ligand Binding

The potential of a receptor to couple to a G protein is most often inferred from the deduced structure of the receptor, i.e., from the fact that it conforms to a seven-transmembrane-domain motif and from the nature of the effectors and/or second messengers regulated. Further studies may reveal that the receptor (or a subpopulation of receptor) binds agonists with high affinity, that this binding is sensitive to GTP or PTX, that agonists stimulate a membrane-associated GTPase, or that PTX inhibits the regulation by agonists of a particular effector. While these features of ligand binding or action are generally good indicators of coupling, they do not permit one to distinguish among families of G proteins regulated (high-affinity binding, GTP-sensitivity) or subtypes of the G_i family (PTX) when applied to mammalian cells. Insect cells afford the opportunity to perform more revealing kinds of experiments in that specific pairings of receptors and G proteins can be created based simply on the choice of recombinant baculoviruses. Thus, it is possible to express a particular receptor and permutation of G protein subunits and determine whether they communicate based on established indices of coupling. One convenient index, as described above, is high-affinity agonist binding.

High concentrations of appropriately modified mammalian G protein subunits can be expressed by insect cells with recombinant baculoviruses (GRABER et al. 1992a, 1992b; INIGUEZ-LLUHI et al. 1992; LABRECQUE et al. 1992; HEPLER et al. 1993; LINDER et al. 1993), and functional heterotrimeric G proteins can be produced by co-infecting with baculoviruses encoding α, β, and γ subunits (HEPLER et al. 1993; BUTKERAIT et al. 1995; KOZASA and GILMAN 1995). Co-infection of insect cells with receptor and combinations of G protein subunits culminates in a functional pairing, as described below. The advantages of insect cells are obvious. First, the pairing of receptors and G proteins is simple – the method for obtaining cells co-expressing the two kinds of proteins is the same as that for generating cells with receptor alone, and no purification of the two is necessary. Second, any combination of receptor and G protein subunits can be introduced. Third, the background of receptors and G proteins endogenous to insect cells is usually inconsequential. An additional benefit relative to in vitro forms of reconstitution is the fact that receptors and G protein subunits are properly inserted into the membrane so that their interaction can be studied in a native environment.

I. Influence of Heterotrimeric G Proteins on Binding of Agonists

BUTKERAIT et al. (1995) used the co-infection strategy to investigate the selectivity of coupling between the human 5-HT$_{1A}$ receptor and several families of G proteins. When Sf9 cells were infected with the 5-HT$_{1A}$ receptor alone (5–30 pmol receptor/mg membrane protein), membranes exhibited a single, low-affinity site for the agonist [^{3}H]8-hydroxy-N,N-dipropyl-2-aminotetralin (8-OH-DPAT; K_d = 7–20 nM). Binding was insensitive to GTP. The low affinity for agonist and insensitivity to GTP indicated an essentially uncoupled receptor. When both the 5-HT$_{1A}$ receptor and G$_{i1}$ (α_{i1}, β_1, and γ_2) were expressed, however, the characteristics of agonist binding changed dramatically. The affinity for [^{3}H]8-OH-DPAT increased (K_d = 1 nM) for a substantial proportion of the receptor, and the binding was sensitive to GTP. The K_d was in agreement with that observed in membranes from hippocampus and from mammalian cells in which the receptor was introduced by transfection. Similar results were achieved with [^{125}I]R-(+)-$trans$-8-hydroxy-2-[N-n-propyl-N-3'-iodo-2'-propenyl)amino]tetralin (8-OH-PIPAT) as the agonist. Binding of the antagonist [^{125}I]4-(2'-methoxyphenyl)-1-[2'(n-2''-pyridinyl)-p-iodobenzamido]ethyl-piperazine (MPPI), however, was unaffected by the presence or absence of G protein, consistent with the agonist-selective effects of G protein coupling.

Similar results were reported for rat D$_2$ dopamine receptors expressed in Sf9 cells. In membranes expressing D$_{2S}$ or D$_{2L}$ receptors alone (>1 pmol/mg membrane protein), agonist binding was not sensitive to GTP. When G$_{i1}$ was co-expressed with the receptors, however, affinity for agonist increased 20- and 90-fold, respectively, as determined by increases in binding of agonist at sub-K_d concentrations. The binding of antagonist was unaltered by co-infection with G protein (BOUNDY et al. 1996). Co-expressing G$_{i1}$ or G$_{i2}$ with the D$_{2S}$ receptor similarly promoted high-affinity, guanine-nucleotide-sensitive binding of agonist when assayed in intact Sf9 cells (GRÜNEWALD et al. 1996b).

Thus, the expression of a receptor with a G protein in Sf9 cells can result in a functional interaction manifested as a coupled phenotype indistinguishable from that observed in mammalian cells. By using different G proteins, the selectivity of interaction can be determined. BUTKERAIT et al. (1995) explored issues of selectivity for the 5-HT$_{1A}$ receptor by varying α subunits in combination with $\beta_1\gamma_2$. Co-expression of the receptor with any member of the G_i family, G_{i1}, G_{i2}, G_{i3}, G_o, or G_z, was associated with an increase in affinity for [^{3}H]8-OH-DPAT, as determined by increases in binding of the agonist at sub-K_d concentrations (Fig. 2). G_s and G_q did not increase affinity for the agonist, though [^{3}H]8-OH-DPAT binding was somewhat sensitive to α_q alone. These results both confirmed the predicted preference of the 5-HT$_{1A}$ receptor for members of the G_i family and indicated that selectivity of receptor–G protein interactions can be maintained in an Sf9 cell expression system.

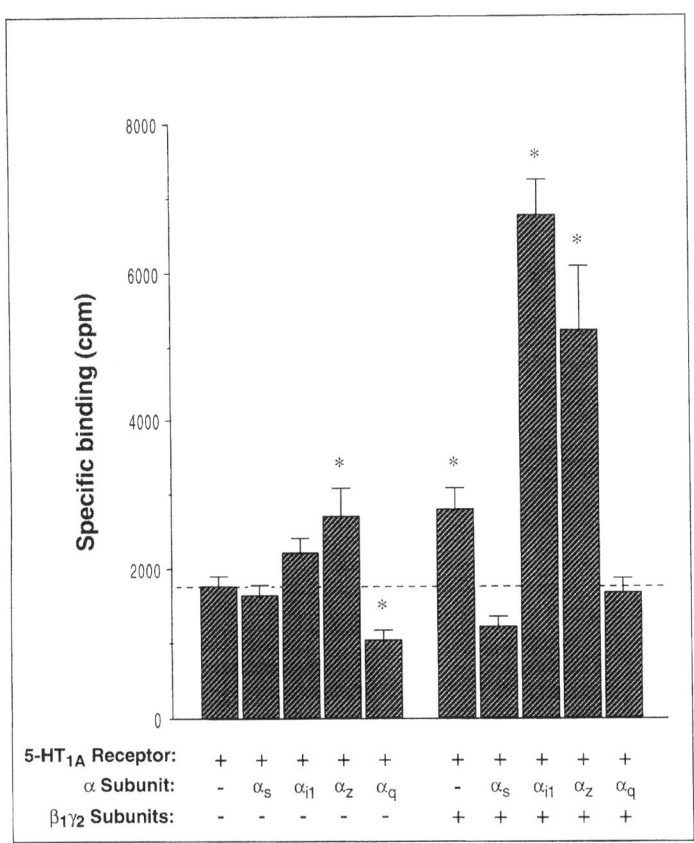

Fig. 2. [^{3}H]-8-hydroxy-N,N-dipropyl-2-aminotetralin (8-OH-DPAT) binding to the 5-hydroxytryptamine-1A (5-HT$_{1A}$) receptor co-expressed with G protein subunits in *Spodoptera frugiperda* (Sf9) cells. Specific binding of [^{3}H]-8-OH-DPAT at 0.5 nM was determined for membranes prepared from Sf9 cells expressing the 5-HT$_{1A}$ receptor, α subunits, and $\beta_1\gamma_2$ as indicated. Data represent the mean ± the standard error of the mean of 5–24 individual experiments assayed in triplicate. As a point of reference, the *dashed line* indicates [^{3}H]-8-OH-DPAT binding where the 5-HT$_{1A}$ receptor is expressed alone. Statistical significance (*$P < 0.01$) was determined using Student's t test (BUTKERAIT et al. 1995)

A permutation of the technique is to study the high-affinity binding of agonist upon reconstitution of receptor-containing Sf9 membranes with purified G proteins. This strategy has been used by several laboratories and lends itself to a quantitative analysis of coupling in relation to receptor:G protein stoichiometry. This kind of relationship cannot be studied simply in the co-infection technique, as receptor and G protein expression cannot be manipulated precisely. However, the technique of membrane reconstitution with purified G proteins is not simple either, as the G proteins must first be purified. This latter technique has been used to study the coupling of the 5-HT$_{1A}$, 5-HT$_{1B}$, and adenosine A$_1$ receptors to members of the G$_i$ family, G$_{i1}$, G$_{i2}$, G$_{i3}$ and G$_o$ (FIGLER et al. 1996; YASUDA et al. 1996; CLAWGES et al. 1997). Reconstituting Sf9 membranes expressing either the 5-HT$_{1A}$ or 5-HT$_{1B}$ receptors with heterotrimeric G$_{i/o}$ (but not G$_t$) increased the affinity of these receptors for [^{3}H]5-HT, as deduced by increases in binding of the agonist at sub-K_d concentrations. Only modest differences in efficacy among the different G proteins were noted in these studies, with a rank order of G$_{i3}$ > G$_{i1}$ > G$_o$ > G$_{i2}$. The A$_1$ adenosine receptor coupled with all of these G proteins essentially equivalently (FIGLER et al. 1996).

II. Influence of Individual G Protein Subunits on Binding of Agonists

The fact that each component of the receptor–G protein complex can be manipulated independently makes it possible to determine the influence of individual G protein subunits on coupling in insect cells. In the co-infection studies discussed above, membranes expressing the 5-HT$_{1A}$ receptor and α_i alone displayed an affinity for agonist that was intermediate between the affinity for receptor alone and that for receptor plus heterotrimeric G protein (BUTKERAIT et al. 1995). Similar results were observed when D$_{2L}$ or D$_{2S}$ receptors were co-expressed with α_{i1}. The affinity of the D$_{2L}$ or D$_{2S}$ dopamine receptors for agonist, as determined by binding of the agonist at sub-K_d concentrations, was increased 2–7-fold upon introduction of α_{i1}, which was approximately one tenth the value observed for $\alpha_{i1}\beta\gamma$. The small changes in affinity for agonists observed upon co-infection of receptor with α subunits alone were consistent with data indicating that α subunits can by themselves interact weakly with receptors (FUNG 1983; KELLEHER and JOHNSON 1988). It is also conceivable that the mammalian α subunits interact with $\beta\gamma$ endogenous to the insect cells.

Co-expression of receptors with $\beta_1\gamma_2$ alone also led to a modest increase in the affinity of receptors for agonists. Curiously, [^{3}H]8-OH-DPAT binding to the 5-HT$_{1A}$ receptor in membranes co-expressing $\beta_1\gamma_2$ was sensitive to guanylylimidodiphosphate. Quite possibly, the mammalian $\beta\gamma$ interacts to some extent with endogenous insect α subunits resembling those of the G$_i$ family to in turn support a modicum of high-affinity binding of agonist. One such α subunit might be the α_o-like protein that MULHERON et al. (1994) demonstrated could be activated by the 5-HT$_{1A}$ receptor. The superimposed expression of high levels of a mammalian α subunit, however, would negate the contributions of endogenous α subunits. The addition of purified α or $\beta\gamma$ subunits to Sf9 membranes expressing A_1 adenosine, 5-HT_{1A}, and 5-HT_{1B} receptors alone did not promote an increase in high-affinity binding of agonists (FIGLER et al. 1996; CLAWGES et al. 1997).

In experiments where different γ subunits were co-infected with α_{i1}, β_1, and the 5-HT_{1A} receptor, γ_1 was found to be significantly less effective in promoting high-affinity binding of an agonist than γ_2, γ_3, γ_5, and γ_7 were (BUTKERAIT et al. 1995). Similarly, a decreased efficacy of $\beta_1\gamma_1$ (compared with $\beta_1\gamma_2$, $\beta_1\gamma_3$, $\beta_2\gamma_2$, and $\beta_2\gamma_3$) was observed upon addition of the purified heterodimers and α_{i2} to Sf9 membranes containing the adenosine A_1 receptor (FIGLER et al. 1996). Additional studies indicated that the deficiency of γ_1 is related to the nature of prenylation (YASUDA et al. 1996). γ_1 is normally farnesylated, whereas the other subunits are geranylgeranylated. A mutation in γ_1 that allows the subunit to be geranylgeranylated greatly enhanced the ability of γ_1 to stabilize high-affinity binding of agonist. In contrast, a mutation of γ_2 causing it to become farnesylated instead of geranylgeranylated reduced its ability to promote high-affinity binding.

III. Characterization of Inverse Agonism

There is a great deal of interest in constitutive activity expressed by GPCRs (Chap. 2). Many receptors, particularly when overexpressed, exhibit a certain level of agonist-independent activation of G proteins and consequent regulation of effectors. This activity is due to the existence of a small amount of R*, which is present through the equilibrium established with R. Inverse agonists decrease constitutive activity, presumably by binding R in preference to R* and thereby drawing the equilibrium toward a form of receptor less able to interact with G proteins. The characterization of inverse agonists by binding has been slow, however, as it has been very difficult to label populations of R selectively in tissues or mammalian expression systems. In particular, the presence of G proteins in these environments precludes obtaining a uniform population of receptor, so the binding of an inverse agonist is complicated by a poorly defined mixture of R, R*, and R*G. However, since receptors expressed in insect cells, in sufficient density, exhibit an essentially uncoupled phenotype, they are ideally suited for the analysis of binding in which the equilibrium between R and R* is not distorted by the presence of a G protein. In studies of the 5-HT_{1A} receptor, the binding of agonists, neutral antagonists, and inverse agonists was explored using uniform populations of coupled and uncoupled receptors (BARR and MANNING 1997). Using low concentrations of a radiolabeled agonist highly selective for R* ([^{125}I]8-OH-PIPAT), the coupled form of receptor was selectively labeled in membranes co-expressing receptor and G protein. The uncoupled form of receptor was labeled in membranes expressing the receptor alone using what was proven in other experiments to be a nearly neutral antagonist ([^{125}I]MPPI). The binding of 5-HT (an agonist), MPPI (an almost neutral antagonist), and spiperone (which was demonstrated to be an inverse agonist) to the coupled and uncoupled forms of receptor was examined in competition binding assays. Whereas the ratio of affinities of serotonin for coupled and uncoupled receptor was approximately 100, that of MPPI was essentially one. These ratios are appropriate for agonists and antagonists, respectively. Of interest, spiperone displaced [^{125}I]MPPI from uncoupled receptors in a competitive manner but displaced [^{125}I]8-OH-PIPAT from coupled receptors in a non-competitive manner. The observation of a conditional non-competitive displacement is consistent with the predictions of WREGGETT and DE LEAN (1984).

D. Reconstitution of Mammalian Receptors and G Proteins: G Protein Activation

The binding of an agonist to a receptor with high affinity is one index of coupling to a G protein. A perhaps more pragmatic index is activation of the G protein, i.e. the ability of an agonist to promote binding of GTP to the α subunit, with consequent release of the subunit from $\beta\gamma$. The advantages that make the Sf9 cells suited to the analysis of coupling through changes in agonist binding apply to the analysis of coupling by G protein activation. These include an essentially null background on which mammalian receptors and G proteins can be expressed, the formation of contacts within a membrane milieu, and the ability to define precisely the proteins to be reconstituted.

Of the assays used to monitor activation of G proteins, that involving the binding of GTPγS has become the most popular. GTPγS is an analogue of GTP that binds G protein α subunits with high affinity, is not hydrolyzed by the α subunit, and can be obtained easily in radiolabeled form. An increasingly common type of assay used particularly with transfected mammalian cells entails the incubation of membranes from the cell of choice with [^{35}S]GTPγS in the presence or absence of ligand. If the conditions are appropriately chosen, the agonist promotes the release of GDP from the α subunit, whereupon [^{35}S]GTPγS binds in its place. The amount of bound [^{35}S]GTPγS can then be quantified by subjection of the membranes to rapid filtration to remove free radiolabels from bound radiolabels. The advantages of the assay are its relative simplicity and the direct, effector-independent measurement of an early event in G protein activation. Weaknesses, however, include an occasionally problematic signal and the inability to distinguish the G proteins activated. Co-expression of receptors and G proteins in insect cells has come to the fore in providing an especially strong and informative assay for G protein activation.

I. Activation Following Co-Expression of Receptor and G Protein

BARR et al. (1997) established an assay for the activation of G proteins co-expressed with various receptors in Sf9 cells. The receptor of interest was paired with a G protein from one of the four families, i.e. α_s, α_z, α_q, or α_{13}, together with $\beta_1\gamma_2$, all of which were introduced by infection with recombinant baculoviruses. Membranes were prepared and incubated with [^{35}S]GTPγS in the presence or absence of agonist 48h following infection. The membranes were then gently solubilized in non-ionic detergent, and the selected α subunit was immunoprecipitated. The amount of [^{35}S]GTPγS bound to the subunit was counted.

Of the four receptors tested (the β_1-adrenergic, 5-HT$_{1A}$, protease-activated 1 (thrombin), and neurokinin$_1$ (NK$_1$) receptors), the β_1-adrenergic receptor demonstrated the most selective coupling. Isoproterenol, working through the β_1-adrenergic receptor, activated G$_s$ alone (Fig. 3). Conversely, G$_s$ was activated only by the β_1-adrenergic receptor. Thus, the pairing between the β_1-adrenergic receptor and G$_s$ was exclusive. The 5-HT$_{1A}$ receptor, as predicted from studies in Sf9 membranes (MULHERON et al. 1994; BUTKERAIT et al. 1995;

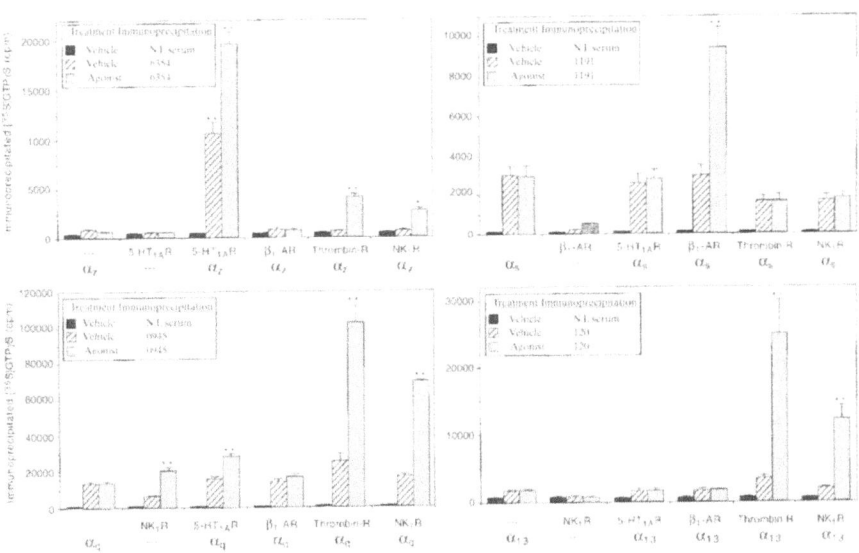

Fig. 3. Receptor-promoted binding of [^{35}S]guanosine 5'-3-O-(thio)triphosphate (GTPγS) to the α subunits of the G$_s$, G$_i$, G$_q$, and G$_{12}$ families. Membranes were prepared from *Spodoptera frugiperda* cells expressing recombinant receptors and G proteins (the α subunit along with $\beta_1\gamma_2$), as indicated. The membranes were incubated with [^{35}S]GTPγS ± agonist in the presence of 3mM free Mg^{2+}, and α subunits were immunoprecipitated with antisera directed against the indicated α subunits or with nonimmune serum. Agonists were serotonin (1μM) for the 5-HT$_{1A}$ receptor, isoproterenol (1μM) for the β_1-adrenergic receptor, the peptide SFLLRN (30μM) for the thrombin receptor, and [Sar^9, Met(O$_2$)^{11}]substance P for the NK$_1$ receptor. Vehicles were ascorbic acid (0.003%) for serotonin and isoproterenol and water for the other agonists. Statistically significant increases from values obtained with $\alpha\beta_1\gamma_2$ without receptor (*far left set of bars in each panel*) are noted (*$P < 0.05$; **$P < 0.01$). $n = 3$ (BARR et al. 1997)

CLAWGES et al. 1997) and its characterization in neuronal tissue and mammalian expression systems (INNIS and AGHAJANIAN 1987; ZGOMBICK et al. 1989), communicated preferentially with the G_i-family representative G_z, mediating 5-HT-promoted binding of [^{35}S]GTPγS to α_z. However, the coupling was not exclusive, as the receptors for thrombin and NK_1 also communicated with G_z to some extent. The latter two receptors nevertheless mediated a more robust activation of G_q. Furthermore, these two receptors alone activated G_{13}.

The one curious feature of these studies was the stimulation by the NK_1 receptor agonist [Sar^9, Met(O_2)^{11}] of [^{35}S]GTPγS binding to a small amount of an α_q-like protein endogenous to insect cells, i.e., a protein observed when the mammalian α_q is omitted from the infection protocol but the α_q-directed antibody is used. Quite possibly, the NK_1 receptor interacts with and activates the insect homologue of G_q, as has been described for other receptors when assayed by effector regulation. Interestingly, none of the other receptors examined activated endogenous insect G proteins discernible under the conditions of the assay despite the fact that immunoreactive insect homologues for α_s and α_{12} have been described. Perhaps these data suggest that α_q is especially highly expressed by Sf9 cells or that the NK_1 baculovirus does not efficiently suppress expression of insect cell proteins during infection.

The fact that endogenous α subunits can be activated by exogenous GPCRs under these conditions emphasizes the advantage of using immunoprecipitation rather than membrane filtration to isolate [^{35}S]GTPγS-bound α subunits. Moreover, the selectivity achieved with immunoprecipitation has a direct bearing on the signal relative to background. This is perhaps more relevant to assays applied to mammalian cells. When filtration is used to separate bound radiolabels from free radiolabels, the membranes trap all proteins that have bound [^{35}S]GTPγS, whether they are the G proteins of interest, other G proteins, low-molecular-weight G proteins, or other proteins altogether. The consequent "background" has a significant impact on the ability to detect an agonist-promoted event. This need not be a severe problem for Sf9 cells, where the G protein of interest is uniquely expressed at high levels. However, the immunoprecipitation step is an essential guarantor of specificity.

For example, G proteins such as G_z, G_{12}, and G_{13} have an intrinsically low rate of GDP/GTP exchange (KOZASA and GILMAN 1995) and, for this reason and perhaps others, they bind less [^{35}S]GTPγS over a period of time in response to an activated GPCR than other G proteins do. Other G proteins, moreover, bind a certain amount of [^{35}S]GTPγS in the absence of agonist/receptor. Given that an α_q-like protein endogenous to Sf9 cells can bind detectable amounts of [^{35}S]GTPγS at the prompting of an activated NK_1 receptor, measurements of α_{13} activation by that same receptor would be imperiled by the "background" contributions of the endogenous G protein in filtration assays. Subunit-specific immunoprecipitations obviate this concern.

Of interest, the co-expression of the 5-HT$_{1A}$ receptor with G_z caused a significant incorporation of [^{35}S]GTPγS into α_z even in the absence of agonist. This was observed for no other pairing of receptor and G protein. As α_z did

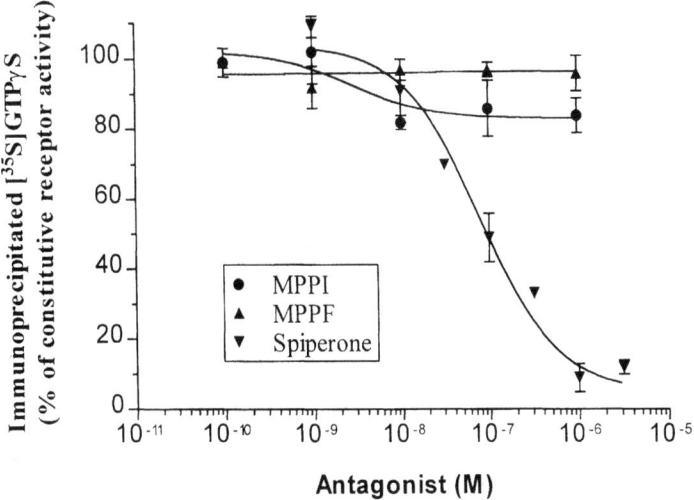

Fig. 4. Suppression of constitutive [^{35}S] guanosine 5'-3-O-(thio)triphosphate (GTPγS) binding by 5-hydroxytryptamine-1A (5-HT$_{1A}$)-receptor antagonists. Membranes were prepared from *Spodoptera frugiperda* (Sf9) cells expressing the 5-HT$_{1A}$ receptor and G$_z$ ($\alpha_z\beta_1\gamma_2$) and then incubated with the indicated concentrations of antagonists (with no agonist present) and [^{35}S]GTPγS as described for Fig. 3. Immunoprecipitation was performed with the α_z-directed antiserum 6354 or non-immune serum. Data are expressed as a percentage of [^{35}S]GTPγS binding (obtained without added antagonist) minus background, as defined by immunoprecipitation with non-immune serum. n = 5–6 (BARR and MANNING 1997)

not bind [^{35}S]GTPγS in the absence of receptor, the agonist-independent activity appeared to be a function of the receptor. That is, the 5-HT$_{1A}$ receptor was constitutively active. The significant agonist-independent activity of the 5-HT$_{1A}$ receptor provided an opportunity to test several 5-HT$_{1A}$ antagonists for inverse-agonist activity (BARR and MANNING 1997). In part, efforts to do so were achieved in the ligand-binding studies described above, but the more compelling demonstration was at the level of G protein activation itself. Of the three antagonists tested – spiperone, 4-(2'-methoxyphenyl)-1-1[2'(n-2''-pyridinyl)-p-fluorobenzamido]ethyl-piperazine (MPPF), and MPPI – spiperone inhibited agonist-independent binding of [^{35}S]GTPγS to α_z almost completely (Fig. 4). MPPI inhibited binding by 10%–20%, and MPPF had no effect. The inhibition of agonist-independent binding of [^{35}S]GTPγS by spiperone was completely antagonized by MPPI. Thus, under the conditions of the assay, spiperone appeared to be a full inverse agonist, MPPI at best a partial inverse agonist, and MPPF a neutral antagonist.

Once again, the signal-to-noise ratio gained by using immunoprecipitation to isolate the [^{35}S]GTPγS-bound α subunit was helpful. Since there was so little background, the wide range of efficacies could be clearly delineated. Agonist-independent activity of the 5-HT$_{1A}$ receptor and the inverse-agonist properties of spiperone have also been observed in transfected Chinese-hamster ovary cells, using [^{35}S]GTPγS binding to membrane as the endpoint (NEWMAN-TANCREDI et al. 1997). The fact that the constitutive activity of this receptor had not been noted using other measures of receptor function emphasizes the advantage of looking at early events in the receptor-signaling cascade. Finally, in the Sf9 cell [^{35}S]GTPγS binding assays, it is interesting to note that the constitutive activity of the 5-HT$_{1A}$ receptor is evident only for G$_z$ and not for G$_q$, although both G proteins are activated by the receptor in the presence of 5-HT. Whether this discrepancy reflects a difference in the strength of signal (5-HT$_{1A}$ activation of α_z exceeded that of α_q) or is instead a qualitative difference in the coupling of the receptor with these two G proteins remains to be determined. The ability to examine each receptor–G protein pair independently makes this system ideally suited for the investigation of these kinds of issues.

One promising use of the [^{35}S]GTPγS-binding assay is the determination of the coupling profile for newly identified GPCRs. In a recent study, two homologues of the human CXCR4 chemokine receptor were identified in mouse thymus, cloned, and inserted into baculoviruses, and their coupling to G$_{i2}$ in Sf9 cells was examined using a co-infection strategy (MOEPPS et al. 1997). A twofold increase in binding of [^{35}S]GTPγS was induced by the chemokine stromal-cell-derived factor 1α in Sf9 membranes co-expressing the homologue murine leukocyte-derived seven-transmembrane-domain receptor (mLESTR)-A or mLESTR-B and G$_{i2}$ (α_{i2}, β_1, and γ_3). Presumably, any receptor for which a ligand has been identified or any receptor that exhibits constitutive activity (or can be made to exhibit it by mutation) can be characterized according to the G proteins it activates.

II. Activation Following Addition of Purified G Protein to Membranes

Reconstitution of Sf9 cell membranes containing a mammalian receptor with a purified G protein provides another setting for evaluating the activation event. The coupling of the 5-HT$_{2C}$ receptor to G$_q$, an interaction long assumed but never directly demonstrated, was examined by HARTMAN and NORTHUP (1996). Membranes from Sf9 cells expressing the 5-HT$_{2C}$ receptor were extracted with urea to remove extrinsic membrane proteins and were reconstituted with purified squid-retinal α subunits and bovine-brain $\beta\gamma$. The use of a chaotropic agent and retinal α subunits avoided the use of detergents for reconstitution, thereby retaining a more natural membrane environment. Serotonin stimulated [^{35}S]GTPγS binding only in membranes reconstituted with α_q, indicating that the receptor was selective in its activation of an added G protein. Three 5-HT$_{2C}$ antagonists – mianserin, ketanserin and mesulergine – competitively inhibited the stimulation of [^{35}S]GTPγS binding induced by 5-HT. In the absence of agonist, mianserin and ketanserin are inverse agonists for the 5-HT$_{2C}$ receptor. Further studies demonstrated that mesulergine also inhibits agonist-independent activity for the 5-HT$_{2C}$ receptor, but only at receptor densities far above those required for detection of inverse-agonist activity for the other antagonists. These results are consistent with the demonstration of 5-HT$_{2C}$ constitutive activity in 293 cells (BARKER et al. 1994) and in whole Sf9 cells (LABRECQUE et al. 1995), where PI hydrolysis is the endpoint. The efficacies of these ligands as inverse agonists as determined by the [^{35}S]GTPγS assay exceeded those as determined by effector activity, perhaps reflecting differences in the endpoint measured, receptor density, coupling efficiency with insect vs squid α_q, or membrane treatment. Regardless, these data, along with those of other reconstitution studies (HEITZ et al. 1995, 1997; YASUDA et al. 1996), indicate that reconstitution of Sf9 membranes expressing receptors coupled to purified G proteins provides a useful system for studying the interactions of receptors with individual G proteins. Furthermore, with the results discussed above (BARR and MANNING 1997), these data suggest that G protein activity in Sf9 cells is an appealing model for detailed investigation of inverse agonism.

III. Limitations and Technical Considerations

The baculovirus/insect cell expression system is a promising model for studying the activation of G proteins by GPCRs. We view the two activation paradigms – co-expression of receptor and G protein versus addition of purified G protein to membranes expressing receptor alone – to be highly complementary. With co-expression, interactions of receptor and G protein occur in a native membrane environment, with the targeting of the G protein achieved at least partly through normal co- and post-translational modifications. Furthermore, no purification of G protein is required, since any combination of G protein α, β, and γ subunits can be introduced with recombinant baculoviruses. Reconstitution with purified G proteins generally requires disruption of the membrane environment; the purification of G proteins, while easier than it used to be, is still time-consuming and often difficult.

The primary advantage of techniques utilizing purified G proteins is the ability to control stoichiometry very precisely; the concentration of each component can be manipulated independently. Therefore, the potency and efficacy of each subunit in supporting high-affinity binding of agonist or agonist-induced activation of G protein can be accurately determined. This information can be especially helpful in exploring fine differences in the interactions of receptors with different G proteins. Control of stoichiometry in co-expression strategies is far more limited. Whereas the relative expression of multiple viruses can be grossly controlled by altering the multiplicity of infection (the number of infectious virus particles per cell), precise manipulations of expression are not possible. Fine quantitative comparisons across different receptor–G protein combinations on the basis of [^{35}S]GTPγS binding is almost impossible. Comparisons must be made using a relative scale, examining the increase over basal activity or using a standard condition as a normalization reference.

E. Conclusions

The insect cell expression system shows great promise for examining fundamental aspects of receptor–G protein communication. By expressing receptor alone or together with various G protein subunits, it is possible to define the interactions in an almost unambiguous fashion. Expression in insect cells can be used for characterizing new receptor ligands in terms of their selectivity for an array of receptors and in terms of their properties as agonists, antagonists, or inverse agonists. Efficacy with respect to one of the earliest events in signaling can be established. Particularly promising is the examination of coupling with G proteins whose lack of well-defined effector systems and slow GDP–GTP exchange rates complicate detection by other means. The expression system is also well suited to the study of novel GPCRs.

References

Aharony D, Little J, Powell S, Hopkins B, Bundell KR, McPheat WL, Gordon RD, Hassall G, Hockney R, Griffin R, Graham A (1993) Pharmacological characterization of cloned human NK-2 (neurokinin A) receptor expressed in a baculovirus/Sf-21 insect cell system. Mol Pharmacol 44:356–363

Alblas J, van Etten I, Khanum A, Moolenaar WH (1995) C-terminal truncation of the neurokinin-2 receptor causes enhanced and sustained agonist-induced signaling. J Biol Chem 270:8944–8951

Arey BJ, Stevis PE, Deecher DC, Shen ES, Frail DE, Negro-Vilar A, Lopez FJ (1997) Induction of promiscuous G protein coupling of the follicle-stimulating hormone (FSH) receptor: a novel mechanism for transducing pleiotropic actions of FSH isoforms. Mol Endocrinol 11:517–526

Axelrod J (1995) Phospholipase A_2 and G proteins. Trends Neurosci 18:64–65

Barker EL, Westphal RS, Schmidt D, Sanders-Bush E (1994) Constitutively active 5-hydroxytryptamine$_{2C}$ receptors reveal novel inverse agonist activity of receptor ligands. J Biol Chem 269:11887–11890

Barr AJ, Manning DR (1997) Agonist-independent activation of G_z by the 5-hydroxytryptamine$_{1A}$ receptor co-expressed in *Spodoptera frugiperda* cells: distinguishing inverse agonists from neutral antagonists. J Biol Chem 272:32979–32987

Barr AJ, Brass LF, Manning DR (1997) Reconstitution of receptors and GTP-binding regulatory proteins (G proteins) in Sf9 cells. J Biol Chem 272:2223–2229

Bernard AR, Kost TA, Overton L, Cavegn C, Young J, Bertrand M, Yahia-Cherif Z, Chabert C, Mills A (1994) Recombinant protein expression in a *Drosophila* cell line: comparison with the baculovirus system. Cytotechnology 15:139–144

Birnbaumer L, Abramowitz J, Yatani A, Okabe K, Mattera R, Graf R, Sanford J, Codina J, Brown AM (1990) Roles of G proteins in coupling of receptors to ionic channels and other effector systems. Crit Rev Biochem Molec Biol 25:225–244

Boundy VA, Luedtke RR, Gallitano AL, Smith JE, Filtz TM, Kallen RG, Molinoff PB (1993) Expression and characterization of the rat D_3 dopamine receptor: pharmacologic properties and development of antibodies. J Pharmacol Exp Ther 264:1002–1011

Boundy VA, Lu L, Molinoff PB (1996) Differential coupling of rat D_2 dopamine receptor isoforms expressed in *Spodoptera frugiperda* insect cells. J Pharmacol Exp Ther 276:784–794

Bourne HR, Sanders DA, McCormick F (1991) The GTPase superfamily: conserved structure and molecular mechanism. Nature 349:117–127

Butkerait P, Zheng Y, Hallak H, Graham TE, Miller HA, Burris KD, Molinoff PB, Manning DR (1995) Expression of the human 5-hydroxytryptamine$_{1A}$ receptor in Sf9 cells. J Biol Chem 270:18691–18699

Chabert C, Cavegn C, Bernard A, Mills A (1994) Characterization of the functional activity of dopamine ligands at human recombinant dopamine D$_4$ receptors. J Neurochem 63:62–65

Chidiac P, Nouet S, Bouvier M (1996) Agonist-induced modulation of inverse agonist efficacy at the β_2-adrenergic receptor. Mol Pharmacol 50:662–669

Conklin BR, Bourne HR (1993) Structural elements of Gα subunits that interact with G$\beta\gamma$, receptors and effectors. Cell 73:631–641

Conklin BR, Farfel Z, Lustig KD, Julius D, Bourne HR (1993) Substitution of three amino acids switches receptor specificity of G$_q\alpha$ to that of G$_i\alpha$. Nature 363:274–276

Clapham DE, Neer EJ (1997) G protein $\beta\gamma$ subunits. Annu Rev Pharmacol Toxicol 37:167–203

Clawges HM, Depree KM, Parker EM, Graber SG (1997) Human 5-HT$_1$ receptor subtypes exhibit distinct G protein coupling behaviors in membranes from Sf9 cells. Biochemistry 36:12930–12938

Debburman SK, Kunapuli P, Benovic JL, Hosey MM (1995) Agonist-dependent phosphorylation of human muscarinic receptors in *Spodoptera frugiperda* insect cell membranes by G protein-coupled receptor kinases. Mol Pharmacol 47:224–233

Dong GZ, Kameyama K, Rinken A, Haga T (1995) Ligand binding properties of muscarinic acetylcholine receptor subtypes (m$_1$–m$_5$) expressed in baculovirus-infected insect cells. J Pharmacol Exp Ther 274:378–384

Figler RA, Graber SG, Lindorfer MA, Yasuda H, Linden J, Garrison JC (1996) Reconstitution of recombinant bovine A$_1$ adenosine receptors in Sf9 cell membranes with recombinant G proteins of defined composition. Mol Pharmacol 50:1587–1595

Fung BK (1983) Characterization of transducin from bovine retinal rod outer segments. I. Separation and reconstitution of the subunits. J Biol Chem 258:10495–10502

Garnovskaya MN, Gettys TW, Van Biesen T, Prpic V, Chuprun JK, Raymond JR (1997) 5-HT$_{1A}$ receptor activates Na$^+$/H$^+$ exchange in CHO-K1 cells through G$_{i\alpha2}$ and G$_{i\alpha3}$. J Biol Chem 272:7770–7776

George ST, Arbabian MA, Ruoho AE, Kiely J, Malbon CC (1989) High-efficiency expression of mammalian β-adrenergic receptors in baculovirus-infected insect cells. Biochem Biophys Res Comm 163:1265–1269

Graber SG, Figler RA, Garrison JC (1992a) Expression and purification of functional G-protein α subunits using a baculovirus expression system. J Biol Chem 267:1271–1278

Graber SG, Figler RA, Kalman-Maltese VK, Robishaw JD, Garrison JC (1992b) Expression of functional G protein $\beta\gamma$ dimers of defined subunit composition using a baculovirus expression system. J Biol Chem 267:13123–13126

Grünewald S, Haase W, Reilander H, Michel H (1996a) Glycosylation, palmitoylation, and localization of the human D$_{2S}$ receptor in baculovirus-infected insect cells. Biochemistry 35:15149–15161

Grünewald S, Reilander H, Michel H (1996b) In vivo reconstitution of dopamine D$_{2S}$ receptor-mediated G protein activation in baculovirus-infected insect cells: preferred coupling to G$_{i1}$ versus G$_{i2}$. Biochemistry 35:15162–15173

Haga K, Uchiyama H, Haga T, Ichiyama A, Kanagawa K, Matuo H (1989) Cerebral muscarinic acetylcholine receptors interact with three kinds of GTP-binding proteins in a reconstitution system of purified components. Mol Pharmacol 35:286–294

Harteneck C, Obukhov AG, Zobel A, Kalkbrenner F, Schultz G (1995) The *Drosophila* cation channel *trpl* expressed in insect Sf9 cells is stimulated by agonists of G-protein-coupled receptors. FEBS Lett 358:297–300

Hartman JL, IV, Northup JK (1996) Functional reconstitution in situ of 5-hydroxytryptamine$_{2c}$ (5HT$_{2c}$) receptors with α_q and inverse agonism of 5HT$_{2c}$ receptor antagonists. J Biol Chem 271:22591–22597

Hayashi MK, Haga T (1996) Purification and functional reconstitution with GTP-binding regulatory proteins of hexahistidine-tagged muscarinic acetylcholine receptors (m2 subtype). J Biochem 120:1232–1238

Heitz F, McClue SJ, Harris BA, Guenet C (1995) Expression of human M2 muscarinic receptors in Sf9 cells: characterisation and reconstitution with G-proteins. J Recept Signal Transduct Res 15:55–70

Heitz F, Nay C, Guenet C (1997) Expression of functional human muscarinic M2 receptors in different insect cell lines. J Recept Signal Transduct Res 17:305–317

Hepler JR, Gilman AG (1992) G-proteins. Trends Biochem Sci 17:383–387

Hepler JR, Kozasa T, Smrcka AV, Simon MI, Rhee SG, Sternweis PC, Gilman AG (1993) Purification from Sf9 cells and characterization of recombinant G$_q$ α and G$_{11}$ α. J Biol Chem 268:14367–14375

Hu Y, Rajan L, Schilling WP (1994) Ca^{2+} signaling in Sf9 insect cells and the functional expression of a rat brain M$_5$ muscarinic receptor. Am J Physiol 266:C1736–1743

Hung DT, Wong YH, Vu TK, Coughlin SR (1992) The cloned platelet thrombin receptor couples to at least two distinct effectors to stimulate phosphoinositide hydrolysis and inhibit adenylyl cyclase. J Biol Chem 267:20831–20834

Iniguez-Lluhi JA, Simon MI, Robishaw JD, Gilman AG (1992) G-protein $\beta\gamma$ subunits synthesized in Sf9 cells. J Biol Chem 267:23409–23417

Innis RB, Aghajanian GK (1987) Pertussis toxin blocks 5-HT$_{1A}$ and GABA$_B$ receptor-mediated inhibition of serotonergic neurons. Eur J Pharmacol 143:195–204

Jansson CC, Karp M, Oker-Blom C, Nasman J, Savola JM, Akerman KE (1995) Two human α_2-adrenoceptor subtypes α_2A-C10 and α_2B-C2 expressed in Sf9 cells couple to transduction pathway resulting in opposite effects on cAMP production. Eur J Pharmacol 290:75–83

Javitch JA, Kaback J, Li X, Karlin A (1994) Expression and characterization of human dopamine D$_2$ receptor in baculovirus-infected insect cells. J Recept Res 14:99–117

Kameyama K, Haga K, Haga T, Moro O, Sadee W (1994) Activation of a GTP-binding protein and a GTP-binding-protein-coupled receptor kinase (β-adrenergic-receptor kinase-1) by a muscarinic receptor m2 mutant lacking phosphorylation sites. Eur J Biochem 226:267–276

Kelleher DJ, Johnson GL (1988) Transducin inhibition of light-dependent rhodopsin phosphorylation: evidence for $\beta\gamma$ subunit interaction with rhodopsin. Mol Pharmacol 34:452–460

Klaiber K, Williams N, Roberts TM, Papazian DM, Jan LY, Miller C (1990) Functional expression of Shaker K$^+$ channels in a baculovirus-infected insect cell line. Neuron 5:221–226

Kleymann G, Boege F, Hahn M, Hampe W, Vasudevan S, Reilander H (1993) Human β_2-adrenergic receptor produced in stably transformed insect cells is functionally coupled via endogenous GTP-binding protein to adenylyl cyclase. Eur J Biochem 213:797–804

Kozasa T, Gilman AG (1995) Purification of recombinant G-proteins from Sf9 cells by hexahistidine tagging of associated subunits. J Biol Chem 270:1734–1741

Kühn B, Schmid A, Harteneck C, Gudermann T, Schultz G (1996) G-proteins of the G$_q$ family couple the H$_2$ histamine receptor to phospholipase C. Mol Endocrinol 10:1697–1707

Kukkonen JP, Nasman J, Ojala P, Oker-Blom C, Akerman KE (1996) Functional properties of muscarinic receptor subtypes Hm1, Hm3 and Hm5 expressed in Sf9 cells using the baculovirus expression system. J Pharmacol Exp Ther 279:593–601

Kusui T, Hellmich MR, Wang LH, Evans RL, Benya RV, Battey JF, Jensen RT (1995) Characterization of gastrin-releasing peptide receptor expressed in Sf9 insect cells by baculovirus. Biochemistry 34:8061–8075

Kwatra MM, Schwinn DA, Schreurs J, Blank JL, Kim CM, Benovic JL, Krause JE, Caron MG, Lefkowitz RJ (1993) The substance P receptor, which couples to $G_{q/11}$, is a substrate of β-adrenergic receptor kinase 1 and 2. J Biol Chem 268:9161–9164

Labrecque J, Caron M, Torossian K, Plamondon J, Dennis M (1992) Baculovirus expression of mammalian G-protein α subunits. FEBS Lett 304:157–162

Labrecque J, Fargin A, Bouvier M, Chidiac P, Dennis M (1995) Serotonergic antagonists differentially inhibit spontaneous activity and decrease ligand binding capacity of the rat 5-hydroxytryptamine type-2C receptor in Sf9 cells. Mol Pharmacol 48:150–159

Laugwitz K-L, Allgeier A, Offermanns S, Spicher K, Van Sande J, Dumont JE, Schultz G (1996) The human thyrotropin receptor: a heptahelical receptor capable of stimulating members of all four G-protein families. Proc Nat Acad Sci USA 93:116–120

Leopoldt D, Harteneck C, Nürnberg B (1997) G-proteins endogenously expressed in Sf9 cells: interactions with mammalian histamine receptors. Naunyn Schmiedebergs Arch Pharmacol 356:216–224

Linder ME, Middleton P, Hepler JR, Taussig R, Gilman AG, Mumby SM (1993) Lipid modifications of G-proteins: α subunits are palmitoylated. Proc Nat Acad Sci USA 90:3675–3679

Lindorfer MA, Sherman NE, Woodfork KA, Fletcher JE, Hunt DF, Garrison JC (1996) G-protein γ subunits with altered prenylation sequences are properly modified when expressed in Sf9 cells. J Biol Chem 271:18582–18587

Liu X, DePasquale JA, Griswold MD, Dias JA (1994) Accessibility of rat and human follitropin receptor primary sequence (R265-S296) in situ. Endocrinology 135:682–691

Mach RH, Luedtke RR, Unsworth CD, Boundy VA, Nowak PA, Scripko JG, Elder ST, Jackson JR, Hoffman PL, Evora PH, Rao AV, Molinoff PB, Childers SR, Ehrenkaufer RL (1993) ^{18}F-labeled benzamides for studying the dopamine D_2 receptor with positron emission tomography. J Med Chem 36:3707–3720

Mazina KE, Strader CD, Tota MR, Daniel S, Fong TM (1996) Purification and reconstitution of a recombinant human neurokinin-1 receptor. J Recept Signal Transduct Res 16:191–207

Miller LK (1988) Baculoviruses as gene expression vectors. Annu Rev Microbiol 42:177–199

Mills A, Allet B, Bernard A, Chabert C, Brandt E, Cavegn C, Chollet A, Kawashima E (1993) Expression and characterization of human D_4 dopamine receptors in baculovirus-infected insect cells. FEBS Lett 320:130–134

Moepps B, Frodl R, Rodewald HR, Baggiolini M, Gierschik P (1997) Two murine homologues of the human chemokine receptor CXCR4 mediating stromal cell-derived factor 1α activation of G_{i2} are differentially expressed in vivo. Eur J Immunol 27:2102–2112

Morris AJ, Scarlata S (1997) Regulation of effectors by G-protein α- and $\beta\gamma$-subunits. Recent insights from studies of the phospholipase c-β isoenzymes. Biochem Pharmacol 54:429–435

Mouillac B, Caron M, Bonin H, Dennis M, Bouvier M (1992) Agonist-modulated palmitoylation of β_2-adrenergic receptor in Sf9 cells. J Biol Chem 267:21733–21737

Mulheron JG, Casañas SJ, Arthur JM, Garnovskaya MN, Gettys TW, Raymond JR (1994) Human 5-HT_{1A} receptor expressed in insect cells activates endogenous G_o-like G-protein(s). J Biol Chem 269:12954–12962

Munoz M, Sautel M, Martinez R, Sheikh SP, Walker P (1995) Characterization of the human Y1 neuropeptide Y receptor expressed in insect cells. Mol Cell Endocrin 107:77–86

Nakamura F, Kato M, Kameyama K, Nukada T, Haga T, Kato H, Takenawa T, Kikkawa U (1995) Characterization of G_q family G-proteins $G_{L1}\alpha$ ($G_{14}\alpha$), $G_{L2}\alpha$ ($G_{11}\alpha$), and $G_q\alpha$ expressed in the baculovirus-insect cell system. J Biol Chem 270:6246–6253

Narayan P, Gray J, Puett D (1996) Expression of functional lutropin/choriogonadotropin receptor in the baculovirus system. Mol Cell Endocrin 117:95–100

Nasman J, Jansson CC, Akerman KE (1997) The second intracellular loop of the α_2-adrenergic receptors determines subtype-specific coupling to cAMP production. J Biol Chem 272:9703–9708

Nebigil CG, Garnovskaya MN, Casañas SJ, Mulheron JG, Parker EM, Gettys TW, Raymond JR (1995) Agonist-induced desensitization and phosphorylation of human 5-HT$_{1A}$ receptor expressed in Sf9 insect cells. Biochemistry 34:11954–11962

Newman-Tancredi A, Conte C, Chaput C, Verriele L, Millan MJ (1997) Agonist and inverse agonist efficacy at human recombinant serotonin 5-HT1A receptors as a function of receptor:G-protein stoichiometry. Neuropharmacology 36:451–459

Ng GY, George SR, Zastawny RL, Caron M, Bouvier M, Dennis M, O'Dowd BF (1993) Human serotonin$_{1B}$ receptor expression in Sf9 cells: phosphorylation, palmitoylation, and adenylyl cyclase inhibition. Biochemistry 32:11727–11733

Ng GY, Mouillac B, George SR, Caron M, Dennis M, Bouvier M, O'Dowd BF (1994a) Desensitization, phosphorylation and palmitoylation of the human dopamine D$_1$ receptor. Eur J Pharmacol 267:7–19

Ng GY, O'Dowd BF, Caron M, Dennis M, Brann MR, George SR (1994b) Phosphorylation and palmitoylation of the human D2$_L$ dopamine receptor in Sf9 cells. J Neurochem 63:1589–1595

Ng GY, Trogadis J, Stevens J, Bouvier M, O'Dowd BF, George SR (1995) Agonist-induced desensitization of dopamine D$_1$ receptor-stimulated adenylyl cyclase activity is temporally and biochemically separated from D$_1$ receptor internalization. Proc Nat Acad Sci USA 92:10157–10161

O'Reilly D, Miller LK, Luckow VA (1994) Baculovirus expression vectors: a laboratory manual. Oxford University Press, Oxford

Obermeier H, Wehmeyer A, Schulz R (1996) Expression of μ-, δ- and κ-opioid receptors in baculovirus-infected insect cells. Eur J Pharmacol 318:161–166

Obosi LA, Hen R, Beadle DJ, Bermudez I, King LA (1997) Mutational analysis of the mouse 5-HT$_7$ receptor: importance of the third intracellular loop for receptor-G-protein interaction. FEBS Lett 412:321–324

Obosi LA, Schuette DG, Europe-Finner GN, Beadle DJ, Hen R, King LA, Bermudez I (1996) Functional characterisation of the Drosophila 5-HT$_{dro1}$ and 5-HT$_{dro2B}$ serotonin receptors in insect cells: activation of a Gα_s-like protein by 5-HT$_{dro1}$ but lack of coupling to inhibitory G-proteins by 5-HT$_{dro2B}$. FEBS Lett 381:233–236

Offermanns S, Laugwitz K-L, Spicher K, Schultz G (1994) G-proteins of the G$_{12}$ family are activated via thromboxane A$_2$ and thrombin receptors in human platelets. Proc Nat Acad Sci USA 91:504–508

Parker EM, Grisel DA, Iben LG, Nowak HP, Mahle CD, Yocca FD, Gaughan GT (1994) Characterization of human 5-HT$_1$ receptors expressed in Sf9 insect cells. Eur J Pharmacol 268:43–53

Parker EM, Kameyama K, Higashijima T, Ross EM (1991) Reconstitutively active G-protein-coupled receptors purified from baculovirus-infected insect cells. J Biol Chem 266:519–527

Pei G, Tiberi M, Caron MG, Lefkowitz RJ (1994) An approach to the study of G-protein-coupled receptor kinases: an in vitro-purified membrane assay reveals differential receptor specificity and regulation by G $\beta\gamma$ subunits. Proc Nat Acad Sci USA 91:3633–3636

Pickering DS, Taverna FA, Salter MW, Hampson DR (1995) Palmitoylation of the GluR6 kainate receptor. Proc Nat Acad Sci USA 92:12090–12094

Pregenzer JF, Alberts GL, Im WB (1997) Agonist-induced [^{35}S]GTPγs binding in the membranes of *Spodoptera frugiperda* insect cells expressing the human D$_3$ dopamine receptor. Neurosci Lett 226:91–94

Quehenberger O, Prossnitz ER, Cochrane CG, Ye RD (1992) Absence of G$_i$ proteins in the Sf9 insect cell. J Biol Chem 267:19757–19760

Raming K, Krieger J, Strotmann J, Boekhoff I, Kubick S, Baumstark C, Breer H (1993) Cloning and expression of odorant receptors. Nature 361:353–356

Reilander H, Boege F, Vasudevan S, Maul G, Hekman M, Dees C, Hampe W, Helmreich EJ, Michel H (1991) Purification and functional characterization of the human β_2-adrenergic receptor produced in baculovirus-infected insect cells. FEBS Lett 282:441–444

Richardson RM, Hosey MM (1992) Agonist-induced phosphorylation and desensitization of human m2 muscarinic cholinergic receptors in Sf9 insect cells. J Biol Chem 267:22249–22255

Rinken A, Kameyama K, Haga T, Engstrom L (1994) Solubilization of muscarinic receptor subtypes from baculovirus-infected Sf9 insect cells. Biochem Pharmacol 48:1245–1251

Robeva AS, Woodard R, Luthin DR, Taylor HE, Linden J (1996) Double tagging recombinant A_1- and A_{2A}-adenosine receptors with hexahistidine and the FLAG epitope. Biochem Pharmacol 51:545–555

Ross SM, Taverna FA, Pickering DS, Wang LY, MacDonald JF, Pennefather PS, Hampson DR (1994) Expression of functional metabotropic and ionotropic glutamate receptors in baculovirus-infected insect cells. Neurosci Lett 173:139–142

Sternweis PC, Smrcka AV (1992) Regulation of phospholipase C by G-proteins. Trends Biochem Sci 17:502–506

Strathmann MP, Simon MI (1991) Gα12 and Gα13 subunits define a fourth class of G-protein a subunits. Proc Natl Acad Sci USA 88:5582–5586

Sugamori KS, Demchyshyn LL, McConkey F, Forte MA, Niznik HB (1995) A primordial dopamine D_1-like adenylyl cyclase-linked receptor from *Drosophila melanogaster* displaying poor affinity for benzazepines. FEBS Lett 362:131–138

Sutkowski EM, Tang WJ, Broome CW, Robbins JD, Seamon KB (1994) Regulation of forskolin interactions with type I, II, V, and VI adenylyl cyclases by $G_s\alpha$. Biochemistry 33:12852–12859

Tang WJ, Gilman AG (1992) Adenylyl cyclases. Cell 70:869–872

Taussig R, Tang WJ, Hepler JR, Gilman AG (1994) Distinct patterns of bidirectional regulation of mammalian adenylyl cyclases. J Biol Chem 269:6093–6100

Vanden Broeck J, Vulsteke V, Huybrechts R, De Loof A (1995) Characterization of a cloned locust tyramine receptor cDNA by functional expression in permanently transformed *Drosophila* S2 cells. J Neurochem 64:2387–2395

Van Dongen AM, Codina J, Olate J, Mattera R, Joho R, Birnbaumer L, Brown AM (1988) Newly identified brain potassium channels gated by the guanine nucleotide binding protein G_o. Science 242:1433–1437

Vasudevan S, Reilander H, Maul G, Michel H (1991) Expression and cell membrane localization of rat M_3 muscarinic acetylcholine receptor produced in Sf9 insect cells using the baculovirus system. FEBS Lett 283:52–56

Vasudevan S, Premkumar L, Stowe S, Gage PW, Reilander H, Chung SH (1992) Muscarinic acetylcholine receptor produced in recombinant baculovirus-infected Sf9 insect cells couples with endogenous G-proteins to activate ion channels. FEBS Lett 311:7–11

Wehmeyer A, Schulz R (1997) Overexpression of δ-opioid receptors in recombinant baculovirus-infected *Trichoplusiani* "High 5" insect cells. J Neurochem 68:1361–1371

Wilkie TM, Yokoyama S (1994) Evolution of the G-protein alpha subunit multigene family. Soc Gen Physiol Ser 49:249–270

Wong SK, Parker EM, Ross EM (1990) Chimeric muscarinic cholinergic: β-adrenergic receptors that activate G_s in response to muscarinic agonists. J Biol Chem 265:6219–6224

Wreggett KA, De Lean A (1984) The ternary complex model. Its properties and application to ligand interactions with the D2-dopamine receptor of the anterior pituitary gland. Mol Pharmacol 26:214–227

Yasuda H, Lindorfer MA, Woodfork KA, Fletcher JE, Garrison JC (1996) Role of the prenyl group on the G-protein γ subunit in coupling trimeric G-proteins to A1 adenosine receptors. J Biol Chem 271:18588–18595

Yatani A, Mattera R, Codina J, Graf R, Okabe K, Padrell E, Iyengar R, Brown AM, Birnbaumer L (1988) The G-protein gated atrial K$^+$ channel is stimulated by three distinct G$_i\alpha$ subunits. Nature 336:680–682

Zgombick JM, Beck SG, Mahle CD, Craddock-Royal B, Maayani S (1989) Pertussis-toxin-sensitive guanine nucleotide-binding protein(s) couple adenosine A$_1$ and 5-hydroxytryptamine$_{1A}$ receptors to the same effector systems in rat hippocampus: biochemical and electrophysiological studies. Mol Pharmacol 35:484–494

CHAPTER 13
Altering the Relative Stoichiometry of Receptors, G Proteins and Effectors: Effects on Agonist Function

G. MILLIGAN

A. Introduction

I. Background

Signal transduction mediated by guanine nucleotide-binding protein (G protein)-coupled mechanisms requires, at a minimum, the contributions of a seven-transmembrane element G protein-coupled receptor (GPCR), the subunits of a heterotrimeric G protein and a G protein-regulated effector, which may be either (1) an enzyme involved in controlling the rate of production or degradation of an intracellular second messenger or (2) an ion channel. Activity of this minimal functional element is regulated directly by the efficacy of ligands that bind to the GPCR at sites overlapping that identified by the natural ligand. Moreover, interaction with agents at other sites on the GPCR can allow allosteric modification of the functions of agonists and antagonists. Activated receptors increase the fraction of the time the G protein spends in the guanosine triphosphate- (GTP; active) compared with the guanosine diphosphate-bound (inactive) state; thus, they control the temporal framework of effector regulation. In native states, regulation of the functional output of the transmembrane signaling cascade will also be controlled by the levels of expression and functional status of a wide range of other proteins that regulate the temporal frame of the activation process. These include regulators of G protein signaling proteins, which decrease the time frame of activation of the G protein and certain kinases and arrestins that function to interfere with productive interactions between GPCRs and G proteins. Although consideration of the contribution of these proteins is absolutely vital to an integrative understanding of intact cell function, relatively little information is currently available on quantitative aspects of modulation of their levels and the control of cellular signaling. As such, the current discussion will focus on how alterations in the relative levels and stoichiometry of each of the minimum elements of the G protein-mediated cascades may regulate the efficiency of signal-transduction events. Simple models based on the concept of "spare receptors" and the location and maximal effect of agonist concentration–response curves provide test systems to assess the usefulness of such ideas in pharmacology.

II. Systems to Modulate GPCR–G Protein–Effector Stoichiometries

Despite intense investigation of the individual components of such G protein-coupled transduction cascades and the basic features of cellular response to the presence of receptor ligands, surprisingly little is actually known about the levels of expression of these proteins in particular cells. Certain pathophysiological conditions are known to regulate levels of expression of GPCRs and G proteins. However, uncertainty about the overall cellular patterns of protein expression associated with disease has resulted in the fact that most studies designed to understand how alterations in the stoichiometry of GPCRs, G proteins and effector enzymes may modify cellular function have been conducted in transfected cell lines or in transgenic animals (AKHTER et al. 1997a).

A range of strategies, some of them introduced relatively recently, are available for cell transfection studies. Transient expression studies are the most popular because of the relatively short time scale required to perform the experiments and the ease of expressing a range of amounts of the protein of interest by varying the amounts of complementary DNA (cDNA) or DNA in the transfections (WISE et al. 1997a). Although useful, such studies are often used primarily to explore potential GPCR–G protein–effector interactions in cells expressing high levels of the signaling proteins. Generation and isolation of a range of individual cell lines derived from the same primary transfection and expressing the protein(s) of interest at a range of levels (LEVY et al. 1993) is more attractive. However, these techniques may be unusable without the ability to identify clones with the hoped-for range of expression levels. This problem may be alleviated by combinations of high-level stable expression and the subsequent use of irreversible (or slowly dissociating) receptor antagonists (MACEWAN and MILLIGAN 1995; VOGEL et al. 1995) to artificially control access of agonist to the expressed GPCR population. The same issues are also relevant to the generation and use of transgenic animals for such studies (MILANO et al. 1994a; BOND et al. 1995). As such, at least for analysis of expression levels and signaling functions of GPCRs, the recent availability of cell lines that induce expression of stably transfected cDNA offers an important new advance (THEROUX et al. 1996; KRUMINS et al. 1997). Theoretically, inducible expression systems based on metallothionein and mammalian steroid promoters have been available for some time. However, in many cases, these allowed a significant degree of expression in the absence of the inducer. More recent systems based on either insect steroid promoters (Biorad) or tetracycline repressors (Clontech) may overcome this problem.

Conceptually, antisense strategies also offer a specifically targeted means to control levels of expression of signaling molecules. This has indeed been of considerable use, particularly (as with transgenic knockout experiments; ROHRER et al. 1996; RUDOLPH et al. 1996; SUSULIC et al. 1996; CAVALLI et al. 1997; OFFERMANS et al. 1997) in studies designed to attribute specific functions to closely related G protein-coupled signaling polypeptides (KLEUSS et al. 1991, 1992, 1993; MOXHAM et al. 1993; MOXHAM and MALBON 1996; CHEN et al.

1997). However, although inducible antisense expression may provide a mechanism, it has yet to find widespread use in the types of quantitative pharmacological evaluation that are the topic of the current chapter.

III. Cellular Distribution of Elements of G Protein-Coupled Signaling Cascades

Although relatively little is known about the absolute levels of expression of the components of G protein-coupled signaling cascades in most cells and tissues, even less is known about their subcellular distribution. Differences in the cellular distribution of the individual polypeptides, potential roles of cytoskeletal elements in limiting free access of the expressed polypeptides to one another and the possibility that specific heterotrimeric $\alpha\beta\gamma$ configurations of G proteins are able to interact selectively with different receptors and effectors add to the complexity of the pattern. Although little is known about the cellular disposition of signaling polypeptides in relation to one another, it is now clear that simple views of their homogeneous distribution at the plasma membrane are incorrect. This situation is likely to be altered rapidly in the coming years as intense effort involving combinations of molecular and cell biology is expended. This has already started to define the spatial localization of systems designed to generate, respond to and destroy cyclic adenosine monophosphate (cAMP; HOUSLAY and MILLIGAN 1997).

The view that the components of G protein-coupled cascades are entirely mobile in the membrane and that signaling efficiency is simply a reflection of the relative concentrations and affinities of interactions between these polypeptides is clearly inadequate to explain a number of observations, including experiments which show that there is a lack of cross-interaction between receptors that apparently couple to the same G protein in NG108-15 cells (GRAESNER and NEUBIG 1993). This could reflect limitations in the lateral mobility of part of the cellular G protein population due either to interactions of these proteins with components of the cytoskeleton or because they are targeted to specific sections of the plasma membrane. It is an inherent necessity that GPCRs and G proteins reside at the plasma membrane for at least a significant part of the time. Recent studies have begun to unravel aspects of how they are targeted there and how they may move within subdomains of the plasma membrane in response to agonist ligands. A series of fascinating studies on the delivery of GPCRs to the plasma membrane have been performed on polarized canine kidney cells. Within the three highly homologous α_2-adrenoceptors, both the α_{2A}-adrenoceptor and the α_{2B}-adrenoceptor are targeted to the basolateral membrane (WOZNIAK and LIMBIRD 1996). However, although the α_{2A}-adrenoceptor is delivered directly to this surface, the α_{2B}-adrenoceptor appears to be initially inserted at random into the apical and basolateral surfaces but is then selectively retained by the basolateral surface (WOZNIAK and LIMBIRD 1996). In contrast, although part of the steady-state α_{2C}-adrenoceptor population has a basolateral plasma-membrane location to which it is delivered directly (at least in these cells), a proportion of the cellular levels of this receptor has an intracellular location (WOZNIAK and LIMBIRD 1996). In contrast to the α_2-adrenoceptors in the same cells, the A_1 adenosine receptor is selectively enriched in the apical membrane (SAUNDERS et al. 1996). Furthermore, disruption of microtubules interfered with the targeting of the A_1 adenosine receptor to the apical surface but did not interfere with the initial apical component of α_{2B}-adrenoceptor distribution (SAUNDERS and LIMBIRD 1997).

The possibility that specialized regions of the plasma membrane may concentrate signaling components has been raised both by observations of a non-uniform distribution of fluorescent agonists and antagonists at GPCRs in both fixed tissue and confocal microscopy studies (McGRATH et al. 1996) and by the use of antibodies raised against either peptide sequences derived from GPCRs or following expression of "epitope-tagged" GPCRs (VON ZASTROW and KOBILKA 1992; MOLINO et al. 1997). There is selective enrichment of signaling polypeptides, including heterotrimeric G proteins, in glycosphingolipid-rich regions named caveolae (LI et al. 1996; SONG et al. 1996). Intriguingly, two recent reports have indicated that the targeting of GPCRs to caveolae may require agonist activation. In cardiac myocytes, addition of a muscarinic acetylcholine receptor agonist resulted in movement of a proportion of the m2 muscarinic acetylcholine receptor population to a caveolar location and the subsequent interaction of the receptor with caveolin-3, a muscle-specific form of caveolin (FERON et al. 1997). A similar occurrence has been reported for the bradykinin B2 receptor (DE WEERD and LEEB-LUNDBERG 1997). Caveolin appears to bind tightly to inactive forms of G proteins and not to mutationally activated forms of G protein α subunits. It could be suggested either that (1) agonist-mediated transfer of a GPCR to the caveolae would act to compete with caveolin for the G protein and, thus, allow it to be activated by the GPCR to initiate signaling or, alternatively, that (2) movement of the GPCR to the caveolae is part of the desensitization response for the GPCR.

The ability to fluorescently label a GPCR offers the potential to examine its distribution and agonist-induced redistribution in intact cells and in real time. One such approach has involved tagging of GPCRs at their C-terminal tails with green fluorescent protein (GFP; BARAK et al. 1997; TARASOVA et al. 1997). Such studies have demonstrated that both a β_2-adrenoceptor–GFP fusion protein and an equivalent construct of the cholecystokinin-A receptor have the capacity to activate adenylyl cyclase and to be redistributed from the plasma membrane to internal structures in response to agonist. Such GFP-tagged GPCRs and analysis of the interactions between fluorescently labeled GPCRs and other signaling polypeptides are likely to be widely used in the near future to explore the details of cellular localization and protein–protein interactions.

For the G proteins, there is certainly evidence that G_s and certain other G proteins can interact directly with tubulin (ROYCHOWDHURY and RASENICK 1994; YAN et al. 1996). Furthermore, signal transduction via the stimulatory arm of adenylyl cyclase is regulated by such structural organization, because microtubule-disrupting agents (such as colchicine and vinblastine) increase agonist-mediated regulation of cAMP production and formation of the $G_s\alpha$–adenylyl cyclase complex (LEIBER et al. 1995).

A further area of potential complication in understanding the importance of stoichiometry of G protein-coupled signaling cascades is the growing evidence that it is the identity of the overall G protein heterotrimer that defines interactions with particular receptors. Due to the availability of a wide range of antisera for individual G protein α subunits, quantification of these polypeptides has become relatively easy. However, until recently, much less emphasis was placed on the identity of the β and γ subunits associated with α subunits, even though considerable genetic variability exists at these loci. The strongest evidence that different $\beta\gamma$ combinations contribute to the specificity of interactions of G proteins with receptors has been derived from electrophysiological studies following injection of antisense deoxyoligonucleotides anticipated to eliminate specific isoforms of β and γ subunits. In the original studies in pituitary GH3 cells (KLEUSS et al. 1991, 1992, 1993), the muscarinic acetylcholine regulation of voltage-operated Ca^{2+} channels was shown to require the $\beta 3$ and $\gamma 4$ isoforms in association with the G_o splice variant $G_{o1}\alpha$, whereas somatostatin regulation of the same (or a very similar) conductance required $G_{o2}\alpha$, $\beta 1$ and $\gamma 3$. If such exquisite selectivity is the norm, then means will have to be sought to establish the absolute quantitative levels of individual heterotrimers present rather than just global levels of α subunits. To date, use of expressed and purified $\beta\gamma$ complexes of defined molecular identity have shown little selectivity in the regulation of effectors (even in relatively physiological settings; WICKMAN et al. 1994). The major exception is that a $\beta 1\gamma 1$ complex (which physiologically appears to be restricted in distribution to photoreceptor-containing cells) tends to show tenfold lower potency compared with other defined $\beta\gamma$ variants. This is despite the activity of many effector systems regulated in this manner, including certain isoforms of adenylyl cyclase, phospholipase C and a number of ion channels (CLAPHAM and NEER 1997). As such, heterotrimer identity may primarily define the selectivity of interactions with GPCRs rather than with effector systems.

B. GPCR–G Protein Fusion Proteins. A Novel Means to Restrict and Define the Stoichiometry of Expression of a GPCR and a G Protein α Subunit

Most schematics which depict G protein-coupled signaling cascades show GPCR and G protein to be present in a ratio of 1:1. This is very different from direct measures of the bulk membrane levels of the proteins (ALOUSI et al. 1991; KIM et al. 1994; POST et al. 1995). However, a highly novel means to ensure that this ratio is produced and that the two proteins must be in proximity following their expression has been the generation of a single polypeptide containing both functionalities. In the first report of such a construct, BERTIN et al. (1994) ligated together cDNA species encoding the human β_2-adrenoceptor and the α subunit of the G protein G_s such that the N-terminus of the G protein was linked directly to the C-terminus of the receptor. This construct was then expressed in S49 cyc⁻ cells (which do not express $G_s\alpha$). Upon addition of the agonist isoproterenol, adenylyl cyclase activity was stimulated. Although potentially compromised by the endogenous expression of a β_2-adrenoceptor in these cells, a marked increase in affinity for the agonist following expression of the fusion protein was taken as evidence that the activation probably occurred via the receptor in the fusion protein (BERTIN et al. 1994). As this construct displayed a poor ability to be desensitized (BERTIN et al. 1994), it has recently been used to attempt to interfere with the proliferation of ras-transformed tumor cells (BERTIN et al. 1997), based on the concept that ras–raf interactions and subsequent stimulation of extracellular signal-regulated protein kinases/mitogen-activated protein kinases can be limited by elevated cAMP levels.

Subsequently, WISE et al. (1997b) adopted a similar strategy to link together the porcine α_{2A}-adrenoceptor and the G protein $G_{i1}\alpha$. This single polypeptide could be detected following expression by an antiserum directed towards an internal epitope within $G_{i1}\alpha$ (WISE et al. 1997b). Cells that do not express G_i-like G proteins are not widely available to function as a null background. Thus, because $G_{i1}\alpha$ is a pertussis toxin-sensitive G protein, the fusion-protein construct was built using a modified form of the G protein, in which the cysteine which acts as substrate for adenosine diphosphate (ADP) ribosylation by pertussis toxin was modified to glycine ($Cys^{351}Gly$). Following expression, the cells could be treated with pertussis toxin, which acted to modify the endogenously expressed G_i-like G proteins. This modification attenuates productive interactions between GPCRs and these G proteins and ensures that any signal obtained must have derived from activation of the receptor-attached G protein. By treating this fusion construct as an agonist-activated enzyme in which the enzyme activity is hydrolysis of GTP, simple enzyme kinetics were used to measure both the K_m for GTP and the V_{max} of the fusion protein GTPase activity stimulated by addition of agonist. Following transient expression of the fusion construct in COS-7 cells, maximally effective levels of the agonist UK14304 produced an increase in V_{max} without alteration of K_m for GTP. Concurrent measurement of levels of expression of the fusion protein by saturation [³H]antagonist-binding studies allowed calculation of a turnover number of about three per minute. Co-expression of G protein $\beta\gamma$ complex (as the β_1 and γ_2 subunits) resulted in doubling of the turnover number (WISE et al. 1997b). This demonstrated both the interaction of the fusion protein with $\beta\gamma$ complex and a role for these subunits in effective function and interaction of GPCR and G protein α subunit. This α_{2A}-adrenoceptor–$Cys^{351}Gly$ $G_{i1}\alpha$ fusion protein has also allowed measurement of agonist efficacy at the α_{2A}-adrenoceptor–$G_{i1}\alpha$ tandem. UK14304 functioned as a partial agonist compared with epinephrine and norepinephrine, as assessed by the capacity of each ligand to stimulate either the V_{max} of the GTPase reaction or the binding of [^{35}S]GTPγS (WISE et al. 1997c).

The rank order of efficacy of agonists at this construct was the same as that measured following individual co-expression of the $α_{2A}$-adrenoceptor and Cys^{351}Gly $G_{i1}α$. The absolute efficacy of partial agonists was greater when examining the co-expressed proteins but, in these experiments, no effort was made to limit the levels of expression of the G protein to 1:1 with the receptor, a situation which, as noted above, is imposed upon the fusion protein. Although it is implied above that the GPCR of the fusion protein is restricted to interacting only with its attached G protein, it has recently been shown that this is not true following stable expression of the $α_{2A}$-adrenoceptor–Cys^{351}Gly-$G_{i1}α$ fusion protein in Rat-1 fibroblasts (BURT et al. 1998). Membranes prepared from such stable cell lines in the absence of prior pertussis toxin treatment resulted in estimates of agonist-induced turnover number that were markedly greater than those produced following transient-expression studies in COS-7 cells. Following pertussis toxin treatment, the rate of agonist-stimulated GTPase activity was markedly reduced but still clearly measurable, with a turnover number now close to that measured following transient transfection. The most obvious interpretation of these experiments was that the fusion protein-constrained GPCR was able to activate both its fusion-protein partner and endogenous G_i-like G proteins. Indeed, if it is assumed that the stimulated GTPase rate of the fusion-protein-linked G protein and the endogenous G proteins are similar, then the agonist-induced V_{max} GTPase rates derived from the receptor-linked G protein (i.e., signal following pertussis toxin treatment) and endogenous G proteins (i.e., signal in the absence of pertussis toxin treatment minus the signal following pertussis toxin treatment) provide a ratio of the number of endogenously expressed G proteins activated per copy of the GPCR–G protein fusion protein (BURT et al. 1998).

The selection of the Cys^{351}Gly mutation of $G_{i1}α$ for construction of the fusion protein was based on the fact that this mutation in $G_{i1}α$ did not prevent interaction with the $α_{2A}$-adrenoceptor (WISE et al. 1997a). However, BAHIA et al. (1998) have recently explored the quantitative effects of the identity of amino acid 351 of $G_{i1}α$ on the capacity of the G protein to act as an agonist-stimulated GTPase following co-expression with the $α_{2A}$-adrenoceptor. Forms of $G_{i1}α$ in which residue 351 was occupied by any of the 20 natural amino acids were tested. A number of these, particularly if the amino acid had a fixed positive or negative charge in its side group, functioned very poorly as agonist-activated GTPases. In contrast, amino acids with branched-chain aliphatic or aromatic side chains functioned well, in many cases noticeably better than wild type (Cys^{351}) $G_{i1}α$. A strong correlation was observed between agonist-induced function and the n-octanol/water partition coefficient of the amino acid, a measurement of hydrophobicity (BAHIA et al. 1998). Interestingly Gly^{351} acted as an outlier from such analyses, functioning more poorly as an $α_{2A}$-adrenoceptor-stimulated GTPase than would have been anticipated. As such, a direct assessment of the relative function of Cys$^{351}G_{i1}α$ and Gly$^{351}G_{i1}α$ was conducted following their incorporation into fusion proteins containing the α_{2A}-adrenoceptor (BAHIA et al., unpublished). As anticipated from the discussion above, the fusion protein containing $Gly^{351}G_{i1}\alpha$ displayed only about 50% of the GTPase turnover number in response to epinephrine compared with parallel studies using the $Cys^{351}G_{i1}\alpha$-containing fusion protein. Such fusion proteins offer unique opportunities to examine many quantitative aspects of the pharmacology of GPCR–G protein interactions, but a number of features of such constructs must be considered carefully before they can be expected to produce insights of direct relevance to native systems.

C. G Protein-Coupled Receptors

Knowledge of the levels of expression of GPCRs in cells is considerably further advanced than for the other components of G protein-linked signaling cascades. This is largely a reflection of the fact that, at least for those receptors that have attracted the attention of the pharmaceutical industry as potential therapeutic targets, there is often a large and detailed chemical base of compounds that can be used (in radiolabeled form) to quantify their levels. As antagonist ligands show little or no ability to discriminate between receptors whether or not they are in contact with their cognate G proteins, the binding of such ligands can usually be modeled by considering the interaction to take place at a single population of non-interacting sites. This is the simplest possible model and, thus, the binding of antagonist ligands produces the best data for measurement of absolute levels of expression of a GPCR. Because agonist ligands are expected to show higher-affinity binding to the G protein-coupled state of a GPCR compared with the uncoupled state and because the differences in affinity can be substantial, useful information about the relative contributions from these two states of the GPCR can be gained from direct binding studies with a radiolabeled agonist and comparison with results obtained using a radiolabeled antagonist. However, for GPCRs for which only [^{3}H]agonist ligands are available, total cellular levels of the receptor can often to difficult to assess accurately. Levels of individual GPCRs vary widely among different cells and tissues, ranging from a few hundred to a few hundred thousand copies per cell. Thus, in transfection studies, it is often unclear if altered signaling of a GPCR attributed to its "overexpression" may have physiological correlates in certain cell types. Furthermore, as many GPCRs are not silent in the absence of ligand (MILLIGAN et al. 1995; MILLIGAN and BOND 1997; Chap. 2), different levels of cellular function may result in the absence of agonist simply because of expression of similar levels of two GPCRs that couple to the same G protein-mediated pathway.

As with many aspects of GPCR biology, the greatest amount of information on the effect of regulation of receptor levels on the effectiveness of transmembrane signaling has been obtained for the β2-adrenoceptor. Following stable expression of the β2-adrenoceptor in Chinese hamster fibroblast cells,

BOUVIER et al. (1988) demonstrated that in clones expressing a wide range of receptor levels (between 80 fmol and 8 pmol/mg membrane protein) increases in the absolute isoprenaline stimulation of adenylyl cyclase activity were observed in membranes of these clones up to levels of expression of some 2 pmol/mg membrane protein. A plateau effect of up to some 5 pmol/mg membrane protein was then observed. However, at higher levels of expression, a decline in the maximal agonist-stimulated adenylyl cyclase activity was recorded. These data appear to indicate that, at levels of the receptor below 2 pmol/mg membrane protein, the β_2-adrenoceptor appears to be the limiting element in achieving maximal agonist stimulation of adenylyl cyclase. As anticipated from simple pharmacological theory, the EC_{50} for isoproterenol stimulation of adenylyl cyclase was reduced in clones expressing higher levels of the receptor, consistent with the notion of "spare receptors" at high levels of expression. One possible explanation of the reduced maximal effectiveness of isoproterenol in membranes of cells expressing above 5 pmol/mg membrane protein of the receptor was that, at such levels, activation of the inhibitory G_i-like G proteins was occurring, as was activation of G_s. Activation of G_i by the β-adrenoceptor has certainly been observed in some studies (XIAO et al 1995).

Interestingly, DAAKA et al. (1997) have recently provided evidence to suggest that the $\beta2$-adrenoceptor only interacts effectively with G_i following protein kinase A (PKA)-mediated phosphorylation of the receptor. Such phosphorylation contributes to heterologous desensitization of this GPCR (FREEDMAN and LEFKOWITZ 1996). This is an attractive idea, as effective interaction with G_i would provide a means to lower cellular cAMP levels and the temporal frame of elevated PKA activity, thus contributing to desensitization of the receptor. Furthermore, levels of constitutive activation of adenylyl cyclase, and thus of cAMP levels, would be anticipated to increase with increasing receptor-expression levels. This did not seem to account for the results of BOUVIER et al. (1988), however, as treatment of the cells with pertussis toxin to eliminate potential interactions of the receptor with G_i failed to result in levels of isoproterenol stimulation of adenylyl cyclase as high as the levels achieved in clones expressing lower levels of receptor. The explanation for these observations still remains unclear. To extend such analyses, WHALEY et al. (1994) performed a careful quantitative study examining the effects of a wide range of levels of expression of the β_2-adrenoceptor (following stable expression in L cells) on both the EC_{50} of epinephrine and the maximal adenylyl cyclase activity that could be achieved by this ligand. These studies demonstrated that agonist EC_{50} declined as receptor number increased and that mathematical predictions that plots of $\log(EC_{50})$ versus $\log$(receptor number) should follow a close to linear relationship were valid (WHALEY et al. 1994). In these cells, plots of epinephrine-stimulated adenylyl cyclase activity against β_2-adrenoceptor number resulted in a rectangular hyperbola in which half-maximal activation of adenylyl cyclase could be obtained with levels of the receptor of only 10–20 fmol/mg membrane protein. As such, close to maximal activation of adenylyl cyclase would be expected with expression of the β_2-adrenoceptor of as little as 50 fmol/mg membrane protein. Because little is known about the levels of expression of adenylyl cyclase in these cells, it is difficult to usefully examine the very different conclusions that might be drawn from the studies of BOUVIER et al. (1988) and of WHALEY et al. (1994) about the limiting element of the stimulatory adenylyl cyclase cascade. To extend this type of study, MACEWAN et al. (1995) used a combination of clones of NG108-15 cells transfected to express differing levels of the β_2-adrenoceptor and an irreversible antagonist to limit access of both full and partial agonists to the expressed receptors. Both the change in EC_{50} for agonist ligands with receptor level and the way partial agonist efficacy varied with receptor level were explored.

Notably, highly similar conclusions were reached whether agonist regulation of adenylyl cyclase activity in cellular membranes or the ability of agonists to promote the high-affinity binding of [^{3}H]forskolin to a complex of $G_s\alpha$ and adenylyl cyclase (Sect. E for a discussion of this assay) in intact cells was measured. Agonist EC_{50} decreased with increasing receptor level, and the efficacy of partial agonists was shown to be higher in cells expressing higher levels of the receptor (MACEWAN et al. 1995). Again, in this cell system, only 30 fmol receptor/mg membrane protein was required to produce half-maximal activation of the adenylyl cyclase cascade in response to a maximally effective concentration of isoproterenol. Equivalent analyses indicated that salbutamol was required to occupy 500 fmol β_2-adrenoceptor/mg membrane protein to cause the same amount of stimulation, while ephedrine was such a poor agonist that it failed to fully activate cellular adenylyl cyclase even in clones expressing the receptor at levels greater than 2 pmol/mg membrane protein. The number of β_2-adrenoceptors required to be occupied by different agonists to produce activation of 50% of the adenylyl cyclase population thus provided a novel relative efficacy measurement in this cell system. As total cellular levels of both $G_s\alpha$ and adenylyl cyclase had previously been measured in these cells (KIM et al. 1994; Sect. E), this was the first system for which a true quantitative description could be provided to examine how signaling efficiency is modulated by alterations in levels of each component of a signaling cascade (MACEWAN et al. 1996).

The results of WHALEY et al. (1994) and MACEWAN et al. (1995) seem to explain why compounds like isoproterenol act as full agonists in virtually all systems. On this basis, the development and use of partial agonists at the β_2-adrenoceptor and other adenylyl cyclase-stimulatory GPCRs might allow selective targeting of therapeutic responses to cells and tissues expressing relatively high levels of the GPCR of interest.

β_2-Adrenoceptors are frequently co-expressed with the other β-adrenoceptor subtypes (β_1 and β_3) in the heart and in adipose tissue, for example. Although most studies have centered on the β_2-adrenoceptor, binding studies with selective ligands have indicated the β_1-adrenoceptor to be the most prevalent subtype in the heart. Despite this, it appears that, at equivalent levels of expression, the β_1-adrenoceptor has a lower ability to activate adenylyl cyclase. To examine the relative functions of co-expressed β_1- and β_2-adrenoceptors in detail, LEVY et al. (1993) isolated clones of L cells co-expressing these two receptors in different ratios. Using selective blockers of the two GPCR subtypes, they were able to demonstrate that the β_2-adrenoceptor was always able to cause greater activation of adenylyl cyclase activity than the β_1-adrenoceptor at full occupancy with isoproterenol. LEVY et al. (1993) described this effect as "receptor efficacy". As with the study of WHALEY et al. (1994), L cells were used as the host. However, different data was obtained regarding the number of receptors required to be expressed and activated to cause full stimulation of the adenylyl cyclase population. In the case of the β_2-adrenoceptor, this was estimated to be 1000 fmol/mg membrane protein while, for the β_1-adrenoceptor, because of its lower measured receptor efficacy, expression at levels close to 3000 fmol/mg membrane protein were insufficient to cause maximal activation (LEVY et al. 1993). The reasons for this discrepancy are not clear. It will be of great interest to ascertain whether "receptor-efficacy" differences for GPCRs that couple to the same basic signal-transduction cascade truly reflect quantitative differences in the ability of the agonist-occupied receptors to cause activation of the same G protein pool or relate to cellular differences in receptor distribution (SAUNDERS et al. 1996; LIMBIRD 1997; MILLIGAN 1998). Such differences would then result in different abilities of intracellular second messenger-regulated kinases to "sample" (HOUSLAY and MILLIGAN 1997) the message generated at different sites of the plasma membrane.

A potentially even more intriguing aspect of "receptor efficacy", which has yet to be explored at a detailed quantitative level, is the idea that different GPCR polymorphisms may produce variations in quality and quantity of signal. In the case of the β_2-adrenoceptor, three distinct polymorphisms have been reported within the human population. These polymorphisms display differences in agonist-mediated regulation and show associations with conditions (such as asthma) where contribution from β_2-adrenoceptor function would be anticipated (GREEN et al. 1993, 1994, 1995; TURKI et al. 1995, 1996). In the case of the β_3-adrenoceptor, a Trp→Arg mutation at amino acid position 64 in the human β_3-adrenoceptor is reportedly associated with morbid obesity (CLEMENT et al. 1995). Although it has not been easy for a number of groups to observe, following expression in Chinese hamster ovary (CHO)-K1 and human embryonic kidney (HEK)-293 cells, this mutant has recently been reported to produce reduced maximal stimulation of adenylyl cyclase compared with the wild type receptor (PIETRI-ROUXEL et al. 1997).

In mammalian heart, the expression level of β_1-adrenoceptor predominates over that of β_2-adrenoceptor. Levels of these receptors have been increased in a cardiac-specific manner by transgenic overexpression. With very high-level overexpression of the β_2-adrenoceptor, the contractility of the heart and adenylyl cyclase activity were maximal in the absence of agonist stimulation (ROCKMAN et al. 1996; KOCH et al. 1998). In contrast, effects of transgenic overexpression of the β_1-adrenoceptor have been less clear-cut (MANSIER et al. 1996). It is unclear if this reflects the poor "receptor efficacy" of this GPCR, the more limited fold overexpression achieved or a combination thereof. By contrast, overexpression of the β_1-adrenoceptor in adipose tissue has been reported to limit diet-induced obesity (SOLOVEVA et al. 1997), presumably by increasing lipolytic activity in response to circulating catecholamine. Chronic congestive heart failure is associated with a reduction in β-adrenoceptor density and a poor signaling capacity of the remaining receptor population. Features of this condition have been reported to be alleviated by introduction of extra β_2-adrenoceptor levels into isolated cardiac myocytes (AKHTER et al. 1997a). As transgenic overexpression of the A_1 adenosine receptor in the mouse heart has been reported to increase myocardial resistance to ischemia (MATHERNE et al. 1997), regulation of the stoichiometry of expression of polypeptides associated with G protein-mediated signaling cascades offers a potentially exciting avenue through which to modify pathophysiological conditions. Because the β-adrenoceptor isoforms display differential desensitization characteristics (MILLIGAN et al. 1994), mutationally modified forms of these receptors that display a reduced ability to be desensitized but maintain high "receptor-efficacy" parameters may offer benefits beyond those achieved by the wild type GPCRs.

Detailed analysis of the effects of regulating levels of the three α_1-adrenoceptor subtypes has recently provided insights into strategies for modulation of the effectiveness of agonists at phosphoinositidase C-linked GPCRs (THEROUX et al. 1996). A number of studies have suggested that the α_1-adrenoceptor subtypes might also display the features of differing "receptor efficacy" (described above for the β_1- and β_2-adrenoceptors). Using a tetracycline-repressible expression system, THEROUX et al. (1996) were able to examine the correlations between levels of GPCR expression and signal output measured as inositol phosphate (IP) generation for each of the human α_1-adrenoceptor subtypes in specific cell clones of HEK-293 cells. With relatively modest (up to 1 pmol/mg protein) levels of expression, none of the individual α_1-adrenoceptor subtypes displayed substantive agonist-independent generation of IPs, and each displayed increasing degrees of signal with increasing expression. However, maximal effects of norepinephrine followed a pattern ($\alpha_{1a} > \alpha_{1b} > \alpha_{1d}$), at equivalent receptor expression levels, for the stimulated generation of IPs and for increases in intracellular Ca^{2+} levels. As the measured responses were markedly variable, only the majority of data derived from such an exhaustive approach was likely to provide suitable correlates. Sustained treatment of cells expressing a constitutively active mutant of the hamster α_{1b}-adrenoceptor (PEREZ et al. 1996; SCHEER et al. 1996) with ligands that function as inverse agonists can result in substantial upregulation of the GPCR. LEE et al. (1997) examined the effects of such upregulation on agonist stimulation of phosphoinositidase C and phospholipase D activities. The potency of phenylephrine to stimulate these activities was unaltered by upregulation of the GPCR, but the maximal effects of the agonist were increased substantially. These results again imply that the capacity of the effector enzymes is not limiting for output via the phosphoinositidase C-linked cascade.

Although the m2 muscarinic acetylcholine receptor is not one generally associated with phosphoinositidase C activation, expression in specific systems can allow such an output. Regulation of GPCR availability to agonist in a stably transfected CHO cell line expressing this receptor can be achieved by pretreatment with the slowly dissociating antagonist quinuclidinyl benzilate (VOGEL et al. 1995). Although agonist stimulation of phosphoinositidase C activity was poor (less than threefold even at relatively high receptor-availability levels), there was a good correlation between receptor levels and signal generation. The EC_{50} values for a range of ligands were essentially unchanged, with receptor level as might be anticipated if receptor activation of the relevant G protein(s) remained limiting in comparison to the levels and activity of phosphoinositidase C. Both of these features were markedly different when agonist-mediated inhibition of forskolin-amplified adenylyl cyclase was measured in the same cells. Here, even at low levels of receptor availability, efficacious agonists produced a maximal degree of inhibition, and it was only with the use of the weak partial agonist pilocarpine that a less-than-maximal degree of inhibition could be observed. This also reached maximal levels as receptor availability increased. As might be anticipated, agonist EC_{50} for cAMP inhibition decreased with increasing receptor level. VOGEL et al. (1995) also observed an ability of high concentrations of agonists to stimulate cAMP production via the porcine m2 muscarinic acetylcholine receptor in these CHO cells. This appeared to be via a direct activation of $G_s\alpha$ rather than via release of $\beta\gamma$ complex from G_i, as pertussis toxin treatment did not prevent the stimulation.

The observation that control of adenylyl cyclase activity by GPCRs (which either produce predominantly stimulatory or inhibitory regulation) is often maximal at low levels of receptor expression tends to suggest either that levels of adenylyl cyclase are generally low in cells or that the coupling efficiency of GPCR and G protein to adenylyl cyclase is very high. Certainly, direct estimates of adenylyl cyclase levels do indicate it to be a poorly expressed protein in many cell systems (Sect. E). As such, transgenic elevation of levels of GPCRs linked to adenylyl cyclase is likely to result primarily in increases in agonist potency and efficacy rather than maximal regulation of cAMP. In at least one case, levels of expression have been high enough to render the system virtually fully active in the absence of agonist (BOND et al. 1995). The effects of directly increasing adenylyl cyclase levels will be discussed later (Sect. E). In contrast, increases in levels of phosphoinositidase C-linked GPCRs frequently translate directly to greater maximal generation of IPs without alteration in agonist potency. However, measurement of this cascade at the level of increases in intracellular $[Ca^{2+}]$ might be anticipated to saturate much more rapidly with increasing GPCR level, as the release of Ca^{2+} from intracellular stores by generated inositol 1,4,5-trisphosphate represents a further amplification step.

D. $G_s\alpha$

One major difficulty with examining the effects of modulation of G protein α-subunit levels on transmembrane signaling efficiency has been the lack of cells with null backgrounds, which prevented truly quantitative studies from being undertaken. Furthermore, the capacity of closely related G proteins to display overlapping functionalities (at least in transfected cell systems) suggests that even cell lines generated from G protein gene knockout studies may not be ideal. It is also true that, at a gross total cell or tissue level, measured levels of G protein are often very high compared with the levels of other signal-transducing polypeptides (ALOUSI et al. 1991; KIM et al. 1994). Indeed, in the central nervous system, the pertussis toxin-sensitive G protein G_o comprises some 1% of the total protein in many brain regions (GIERSCHIK et al. 1986; MILLIGAN et al. 1987). As such, further overexpression might be anticipated to have limited benefits in understanding the effects of quantitative variation of G protein levels in regulating the efficiency of signal transduction.

The high levels of G proteins in cells and tissues have allowed both semi-quantitative immunoblotting and enzyme-linked immunosorbent assays (ELISA) to provide good estimates of expression levels (MILLIGAN 1993). Despite the high levels of expression, there is good evidence indicating that not all of the cellular population of specific G proteins is equally accessible to individual GPCRs, which may mediate the same primary function (NEUBIG 1994). For example, it has been suggested that, in NG108-15 cells, the endogenously expressed δ opioid, M4 muscarinic acetylcholine and α_{2B}-adrenoceptors utilize non-overlapping pools of G_i (GRAESNER and NEUBIG 1993). Furthermore, there are clearly non-plasma membrane pools of G proteins; these may have functions other than trans-plasma membrane signal transduction (DENKER et al. 1996; GIESBERTS et al. 1997; HAMILTON and NATHANSON 1997). Strong evidence has accrued for the interaction of at least some G proteins with elements of the cellular cytoskeleton (IBARRONDO et al. 1995; ADOLFSSON et al. 1996; COTE et al. 1997). Definition and understanding of the cellular architecture of G protein expression remains a major challenge for the future. Estimates of R:G interaction ratios vary depending on whether they are based on ligand-binding studies that estimate the fraction of a GPCR in a high-affinity agonist-binding state or on measurement of agonist stimulation of the GTPase activity of provided G proteins. Perhaps for the reasons noted above, transgenic overexpression of G protein α subunits has been limited in use.

An exception to this has been $G_s\alpha$. $G_s\alpha$ can be expressed as a number of splice variants (BRAY et al. 1986). The primary difference between the individual forms is the presence or absence of the information encoded by exon 3 of the $G_s\alpha$ gene. The major isoform of $G_s\alpha$ in mouse heart appears to be a short variant. Transgenic overexpression of a short splice variant of $G_s\alpha$ in a cardiac-specific manner in mice has been achieved (GAUDIN et al. 1995). The degree of total overexpression of $G_s\alpha_{short}$ appeared modest (2.8-fold). However, estimates of levels of expression of $G_s\alpha$ in (rat) ventricular myocytes are higher than in many other systems and indicate the presence of some 3.5×10^6 copies per cell (40 pmol/mg membrane protein) of the short isoform and some 1.2×10^6 copies of the long isoform (POST et al. 1995). Thus, if similar levels of expression are found in the mouse, then at an absolute level, a 2.8-fold overexpression of $G_s\alpha_{short}$ would correspond to a high level of protein (72 pmol/mg of membrane protein). Despite this level of G protein overexpression, the functional effects on short-term signaling were limited. No alterations were reported in membranes in either the basal adenylyl cyclase activity or when this activity was stimulated by GTP + isoproterenol, GTPγS, NaF or forskolin (GAUDIN et al. 1995). Two features of the system were, however, modulated. A higher proportion of the population of β-adrenoceptors were present in the state of high-affinity for agonist, and there was a reduced lag time before stimulation of adenylyl cyclase activity was enhanced by the poorly hydrolyzed analogue of GTP, Gpp[NH]p (GAUDIN et al. 1995). Both of these features might be anticipated, based on the law of mass-action. However, if the expressed excess $G_s\alpha_{short}$ was correctly targeted following expression, then a greater fraction of the receptor population in the G_s-coupled state might have implied that more $G_s\alpha$ activation would occur in the absence of receptor agonist. This might have been expected to result in a greater degree of basal adenylyl cyclase activity. Subsequently, it was reported that there are both increased mortality and features diagnostic of dilated cardiomyopathy in the $G_s\alpha$-overexpressing animals compared with controls (IWASE et al. 1997), and there is an increase in the efficacy of the β-adrenoceptor–$G_s\alpha$–adenylyl cyclase cascade (IWASE et al. 1996).

Overexpression of $G_s\alpha$ in stable cell lines has also been reported. S49 cyc⁻ lymphoma cells are a variant of S49 cells, which do not express $G_s\alpha$ and, thus, provide the type of null background alluded to earlier (Chap. 3.2). LEVIS and BOURNE (1992) transfected these cells with a modified version of the long isoform of $G_s\alpha$ in which the sequence derived from exon 3 was modified to encode a sequence identified by monoclonal antibody 12CA5. This modified form of the polypeptide was shown to be functional, but no absolute quantitation of the levels of expression were provided. Use of such a system would seem highly suitable for assessing how much $G_s\alpha$ would be required to achieve maximal and half-maximal regulation of adenylyl cyclase. This issue has recently been addressed by BARBER and colleagues (KRUMINS and BARBER 1997; KRUMINS et al. 1997). By generating cell lines following electroporation of S49 cyc⁻ lymphoma cells with a long-isoform $G_s\alpha$ construct from which expression could be induced by addition of dexamethasone, KRUMINS et al. (1997) noted that induction of $G_s\alpha$ allowed increasing levels of epinephrine-stimulated adenylyl cyclase activity, which reached a maximal level. They also noted that agonist EC_{50} decreased as levels of $G_s\alpha$ increased, although this effect was relatively small (less than twofold). In concert with the development of agonist stimulation of adenylyl cyclase activity, dexamethasone treatment of the transfected S49 cyc⁻ lymphoma cells also resulted in an increased magnitude of GTPγS shifts in the competition curves of epinephrine binding versus [^{125}I]iodocyanopindolol binding to the β_2-adrenoceptor. This indicated an increasing presence of a high-affinity agonist-binding site, which is likely to reflect GPCR–G protein interactions (KRUMINS et al. 1997). The combination of these results have been interpreted as providing evidence of a shuttle model rather than a $G_s\alpha$–adenylyl cyclase pre-coupled model for $G_s\alpha$ activation of adenylyl cyclase (KRUMINS et al. 1997).

In a similar vein, in membranes from hearts of mice transgenically overexpressing the receptor, GURDAL et al. (1997) have reported more effective interactions between the β_2-adrenoceptor and $G_s\alpha$ in the absence of agonist. In contrast, MULLANEY and MILLIGAN (1994) utilized the hemagglutinin-tagged variant construct of the long isoform of $G_s\alpha$ and expressed it stably in the genetic background of NG108-15 cells. The long isoform provides greater than 85% of the overall steady-state levels of endogenous $G_s\alpha$ in these cells and has been shown to be present at levels of 1.2×10^6 copies/cell (KIM et al. 1994). Although not altering the number of amino acids in the polypeptide chain, the modifications introduced by LEVIS and BOURNE (1992) alter the mobility of the epitope-tagged polypeptide such that it migrates more slowly through sodium dodecyl sulfate polyacrylamide-gel electrophoresis (MULLANEY and MILLIGAN 1994). This allowed concurrent detection of the introduced, modified G protein and the endogenously expressed wild type long isoform of $G_s\alpha$ by immunoblotting with an antiserum directed against the C-terminal section of $G_s\alpha$. In clone BST15, the total level of $G_s\alpha_{long}$ was increased to 2×10^6 copies/cell. Both of these forms of $G_s\alpha_{long}$ were shown to be regulated by agonists at the IP prostanoid receptor (MULLANEY and MILLIGAN 1994), but regulation of adenylate-cyclase activity in these cells was not different compared with that in membranes from the parental control (MULLANEY et al. 1996), either in maximal amplitude or in EC_{50} for a prostanoid agonist. Forskolin, NaF and the poorly hydrolyzed analogue of GTP Gpp[NH]p each produced equivalent stimulations of adenylyl cyclase activity in membranes from clone BST15 and NG108-15 cells. Although these results may seem distinct from those of KRUMINS et al. (1997), the difference in level of expression between parental NG108-15 cells and clone BST15 is sufficiently small that any predicted effects might also be expected to be very limited. Furthermore, as the levels of expression of $G_s\alpha$ in the parental NG108-15 cells are so large in comparison to levels of adenylyl cyclase (Sect. E), it was not surprising that addition of further $G_s\alpha$ produced little enhancement of function.

Apart from the S49 cyc$^-$ lymphoma cell line, strategies that result in a reduction in cellular levels of $G_s\alpha$ offer the potential ability to examine whether this can restrict the maximal effectiveness of the adenylyl cyclase cascade. Sustained treatment of cells with cholera toxin is known to result in a major reduction in cellular $G_s\alpha$ levels, as persistent activation of this polypeptide (either due to cholera-toxin-catalyzed ADP-ribosylation or via mutagenesis) results in enhanced degradation of the polypeptide (CHANG and BOURNE 1989; MILLIGAN et al. 1989; MACLEOD and MILLIGAN 1990). Reduction of $G_s\alpha$ levels by 90% in GH3 cells via sustained treatment with cholera toxin has, however, been reported to have little effect on GTP-stimulated adenylyl cyclase activity (CHANG and BOURNE 1989), again implying a considerable cellular steady-state molar excess over the amount required to maximally activate the cascade. In contrast, MACLEOD and MILLIGAN (1990) have shown that large reductions in $G_s\alpha$ levels in neuroblastoma × glioma hybrid cells can result in a reduced ability of forskolin to stimulate adenylyl cyclase activity without altering Mn^{2+}-stimulated activity (which provides an estimate of levels of the adenylyl cyclase catalytic moiety in the absence of G protein regulation). Thus, in some systems, it is possible to reduce levels of G_s sufficiently to alter the effectiveness of this cascade. However, the presence of cholera toxin will ensure that the remaining $G_s\alpha$ is in an active, ADP-ribosylated state and will thus limit any prospect of examining agonist regulation of the cascade under such conditions. Antisense strategies, probably driven using an inducible promoter system (Sect. A), again offer a direct means to explore this question. Antisense elimination of $G_s\alpha$ with the aim of assessing whether reduction in levels of this polypeptide will result in reduced receptor-mediated activation of adenylyl cyclase activity has, in fact, been employed. In GH3 and related pituitary cell lines, thyrotropin-releasing hormone (TRH) has, in some studies, been reported to cause both stimulation of adenylyl cyclase activity and its more established stimulation of activity of phosphoinositidase C. PAULSSEN et al. (1992) demonstrated that antisense reduction of $G_s\alpha$ levels in these cells resulted in a reduced ability of TRH to stimulate cAMP production but that this did not alter the ability of TRH to stimulate phosphoinositidase C activity. Such studies again suggest that it is possible to reduce cellular $G_s\alpha$ levels to an extent sufficient to interfere with maximal activation of the adenylyl cyclase cascade. Other studies have also targeted antisense elimination of $G_s\alpha$ (WANG et al. 1992). Although such studies have reported substantial reductions of cellular $G_s\alpha$, the primary function has not been to determine the effect of this on the regulation of adenylyl cyclase function. This is an area that would benefit greatly from further detailed quantitative analysis.

Transgenic overexpression of the phosphoinositidase C-linked G protein $G_q\alpha$ in the hearts of mice has also been reported (D'ANGELO et al. 1997). Interestingly, as with transgenic overexpression of a constitutively active mutant of the α_{1B}-adrenoceptor (MILANO et al. 1994b), overexpression of this G protein α subunit at levels four times above normal resulted in increased heart weight and myocyte size and decreased β-adrenoceptor function. All of these are features associated with chronic cardiac failure. By contrast, cardiac overexpression of the wild type α_{1B}-adrenoceptor resulted in only depressed signaling via the β-adrenoceptor (AKHTER et al. 1997b) rather than myocardial hypertrophy. Studies again suggest the possibility that increases in levels of signaling elements proximal to phosphoinositidase C can result in enhanced effectiveness of this cascade. The apparent redundancy of function of the closely related phosphoinositidase C-linked G proteins $G_q\alpha$ and $G_{11}\alpha$ (OFFERMANNS et al. 1994) has limited the conclusions that can be drawn unambiguously from $G_q\alpha$ knockout mice (OFFERMANNS et al. 1997). However, as platelets and other hemopoietically derived cells lack expression of $G_{11}\alpha$ (MILLIGAN et al. 1993; JOHNSON et al. 1996), the lack of response of platelets from $G_q\alpha$ knockout mice to a series of platelet activators provided evidence for a direct role (OFFERMANNS et al. 1997). Perhaps surprisingly, therefore, murine embryonic stem cells lacking $G_q\alpha$ actually display markedly enhanced phosphoinositidase C response to bradykinin (RICUPERO et al. 1997), which is returned to normal levels by expression of $G_q\alpha$ in these cells. The basis of these observations remains unexplained, though they are clearly of great interest.

E. Effector Enzymes: Adenylyl Cyclase

Following cDNA cloning of the first molecularly defined adenylyl cyclase (type 1) by the laboratories of REED and GILMAN (KRUPINSKI et al. 1989), a steady increase in numbers of family members has been recorded. Nine types have now been defined (TAUSSIG and GILMAN 1995; HOUSLAY and MILLIGAN 1997). These share common structural features, based on two groups of six putative transmembrane spanning elements connected by a long intracellular loop; in addition, the N- and C-termini of all the family members are predicted to be intracellular. All of the subtypes of this family are believed to be positively regulated by $G_s\alpha$, although the quantitative details of this may vary (HARRY et al. 1997). Individual family members differ in their modes of regulation by other means, including G protein $\beta\gamma$ complexes, Ca^{2+}/calmodulin, protein kinase C-mediated phosphorylation and $[Ca^{2+}]_i$ levels. Recently, the crystal structure of the C2 domain of adenylyl cyclase type II has been solved (ZHANG et al. 1997), allowing new insights into the catalytic mechanism and regulation of this enzyme, and rapid progress is now being made in this area (DESSAUER et al. 1997; YAN et al. 1997a, 1997b). Until recently, antisera able to selectively identify individual adenylyl cyclase isoforms have not been available and, indeed, commercially available antisera of reasonable titer and specificity remain difficult to obtain. This has resulted in the restriction of analysis of the tissue and cellular distribution of the adenylyl cyclase isoforms to detection of relevant messenger RNA (mRNA) by Northern blotting or reverse-transcriptase polymerase chain reaction (RT-PCR). These approaches cannot, however, provide estimates of levels of expression of the individual polypeptides. As a first approach to this, a number of workers have attempted to utilize the ability of $[^3H]$forskolin to bind to the activated complex of G_s and adenylyl cyclase (GSAC) with high affinity (SEAMON and DALY 1985; YAMASHITA et al. 1986; ALOUSI et al. 1991; KIM et al. 1994, 1995; POST et al. 1995; MACEWAN et al. 1996; STEVENS and MILLIGAN 1998). If sufficient $G_s\alpha$ is available to interact with the total cellular population of adenylyl cyclase, then such an approach provides a potential means to quantify levels of adenylyl cyclase expression. In normal circumstances, the GSAC might be anticipated to be a transitory complex, as the intrinsic GTPase activity of GTP-liganded $G_s\alpha$ would act to destabilize and inactivate the complex. However, when using cell membranes where quasi-persistent activation of $G_s\alpha$ can be achieved by the addition of either poorly hydrolyzed analogues of GTP or AlF_4^-, a binding isotherm can be generated to allow analysis of the population of GSAC (KIM et al. 1994).

In all reported studies, levels have been substantially lower than for $G_s\alpha$. In membranes of NG108-15 cells, levels of $G_s\alpha$ (1.2×10^6 copies/cell) have been reported to be some 70-fold higher than the levels of GSAC that can be formed by addition of maximally effective concentrations of Gpp[NH]p (KIM et al. 1994). It has been possible to adapt these approaches to intact cells where specific [^{3}H]forskolin binding is driven in an agonist-dependent manner. In intact S49 lymphoma cells maintained at low temperature to prevent the dissociation and inactivation of GSAC once it formed, maximal agonist occupancy of the β_2-adrenoceptor resulted in an estimated B_{max} for binding of [^{3}H] forskolin of 3000 sites/cell (ALOUSI et al. 1991). If it is assumed that the specific binding of [^{3}H]forskolin to GSAC occurs at a single site (although there have been suggestions that there may be two binding sites for forskolin on each adenylyl cyclase molecule; ZHANG et al. 1997), then the total number of GSAC complexes that can be formed in these cells is also substantially lower than the number of copies of $G_s\alpha$ (100000; RANSNAS and INSEL 1988).

Agonist regulation of [^{3}H]forskolin binding was not observed in variants of S49 lymphoma cells that either lacked $G_s\alpha$ expression (cyc$^-$), expressed a form of this polypeptide that failed to interact with the receptor (unc) or where $G_s\alpha$ failed to become activated properly in response to agonist (H21a; ALOUSI et al. 1991). Use of similar whole-cell [^{3}H]forskolin-binding experiments in NG108-15 cells transfected to express varying levels of the β_2-adrenoceptor resulted in isoproterenol-stimulation of specific [^{3}H]forskolin binding, in which the maximal levels achieved were similar in cells expressing 25,000 copies/cell and 400,000 copies/cell of this GPCR (KIM et al. 1995). However, as with direct measurements of adenylyl cyclase activity, the ability of isoproterenol to produce this effect was markedly greater in the cells with higher GPCR levels (KIM et al. 1995). In contrast, concentration–effect curves for stimulation of [^{3}H]forskolin binding by an agonist at the endogenously expressed IP prostanoid receptor was not different in these cell lines (KIM et al. 1995). It is also of interest to note that agonist-independent constitutive activity could be recorded using the intact-cell [^{3}H]forskolin-binding assay. The level of specific binding of [^{3}H]forskolin measured in the absence of ligand was greater in the cells expressing high levels of the β_2-adrenoceptor compared with either those with low levels or the parental, untransfected cells (KIM et al. 1995). This agonist-independent [^{3}H]forskolin binding was partially reversed by addition of propranolol, which had no effect on basal [^{3}H]forskolin binding in cells with low levels of the β_2-adrenoceptor (KIM et al. 1995). These results paralleled those obtained earlier with membranes from these cell lines when examining the regulation of adenylyl cyclase activity (ADIE et al. 1994a, 1994b). This basic assay has also been used to examine the efficacy of β-adrenoceptor partial agonists in both S49 lymphoma cells (ALOUSI et al. 1991) and in NG108-15 cell lines transfected to express varying levels of the β_2-adrenoceptor (MACEWAN et al. 1995). As anticipated from receptor theory (and in line with direct adenylyl cyclase-activity measurements), higher levels of expression of this receptor resulted in increased efficacy of partial agonists (MACEWAN et al. 1995).

Although much of the evidence detailed indicates that adenylyl cyclase levels represent the limiting element for maximal potential cAMP generation in many cell types, little has been done at a quantitative level to test this directly. If specific binding of [^{3}H]forskolin represents a useful means to examine quantitative aspects of the expression of adenylyl cyclase and its regulation by agonists at GPCRs, it is necessary to demonstrate that it could be used to measure the degree of overexpression of adenylyl cyclase following stable transfection of an adenylyl cyclase isoform into a heterologous cell system. NG108-15 cells and clones derived from them do not express adenylyl cyclase type II, as assessed by an inability to detect relevant mRNA using RT-PCR (MACEWAN et al. 1996). Following transfection of β_2-adrenoceptor-expressing NG108-15 cells with a cDNA encoding adenylyl cyclase type II, individual clones positive for this mRNA were examined for both guanine, nucleotide-stimulated [^{3}H]forskolin binding in membrane preparations and isoproterenol-stimulated [^{3}H]forskolin binding in whole cells. In each case, higher levels of stimulated [^{3}H]forskolin binding were observed than were observed in the parental cells, and the degree of extra binding correlated with the levels of detected adenylyl cyclase type II mRNA (MACEWAN et al. 1996). In the intact cell studies, both basal [^{3}H]forskolin binding and the maximal response to isoproterenol addition increased substantially. However, even in these clones, adenylyl cyclase appeared to remain the limiting element, as the EC_{50} for the agonist was unaffected by increases in levels of adenylyl cyclase of up to eightfold (MACEWAN et al. 1996).

Partial agonists at the β_2-adrenoceptor did not show altered efficacy with enhanced adenylyl cyclase levels. Perhaps surprisingly, agonists at the A2 adenosine and secretin receptors, which are expressed endogenously at low levels in NG108-15 cells, displayed similar maximal effects when compared with isoproterenol in the adenylyl cyclase type II-overexpressing cells and in the clones with lower adenylyl cyclase levels. This was despite their inability to cause full activation of adenylyl cyclase in the parental cells without overexpression of adenylyl cyclase (MACEWAN et al. 1996). This may relate to the type of compartmentalization described for G_i-linked GPCRs in the cells (GRAESNER and NEUBIG 1993) and clearly requires further analysis.

It would also be of considerable interest to expand this type of analysis to other isoforms of adenylyl cyclase and to other cell backgrounds. No reports on transgenic overexpression of this element have yet been reported. If less than a maximal activation of adenylyl cyclase is required to produce maximal contractile responses in the heart, as seems evident in the studies from MILANO et al. (1994a), then overexpression of adenylyl cyclase in heart might not be anticipated to result in quantitative physiological benefit.

F. Conclusions

Emerging information on the cellular disposition of the polypeptides of signal-transduction cascades should soon allow a concerted effort to understand how targeted alterations in individual components will effect the overall effectiveness of information transfer from receptor to effector.

Acknowledgments. Work in the author's laboratory in this area is supported by the Medical Research Council and the Biotechnology and Biological Sciences Research Council.

References

Adie EJ, Milligan G (1994a) Regulation of basal adenylate-cyclase activity in neuroblastoma × glioma hybrid, NG108-15, cells transfected to express the human β_2-adrenoceptor: evidence for empty receptor stimulation of the adenylate-cyclase cascade. Biochem J 303:803–808

Adie EJ, Milligan G (1994b) Agonist regulation of cellular G_s α-subunit levels in neuroblastoma × glioma hybrid NG108-15 cells transfected to express different levels of the human β_2-adrenoceptor. Biochem J 300:709–715

Adolfsson JL, Ohd JF, Sjolander A (1996) Leukotriene D4-induced activation and translocation of the G-protein αi3-subunit in human epithelial cells. Biochem Biophys Res Commun 226:413–419

Akhter SA, Skaer CA, Kypson AP, McDonald PH, Peppel KC, Glower DD, Lefkowitz RJ, Koch WJ (1997a) Restoration of β-adrenergic signaling in failing cardiac ventricular myocytes via adenoviral-mediated gene transfer. Proc Natl Acad Sci USA 94:12100–12105

Akhter SA, Milano CA, Shotwell KF, Cho MC, Rockman HA, Lefkowitz RJ, Koch WJ (1997b) Transgenic mice with cardiac overexpression of α1B-adrenergic receptors. In vivo α1-adrenergic-receptor-mediated regulation of β-adrenergic signaling. J Biol Chem 272:21253–21259

Alousi AA, Jasper JR, Insel PA, Motulsky HJ (1991) Stoichiometry of receptor–G_s–adenylate-cyclase cascade. FASEB J 5:2300–2303

Bahia DS, Wise A, Fanelli F, Lee M, Rees S, Milligan G (1998) Hydrophobicity of residue^{351} of the G-protein $G_{i1}\alpha$ determines the extent of activation by the α_{2A}-adrenoceptor. Biochemistry 37:11555–11562

Barak LS, Ferguson SSG, Zhang J, Martenson C, Meyer T, Caron MG (1997) Internal trafficking and surface mobility of a functionally intact β_2-adrenergic-receptor-green-fluorescent-protein conjugate. Mol Pharmacol 51:177–184

Bertin B, Freissmuth M, Jockers R, Strosberg AD, Marullo S (1994) Cellular signaling by an agonist-activated receptor/Gsα fusion protein. Proc Natl Acad Sci USA 91:8827–8831

Bertin B, Jockers R, Strosberg AD, Marullo S (1997) Activation of a β2-adrenergic receptor/Gsα fusion protein elicits a desensitization-resistant cAMP signal capable of inhibiting proliferation of two cancer cell lines. Receptors Channels 5:41–51

Bond RA, Leff P, Johnson TD, Milano CA, Rockman HA, McMinn TR, Apparsundaram S, Hyek MF, Kenakin TP, Allen LF, Lefkowitz RJ (1995) Physiological effects of inverse agonists in transgenic mice with myocardial overexpression of the β_2-adrenoceptor. Nature 374:272–276

Bouvier M, Hnatowich M, Collins S, Kobilka BK, DeBlasi A, Lefkowitz RJ, Caron MG (1988) Expression of a human cDNA encoding the β2-adrenergic receptor in Chinese-hamster fibroblasts (CHW): functionality and regulation of the expressed receptors. Mol Pharmacol 33:133–139

Bray P, Carter A, Simons C, Guo V, Puckett C, Kamholz J, Spiegel A, Nirenberg M (1986) Human cDNA clones for four species of Gαs signal transduction protein. Proc Natl Acad Sci USA 83:8893–8897

Burt AR, Sautel M, Wilson MA, Rees S, Wise A, Milligan G (1998) Agonist occupation of an α_{2A}-adrenoceptor-G$_{i1}\alpha$ fusion protein results in activation of both receptor-linked and endogenous G proteins. Comparisons of their contributions to GTPase activity and signal transduction and analysis of receptor-G protein activation stoichiometry. J Biol Chem 273:10367–10375

Cavalli A, Lattion AL, Hummler E, Nenniger M, Pedrazzini T, Aubert JF, Michel MC, Yang M, Lembo G, Vecchione C, Mostardini M, Schmidt A, Beermann F, Cotecchia S (1997) Decreased blood pressure response in mice deficient of the α1b-adrenergic receptor. Proc Natl Acad Sci USA 94:11589–11594

Chang FH, Bourne HR (1989) Cholera toxin induces cAMP-independent degradation of G$_s$. J Biol Chem 264:5352–5357

Chen JF, Guo JH, Moxham CM, Wang HY, Malbon CC (1997) Conditional, tissue-specific expression of Q205L Gαi2 in vivo mimics insulin action. J Mol Med 75:283–289

Clapham DE, Neer EJ (1997) G protein $\beta\gamma$ subunits. Annu Rev Pharmacol Toxicol 37:167–203

Clement K, Vaisse C, Manning BS, Basdevant A, Guy-Grand B, Ruiz J, Silver KD, Shuldiner AR, Froguel P, Strosberg AD (1995) Genetic variation in the β3-adrenergic receptor and an increased capacity to gain weight in patients with morbid obesity. N Engl J Med 333:352–354

Cote M, Payet MD, Dufour MN, Guillon G, Gallo-Payet N (1997) Association of the G-protein α(q)/α11-subunit with cytoskeleton in adrenal glomerulosa cells: role in receptor-effector coupling. Endocrinology 138:3299–3307

Daaka Y, Luttrell LM, Lefkowitz RJ (1997) Switching of the coupling of the β2-adrenergic receptor to different G-proteins by protein kinase A. Nature 390:88–91

D'Angelo DD, Sakata Y, Lorenz JN, Boivin GP, Walsh RA, Liggett SB, Dorn GW 2nd (1997) Transgenic Gαq overexpression induces cardiac contractile failure in mice. Proc Natl Acad Sci USA 94:8121–8126

De Weerd WFC, Leeb-Lundberg LMF (1997) Bradykinin sequesters B2 bradykinin receptors and the receptor-coupled Gα subunits Gα_q and Gα_i in caveolae in DDT1 MF2 smooth-muscle cells. J Biol Chem 272:17858–17866

Denker SP, McCaffery JM, Palade GE, Insel PA, Farquhar MG (1996) Differential distribution of α subunits and $\beta\gamma$ subunits of heterotrimeric G-proteins on Golgi membranes of the exocrine pancreas. J Cell Biol 133:1027–1040

Dessauer CW, Scully TT, Gilman AG (1997) Interactions of forskolin and ATP with the cytosolic domains of mammalian adenylyl cyclase. J Biol Chem 272:22272–22277

Feron O, Smith TW, Michel T, Kelly RA (1997) Dynamic targeting of the agonist-stimulated m$_2$ muscarinic acetylcholine receptor to caveolae in cardiac myocytes. J Biol Chem 272:17744–17748

Freedman NJ, Lefkowitz RJ (1996) Desensitization of G-protein-coupled receptors. Rec Progr Horm Res 51:319–353

Gaudin C, Ishikawa Y, Wight DC, Mahdavi V, Nadal-Ginard B, Wagner TE, Vatner DE, Homcy CJ (1995) Overexpression of Gsα protein in the hearts of transgenic mice. J Clin Invest 95:1676–1683

Gierschik P, Milligan G, Pines M, Goldsmith P, Codina J, Klee W, Spiegel A (1986) Use of specific antibodies to quantitate the guanine nucleotide-binding protein Go in brain. Proc Natl Acad Sci USA 83:2258–2262

Giesberts AN, van Ginneken M, Gorter G, Lapetina EG, Akkerman JW, van Willigen G (1997) Subcellular localization of α-subunits of trimeric G-proteins in human platelets. Biochem Biophys Res Commun 234:439–444

Graeser D, Neubig RR (1993) Compartmentation of receptors and guanine nucleotide-binding proteins in NG108-15 cells: lack of cross-talk in agonist binding among the α2-adrenergic, muscarinic, and opiate receptors. Mol Pharmacol 43:434–443

Green SA, Cole G, Jacinto M, Innis M, Liggett SB (1993) A polymorphism of the human β2-adrenergic receptor within the fourth transmembrane domain alters ligand binding and functional properties of the receptor. J Biol Chem 268:23116–23121

Green SA, Turki J, Innis M, Liggett SB (1994) Amino-terminal polymorphisms of the human β2-adrenergic receptor impart distinct agonist-promoted regulatory properties. Biochemistry 33:9414–9419

Green SA, Turki J, Bejarano P, Hall IP, Liggett SB (1995) Influence of β2-adrenergic receptor genotypes on signal transduction in human airway smooth-muscle cells. Am J Respir Cell Mol Biol 13:25–33

Gurdal H, Bond RA, Johnson MD, Friedman E, Onaran HO (1997) An efficacy-dependent effect of cardiac overexpression of β2-adrenoceptor on ligand affinity in transgenic mice. Mol Pharmacol 52:187–194

Hamilton SE, Nathanson NM (1997) Differential localization of G-proteins, Gαo and Gα-1, -2, and -3, in polarized epithelial MDCK cells. Biochem Biophys Res Comm 234:1–7

Harry A, Chen Y, Magnusson R, Iyengar R, Weng G (1997) Differential regulation of adenylyl cyclases by Gα_s. J Biol Chem 272:19017–19021

Houslay MD, Milligan G (1997) Tailoring cAMP-signalling responses through isoform multiplicity. Trends Biochem Sci 22:217–224

Ibarrondo J, Joubert D, Dufour MN, Cohen-Solal A, Homburger V, Jard S, Guillon G (1995) Close association of the α subunits of G_q and G11 G-proteins with actin filaments in WRK1 cells: relation to G-protein-mediated phospholipase C activation. Proc Natl Acad Sci USA 92:8413–8417

Iwase M, Bishop SP, Uechi M, Vatner DE, Shannon RP, Kudej RK, Wight DC, Wagner TE, Ishikawa Y, Homcy CJ, Vatner SF (1996) Adverse effects of chronic endogenous sympathetic drive induced by cardiac GSα overexpression. Circ Res 78:517–524

Iwase M, Uechi M, Vatner DE, Asai K, Shannon RP, Kudej RK, Wagner TE, Wight DC, Patrick TA, Ishikawa Y, Homcy CJ, Vatner SF (1997) Cardiomyopathy induced by cardiac Gsα overexpression. Am J Physiol 272:H585–H589

Johnson GJ, Leis LA, Dunlop PC (1996) Specificity of Gαq and Gα11 gene expression in platelets and erythrocytes. Expressions of cellular differentiation and species differences. Biochem J 318:1023–1031

Kim GD, Adie EJ, Milligan G (1994) Quantitative stoichiometry of the proteins of the stimulatory arm of the adenylyl cyclase cascade in neuroblastoma × glioma hybrid NG108-15 cells. Eur J Biochem 219:135–143

Kim GD, Carr IC, Milligan G (1995) Detection and analysis of agonist-induced formation of the complex of the stimulatory guanine nucleotide-binding protein with adenylate cyclase in intact wild-type and β2-adrenoceptor-expressing NG108-15 cells. Biochem J 308:275–281

Kleuss C, Hescheler J, Ewel C, Rosenthal W, Schultz G, Wittig B (1991) Assignment of G-protein subtypes to specific receptors inducing inhibition of calcium currents. Nature 353:43–48

Kleuss C, Scherubl H, Hescheler J, Schultz G, Wittig B (1992) Different β-subunits determine G-protein interaction with transmembrane receptors. Nature 358:424–426

Kleuss C, Scherubl H, Hescheler J, Schultz G, Wittig B (1993) Selectivity in signal transduction determined by γ subunits of heterotrimeric G-proteins. Science 259:832–834

Koch WJ, Lefkowitz RJ, Milano CA, Akhter SA, Rockman HA (1998) Myocardial overexpression of adrenergic receptors and receptor kinases. Adv Pharmacol 42:502–506

Krumins AM, Barber R (1997) Examination of the effects of increasing G_s protein on β2-adrenergic receptor, G_s, and adenylyl cyclase interactions. Biochem Pharmacol 54:61–72

Krumins AM, Lapeyre JN, Clark RB, Barber R (1997) Evidence for the shuttle model for Gsα activation of adenylyl cyclase. Biochem Pharmacol 54:43–59

Krupinski J, Coussen F, Bakalyar HA, Tang WJ, Feinstein PG, Orth K, Slaughter C, Reed RR, Gilman AG (1989) Adenylyl cyclase amino acid sequence: possible channel-or transporter-like structure. Science 244:1558–1564

Lee TW, Cotecchia S, Milligan G (1997) Up-regulation of the levels of expression and function of a constitutively active mutant of the hamster α_{1B}-adrenoceptor by ligands that act as inverse agonists. Biochem J 325:733–739

Leiber D, Jasper JR, Alousi AA, Martin J, Bernstein D, Insel PA (1993) Alteration in G_s-mediated signal transduction in S49 lymphoma cells treated with inhibitors of microtubules. J Biol Chem 268:3833–3837

Levis MJ, Bourne HR (1992) Activation of the α subunit of G_s in intact cells alters its abundance, rate of degradation, and membrane avidity. J Cell Biol 119:1297–1307

Levy FO, Zhu X, Kaumann AJ, Birnbaumer L (1993) Efficacy of β_1-adrenergic receptors is lower than that of β_2-adrenergic receptors. Proc Natl Acad Sci USA 90: 10798–10802

Li S, Couet J, Lisanti MP (1996) Src tyrosine kinases, $G\alpha$ subunits and H-Ras share a common membrane-anchored scaffolding protein, caveolin. J Biol Chem 271: 29182–29190

MacEwan DJ, Kim GD, Milligan G (1995) Analysis of the role of receptor number in defining the intrinsic activity and potency of partial agonists in neuroblastoma × glioma hybrid NG108-15 cells transfected to express differing levels of the human β_2-adrenoceptor. Mol Pharmacol 48:316–325

MacEwan DJ, Kim GD, Milligan G (1996) Agonist regulation of adenylate-cyclase activity in neuroblastoma × glioma hybrid NG108-15 cells transfected to co-express adenylate cyclase type II and the β_2-adrenoceptor. Evidence that adenylate cyclase is the limiting component for receptor-mediated stimulation of adenylate-cyclase activity. Biochem J 318:1033–1039

Macleod KG, Milligan G (1990) Biphasic regulation of adenylate cyclase by cholera-toxin in neuroblastoma × glioma hybrid cells is due to the activation and subsequent loss of the α subunit of the stimulatory GTP-binding protein (GS). Cell Signal 2:139–151

Mansier P, Medigue C, Charlotte N, Vermeiren C, Coraboeuf E, Deroubai E, Ratner E, Chevalier B, Clairambault J, Carre F, Dahkli T, Bertin B, Briand P, Strosberg D, Swynghedauw B (1996) Decreased heart-rate variability in transgenic mice over-expressing atrial β_1-adrenoceptors. Am J Physiol 271:H1465-H1472

Matherne GP, Linden J, Byford AM, Gauthier NS, Headrick JP (1997) Transgenic A1 adenosine receptor overexpression increases myocardial resistance to ischemia. Proc Natl Acad Sci USA 94:6541–6546

McGrath JC, Arribas S, Daly CJ (1996) Fluorescent ligands for the study of receptors. Trend Pharmacol Sci 17:393–399

Milano CA, Allen LF, Rockman HA, Dolber PC, McMinn TR, Chien KR, Johnson TD, Bond RA, Lefkowitz RJ (1994a) Enhanced myocardial function in transgenic mice overexpressing the β2-adrenergic receptor. Science 264:582–586

Milano CA, Dolber PC, Rockman HA, Bond RA, Venable ME, Allen LF, Lefkowitz RJ (1994b) Myocardial expression of a constitutively active α1B-adrenergic receptor in transgenic mice induces cardiac hypertrophy. Proc Natl Acad Sci USA 91:10109–10113

Milligan G (1993) Qualitative and quantitative characterization of the distribution of G-protein α subunits in mammals. Handb Exp Pharm 108/II:45–64

Milligan G (1998) New aspects of G-protein-coupled receptor signalling and regulation. Trends Endocrinol Metabolism 9:13–19

Milligan G, Bond RA (1997) Inverse agonism and the regulation of receptor number. Trends Pharmacol Sci 18:468–474

Milligan G, Streaty RA, Gierschik P, Spiegel AM, Klee WA (1987) Development of opiate receptors and GTP-binding regulatory proteins in neonatal rat brain. J Biol Chem 262:8626–8630

Milligan G, Unson CG, Wakelam MJ (1989) Cholera-toxin treatment produces down-regulation of the α-subunit of the stimulatory guanine-nucleotide-binding protein (G_s). Biochem J 262:643–649

Milligan G, Mullaney I, McCallum JF (1993) Distribution and relative levels of expression of the phosphoinositidase C-linked G-proteins $G_q\alpha$ and $G_{11}\alpha$: Absence of $G_{11}\alpha$ in human platelets and haemopoietically derived cell lines. Biochim Biophys Acta 1179:208–212

Milligan G, Svoboda P, Brown CM (1994) Why are there so many adrenoceptor subtypes? Biochem Pharmacol 48:1059–1071

Milligan G, Bond RA, Lee M (1995) Inverse agonism: pharmacological curioisity or potential therapeutic strategy? Trends Pharmacol Sci 16:10–13

Molino M, Bainton DF, Hoxie JA, Coughlin SR, Brass LF (1997) Thrombin receptors on human platelets. Initial localization and subsequent redistribution during platelet activation. J Biol Chem 272:6011–6017

Moxham CM, Malbon CC (1996) Insulin action impaired by deficiency of the G-protein subunit Giα2. Nature 379:840–844

Moxham CM, Hod Y, Malbon CC (1993) Giα2 mediates the inhibitory regulation of adenylyl cyclase in vivo: analysis in transgenic mice with Giα2 suppressed by inducible antisense RNA. Dev Genet 14:266–273

Mullaney I, Milligan G (1994) Equivalent regulation of wild-type and an epitope-tagged variant of Gsα by the IP prostanoid receptor following expression in neuroblastoma × glioma hybrid, NG108-15, cells. FEBS Lett 353:231–234

Mullaney I, Carr IC, Milligan G (1996) Overexpression of G(s)α in NG108-15, neuroblastoma × glioma cells: effects on receptor regulation of the stimulatory adenylyl cyclase cascade. FEBS Lett 397:325–330

Neubig RR (1994) Membrane organization in G-protein mechanisms. FASEB J 8:939–946

Offermanns S, Heiler E, Spicher K, Schultz G (1994) G_q and G11 are concurrently activated by bombesin and vasopressin in Swiss 3T3 cells. FEBS Lett 349:201–204

Offermanns S, Toombs CF, Hu YH, Simon MI (1997) Defective platelet activation in Gα(q)-deficient mice. Nature 389:183–186

Paulssen RH, Paulssen EJ, Gautvik KM, Gordeladze JO (1992) The thyroliberin receptor interacts directly with a stimulatory guanine-nucleotide-binding protein in the activation of adenylyl cyclase in GH3 rat pituitary tumour cells. Evidence obtained by the use of antisense RNA inhibition and immunoblocking of the stimulatory guanine-nucleotide-binding protein. Eur J Biochem 204:413–418

Perez DM, Hwa J, Gaivin R, Mathur M, Brown F, Graham RM (1996) Constitutive activation of a single effector pathway: evidence for multiple activation states of a G-protein-coupled receptor. Mol Pharmacol 49:112–122

Pietri-Rouxel F, St John Manning B, Gros J, Strosberg AD (1997) The biochemical effect of the naturally occurring Trp64→Arg mutation on human β3-adrenoceptor activity. Eur J Biochem 247:1174–1179

Post SR, Hilal-Dandan R, Urasawa K, Brunton LL, Insel PA (1995) Quantification of signalling components and amplification in the β-adrenergic-receptor–adenylate-cyclase pathway in isolated adult rat ventricular myocytes. Biochem J 311:75–80

Ransnas LA, Insel PA (1988) Quantitation of the guanine-nucleotide-binding regulatory protein G_s in S49 cell membranes using antipeptide antibodies to αs. J Biol Chem 263:9482–9485

Ricupero DA, Polgar P, Taylor L, Sowell MO, Gao Y, Bradwin G, Mortensen RM (1997) Enhanced bradykinin-stimulated phospholipase C activity in murine embryonic stem cells lacking the G-protein αq-subunit. Biochem J 327:803–809

Rockman HA, Koch WJ, Milano CA, Lefkowitz RJ (1996) Myocardial β-adrenergic receptor signaling in vivo: insights from transgenic mice. J Mol Med 74:489–495

Rohrer DK, Desai KH, Jasper JR, Stevens ME, Regula DP Jr, Barsh GS, Bernstein D, Kobilka BK (1996) Targeted disruption of the mouse β_1-adrenergic receptor gene: developmental and cardiovascular effects. Proc Natl Acad Sci USA 93:7375–7380

Roychowdhury S, Rasenick MM (1994) Tubulin–G-protein association stabilizes GTP binding and activates GTPase: cytoskeletal participation in neuronal signal transduction. Biochemistry 33:9800–9805

Rudolph U, Spicher K, Birnbaumer L (1996) Adenylyl cyclase inhibition and altered G-protein-subunit expression and ADP-ribosylation patterns in tissues and cells from Gi2$\alpha^{-/-}$ mice. Proc Natl Acad Sci USA 93:3209–3214

Saunders C, Limbird LE (1997) Disruption of microtubules reveals two independent apical targeting mechanisms for G-protein-coupled receptors in polarized renal epithelial cells. J Biol Chem 272:19035–19045

Saunders C, Keefer JR, Kennedy AP, Wells JN, Limbird LE (1996) Receptors coupled to pertussis toxin-sensitive G-proteins traffic to opposite surfaces in Madin-Darby canine kidney cells. A$_1$ adenosine receptors achieve apical and α_{2A}-adrenergic receptors achieve basolateral localization. J Biol Chem 271:995–1002

Scheer A, Fanelli F, Costa T, De Benedetti PG, Cotecchia S (1997) The activation process of the α1B-adrenergic receptor: potential role of protonation and hydrophobicity of a highly conserved aspartate. Proc Natl Acad Sci USA 1997 94:808–813

Seamon KB, Daly JW (1985) High-affinity binding of forskolin to rat-brain membranes. Adv Cyclic Nucleotide Protein Phosphorylation Res 19:125–135

Soloveva V, Graves RA, Rasenick MM, Spiegelman BM, Ross SR (1997) Transgenic mice overexpressing the β_1-adrenergic receptor in adipose tissue are resistant to obesity. Mol Endocrinol 11:27–38

Song KS, Li S, Okamoto T, Quilliam LA, Sargiacomo M, Lisanti MP (1996) Co-purification and direct interaction of ras with caveolin, an integral membrane protein of caveolae microdomains. Detergent-free purification of caveolae membranes. J Biol Chem 271:9690–9697

Stevens PA, Milligan G (1998) Efficacy of inverse agonists in cells overexpressing a constitutively active β_2-adrenoceptor and type II adenylyl cyclase. Br J Pharmacol 123:335–343

Susulic VS, Frederich RC, Lawitts J, Tozzo E, Kahn BB, Harper ME, Himms-Hagen J, Flier JS, Lowell BB (1995) Targeted disruption of the β3-adrenergic-receptor gene. J Biol Chem 270:29483–29492

Tarasova NI, Stauber RH, Choi JK, Hudson EA, Czerwinski G, Miller JL, Pavlakis GN, Michejda CJ, Wank SA (1997) Visualization of G-protein-coupled receptor trafficking with the aid of the green fluorescent protein. Endocytosis and recycling of cholecystokinin receptor type A. J Biol Chem 272:14817–14824

Taussig R, Gilman AG (1995) Mammalian membrane-bound adenylyl cyclases. J Biol Chem 270:1–4

Theroux TL, Esbenshade TA, Peavy RD, Minneman KP (1996) Coupling efficiencies of human α_1-adrenergic-receptor subtypes: titration of receptor density and responsiveness with inducible and repressible expression vectors. Mol Pharmacol 50:1376–1387

Turki J, Pak J, Green SA, Martin RJ, Liggett SB (1995) Genetic polymorphisms of the β2-adrenergic receptor in nocturnal and non-nocturnal asthma. Evidence that Gly16 correlates with the nocturnal phenotype. J Clin Invest 95:1635–1641

Turki J, Lorenz JN, Green SA, Donnelly ET, Jacinto M, Liggett SB (1996) Myocardial signaling defects and impaired cardiac function of a human β2-adrenergic receptor polymorphism expressed in transgenic mice. Proc Natl Acad Sci USA 93:10483–10488

Vogel WK, Mosser VA, Bulseco DA, Schimerlik MI (1995) Porcine m2-muscarinic-acetylcholine-receptor–effector coupling in Chinese-hamster ovary cells. J Biol Chem 270:15485–15493

Wang HY, Watkins DC, Malbon CC (1992) Antisense oligodeoxynucleotides to GS-protein α-subunit sequence accelerate differentiation of fibroblasts to adipocytes. Nature 358:334–337

Whaley BS, Yuan N, Birnbaumer L, Clark RB, Barber R (1994) Differential expression of the β-adrenergic receptor modifies agonist stimulation of adenylyl cyclase: a quantitative evaluation. Mol Pharmacol 45:481–489

Wise A, Watson-Koken M-A, Rees S, Lee M, Milligan G (1997a) Interactions of the α_{2A}-adrenoceptor with multiple G_i-family G-proteins: studies with pertussis toxin-resistant G-protein mutants. Biochem J 321:721–728

Wise A, Carr IC, Milligan G (1997b) Measurement of agonist-induced guanine nucleotide turnover by the G-protein $G_{i1}\alpha$ when constrained within an α_{2A}-adrenoceptor-$G_{i1}\alpha$ fusion protein. Biochem J 325:17–21

Wise A, Carr IC, Groarke AD, Milligan G (1997c) Measurement of agonist efficacy using an α_{2A}-adrenoceptor–$G_{i1}\alpha$ fusion protein. FEBS Lett 419:141–146

Wozniak M, Limbird LE (1996) The three α_2-adrenergic receptor subtypes achieve basolateral localization in Madin-Darby canine kidney II cells via different targeting mechanisms. J Biol Chem 271:5017–5024

Xiao R, Ji X, Lakatta EG (1995) Functional coupling of the β2-adrenoceptor to a pertussis-toxin-sensitive G-protein in cardiac myocytes. Mol Pharmacol 47:322–329

Yamashita A, Kurokawa T, Higashi K, Danura T, Ishibashi S (1986) Forskolin stabilizes a functionally coupled state between activated guanine nucleotide-binding stimulatory regulatory protein, Ns, and catalytic protein of adenylate-cyclase system in rat erythrocytes. Biochem Biophys Res Commun 137:190–194

Yan K, Greene E, Belga F, Rasenick MM (1996) Synaptic membrane G-proteins are complexed with tubulin in situ. J Neurochem 66:1489–1495

Yan SZ, Huang ZH, Rao VD, Hurley JH, Tang WJ (1997a) Three discrete regions of mammalian adenylyl cyclase form a site for Gsα activation. J Biol Chem 272:18849–18854

Yan SZ, Huang ZH, Shaw RS, Tang WJ (1997b) The conserved asparagine and arginine are essential for catalysis of mammalian adenylyl cyclase. J Biol Chem 272:12342–12349

Zhang G, Liu Y, Ruoho AE, Hurley JH (1997) Structure of the adenylyl cyclase catalytic core. Nature 386:247–253

CHAPTER 14
The Study of Drug–Receptor Interaction Using Reporter Gene Systems in Mammalian Cells

D.M. IGNAR and S. REES

A. Reporter Systems

I. What Is a Reporter Gene System?

"Reporter gene" is the term used to describe a plasmid containing either an inducible or constitutive promoter element that controls the expression of a readily measurable enzyme or other protein. The reporter protein typically has a unique activity or structure to enable it to be distinguished from other proteins present. The choice of reporter gene, which is often an enzyme, is primarily influenced by the availability of a simple, usually colorimetric, fluorescent or luminescent assay of the activity of the protein product. The activity of the reporter protein provides an indirect measurement of the transcriptional activity of the promoter sequence. Reporter gene assays have been used for the characterization of the sequences and transcription factors that control gene expression at the transcriptional level and for the characterization of receptor-mediated signal transduction through the measurement of alterations in gene expression caused by receptor signaling. As such, reporter gene assays have been developed for drug screening and analytical pharmacology in mammalian cell-based assays.

Inducible reporter genes used in studies of signal transduction are inactive or weakly active with respect to transcription until the transcription factor protein(s) that bind to and activate the promoter are themselves activated as a consequence of receptor-mediated signal transduction. The activated promoter drives the transcription of a reporter gene, which is then quantified by assaying the messenger RNA, the protein product itself or the enzymatic activity of the protein product. A constitutive reporter gene contains a promoter that is constantly active, resulting in continuous expression of the reporter gene protein product. Constitutive reporter genes have been used in the analysis of cell tracking, intracellular trafficking of proteins and signal transduction.

II. Detection Methods

1. Enzymatic

Factors influencing the choice of reporter protein include cost, sensitivity, safety (radioactivity detection), stability of the reporter gene response and simplicity of the assay. For these reasons, many widely used reporter genes encode enzymes for which simple, non-radioactive assay procedures have been developed. Presently, the most commonly used enzymes include firefly (*Photinus pyralis*) luciferase (DE WET et al. 1987), *Renilla reniformis* (sea pansy) luciferase (LORENZ et al. 1991), secreted placental alkaline phosphatase (SEAP; HENTHORN et al. 1988) and β-galactosidase (CHEN et al. 1995). The development of increasingly facile, sensitive, cost-effective assays for many of these enzymes has expedited the use of reporter gene systems in analytical pharmacology and drug discovery. Assay systems for many of these enzymes, and vectors containing the reporter genes, are available commercially. Detailed discussion of each of these reporter genes and their assay systems is beyond the scope of this chapter and, thus, the reader is referred to reviews (ALAM and COOK 1990; BRONSTEIN et al. 1994; SUTO and IGNAR 1997).

Within this chapter, we will provide a number of examples of the use of the bioluminescent enzyme firefly luciferase as a reporter enzyme. The main advantages of firefly luciferase are attomolar detection sensitivity, broad dynamic range and facile assay methods that stabilize the luminescent signal for several hours (WILLIAMS et al. 1989; ROELANT et al. 1996). The relatively short half-life of luciferase (~4h) makes it an ideal reporter for mammalian cell lines. A reporter gene product that possesses a long half-life can accumulate within the cell as a consequence of basal promoter activity, which can result in a decrease in signal-to-noise ratio in the assay.

In contrast to firefly luciferase, SEAP is secreted from the cell such that reporter gene proteins present in the medium can simply be removed before an experiment. This property greatly facilitates the measurement of a decrease in transcriptional activity. A further advantage of SEAP is that the reporter activity can be measured without cell lysis. The same is true of *Renilla* luciferase, which utilizes the cell-permeable substrate coelenterazine. The colorimetric assays for SEAP and β-galactosidase are inexpensive and facile but not particularly sensitive, and they have a limited dynamic range. Recently, chemiluminescent assay systems for SEAP and β-galactosidase have been developed; these systems are as sensitive as luciferase assays. In addition, there are commercially available dual systems for assaying more than one reporter gene activity when assessment of more than one signaling event is desirable. For example, assay kits are available from a number of suppliers for the sequential assay of firefly and *Renilla* luciferases within the same reaction well.

A new β-lactamase reporter gene detection system having the advantage of highly sensitive real-time measurement of transcription in individual living cells has recently been introduced (ZLOKARNIK et al. 1998). A membrane-permeant fluorogenic substrate ester composed of 7-hydroxy coumarin and fluorescein attached to a cephalosporin backbone is used to assess transcription. As a consequence of a fluorescence resonance-energy transfer (FRET) event between the 7-hydroxy coumarin and the fluorescein groups in the substrate, the substrate fluoresces green (520 nm) upon excitation at 409 nm. After attack by β-lactamase, expressed as an inducible reporter protein, the fluorescein molecule is released; this disrupts the resonance energy transfer from coumarin to fluorescein, resulting in a large shift in emission wavelength to blue (447 nm). The half-life of this protein is similar to that of firefly luciferase in mammalian cells, which is advantageous for stable cell line generation, as discussed above.

2. Non-Enzymatic

In addition to the enzymatic reporter genes, the *Aequorea victoria* bioluminescent proteins aequorin and green fluorescent protein (GFP) have been used as reporter systems (CHALFIE et al. 1994). The calcium-sensitive photoprotein aequorin has been constitutively expressed as a reporter of calcium signaling in mammalian cells. In the presence of intracellular calcium and the cofactor coelenterazine, an oxidation reaction results in the generation of light that can be detected by conventional luminometry. In contrast to all other reporter proteins, GFP fluorescence is non-enzymatic and requires no additional cofactors (CHALFIE et al. 1994). Upon excitation with blue light at a λ_{max} of 396 nm, the protein emits green light with a λ_{max} of 509 nm. GFP has been widely used as a marker of protein expression (CHALFIE et al. 1994), to follow intracellular movement of proteins (BARAK et al. 1997) and in the construction of FRET assays (HEIM and TSIEN 1995).

Other reporter proteins include human growth hormone and chloramphenicol acetyltransferase, for which enzyme-linked immunosorbent assays have been developed in order to measure the amount of protein present. The use of these proteins will not be described here.

III. Measurement of Intracellular Signaling

1. Inducible Reporter Genes

The use of reporter genes to study drug–receptor interactions in mammalian cells has relied upon the characterization and subsequent assay of the signal transduction events activated as a consequence of receptor stimulation. Inducible reporter genes have been developed to monitor these signal transduction events. Generally, two types of inducible reporter genes have been used for the measurement of intracellular signaling.

a) Responsive Promoters

Reporter gene assays were first used in experiments designed to characterize promoter elements. In such experiments, a reporter gene, such as luciferase, was expressed off a natural promoter sequence, and the ability of extracellular stimuli to activate the promoter was studied using luciferase luminescence as the readout. Such studies were used to identify DNA sequence motifs (termed transcription factor binding sites or hormone response elements) required for promoter activity. Simultaneous studies led to the identification of the transcription factor proteins, which are able to bind to these DNA sequence motifs to promote gene expression. For example, reporter genes containing the consensus cyclic adenosine monophosphate (cAMP) response element (CRE) have been developed for the assessment of agonist activation of G protein-coupled receptors (GPCRs), which couple to heterotrimeric G proteins of the $G\alpha_s$ or $G\alpha_i$ families (STRATOWA et. al. 1995a; GEORGE et al. 1997). Agonist activity at a $G\alpha_s$-coupled GPCR results in the activation of the effector enzyme adenylyl cyclase to cause a subsequent elevation of intracellular cAMP levels. The consequence of an increase in intracellular cAMP is activation of protein kinase A (PKA), which subsequently phosphorylates and activates members of the CREB (CRE-binding protein) family of transcription factors. As shown in Fig. 1, activated CREB causes alterations in gene expression following binding to the CRE (TGACGTCA) found within the promoter element of many genes (STRATOWA et al. 1995a). As such, the use of reporter genes containing CRE elements within the promoter sequence has allowed the development of reporter assays for GPCRs that regulate cAMP.

A number of natural promoter elements have been used in reporter gene constructs to detect receptor signaling, such as the c-fos promoter, which contains several transcription factor binding sites including a CRE element (HILL and TREISMAN 1995) and the intracellular adhesion molecule (ICAM) promoter, which contains two activating protein-1 (AP-1) sites (WEYER et al. 1995). In this latter study, an ICAM promoter-driven luciferase reporter construct was used to characterize the 5-HT_{2C} serotonin receptor.

Natural promoter elements often contain many different transcription factor binding sites. In addition to a CRE, the c-fos promoter contains an AP-1 site and a serum response element (SRE) and, thus, will respond to PKA, protein kinase C (PKC), and mitogen activated protein kinase (MAPK) signaling events (HILL and TREISMAN 1995). In order to increase the specificity of reporter constructs and also to increase the signal-to-noise ratio obtained in a reporter gene assay, a number of synthetic, responsive promoters have been generated. Such promoters usually consist of multiple copies of a specific transcription factor binding site placed upstream of a minimal mammalian promoter element. The minimal promoter element is transcriptionally silent and contains only the RNA polymerase binding site. This synthetic promoter is then used to promote the expression of a reporter gene, such as luciferase (STRATOWA et al. 1995; Table 1). The structure of a synthetic CRE–luciferase reporter construct is shown in Fig. 2. This construct contains six copies of the consensus CRE placed upstream of the minimal herpes simplex virus thymidine kinase promoter element. To optimize the signal to noise obtained with the reporter, the precise position and number of copies of the response element used in such a promoter is derived from empirical study and, usually, many copies of the response element are required (STRATOWA et al. 1995 and references therein).

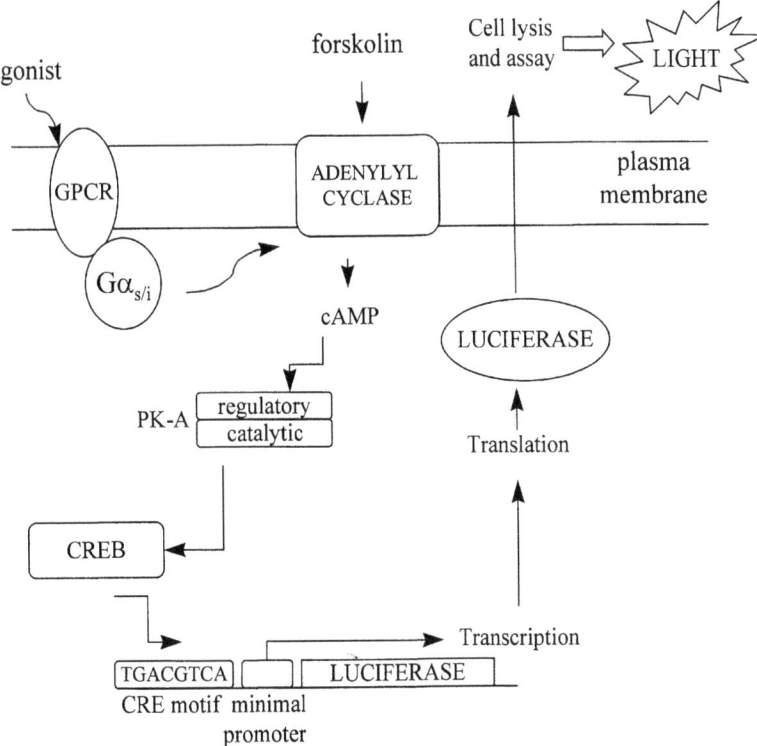

Fig. 1. Schematic representation of a cyclic adenosine monophosphate (cAMP) reporter gene cell line used as a screening system for agonist activity at $G\alpha_s$-coupled receptors. Agonist binding to a receptor capable of functional interaction with $G\alpha_s$ results in the generation of an increase in intracellular cAMP, resulting in the subsequent activation of protein kinase A (PKA). Activated PKA is able to phosphorylate members of the cAMP response element (CRE) binding (CREB) family of transcription factors. Activated CREB is able to bind to CRE to promote the expression of the firefly luciferase reporter gene. Luciferase luminescence provides a readout of agonist activity at the receptor

Many other mammalian response elements have been identified and used in reporter constructs (Table 1). Commonly used examples include the SRE, used to monitor MAPK activation (TREISMAN 1995), the AP-1 binding site (12-O-tetradecanoyl phorbol-13-acetate response element), used to detect PKC activation (STRATOWA et al. 1995b), the sis-inducible element, used to detect growth factor and cytokine signaling (SADOWSKI et al. 1993), and nuclear hormone response elements (KLEIN-HITPAB 1986; DIAMOND et al. 1990).

b) Chimeric Transcription Factors

Chimeric transcription factor reporter genes have been used to characterize receptor activation of the MAPK cascade. Many cell surface receptors, including $G\alpha_i$- and $G\alpha_{q/11}$-coupled GPCRs, are capable of the regulation of this signal

Table 1. Transcription-factor response elements used in reporter constructs

Response element	Transcription factor	Receptors
CRE	CREB, ATF	$G_{\alpha s}$ coupled $G_{\alpha i}$ coupled
TRE (AP-1 site)	c-Jun	$G_{\alpha q}$ coupled Growth factors
SRE	Elk-1, Sap-1	$G_{\alpha q}$ coupled $G_{\alpha i}$ coupled $G_{12/13}$ coupled Growth factors
SIE	ISGF3α	Growth factors Cytokines
NF-AT	NF-AT	$G_{\alpha q}$ coupled Growth factors Cytokines
NRE	Steroid receptors	Estrogen Progesterone Glucocorticoid

AP-1, activating protein 1; *ATF*, activating transcription factor; *CRE*, cyclic adenosine monophosphate response element; *CREB*, CRE-binding protein; *ISGF*, interferon-stimulated gene factor; *NF-AT*, nuclear factor of activated T cells; *NRE*, nuclear hormone response element; *SIE*, sis-inducible element; *SRE*, serum response element; *TRE*, 12-*O*-tetradecanoyl phorbol-13-acetate response element.

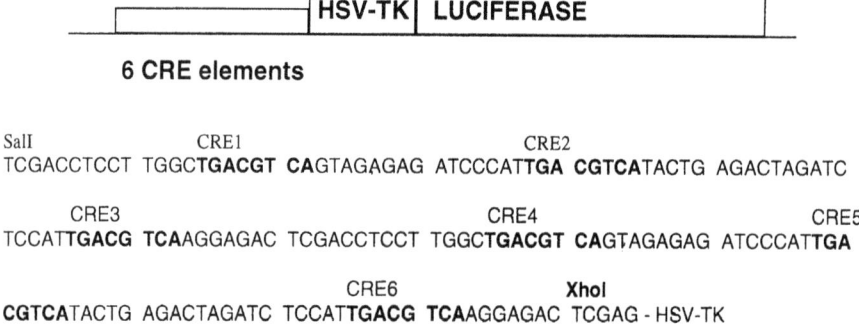

```
                         ┌──────────┬──────────────────────────┐
─────────────────────────┤  HSV-TK  │       LUCIFERASE         │
                         └──────────┴──────────────────────────┘
      6 CRE elements

SalI              CRE1                          CRE2
TCGACCTCCT TGGCTGACGT CAGTAGAGAG ATCCCATTGA CGTCATACTG AGACTAGATC

      CRE3                          CRE4                      CRE5
TCCATTGACG TCAAGGAGAC TCGACCTCCT TGGCTGACGT CAGTAGAGAG ATCCCATTGA

                 CRE6                   XhoI
CGTCATACTG AGACTAGATC TCCATTGACG TCAAGGAGAC TCGAG - HSV-TK
```

Fig. 2. Structure of an inducible reporter gene. The synthetic promoter consists of multiple copies of a transcription factor binding site, in this case six copies of the consensus cyclic adenosine monophosphate response element (CRE; *bold*) linked to a minimal promoter element, such as the herpes simplex virus thymidine kinase promoter (linkage occurs at amino acids +110 to +101 of the promoter). The inducible promoter is used to drive expression of a luciferase reporter gene

transduction cascade (POST and BROWN 1996). Eukaryotic transcription factors generally contain two domains: a regulatory domain and a DNA binding and transactivation domain. The regulatory domain contains a number of serine, threonine or tyrosine residues, which act as substrates for phosphorylation by upstream kinases that are activated as a consequence of the stimulation of cell surface receptors. The DNA binding and transactivation domain is able to recognize and bind to a specific transcription factor binding site, resulting in the activation of any promoter containing that sequence. As such, it is possible to construct chimeric proteins consisting of the regulatory domain of one transcription factor fused to the DNA binding and transactivation domain of a second. As a consequence of binding to a specific transcription factor binding site termed the Gal4 upstream activating sequence (UAS), the Gal4 transcription factor, which is derived from the budding yeast *Saccharomyces cerevisiae*, activates two genes involved in galactose metabolism. Several groups have constructed chimeric transcription factors consisting of the regulatory domain of a mammalian transcription factor linked to the DNA binding and transactivation domain of yeast Gal4 (KORTENJMAN et al. 1994; STRAHL et al. 1996). The mammalian transcription factor regulatory domain is phosphorylated as a consequence of a signal transduction event. This enables the chimeric protein to recognize and bind to the Gal4 UAS, resulting in the activation of a firefly luciferase reporter construct containing multiple copies of the Gal4 UAS in the promoter element (summarized in Fig. 3). Gal4-based reporter gene assays have been developed to report on the activity of many transcription factors, including elk-1, sap1a, activating transcription factor 2 and c-Jun (KORTENJMAN et al. 1994; STRAHL et al. 1996). For example, the Gal4/Elk-1 chimeric transcription factor consists of amino acids 1–147 of Gal4 (which encompasses the DNA binding domain) fused to the regulatory domain of Elk-1 (amino acids 83 to 428; GILLE et al. 1995).

Activation of the MAPK pathway following GPCR, tyrosine kinase or other cell surface receptor stimulation results in the activation of the MAPK enzymes, ERK1 and ERK2 (extracellular signal regulated kinases 1 and 2, also known as p42 and p44 MAPK, respectively). As shown in Fig. 3, activated ERK is able to phosphorylate and activate a Gal4/Elk-1 chimeric transcription factor, with subsequent induction of luciferase expression (KORTENJMANN et. al. 1994). In contrast to conventional MAPK assays (such as peptide phosphorylation assays), or immunoprecipitation and Western blotting to identify phosphorylated MAPK (FUKUDA et al.1997), the reporter gene assay allows a rapid detection of the activation of MAPK enzymes in whole cells and allows the identification of the transcription factors that are regulated as a result of MAPK activity.

2. Reporter Proteins

The *Aequorea victoria* photoprotein aequorin has been used for many years as a reporter of changes in intracellular calcium concentration in mammalian

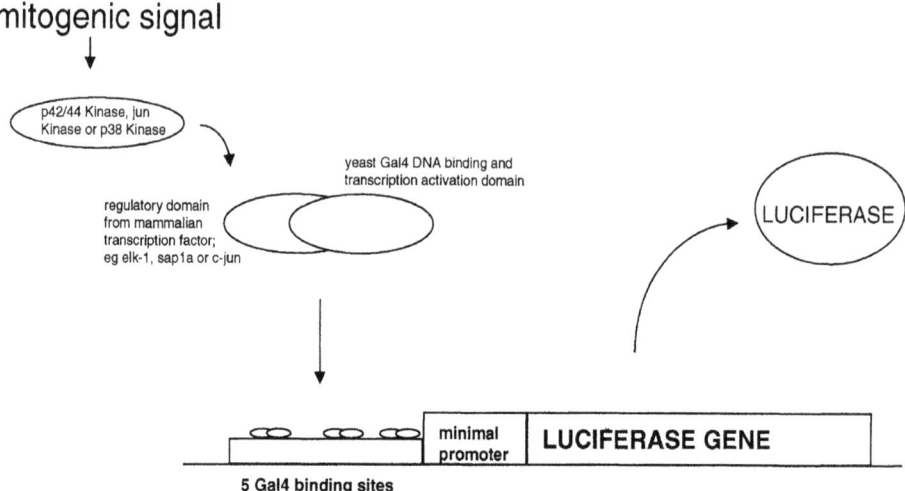

Fig. 3. Principles of a mammalian transcription factor/Gal4 chimeric reporter system. A fusion protein consisting of the regulatory domain from a mammalian transcription factor and the DNA-binding domain of the yeast Gal4 protein is constitutively expressed in mammalian cells. This chimeric protein is phosphorylated by an upstream kinase as a consequence of a signal transduction event, resulting in its activation. Once activated, it is able to bind to the Gal4 response element, resulting in activation of a reporter gene in which luciferase expression is under the transcriptional control of a Gal4-responsive synthetic promoter

cells or *Xenopus* oocytes (ASHLEY and CAMPBELL 1979). Aequorin is a 21-kDa photoprotein that forms a bioluminescent complex when linked to the chromophore cofactor coelenterazine (BRINI et al. 1995). Following the binding of Ca^{2+} to this complex, an oxidation reaction of coelenterazine results in the production of apoaequorin, coelenteramide, CO_2 and light with an emission λ_{max} of 395 nm. Aequorin luminescence has very rapid "flash"-type kinetics; for this reason, a luminometer equipped with injectors is required for detection. Loading of cells has traditionally involved microinjection of purified aequorin protein, which has limited the usefulness of this reporter system. In recent years, cloning of aequorin complementary DNA (cDNA) has allowed expression of this protein, both transiently and stably, in a range of cell types and has greatly expanded the utility of aequorin as a reporter.

Agonist binding at a GPCR that couples to a $G\alpha_{q/11}$ family G protein results in the activation of the phosphoinositidases of the phospholipase $C\beta$ class, which catalyze the formation of the second messenger metabolites *sn*-1-2-diacylglycerol and inositol (1,4,5)-trisphosphate. This is followed by the release of calcium from intracellular stores and the activation of PKC (NEER and CLAPHAM 1989; Fig. 3). The increase in cytoplasmic calcium following agonist binding can be detected by the generation of aequorin luminescence in mammalian cells that constitutively express this reporter (STABLES et al. 1997). Aequorin has been used to report agonist activation of a number of

GPCRs, including a histamine receptor (BRINI et al. 1995), the 5-HT$_{2A}$ serotonin receptor (WEYER et al. 1993), the V$_{1A}$ vasopressin receptor (BUTTON and BRONSTEIN 1993) and the α_1-adrenoceptor (BUTTON and BRONSTEIN 1993; BRINI et al. 1995). Furthermore, aequorin has been co-expressed with the G protein α subunit Gα_{16} to generate a generic screening system for agonist activation of several GPCRs (STABLES et al. 1997).

To expand the range of uses of aequorin, a number of modified aequorins have been constructed in which expression of the protein is targeted to particular cellular compartments in order to measure calcium changes within those compartments (DEGIORGI et al. 1996). This includes aequorin targeted to the mitochondria (RIZZUTO et al. 1992), nucleus (BRINI et al. 1993) and endoplasmic recticulum (BRINI et al. 1997). Furthermore, the construction of fusion proteins with aequorin has facilitated the analysis of local calcium changes. A calcium-sensitive adenylyl cyclase/aequorin fusion protein has been used to report the changes in intracellular calcium concentration that regulate this calcium-sensitive enzyme (NAKAHASHI et al. 1997).

The *A. victoria* GFP is a 238 amino acid photoprotein that emits green light upon fluorescent excitation. Unlike other bioluminescent reporter molecules, no additional substrates or cofactors are required for light emission (CHALFIE et al. 1994). GFP fluorescence is stable and has been measured non-invasively in living cells of many species, including mammalian cells, *Drosophila*, *Caenorhabditis elegans*, yeast and *Escherichia coli*. The use of GFP as an inducible reporter gene has been limited due to the brightness of the protein, which, while readily detectable by fluorescence microscopy or fluorescence-activated cell-sorting analysis, is not easily detectable in a plate fluorimeter. Availability of the cDNA sequence for GFP has resulted in the generation and characterization of several GFP mutants with enhanced fluorescence emission. The active chromophore within GFP is a cyclic hexapeptide spanning amino acids 64–69. Mutation of the serine at amino acid 65 to threonine has resulted in the generation of a protein with a sixfold increase in the intensity of fluorescence emission. Furthermore, the presence of the Ser65Thr mutation and the mutation of the phenylalanine residue at position 64 to leucine results in a 35-fold increase in fluorescence intensity in mammalian cells. In order to facilitate protein folding in mammalian cells, variants of the protein for which codon usage within the cDNA has been optimized for human cell expression are now available (ZOLOTUKHIN et al. 1996).

A number of mutants of GFP with altered excitation or emission characteristics have also been identified. For example mutation of the tyrosine residue at position 66 to histidine generates a protein with blue fluorescence emission [the so-called blue fluorescent protein (BFP)], with a λ_{max} for excitation of 458 nm and for emission of 480 nm (HEIM et al 1994; HEIM and TSIEN 1995). The majority of these variants of GFP protein are now commercially available.

GFP has been widely used in fusion proteins to assess intracellular protein trafficking (FERRER et al. 1997) and subcellular localization of recombinantly expressed proteins (WANG and HAZELRIGG 1994). For example, in an elegant study, a fusion protein between the β_2-adrenoceptor and GFP has been used to monitor receptor expression, localization at the plasma membrane and internalization following agonist stimulation (BARAK et al. 1997). Several groups have investigated the use of GFP and BFP as partners for FRET. In such studies, excitation of BFP at 368nm causes emission of light at 445nm, which excites a Ser65Cys mutant of GFP to generate light emission at 509nm (HEIM and TSIEN 1996). A fusion protein consisting of GFP and BFP with a linker sequence containing a trypsin-cleavage site between the two fluorescent proteins was constructed. When excited at 368nm, the fusion protein emits light at 509nm. Upon treatment with trypsin, the fusion protein was cleaved, with the result that fluorescence excitation at 368nm then generated fluorescence emission at 445nm (HEIM and TSIEN 1996). Such FRET partners may be used as reporters of protease activity. In a similar report, Miyawaki and colleagues described the construction of a BFP–calmodulin–GFP fusion construct and its use as a non-enzymatic reporter of calcium concentration in mammalian cells (MIYAWAKI et al. 1997).

IV. Transient Versus Stable Expression of Reporter Genes

1. Transient Expression

The pharmacological analysis of GPCRs has been greatly facilitated by the ability to create mammalian cell lines either transiently or stably expressing the receptor of interest (HOYER et al. 1994). Reporter vectors can be either transiently transfected into mammalian cells that endogenously express the receptor of interest or co-transfected into a heterologous cell line along with an expression vector for the receptor. Pharmacological experiments are generally performed within 48h of transfection. Examples of experiments for which transient transfection is advantageous are:

1. For studies comparing the regulation of different promoter elements.
2. For optimization of the number of response elements in a reporter construct.
3. For testing the ability of exogenously expressed receptors to activate a particular response element. Transient transfections are labor intensive, use large quantities of plasmid stocks and can be expensive depending on the transfection reagent.

2. Stable Expression

Cell lines stably expressing the receptor and reporter genes are often created if experiments will require repeated use of a single reporter/receptor combination. There are several ways to accomplish the generation of stable lines. In the two-step model, stable cell lines that express only the reporter gene are created. The optimal stable host reporter cell line continues to express the reporter gene for at least several months and exhibits a high signal-to-noise ratio on activation of the reporter gene as a result of stimulation of endogenous receptors or enzymes. Receptors of interest can then be stably transfected into the host reporter gene cell line. Alternatively, the receptor and reporter construct can be co-transfected. This is a good approach when using a cell line for which known endogenous response with which to search for stable lines expressing the reporter gene exists. In this model, activation of the co-transfected receptor can be used to identify useful stable cell lines. Finally, if a receptor is difficult to express, it may be expeditious to simply transfect the reporter gene into a cell line with endogenous expression of the receptor of interest. Compared to transient transfection assays, pharmacological assays using stable cell lines are less expensive, more facile, higher capacity and have a lower degree of variability.

B. Use of Reporter Gene Systems in Pharmacology

I. Drug Discovery

Reporter gene assays have been used for primary screening efforts in both transient transfection and stable cell line formats. In our laboratories, stable cell lines expressing the receptor of interest are selected based on receptor-mediated activation of the reporter gene rather than protein expression. Both receptor binding assay and reporter assay screens can be executed using the same stable cell line. Reporter assays are most useful when screening for agonist ligands; however, these assays can be used to screen for an antagonist by including a submaximal concentration of agonist in each assay well. The stable reporter lines are useful for secondary pharmacological characterization of ligands discovered through the screening process (see below). In addition, stable reporter cell lines can be created for receptor subtypes related to the target of interest for use in selectivity characterization.

Firefly luciferase reporter gene assays have proven to be facile, sensitive, inexpensive, high-capacity screens. The stable cell line screens have been run in 96-well and 384-well format in our experiments, although much higher density formats are possible. Currently, one scientist can assay approximately 8000 or 30,000 compounds per day in the 96-well or 384-well format, respectively. These assays can be automated on a rail system if the appropriate incubation equipment is available.

II. Receptor Pharmacology

Reporter gene assays can be used for pharmacological studies of any type of receptor that modulates gene transcription either directly or through intracellular signaling pathways. This statement assumes that an appropriate transcriptional response element or natural promoter that is responsive to activation of the target receptor has already been discovered. The remainder of this chapter will focus on studies performed in our laboratories with G protein-coupled receptors and integrin receptors.

1. Use of Reporter Gene Assays to Assess Receptor Agonism

a) Example 1. 6CRE-Luciferase as a Reporter for $G\alpha_s$-Coupled Receptor Signaling

CRE–luciferase reporter genes are responsive to any signaling event that results in the phosphorylation and activation of the CREB family of transcription factors. Thus, CRE reporter genes can be used to report on GPCR-mediated alterations in the concentration of intracellular cAMP (HIMMLER et al. 1993; Fig. 1). However, CRE–luciferase reporter genes can be regulated by other signal transduction events in certain cell types. Both calcium/calmodulin-dependent protein kinase II and rsk2 can activate CREB transcription factors in response to receptor-mediated elevation of intracellular calcium levels and activation of the MAPK enzymes (MATTHEWS et al. 1994; XING et al. 1996). In our laboratories, the 6CRE–luciferase reporter gene (Fig. 2) is routinely used to assess agonist activity on a range of $G\alpha_s$- and $G\alpha_i$-coupled GPCRs including members of the $G\alpha_s$-coupled secretin receptor family (Fig. 4). The 6CRE–luciferase reporter gene construct described in Fig. 2 was transfected into Chinese hamster ovary (CHO) cells, and a stable cell line was selected based on the response of the reporter gene to forskolin (which directly activates adenylyl cyclase) and calcitonin. A calcitonin receptor is expressed endogenously in CHO cells. This cell line then served as a host for the subsequent transfection and expression of several secretin family GPCRs. Concentration–response curves (constructed using the human peptide as ligand) and the generation of luciferase luminescence as the readout revealed that the EC_{50} for calcitonin at the CHO calcitonin receptor was 686 pM, the EC_{50} for PTH (parathyroid hormone) at the human PTH receptor was 73 pM, the EC_{50} for CRF (corticotropin-releasing factor) at the human CRF1 receptor was 13 nM, and the EC_{50} for PACAP (pituitary adenylate cyclase-activating peptide) 1–38 at the human PACAP receptor was 260 pM (Fig. 4). These values are comparable to those reported for induction of intracellular cAMP at these receptors (CHEN et al. 1993; HUANG et al. 1996; PISEGNA and WANK 1996; GEORGE et al. 1997). The data in Fig. 4 are expressed as percent maximal response; however, the fold stimulation by the peptides in each cell line can vary widely. Interestingly, the endogenous forskolin and calcitonin fold responses also vary in cell lines derived from host reporter lines that have been transfected with various GPCRs.

b) Example 2. Aequorin as a Reporter for $G\alpha\alpha_{q/11}$-Coupled Receptor Signaling

To demonstrate the ability of aequorin to report agonist activation of GPCRs that couple to G proteins of the $G\alpha_{q/11}$ family, an aequorin reporter gene con-

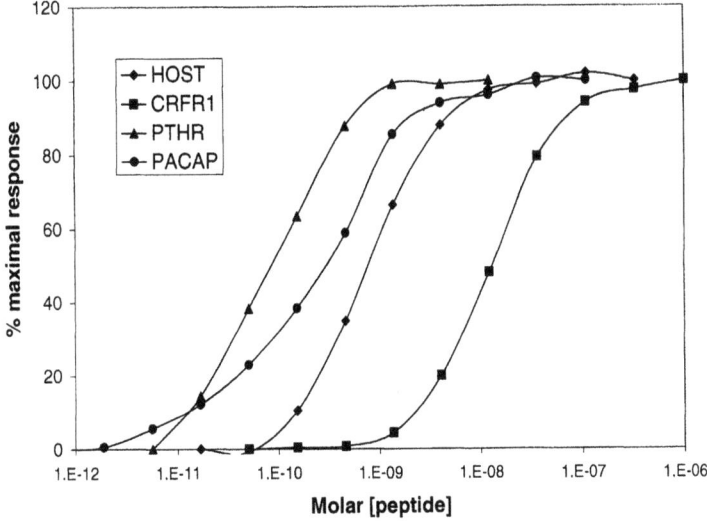

Fig. 4. Agonist stimulation of cyclic-adenosine-monophosphate response element 6 (6CRE)–luciferase reporter gene for several $G_{\alpha s}$-coupled G-protein-coupled receptors. A stable reporter gene host cell line was created by transfecting wild type Chinese hamster ovary cells with the 6CRE–luciferase vector (Fig. 2) and selecting the optimal clone based on endogenous responses to forskolin and calcitonin. The host line was then transfected with expression vectors for human corticotropin-releasing-factor receptor 1 (CRFR1), parathyroid hormone (PTH) and pituitary adenylate-cyclase-activating peptide (PACAP) receptors. The host line and the three lines expressing CRFR1, PTH receptor and PACAP receptor were placed in 96-well viewplates (Packard). On the day before the experiment, the medium was removed and replaced with 50 μl of serum-free Dulbecco's modified Eagle's medium/F-12. The cells were treated with the indicated concentrations of human calcitonin (host line), CRF, PTH 1–34 or PACAP 1–38 peptides in a 50-μl volume, and the plates were incubated for 4 h at 37°C. One hundred microliters of LucLite (Packard) solution was added to the wells and, after 20 min, the luciferase activity was quantified in a Topcount microplate scintillation and luminescence counter (Packard)

taining a mitochondrial targeted aequorin cDNA expressed from a constitutive cytomegalovirus promoter was transiently transfected into a number of cell lines expressing such receptors. In CHO cells stably expressing the $G\alpha_{q/11}$-coupled human endothelin ET_A, angiotensin AT_1, thyrotrophin-releasing hormone (TRH), neurokinin NK_1, and purinergic P2y receptors, transient expression of aequorin followed by the application of a near-maximal concentration of the appropriate agonist caused a large stimulation of aequorin luminescence (Fig. 5a). The increase in luminescence varied from 50-fold for the NK_1 receptor activated by agonist to 250-fold for the ET_A receptor activated by endothelin 1. A concentration–response curve constructed for angiotensin II activity at the AT_1 receptor revealed the EC_{50} for this agonist to be 100 ± 69 nM (Fig. 5b). Similar concentration–response curves were

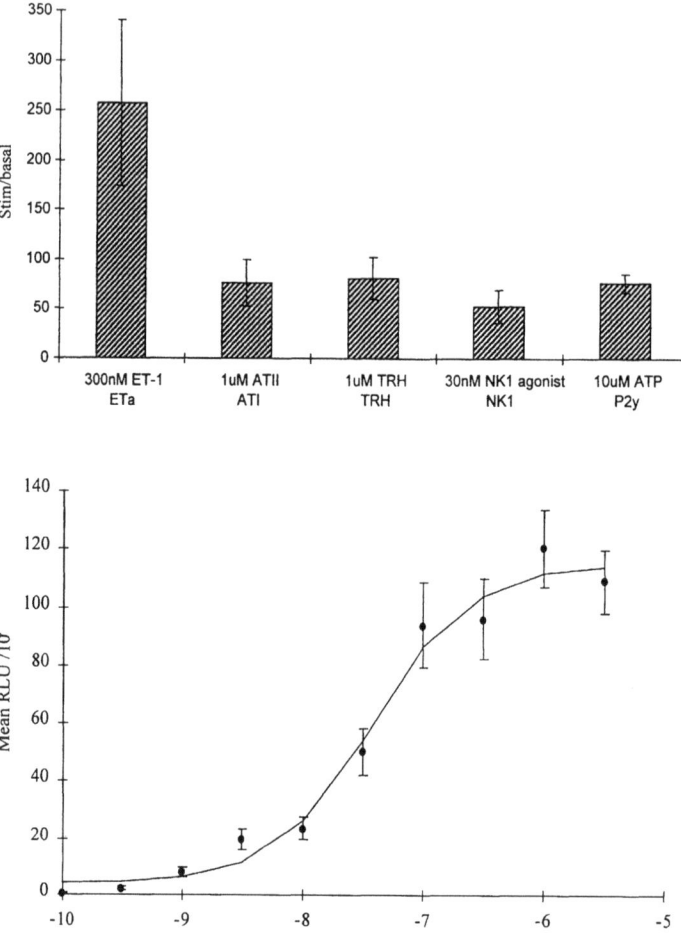

Fig. 5. Agonist stimulation of aequorin luminescence for a range of $G\alpha_{q/11}$-coupled G-protein-coupled receptors. **a** A plasmid in which expression of aequorin is under the transcriptional control of the constitutive cytomegalovirus promoter was transiently transfected into Chinese hamster ovary (CHO) cells stably expressing the human endothelia ET_A, the human angiotensin AT_1, the human neurokinin NK_1, rat thyrotrophin-releasing hormone and hamster P2y receptors. Each receptor was stimulated with a maximal concentration of the appropriate agonist as indicated, and aequorin luminescence was detected in a Dynatech ML3000 luminometer. The data was pooled from 3–5 experiments, with each performed in triplicate, and was expressed as fold stimulation over basal luminescence in response to near-maximal concentrations of agonist. **b** CHO cells stably expressing the AT_1 angiotensin receptor were transiently transfected with an aequorin reporter plasmid. A concentration–response curve for the agonist AT-II was constructed. Data are from a representative of three experiments performed in triplicate and are expressed in relative light units

constructed for endothelin 1 at the ET_A receptor ($EC_{50} = 59 \pm 13$ nM), TRH at the TRH receptor ($EC_{50} = 264 \pm 27$ nM), NK_1 agonist at the NK_1 receptor ($EC_{50} = 8.21 \pm 2.8$ nM) and adenosine triphosphate at the P2y receptor ($EC_{50} = 5.3 \pm 2.5\,\mu$M; data not shown).

c) Example 3. The Use of a Gal4/Elk-1 Chimera to Report Opioid-Receptor-Like Receptor-1 Activation of MAPK

Stimulation of many GPCRs causes a rapid elevation in the activity of a family of closely related serine–threonine kinases known as the MAPKs (POST and BROWN 1996). A recent report by FUKUDA et al. (1997) showed that nociceptin stimulation of the opioid-receptor-like receptor 1 (ORL1 receptor) expressed in CHO cells caused a rapid increase in the activity of the MAPK enzymes ERK1and ERK2. In our laboratory, we have demonstrated ORL1 activation of these enzymes using reporter gene assays. CHO cells stably expressing the ORL1 receptor were transiently transfected with a Gal4/Elk-1 chimeric transcription factor together with a second plasmid containing the firefly luciferase reporter gene under the transcriptional control of a Gal4-responsive promoter (KORTENJMANN et. al. 1994; Fig. 3). Nociceptin caused an increase in Gal4/Elk-1 activity in a concentration-dependent manner, with an EC_{50} of 0.49 nM (range 0.15–1.63 nM; Fig. 7). The EC_{50} for regulation of Gal4/Elk-1 is not significantly different from that for regulation of ERK1/ERK2 determined in a peptide phosphorylation assay in our laboratory (0.28 nM, range 0.17–0.45 nM) or the EC_{50} reported by FUKUDA et al. for nociceptin activation of ERK1/ERK2 (FUKUDA et al. 1997).

2. Use of Reporter Gene Assays to Evaluate Receptor Antagonism

Reporter gene assays can be used to evaluate functional antagonism. A CHO cell line stably transfected with the 6CRE–luciferase cAMP reporter construct (Fig. 1) was described in Sect. B.II.1.a. As shown in Fig. 8, concentration–response curves for calcitonin are progressively shifted to the calcitonin receptor by increasing concentrations of a peptide antagonist. The pK_B calculated from the reporter data is 8.57. The pK_I determined from receptor binding studies using CHO cell membranes is 8.55 (data not shown). Thus, the assessment of the potency of the antagonist in the reporter assay is comparable to that of the classical receptor binding assay.

3. Simultaneous Detection of Multiple Signals

a) Example 1. Dual Reporter Assays

A number of reagents are now commercially available for the sequential measurement of two reporter genes; for example, it is possible to assay firefly and *Renilla* luciferase sequentially in the same assay sample. In this assay, firefly luciferase is assayed first. Following this measurement, the *Renilla* luciferase

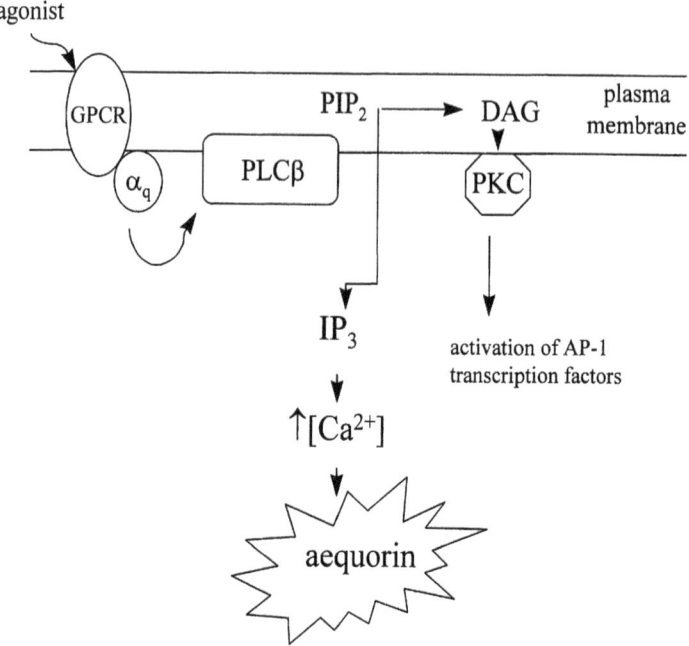

Fig. 6. Construction of an aequorin reporter gene cell line. The aequorin protein is constitutively expressed in mammalian cells. Agonist binding to a receptor which is capable of coupling to a G-protein of the $G\alpha_{q/11}$ family results in the generation of an increase in intracellular calcium concentration, as described in the text. In the presence of calcium and the cofactor coelenterazine, aequorin luminescence is generated from aequorin expressed in the same cell. DAG, diacylglycerol; GDP, guanosine diphosphate; GTP, guanosine triphosphate; IP_3, inositol (1,4,5)-triphosphate; PIP_2, phosphatidylinositol (4,5)-bisphosphate; PKC, protein kinase C; PLCβ, phospholipase Cβb

assay reagent is added to the same assay samples. The *Renilla* luciferase assay reagent quenches the firefly luciferase reaction and initiates the *Renilla* luciferase signal to produce a second luminescence signal to allow detection of this reporter. Dual reporter assays allow the analysis of two signal transduction events in the same cell using two different reporter genes specific to two signal transduction events. Furthermore, in our laboratory, we have used the dual luciferase assay to simultaneously characterize two GPCRs that signal through activation of the G protein α subunit ($G\alpha_s$) in the same assay well. In these experiments, CHO cells stably transfected with a cAMP-responsive firefly luciferase reporter were further transfected with the human vasopressin V_2 receptor. Similarly, CHO cells stably transfected with a cAMP-responsive *Renilla* luciferase reporter were further transfected with the human $β_2$-adrenoceptor. The two cell lines were mixed in individual wells of a 96-well plate, and dose–response curves for the V_2 agonist vasopressin and the $β_2$ agonist isoprenaline were constructed using firefly luminescence as the readout for vasopressin activity and *Renilla* luminescence as the readout for isoprenaline

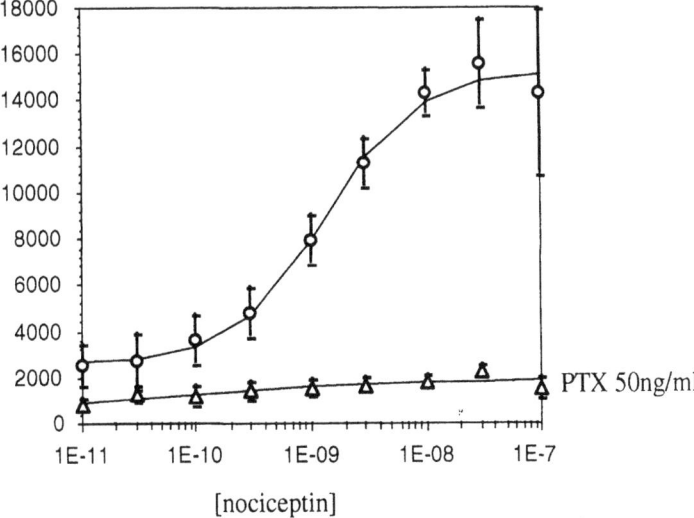

Fig. 7. Nociceptin activation of extracellular signal-regulated kinase using a Gal4/elk-1 reporter gene. Chinese hamster ovary cells stably expressing opioid-receptor-like receptor 1 were transiently transfected with a Gal4/Elk-1 chimeric transcription factor together with a second plasmid containing the firefly luciferase reporter gene under the transcriptional control of a Gal4-responsive promoter. A concentration–response curve to nociceptin was constructed in the absence (●) and the presence (▲) of pre-treatment for 18h with 50ng/ml pertussis toxin. All values are the mean counts per second of at least three experiments performed in duplicate ± the standard error of the mean

activity. Following the addition of firefly luciferase assay reagent, vasopressin was seen to activate the V_2 receptor with an EC_{50} of 11 nM (Fig. 9). As expected, vasopressin did not stimulate *Renilla* luminescence. Following the subsequent addition of *Renilla* luciferase assay reagent to the same cells, the dose–response curve for vasopressin was flattened, and a dose–response curve for isoprenaline was revealed, with an EC_{50} of 0.11 nM (Fig. 9). The use of dual reporter assays for compound screening allows the simultaneous screening of two receptors, with significant savings in compound use, time and cost.

b) Example 2. Combination of Reporter Assays with Other Assay Types

We have developed a novel method for detection of two or more signaling events in a single assay (IGNAR and YINGLING 1997). The LuFLIPRase assay combines reporter gene assays with the measurement of intracellular ion fluxes via fluorescent dyes using the fluorometric imaging plate reader (FLIPR, Molecular Devices). Either inducible reporter genes or reporter proteins can be used in this system. The method is useful for the study of a single receptor that couples to more than one signal transduction event. An example of this is shown in Fig. 10, in which a splice variant of the human PACAP recep-

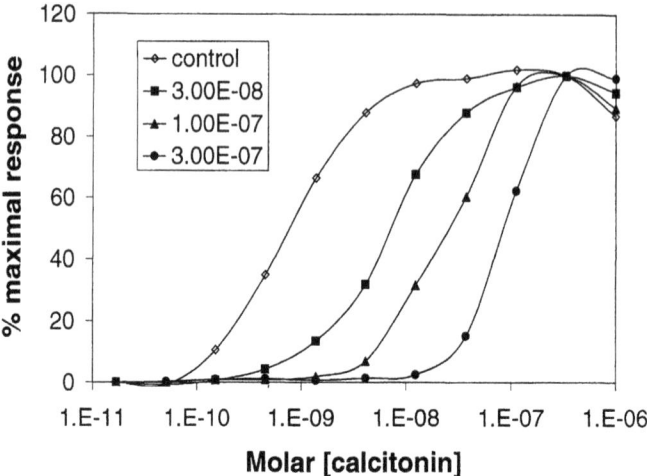

Fig. 8. Antagonism of $G_{\alpha s}$ coupled response using the cyclic-adenosine-monophosphate response element 6 (6CRE)–luciferase reporter gene. Host 6CRE–luciferase reporter gene-expressing Chinese hamster ovary cells were placed in 96-well viewplates (Packard) in Dulbecco's modified Eagle's medium (DMEM)/F-12 medium containing 10% fetal bovine serum. The medium was replaced with serum-free DMEM/F-12 on the day before the experiment. The cells were pre-treated with the indicated concentrations of a peptide antagonist of the calcitonin receptor. Thirty minutes later, human synthetic calcitonin was added to the wells, and the plates were incubated for 4h at 37°C. One hundred microliters of LucLite solution was added to the wells and, after 20 min, luciferase activity was quantified in a Topcount microplate scintillation and luminescence counter (Packard)

tor expressed in 6CRE–luciferase CHO cells was treated with PACAP 1–38, resulting in activation of the CRE reporter gene and mobilization of intracellular calcium. It has been reported that the PACAP receptor couples to both adenylyl cyclase and phosphatidylinositol (PI) lipid metabolism and that the potency of cyclase activation is at least ten times greater than that of PI metabolism (PISEGNA and WANK 1996), similar to the data presented in Fig. 10. Thus, this assay may be useful in the study of agonist trafficking with GPCRs that couple to more than one G protein.

As described above for the dual reporter gene assay (Sect. B.II.3.a), the LuFLIPRase assay can also be used for the simultaneous screening of two GPCRs. In the LuFLIPRase assay, activation of each receptor would be linked to a different detection system. For instance, one cell line would contain a luciferase reporter gene responsive to one receptor and the second cell line would contain a receptor coupled to an intracellular ion flux measurable with fluorescent dye in the FLIPR. Two or more recombinant cell lines can be mixed together in a single well such that screening for ligands at multiple receptors can be prosecuted concurrently, which decreases expense and personnel expenditure. This is a versatile technique that can be adapted to the

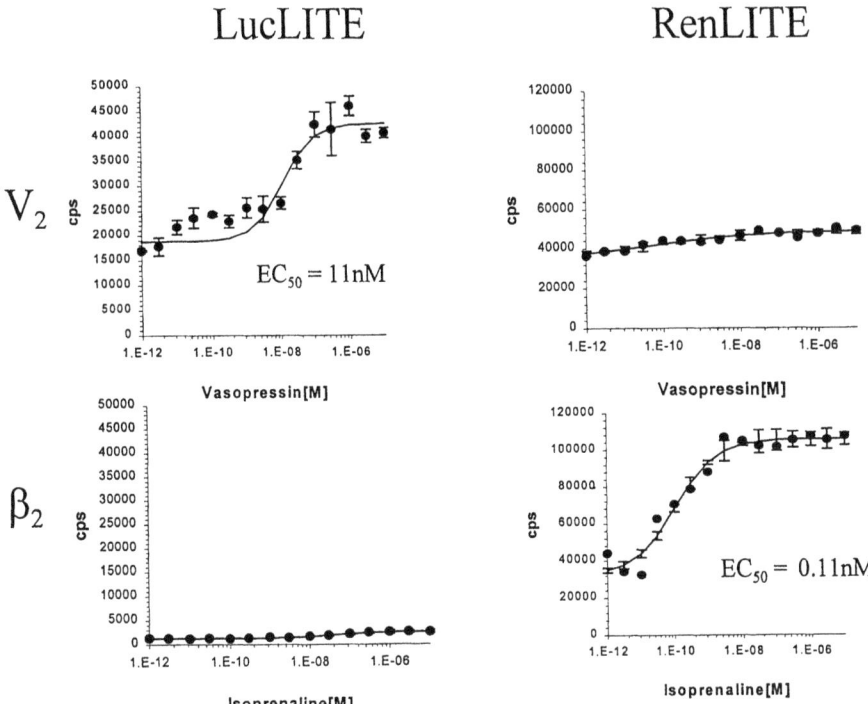

Fig. 9. Characterization of the human vasopressin V_2 receptor and β_2-adrenoceptor using a dual luciferase assay. Chinese hamster ovary (CHO) cells stably expressing the vasopressin V_2 receptor and a firefly luciferase reporter gene and CHO cells stably expressing the β_2-adrenoceptor and a *Renilla* luciferase reporter gene were mixed and placed into individual wells of a 96-well plate. Concentration–response curves to vasopressin and isoprenaline were constructed using firefly luminescence as the readout for vasopressin activity, and *Renilla* luminescence as the readout for isoprenaline activity (see text for details). All values are expressed as mean counts per second of at least three experiments performed in duplicate ± the standard error of the mean

study of almost any receptor-regulated signaling system through the use of reporter genes and fluorescent dyes.

4. Measurement of Constitutive Activity

The phenomenon of constitutive signaling activity of GPCRs (LEFKOWITZ et al. 1993) can be assessed using transient transfection reporter gene assays. The use of a PKC-responsive promoter construct to detect constitutive signaling from the thyrotropin-releasing hormone (TRH) receptor was recently reported (JINSI-PARIMOO and GERSHENGORN 1997). COS-1 cells were transfected with increasing amounts of an expression vector for the TRH receptor. The constitutive activity detected by the reporter gene correlated with the amount of receptor expression. In contrast to the reporter gene data, the measurement of inositol phosphate formation was not sensitive enough to pick up

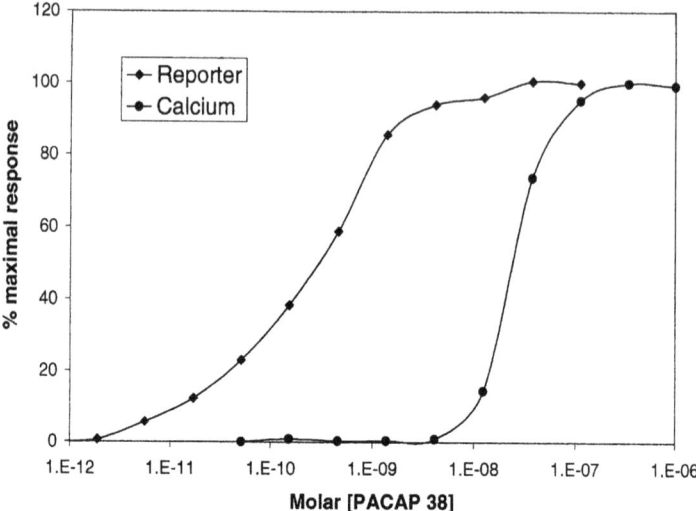

Fig. 10. Concurrent assay of multiple signal transduction events regulated by a single receptor. Host cyclic-adenosine-monophosphate response element-6–luciferase reporter gene-expressing Chinese hamster ovary cells were stably transfected with a splice variant of the human pituitary adenylate-cyclase-activating-peptide (PACAP) receptor. The cells were placed in viewplates in Dulbecco's modified Eagle's medium (DMEM)/F-12 medium containing 10% fetal bovine serum. On the day before the experiment, the medium was removed and replaced with serum-free DMEM/F-12. On the day of the experiment, the cells were loaded with Calcium Green-1 (Molecular Probes) before treatment with the indicated concentrations of PACAP 1–38 peptide. Calcium Green-1 exhibits an increase in fluorescence intensity after binding intracellular calcium ions. The addition of PACAP 1–38 peptide was performed in the fluorometric-imaging plate reader (FLIPR), and readings were taken every second for 60s ($\lambda_{excitation} = 488$ nm). The maximal change in fluorescence during this period was used to create the calcium-concentration–response curve for the calcium signal. After quantification of the calcium signal on the FLIPR, the cells were incubated for 4h at 37°C. One hundred microliters of LucLite was added to the wells and, after 20min, luciferase reporter gene activity was quantified in a Topcount microplate scintillation and luminescence counter

any change in receptor signaling. We have performed similar studies with secretin family GPCRs in CHO cells expressing the 6CRE–luciferase reporter gene. Each GPCR seemed to have a different inherent propensity for constitutive signaling. The human parathyroid hormone receptor (PTHR) demonstrated considerable constitutive activity when only 0.5ng of PTHR expression vector was transfected. Transfection of 50ng of human glucagon receptor expression vector was required to obtain a similar result. The human glucagon-like peptide (GLP)-1 receptor did not exhibit constitutive activity after transfection of up to 50ng of GLP-1 receptor expression vector. The ability to detect constitutive activity is useful in the study and discovery of ligands that induce inverse agonism.

5. Measurement of Efficacy

Although the assessment of efficacy in recombinant systems is difficult to validate regardless of assay type, we have compared the 6CRE-luciferase reporter assay to other, more classical measurements of functional response with regard to estimation of efficacy. We have compared the potency of agonists in the 6CRE-luciferase reporter assay, a cAMP assay and the microphysiometer in the 6CRE-luciferase CHO cells expressing human GLP-1 or glucagon receptors. In both cases, the potency of the peptide agonist was highest in the microphysiometer assay and lowest in the cAMP assay, with the reporter assay in the intermediate range. A similar observation was made by GEORGE et al. (1997) using the endogenous calcitonin response in CHO cells in a comparison of a 6CRE-luciferase reporter assay and measurement of cAMP levels. Reporter assays can also be used to differentiate selective agonist potency. The rank order of potency of several calcitonin peptides from different species on the CHO cell calcitonin receptor was very similar in the 6CRE-luciferase reporter assay and in the microphysiometer.

The ability to discern partial agonists from full agonists is critical to the evaluation of efficacy. We have discovered small molecule partial agonists for several receptors in high-throughput reporter screens. Interestingly, we have observed that the activity of a partial agonist in the reporter assay is usually a much higher percentage of the full agonist activity than in the cAMP assay.

A potential explanation for this phenomenon is that the amount of cAMP produced by a full agonist far exceeds the amount necessary for maximal activation of available PKA or transcriptional activation of the reporter gene. Thus, the assay chosen for analysis of efficacy should assess the signaling event that regulates the physiological function of interest if that information is known.

6. Assessment of New Signaling Pathways

Reporter gene assays can be utilized in the discovery of previously unknown signaling or transcriptional mechanisms regulated by a receptor. These studies require only an agonist to activate the receptor and various reporter gene constructs to perform studies of a wide range of signaling interactions that may yield unexpected results. An example of this is the discovery of steroid hormone-independent activation of nuclear steroid receptors by ligands for receptors localized at the cell surface, such as dopamine and peptide growth factors (POWER et al. 1991; IGNAR-TROWBRIDGE et al. 1993, 1996). In the case of the peptide growth factor studies, the reporter gene assays were instrumental in supporting the hypothesis of cross-talk between peptide growth factor receptors and the estrogen receptor, a proposal which was originally conceived from in vivo observations (IGNAR-TROWBRIDGE et al. 1992). Thus, reporter gene assays can be powerful tools for the elucidation of pharmacological mechanisms.

Acknowledgements. The authors would like to thank Nicola Bevan, Sue Brown, Aaron Goetz, Sarah Scott and Jenny Stables for allowing their data to be presented in this chapter.

References

Alam J, Cook JL (1990) Reporter genes: application to the study of mammalian gene transcription. Anal Biochem 188:245–254

Ashley CC, Campball AK (eds) (1979) The detection and measurement of free Ca^{2+} in cells. Elsevier/North Holland, Amsterdam

Barak LS, Stephen SG, Ferguson JZ, Caron MG (1997) A β-arrestin/green fluorescent protein biosensor for detecting G-protein-coupled receptor activation. J Biol Chem 272:27497–27500

Brini M, Murgia M, Pasti M, Picard D, Pozzan T, Rizzuto R (1993) Nuclear calcium concentration measured with specifically targeted recombinant aequorin EMBO J 12:4813–4819

Brini M, Marsault R, Bastianutto C, Alvarez J, Pozzan T, Rizzuto R (1995) Transfected aequorin in the measurement of cytosolic Ca2+ concentration ([Ca2+]c). A critical evaluation. J Biol Chem 270:9896–9903

Brini M, De Giorgi F, Murgia M, Marsault R, Massimino ML, Cantini M, Rizzuto R, Pozzan T (1997) Subcellular analysis of calcium homeostasis in primary cultures of skeletal muscle myotubes. Mol Biol Cell 8:129–143

Bronstein I, Fortin J, Stanley PE, Stewart GSAB, Kricka LJ (1994) Chemiluminescent and bioluminescent reporter gene assays. Anal Biochem 219:169–181

Button D, Brownstein M (1993) Aequorin-expressing mammalian cell lines used to report Ca2+ mobilization. Cell Calcium 14:663–671

Chalfie M, Tu Y, Euskirchen G, Ward WW, Prasher DC (1994) Green fluorescent protein as a marker for gene expression. Science 263:802–804

Chen R, Lewis KA, Perrin MH, Vale WW (1993) Expression cloning of a human corticotropin-releasing factor receptor. Proc Natl Acad Sci, USA 90:8967–8971

Chen W, Shields TS, Stork PJS, Cone RD (1995) A colorimetric assay for measuring activation of G_s- and G_q-coupled signaling pathways. Anal Biochem 226:349–354

De Wet JR, Wood KV, DeLuca M, Helsinki DR, Subramani S (1987) Firefly luciferase gene: structure and expression in mammalian cells. Mol Cell Biol 7:725–737

Delagrave S, Hawtin RE, Silva CM, Yang MM, Youvan DC (1995) Red-shifted excitation mutants of the green fluorescent protein. Biotechnology 13:151–154

Di Giorgi F, Brini M, Bastianutto C, Marsault R, Montero M, Pizzo P, Rossi R, Rizzuto R (1996) Targeting aequorin and green fluorescent protein to intracellular organelles. Gene 173:113–117

Diamond MI, Miner JN, Yoshinaga SK, Yamamoto KR (1990) Transcription factor interactions: selectors of positive or negative regulation from a single DNA element. Science 249:1266–1272

Ferrer JC, Baque S, Guinovart JJ (1997) Muscle glycogen synthase translocates from the cell nucleus to the cytosol in response to glucose. FEBS Lett 415:249–252

Fukuda K, Shed T, Morikawa H, Kato S, Mori K (1997) Activation of mitogen-activated protein kinase by the nociceptin receptor expressed in Chinese hamster ovary cells. FEBS Lett 412:290–294

George SE, Bungay PJ, Naylor LH (1997) Evaluation of a CRE-directed luciferase reporter gene assay as an alternative to measuring cAMP accumulation. J Biomolecular Screening 2:235–240

Gille H, Kortenjann M, Thomae O, Moomaw C, Slaughter C, Cobb MH, Shaw PE (1995) ERK phosphorylation potentiates Elk-1 mediated ternary complex formation and transactivation. EMBO J 14:951–962

Heim R, Tsien RY (1995) Engineering green fluorescent protein for improved brightness, longer wavelengths and fluorescence resonance energy transfer. Current Biol 6:178–182

Heim R, Prasher DC, Tsien RY (1994) Wavelength mutations and post-translational oxidation of green fluorescent protein. Proc Natl Acad Sci USA 91:12501–12504

Heim R, Cubitt AB, Tsien RY (1995) Improved green fluorescence. Nature 373:663–664

Henthorn P, Zervos P, Raducha M, Harris H, Kadesch T (1988) Expression of a human placental alkaline phosphatase gene in transfected cells: use as a reporter for studies of gene expression. Proc Natl Acad Sci USA 85:6342–6346

Hill CS, Treisman R (1995) Differential activation of c-fos promoter elements by serum, lysophosphatidic acid, G proteins and polypeptide growth factors. EMBO J 14:5037–5047

Himmler A, Stratowa C, Czernilofsky AP (1993) Functional testing of human dopamine D_1 and D_2 receptors expressed in stable cAMP-responsive luciferase reporter cell lines. J Receptor Res 13:79–94

Hoyer D, Clarke DE, Fozard JR, Hartig PR, Martin GR, Mylechareme EJ, Saxena PR, Humphrey PPA (1994) The IUPHAR classification of receptors for 5-hydroxytryptamine (serotonin). Pharmacol Rev 46:157–204

Huang Z, Chen Y, Pratt S, Chen TH, Bambino T, Nissenson RA, Shoback DM (1996) The N-teminal region of the third intracellular loop of the parathyroid hormone (PTH/PTH-related peptide receptor is critical for coupling to cAMP and inositol phosphate/Ca^{2+} signal transduction pathways. J Biol Chem 271:33382–33389

Ignar DM, Yingling J (1997) Assay methods for the measurement of two or more signaling events. US patent application PU3166US1

Ignar-Trowbridge DM, Nelson KG, Bidwell MC, Curtis SW, Washburn TF, McLachlan JA, Korach KS (1992) Coupling of dual signaling pathways: epidermal growth factor action involves the estrogen receptor. Proc Natl Acad Sci USA 89:4658–4662

Ignar-Trowbridge DM, Teng CT, Ross KA, Parker MG, Korach KS, McLachlan JA (1993) Peptide growth factors elicit estrogen receptor-dependent transcriptional activation of an estrogen-responsive element. Mol Endocrinol 7:992–998

Ignar-Trowbridge DM, Pimentel M, Parker MG, McLachlan JA, Korach KS (1996) Peptide growth factor cross-talk with the estrogen receptor requires the A/B domain and occurs independently of protein kinase C or estradiol. Endocrinology 137:1735–1744

Jinsi-Parimoo A, Gershengorn MC (1997) Constitutive activity of native thyrotropin-releasing hormone receptors revealed using a protein kinase C-responsive reporter gene. Endocrinology 138:1471–1475

Klein-Hitpab L, Schorpp M, Wagner U, Ryffel GU (1986) An estrogen-responsive element derived from the 5' flanking region of the *Xenopus vitellogenin* A2 gene functions in transfected human cells. Cell 46:1053–1061

Kortenjann M, Thomas O, Shaw PE (1994) Inhibition of v-raf-dependent c-fos expression and transformation by a kinase-defective mutant of the mitogen-activated protein kinase ERK2. Mol Cell Biol 14:4815–4824

Lefkowitz RJ, Cotecchia S, Samama P, Costa T (1993) Constitutive activity of receptors coupled to guanine-nucleotide-regulatory proteins. Trends Pharmacol Sci 14:303–307

Lorenz WW, McCann RO, Longiaru M, Cormier MJ (1991) Isolation and expression of a cDNA encoding *Renilla reniformis* luciferase. Proc Natl Acad Sci 88:4438–4442

Matthews RP, Guthrie CR, Wailes LM, Zhao X, Means AR, McKnight GS (1994) Calcium/calmodulin-dependent protein kinase types II and IV differentially regulate CREB-dependent gene expression. Mol Cell Biol 14:6107–6116

Miyawaki A, Llopis J, Heim R, McCaffery JM, Adams JA, Ikura M, Tsien RY (1997) Fluorescent indicators for calcium based on green fluorescent proteins and calcium. Nature 388:834–835

Nakahashi Y, Nelson E, Fagan K, Gonzales E, Guillou JL, Cooper DMF (1997) Construction of a full-length calcium-sensitive adenylyl cyclase/aequorin chimera. J Biol Chem 272:18093–18097

Neer EJ, Clapham DE (1989) Roles of G-protein subunits in transmembrane signalling. Nature 333:129–134

Pisegna JR, Wank SA (1996) Cloning and characterization of the signal transduction of four splice variants of the human pituitary adenylate-cyclase-activating polypeptide receptor. J Biol Chem 271:17267–17274

Post GR, Brown JH (1996) G-protein-coupled receptors and signaling pathways regulating growth responses. FASEB J 10:741–749

Power RF, Mani SK, Codina J, Conneely OM, O'Malley BW (1991) Dopaminergic and ligand-independent activation of steroid hormone receptors. Science 254:1636–1639

Rizzuto R, Simpson AWM, Brini M, Pozzan T (1994) Rapid changes of mitochondrial calcium revealed by specifically targeted recombinant aequorin. Nature 358:325–327

Roelant CH, Burns DA, Scheirer W (1996) Accelerating the pace of luciferase reporter gene assays. Biotechniques 20:914–918

Sadowski HB, Shaui K, Darnell JE Jr, Gilman MZ (1993) A common nuclear signal transduction pathway activated by growth factor and cytokine receptors. Science 261:1739–1743

Stables J, Green A, Marshall F, Fraser N, Knight E, Sautel M, Milligan G, Lee M, Rees S (1997) A bioluminescent assay for agonist activity at potentially any G-protein coupled receptor. Anal Biochem 252:115–126

Strahl T, Gille H, Shaw PE (1996) Selective response of ternary complex factor Sap1a to different mitogen-activated protein kinase subgroups. Proc Natl Acad Sci USA 93:11563–11568

Stratowa C, Himmler A, Czernilofsky AP (1995a) Use of a luciferase reporter system for characterizing G-protein-linked receptors. Curr Biol 6:574–581

Stratowa C, Machat H, Burger E, Himmler A, Schafer R, Spevak W, Wryer U, Wiche-Castanon M, Czernilofsky AP (1995b) Functional characterization of the human neurokinin receptors NK1, NK2, and NK3 based on a cellular assay system. J Receptor Signal Transduction Res 15:617–30, 1995

Suto CM, Ignar DM (1997) Selection of an optimal reporter gene for cell-based high throughput screening assays. J Biomolecular Screening 2:7–9

Treisman R (1995) Journey to the surface of the cell: Fos regulation and the SRE EMBO J 14:4905–4913

Wang S, Hazelrigg T (1994) Implications for bcd mRNA localization from spatial distribution of exu protein in *Drosophila* oogenesis. Nature 369:400–403

Weyer U, Schafer R, Himmler A, Mayer SK, Burger E, Czernilofsky AP, Stratowa C (1993) Establishment of a cellular assay system for G-protein coupled receptors: coupling of human NK2 and 5-HT2 receptors to phospholipase C activates a luciferase reporter gene. Receptors and Channels 1:193–200

Williams TM, Burlein JE, Ogden S, Kricka LJ, Kant JA (1989) Advantages of firefly luciferase as a reporter gene: application to the interleukin-2 gene promoter. Anal Biochem 176:28–32

Xing J, Ginty DD, Greenberg ME (1996) Coupling of the RAS-MAPK pathway to gene activation by RSK2, a growth factor-regulated CREB kinase Science 273:959–963

Zlokarnik G, Neglulescu PA, Knapp TE, Mere L, Burres N, Feng L, Whitney MW, Roemer K, Tisane RY (1998) Quantitation of transcription and clonal selection of single living cells with β-lactamase a reporter. Science 279:84–88

Zolotukhin S, Potter M, Hauswirth WW, Guy J, Nuzyczka N (1996) A "humanised" green fluorescent protein cDNA adapted for high-level expression in mammalian cells. J Virol 70:4646–4654

CHAPTER 15
Melanophore Recombinant Receptor Systems

C.K. JAYAWICKREME and M.R. LERNER

A. Introduction

This chapter describes the development and application of a *Xenopus laevis* melanophore recombinant system to study receptor-ligand interactions. Many poikilothermic vertebrates possess the ability to change their skin color rapidly. The cells responsible for these variations in appearance are called chromatophores and depending on the animal, a variety of chromatophores, including melanophores, xanthophores, erythrophores, and iridophores, contribute to the process. Color changes can be mediated by several stimuli including environmental agents, hormonal regulation and neuronal activity (ABE et al. 1969a,b; LERNER et al. 1988; LERNER 1994).

Xenopus laevis melanophores are derived from the neural crest during the course of development. Their dark brown pigment, melanin, is contained in intracellular membrane bound organelles called melanosomes which can be translocated along a microtubule network such that they are either collected at a central location near the nucleus or dispersed evenly throughout an individual cell (THALER and HAIMO 1992; LERNER 1994). The melanophores can rapidly switch their melanosomes between the two states, and an example of skin color control by stimulation of melanophore G protein-coupled receptors (GPCRs) can be seen with almost any type of frog. For example, a frog given melatonin lightens, while one which receives melanocyte stimulating hormone (MSH) darkens (POTENZA and LERNER 1992). These two hormones, respectively, stimulate receptors that lower or raise intracellular second-messenger levels of cAMP (Fig. 1). In these cases, cell color reflects cAMP levels because it controls the molecular motor(s) responsible for positioning pigment within the cell (NILSSON and WALLIN 1997; RODIONOV et al. 1991; ROGERS et al. 1997; ROGERS and GELFAND VI 1998; ROLLAG and ADELMAN 1993; RUBINA et al. 1999). The successful generation of in vitro cultures of these cells (DANIOLOS et al. 1990) has facilitated further characterization of melanophore cell signaling pathways and has made possible their use as a cell based reporter system (JAYAWICKREME and KOST 1997). Figure 1 demonstrates the extent to which a change in the state of pigment disposition within melanophores can be appreciated at a macroscopic level; the in vitro color change of the cultured melanophores (Fig. 1A) is compared with the in vivo color change of melanophores in *Xenopus laevis* frogs (Fig. 1B).

While some signaling cascades that can lead to melanosome translocation may remain to be discovered, among the studies conducted to date it has been shown that in *Xenopus laevis* melanophores, melanosome dispersion can be affected via activation of adenylyl cyclase (DANIOLOS et al. 1990; POTENZA et al. 1992) or phospholipase C (GRAMINSKI et al. 1993; SUGDEN and ROWE 1992), while melanosome aggregation results from the inhibition of adenylyl cyclase (POTENZA et al. 1994; MCCLINTOCK et al. 1993). It would be interesting to investigate what additional signaling events may affect melanosome translocation.

The response to various GPCR ligands suggested the potential usefulness of melanophores as a system to study ligand-GPCR interactions. GPCRs are

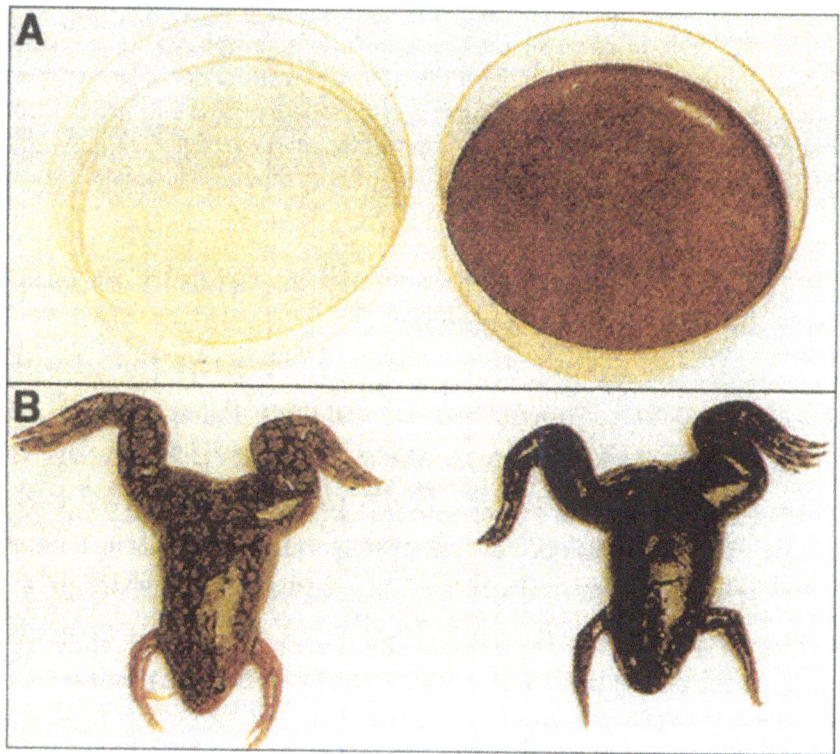

Fig. 1A,B. Macroscopic view of pigment translocation in melanophores (POTENZA and LERNER 1992). **A** The 'in vitro' color change of the cultured melanophores. Two dishes (100 mm) plated with equal number of melanophore cells treated with either 1 nmol/l melatonin (*left*) or 100 nmol/l α-MSH (*right*) for 30 min; **B** 'in vivo' change of melanophores in *Xenopus laevis* frogs. Frogs (~20 g) were subcutaneously injected with 0.5 ml of either 4 μmol/l α-MSH in 70% phosphare buffered saline (PBS) or 40 μmol/l melatonin in 70% PBS and photographed after 60–90 min following injection of drugs

one of the largest classes of cell surface receptors and they are positioned to regulate many critically important biological functions. From the viewpoints of both basic biomedical research and clinical practice, there is tremendous interest in exploring and understanding ligand interactions with GPCRs. The recent availability of clones for numerous GPCRs has engendered demand for rapid, functional GPCR assays. Thus studies were initiated to develop a recombinant system to investigate ligand-GPCR interactions using melanophores (LERNER et al. 1993; LERNER 1994). To date, a large number of exogenous G protein-coupled receptors (GPCRs) and a limited number of receptor tyrosine kinases have been successfully expressed and studied in melanophore cells.

Pigment translocation in melanophores can easily be detected within a few minutes following the activation of effector molecules (Fig. 2) thereby providing a fast, sensitive and versatile reporter technology. The assay has several attractive features: (1) it works with a broad range of receptors whose activation either raise or lower intracellular cAMP levels or that raise intracellular DAG levels; (2) it is rapid – experiments are completed within 30–60 min of adding ligand; (3) it is read directly from living cells without the need to make cell extracts or to add expensive chemical developers; (4) because the readout uses a natural system, it does not require transcription of a reporter gene; and (5) the assay has the flexibility of being used in either well or open lawn formats.

B. Cellular Signaling in Melanophores

Recombinant melanophore receptor assay is based on its ability to translocate melanosomes in response to external stimuli. This chapter describes the signaling events identified to be involved in the translocation of melanosomes within the cell.

I. Signaling Pathways

In melanophores, coupling of GPCRs to translocate melanosomes is the most well characterized signaling pathway (LERNER 1994; SAMMAK et al. 1992; SCHEENEN et al. 1994a,b; ROGERS et al. 1998; COZZI and ROLLAG 1992). Signal transduction mediated by recombinant GPCRs expressed in melanophores is initiated by coupling to endogenous G proteins. Figure 3 illustrates the signaling pathways for GPCRs that leads to activation of PKA and PKC or inhibition of PKA (REILEIN et al. 1998). This pathway is the same as shown for many mammalian cell lines. When a ligand binds to a G_s-coupled receptor, it causes the α-subunits of G_s proteins to dissociate and activate the adenylyl cyclase which in turn activates PKA. This results in the initiation of phosphorylation events which cause the melanosomes to disperse. It has been shown that increasing cAMP is associated with phosphorylation of a 53-kDa/57-kDa protein and centrifugal pigment translocation (DE GRANN et al. 1985a,b;

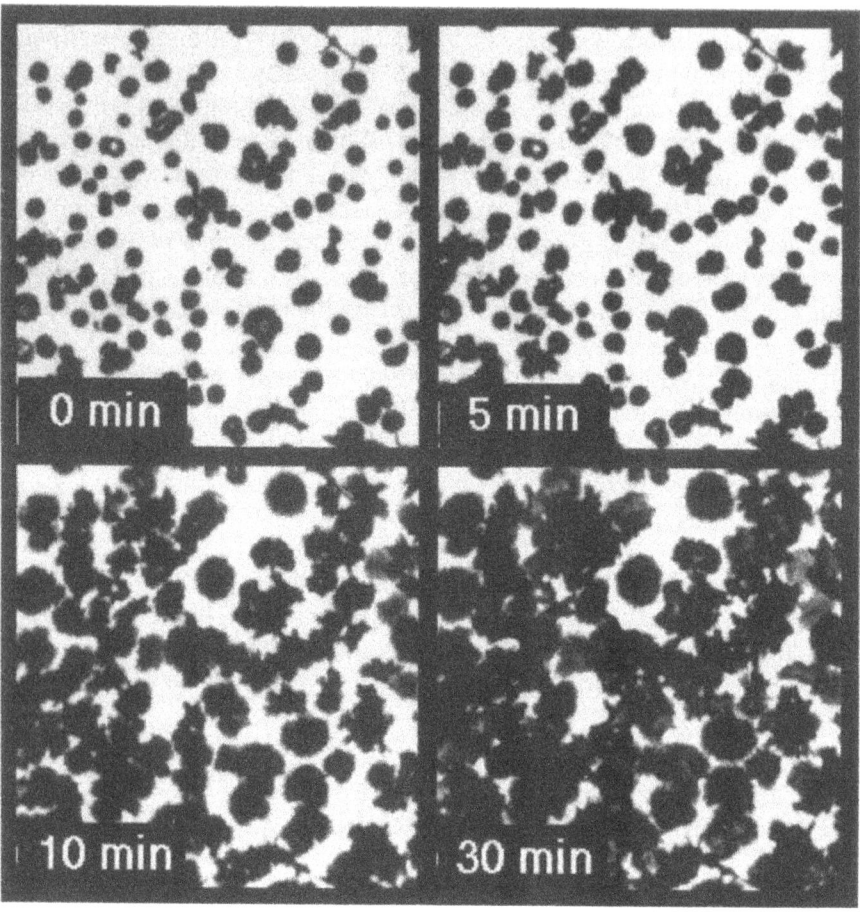

Fig. 2. Melanophore pigment translocation with time. Melanophore cells were treated with 1 nmol/l melatonin for 60 min and then with 10 nmol/l α-MSH. Cell images were captured at: 0 min; 5 min; 10 min; 30 min after the addition of α-MSH. Melatonin at a concentration of 1 nmol/l causes a fall in intracellular cAMP levels leading to centripetal melanosome movement until the organelles are collected at a central location around nucleus as seen in the 0 min image. Treatment with α-MSH at a concentration of 10 nmol/l overrides the effect of melatonin by raising cAMP which induces centrifugal pigment movement, resulting in an even distribution of pigment granules throughout the cytoplasm as seen in 5 min, 10 min, and 30 min images

ROZDZAI and HAIMO 1986). When a G_i-coupled receptor is activated, the dissociated α-subunits inhibit adenylyl cyclase which in turn reverses the pigment dispersion process to result in aggregation. When a G_q-coupled receptor is activated, the α-subunits of the G_q proteins activates phospholipase C, which in turn activate PKC that then initiates phosphorylation events to cause melanosome dispersion.

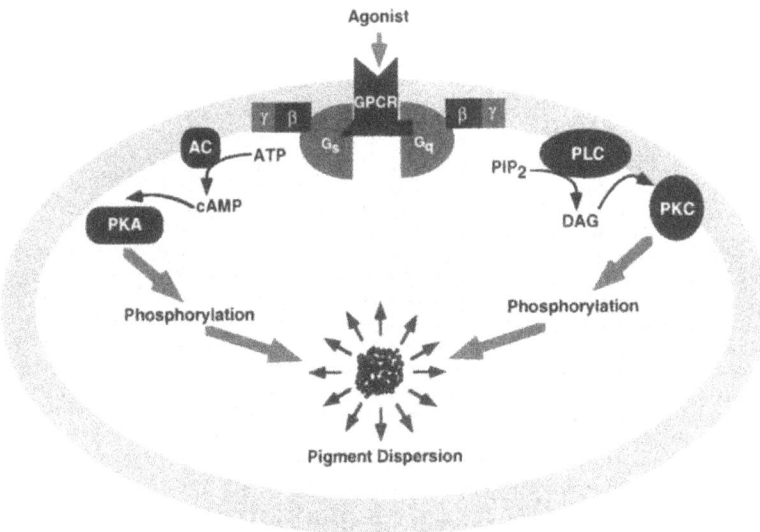

Fig. 3. GPCR signal transduction in melanophores. When a ligand binds to a GPCR the affinity for receptor-G protein interaction increases and this causes the α-subunit of the G protein to dissociate and activate the effector molecules. The α-subunit of the G_s and G_q proteins activate adenylyl cyclase and phospholipase-C respectively, which in turn activate PKA and PKC respectively. The activation of both PKA and PKC results in the initiation of phosphorylation events which cause the melanosomes to disperse. When a G_i-coupled receptor is activated, the dissociated α-subunit inhibits the adenylyl cyclase which in turn reverses the pigment dispersion process to result in aggregation. Thus the activation of recombinant GPCRs in melanophores will lead to pigment dispersion or aggregation depending upon its coupling

Thus the activation of recombinant GPCRs in melanophores leads to pigment dispersion or aggregation depending upon their G protein coupling. Pigment translocation in melanophores can be detected within a few minutes following the activation of effector molecules (Fig. 2). Figure 4 illustrates the general response of melanophore cells to ligands for GPCRs which couple through either G_s, G_q, or G_i proteins with specific emphasis on human β_2-adrenergic receptor function. The left half of the figure shows aggregated cells whereas the right counterpart shows the same cells in a dispersed state.

II. Endogenous Receptor Signaling

The ability of melanophore to translocate pigment granules in response to endogenous ligand has been known for a long time (LERNER 1961). Early studies led to the discovery of such therapeutically important molecules such

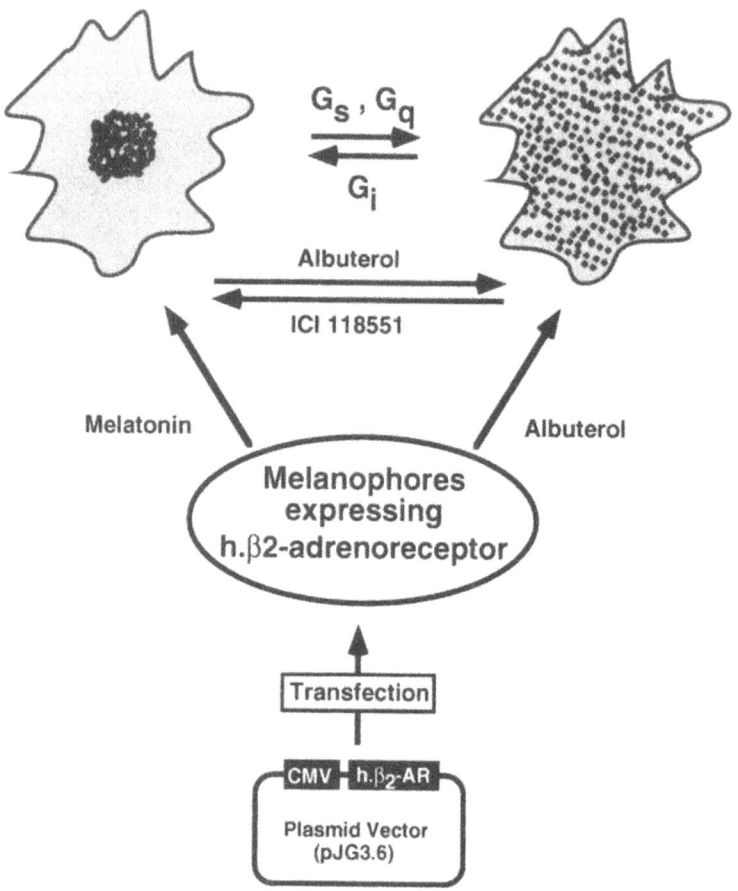

Fig. 4. Schematic illustration of melanosome distribution in recombinant melanophores upon activation of GPCRs. The *top left half* of the figure shows an aggregated cell, whereas the *right half* shows the same cell in a dispersed state. Before a screen, the cells are preset to an aggregated state for G_s/G_q-coupled receptors or to a dispersed state for G_i-coupled receptors. The aggregated state is normally achieved by activating the endogenous G_i-coupled melatonin receptor, whereas the dispersed state could be obtained by exposing the cells to light to activate the endogenous melanophore photoreceptor. The figure illustrates how the coupling of the human β_2-adrenoreceptor is studied in melanophores. First, the cDNA encoding the receptor is subcloned into a plasmid vector (e.g., pJG3.6) downstream of the CMV promoter and then transfected into melanophores. Before a screen, the cells are treated with melatonin to preset them to an aggregated state. Treatment with the β_2-adrenoreceptor agonist albuterol activates PKA, causing melanosome dispersion. Addition of the β_2-adrenoreceptor antagonist ICI 118551 blocks activation, causing melanosomes aggregation

as melatonin and α-melanocyte stimulating hormone and the identification of the receptors these molecules act on. Since then several additional GPCRs endogenous to melanophores have been characterized.

In addition, melanophores contain many distinct G proteins whose presence facilitate the functional expression of numerous exogenous transmembrane receptors. At least 13 different G_α proteins, consisting of the α subunits of G_s, G_q, G_i, G_o, and G_z proteins, have been reported (KARNE 1991) to occur in melanophores. In addition, the presence of at least 8 endogenous GPCRs in melanophores has also been documented. These include receptors for melatonin, α-MSH, endothelin-3, serotonin, VIP, oxitocin, isoproterenol, and β-CGRP. Among these receptors, molecular cloning of the endotheline-3 (KARNE et al. 1993) and melatonin receptors (EBISAWA et al. 1994) has been accomplished. In addition, pharmacological characterization of *Xenopus laevis* β-adrenergic type (POTENZA and LERNER 1992), serotonin (POTENZA and LERNER 1994), melatonin (TEH and SUGDEN 1999), and VIP receptors (MAROTTI et. al. 1999) have been reported.

1. Melatonin 1c Receptor

One of the earliest described actions of melatonin was its ability to cause melanosome aggregation in dermal melanophores of amphibians (LERNER 1961). The action is mediated through a high-affinity melatonin receptor (EC50~50–200 pM) (POTENZA et al. 1994a; TEH and SUGDEN 1999) that is coupled to an inhibitory G protein (G_i) whose ability to induce phosphoinositide hydrolysis has been described (MULLINS et al. 1993). The presence of the melatonin receptor is a key component for the melanophore bio-assay to function in G_s or G_q-coupled receptor screen format, as it allows for the presetting of cells to a melanosome aggregated state by treating the cells with melatonin. Extensive pharmacological characterization of this receptor with various synthetic analogs has been described (JOCKERS et al. 1997; PICKERING et al 1996; SUGDEN 1991, 1992; TEH and SUGDEN 1999) and Steven Reppert's laboratory reported its molecular cloning (EBISAWA et al. 1994). The cDNA encodes a protein of 420 amino acids, which contains seven hydrophobic segments. The studies on the cloned receptor in CHO cells confirmed its high affinity to melatonin and its belonging to GPCR superfamily. Moreover, the knowledge on this xenopus receptor sequence allowed the subsequent cloning of melatonin receptor cDNA from other species, including humans.

2. α-Melanocyte Stimulating Hormone (MSH) Receptor

α-MSH was one of the first peptide hormones that was studied as a crude extract of hog pituitary gland and shown to have activity on frog skin (LEE and LERNER 1956). The frog skin assay was then used to isolate the pure α-MSH peptide (LEE and LERNER 1956; HARRIS and LERNER 1957). Since then a fair amount of work has been devoted to understand its biological, physiological, pharmacological, and therapeutic aspects. The α-MSH receptor is coupled through a G_s protein. As the name implies, this receptor is well known to cause pigment dispersion in melanophores upon stimulation with α-MSH (EC_{50} of 0.5–1 nmol/l). The identification of both synthetic peptide agonists and antagonists to this receptor has also been described (JAYAWICKREME et al. 1994b; QUILLAN et al. 1995).

3. Endothelin-C Receptor

Endothelin peptides have been demonstrated to induce pigment dispersion in melanophores (KARNE et al. 1993). Moreover, the ability of ET-3 specifically to cause pigment dispersion (compare EC_{50} for ET-3 of 24 nmol/l with $\geq$10 μmol/l for ET-1 or ET-2) shows that these cells express the ET_c receptor subtype (KARNE et al. 1993). A cDNA encoding for ET_c receptor has been isolated from a melanophore cDNA library (KARNE et al. 1993). The cDNA encodes a protein of 424 amino acids with the predicted heptahelical structure common to the GPCR super family. This receptor showed a concentration dependent signal desensitization wherein the higher the ligand concentration, the faster the rate of desensitization.

4. Serotonin Receptor

A serotonin receptor endogenous to melanophores has been reported (POTENZA and LERNER 1994b). Serotonin increased intracellular levels of cAMP and induced pigment dispersion in the cells with an EC_{50} of 56 nmol/l (POTENZA and LERNER 1994b). In terms of its ability to increase cAMP, the receptor seems to be most closely related to 5HT4R, 5HT6R, and 5HT7R. However, a series of serotonin receptor ligands have been evaluated as agonists or antagonists at this receptor and the pharmacological profile suggests the presence of a receptor which shares some properties with but appears different from other previously described serotonin receptors (POTENZA and LERNER 1994b).

5. β-Adrenoreceptor

The characterization of a β-adrenoreceptor endogenous to melanophores has been described (POTENZA and LERNER 1992). The treatment of cells with isoproterenol or norepinephrine increased the intracellular levels of cAMP and induced pigment dispersion with an EC_{50} of 380 nmol/l and 347 nmol/l respectively (POTENZA and LERNER 1992). Timolol, a non-selective β-adrenoreceptor antagonist has shown the highest activity (an IC_{50} of 53.7 nmol/l) in blocking the norepinehrine induced pigment dispersion. The study demonstrated that endogenous β-adrenoreceptor prefers agonists which are most selective to β_1-adrenoreceptors. However, in terms of blocking the activity, non-selective antagonist seems to be most effective (POTENZA and LERNER 1992).

6. VIP Receptor

A receptor for VIP-related peptides has been functionally characterized in melanophores (MAROTTI et al. 1999). Its activation stimulated intracellular cAMP accumulation and pigment dispersion, suggesting a G_s protein mediated response. Helodermin, with an EC_{50} of 46.5 pmol/l (MAROTTI et al. 1999), was the most potent activator of followed by PACAP38>VIP>PACAP 27. A similar order of potencies has been observed for the peptides to induce cAMP accumulation. The responses to VIP agonists were selectively inhibited by the VIP antagonists PACAP-(6–27) and (N-Ac-Tyr1-D-Phe2)-GRF(1–29)-NH2.

In addition, evidence for the presence of a photoreceptor has been reported (DANIOLOS et al. 1990; MIYASHITA et at. 1996; MORIYA et al. 1996). The action spectrum showed a peak response to visible light at 460 nm with a fourfold increase in intracellular cAMP levels. The molecular cloning of an opsin, melanopsin, in *Xenopus laevis* melanophores has been reported (PROVENCIO et al. 1998). Its deduced amino acid sequence shares greatest homology with cephalopod opsins. The predicted secondary structure of melanopsin indicates the presence of a long cytoplasmic tail with multiple putative phosphorylation sites, suggesting that this opsin's function may be finely regulated.

In addition the presence of receptors responding to oxytocin and β-CGRP has also been reported (McCLINTOCK et al. 1996). Further characterization of these receptors are yet to be accomplished.

C. Melanophore Assay Technology

Like most recombinant assays, the main features of the melanophore technology includes cell culture, receptor expression and signal detection. This chapter describes the main features of the assay technology.

I. Cell Culture and Related Techniques

Melanophore cell culture and related techniques have now become common practices in many academic and pharmaceutical industry laboratories. While different labs may have slightly varied optimal procedures, the following paragraphs will describe the generally used practices.

1. Preparation of Cultures of Melanophore

The preparation of cultures of *Xenopus laevis* melanophore has been successfully achieved from tadpole dermal cells (DANIOLOS et al. 1990; AKIRA and IDE 1987; FUKUZAWA and IDE 1983; KONDO and IDE 1983; IDE 1974; SELDENRIJK et al. 1979). Once a healthy immortalized melanophore culture has been established, the cells can be propagated for years while preserving the desired properties.

Approximately 1–2 months after an initial primary culture is established from tadpoles, colonies of melanophores are evident. These colonies are isolated from contaminating cells, primarily fibroblasts, by trypsinization followed by Percoll density-gradient centrifugation. Repetitive culturing and Percoll density-gradient centrifugation leads to the generation of pure cultures of melanophores. It has been reported (DANIOLOS et al. 1990; HALABAN et al. 1986; SIEBER-BLUM and CHOKSHI 1985) that addition of mitogenic agents to cultures at this stage improves cell proliferation. Also, it has been shown, that at least at the initial stages, the proliferation of *Xenopus laevis* melanophores is enhanced by the presence of those mitogenic agents required by normal human and murine melanocytes such as dbcAMP, α-MSH, insulin, human placental extract, and IBMX (DANIOLOS et al. 1990). Insulin is a particularly effective mitogenic agent. When the melanophore lines are well established these mitogenic agents are not necessary to keep in continuous cultures.

2. Continuous Culturing of *Xenopus laevis* Melanophores

Once a healthy immortalized melanophore culture has been established, continuous propagation of these cells is straightforward. The cells are grown at room temperature or optimally at 27°C in closed containers. They are typically cultured in 0.7 × Leibovitz L-15 media supplemented with serum. The culture of melanophores in media conditioned with *Xenopus laevis* fibroblasts has tremendously improved the quality of the cells. During the conditioning of media, growth factors are released into the media from fibroblasts. The preparation of conditioned media is typically achieved by pre-exposing the 0.7 × Leibovitz L-15 medium supplemented with 20% fetal bovine serum to fibroblast cells. Typically, collection of two crops derived from 25%–40% and 70%–80% confluence of fibroblasts is the most commonly used procedure. The quality of the continuous cultures of melanophores may vary from laboratory to laboratory depending upon the quality of the media employed to culture the cells.

3. Receptor Expression

To date more than 100 different GPCRs have been expressed in melanophores. The expression of most of these foreign receptors has been accomplished by subcloning the receptor cDNA into a vectors driven by the CMV immediate early gene promoter. While the usefulness of the other promoters has not been extensively evaluated, most commercially available CMV promoter driven vectors work well. The construction of the pJG3.6 vector (GRAMINSKI et al. 1993), containing the CMV promoter, has been facilitated higher transfection efficiencies of GPCRs in melanophores and this vector remains the favorite one and transfection efficiencies as high as 80%–90% have been observed using electoporation methods.

Among the various transfection techniques, electroporation seems to be the most commonly used procedure for the efficient expression of exogenously introduced genes. Though it is not widely used, the successful use of recombinant vaccinia virus for the same purpose has also been described (POTENZA and LERNER 1991). For electroporation, a suspension of melanophore cell in PBS buffer is mixed with the plasmid DNA of receptor of interest. Soon after the electroporation it is important to transfer cells to a tube of conditioned media for quick recovery before plating.

4. Preparation of Stable Melanophore Lines Expressing Exogenous Receptors

Similar techniques that have been described for mammalian cells can easily be adapted for the generation of stable melanophore cell lines expressing foreign receptors. Most commonly used procedures include the co-transfection of plasmid coding for receptors of interest together with a plasmid vector coding for a neomycin resistant gene or subcloning of the receptor cDNA into a vector which encodes neomycin resistant gene followed by transfection. Then, 1–3 days after transfection, cells are treated with G418 (250–500 µg/ml) to select cell colonies of interest.

II. Signal Detection

Melanophore assays are based on the ability to monitor pigment distribution within the cells. Pigment translocation in melanophores can be detected by either measuring the change in light transmittance through the cells (POTENZA and LERNER 1992; GRAMINSKI et al. 1993) or by imaging the cell responses (McCLINTOCK et al. 1993; JAYAWICKREME et al. 1994a). Changes in light transmittance can easily be measured using a microtiter plate reader, while cell imaging either at single cell resolution or at a macro level where cumulative response arising from a collection of cells is achieved using a CCD camera. Both states of intracellular pigment distribution (dispersion or aggregation) are easily detectable.

1. Transmittance Reading

The 96-well microtiter plate format (POTENZA and LERNER 1992; GRAMINSKI et al. 1993), which measure the change in light transmittance through the cells, is the most widely used format in studying ligand-receptor interactions (Fig. 5). In this arrangement, the recombinant melanophores are plated in 96-well plates and ligand induced melanosome translocations are quantified by measuring the change in light transmittance using a microtiter plate reader. Many commercially available plate readers are suitable for this purpose, although a one with the ability to take several readings within a well is preferred. The ligand induced responses are quantified by calculating $(1-T_f/T_i)$, where T_i is the photo transmission immediately before drug addition and T_f is the photo transmittance at the end point. Figure 5 shows an example of measuring human β_2-adrenoreceptor responses in melanophore using a 96-well microtiter plate.

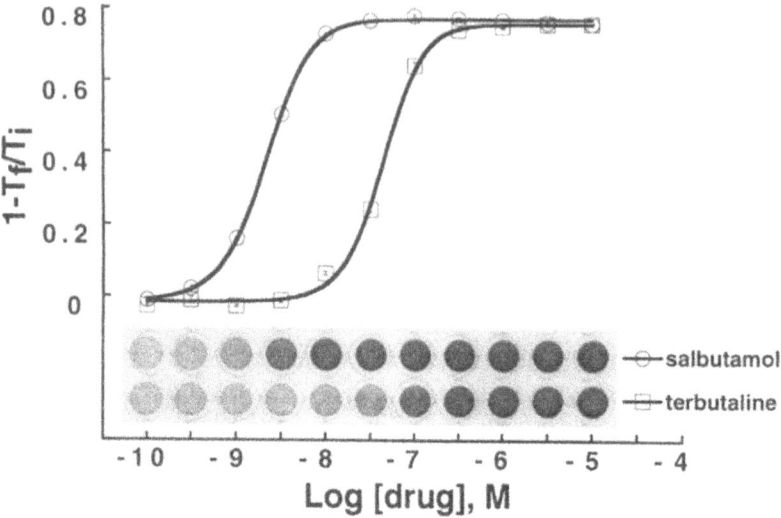

Fig. 5. Measurement of melanophore response in a 96-well microtiter plate. Melanophore cells expressing human β_2-adrenoreceptor were plated in a 96-well plate (GRAMINSKI et al. 1993). Cells were pretreated with 1 nmol/l melatonin for 60 min and were then treated with either salbutamol or terbutaline for 30 min. The transmittance through the wells was measured before (T_i) and after (T_f) drug addition using a plate reader. To generate the dose response curves, change in transmittance, $1-T_f/T_i$, was plotted against the drug concentration. The *inset* shows a photograph of two 96-well plate rows from which the data were used to generate dose response curves

In addition, the recent advancement in 384-well and 1536-well technology should facilitate the melanophore bio-assays in these formats.

2. Digital Imaging

Digital imaging is another attractive way of detecting ligand mediated melanophore responses (McCLINTOCK et al. 1993; JAYAWICKREME et al. 1994a). The technology has been successfully used to detect the responses at single cell resolution or to detect cumulative responses arising from batches of cells. The latter methodology is often the main detection procedure during the melanophore lawn format screens (JAYAWICKREME et al. 1998a,b) to identify responses arising from compounds. In this method the camera, which is controlled by a computer is mounted at a suitable distance away from the cell plates for appropriate imaging. The resolution of the camera should be selected depending on the size of the area needing to be imaged. Image subtraction of original and end point images are used to enhance the signal. When imaging needs to be done at single cell resolution, often the camera is mounted through a microscope to look at a field of cells (McCLINTOCK et al. 1993).

However, with new high resolution $4k \times 4k$ cameras one can even capture images at single cell resolution even without using a microscope. In this method, the translocation of melanosomes within thousands of individual cells are simultaneously tracked by capturing gray scale video images before and after receptor activation. A successful application of this technology to study human β_2-adrenergic receptors which stimulate adenylyl cyclase, murine substance P and bombesin receptors which stimulate PLC, and a human D_2 receptor which inhibits adenylyl cyclase has been reported (MCCLINTOCK et al. 1993). The rapid data-handling ability of video technology should provide a bioassay useful for cloning novel GPCRs by screening cDNA libraries for clones encoding new receptors.

III. Screening Formats

Melanophore technology provides attractive ways to conduct high-throughput screens in drug discovery research. Its ability to screen G_s-, G_q-, or G_i-coupled receptors in the same cell background with pigment translocation as the reporter readout is a feature unique to the melanophore system. Due to the presence of the natural reporter readout, unlike in other widely used recombinant functional high-throughput screen methods, no chemical developers or dyes need to be added to detect the final end point readout. In addition, its fast, stable, and on-time readout make the melanophore system more versatile allowing it to be used in lawn format. Melanophore technology therefore offers two attractive screening platforms, well format and open lawn format, for high-throughput screens.

1. Microtiter Well Format

The 96-well format seems to be the most widely used microtiter well format for screening. Melanophore cells either transiently or stably expressing GPCRs are plated into 96-well plates (Fig. 5). Cell plating is done in CFM at least 6h before a screen to allow sufficient time for cell to attach. Just before a screen CFM is replaced by an assay buffer (typically $0.7 \times$ Leibovitz L-15 supplemented with 0.1% BSA) which is free of serum. Compounds to be screened can be added as a concentrated solution or are diluted in assay buffer. For G_s or G_q assays compounds are pre-treated with 1 nmol/l melatonin to preset the cells to a fully aggregated state. For a G_i assay a pretreatment is generally not required as cells are usually in a dispersed state. However, if the cells need to be preset to a dispersed state, exposure to visible light for 30 min is sufficient.

The screen technology can be automated for high throughput screens with commercially available instrumentation. Scientists at Glaxo WellcomeInc. have successfully used automated high-throughput (unpublished data) technology for drug discovery research to screen G_s-, G_q-, and G_i-coupled receptors. The recent advancement in 384-well and 1536-well format technology

2. Lawn Format

The timely response of melanophore allows for the flexibility of screening compounds on an open lawn of cells. This technology is attractive, especially when large collections of compounds need to be screened. This may be the most efficient way to screen millions of synthetic peptide analogs for their individual functional responses. The development of recombinant melanophore technology for screening synthetic combinatorial peptide libraries using the multi-use peptide library (MUPL) concept has been documented (JAYAWICKREME et at. 1994a,b). Figure 6 shows a schematic illustration of preparing MUPLs. Figure 7 shows responses arising from two α-MSH beads on a lawn of melanophore cells. In this method recombinant melanophores are plated as a lawn of cells on a petri dish and a filter containing partially cleaved synthetic peptide beads is applied over the cells. In this multi-use combinatorial peptide library (MUPL) technique, peptides are liberated from their supports in a dry state so that the problem of signal interference due to mixing of peptide molecules, particularly agonists and antagonists, is avoided. In addition, the peptides are released from their supports in a controlled manner so that fractions are available for multiple independent tests thus eliminating the need for iterative library analysis and resynthesis.

IV. Receptor Cloning and Mutagenesis Studies

Besides its use for comparing how chemicals interact with GPCRs, identifying ligands for orphan receptors, and its applications in drug discovery research, the melanophores should be useful for cloning receptors and for determining how ligands affect the large number of receptor variants generated by site-directed mutagenesis. As the human genome project is getting closer to its completion, its usefulness for cloning may decline while the utility for mutagenesis may become more demanding.

Melanophore imaging technology at single cell resolution provides an ideal means for both cloning and mutagenesis studies. It has been demonstrated that, by imaging fields containing thousands of cells, the presence of a plasmid coding for a receptor could be detected when its frequency was one per 10,000 plasmids transfected. An extensive study has been conducted to illustrate its capability for cloning either G_s, G_q, or G_i receptors (McCLINTOCK et al. 1993). In this study it was demonstrated the melanophore systems ability to detect responses arising from the one plasmid of G_s-coupled β_2-adrenoreceptor diluted with 10,000 plasmids from a liver cDNA library, one plasmid of G_i-coupled D_2 receptor diluted with 10,000 plasmids from a rat olfactory bulb library, or one plasmid of G_q-coupled NK-1 receptor diluted

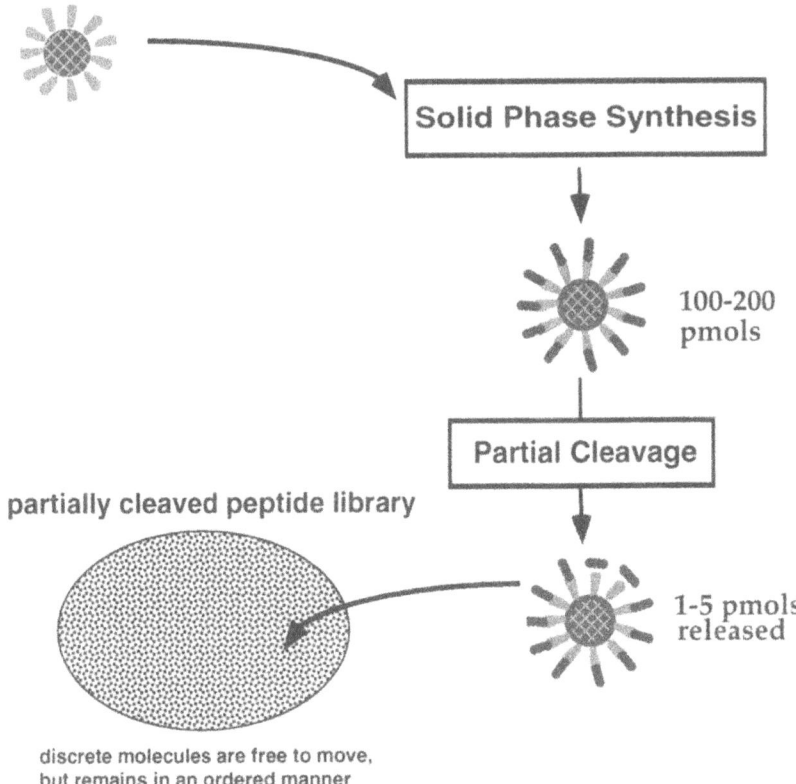

Fig. 6. Preparation of multi-use peptide libraries (MUPLs). Briefly, in this method (JAYAWICKREME et al. 1998a,b), peptides are constructed on a solid support and then partially cleaved in a dry state using gaseous TFA followed by a neutralization with gaseous NH_3. Since the peptides are liberated in a dry state they remain as separate independent entities after non-covalently bound to the bead. In addition, since the peptides are released from their supports in a controlled manner, fractions are available for multiple independent tests

with 10,000 plasmids from a rat brain cDNA library. Similarly one could detect the effect of a ligand on individual mutants in a library of receptor mutants prepared by side-directed mutagenesis.

D. Receptor Studies and Applications

Initial studies in Michael Lerner's laboratory led to the successful development of the melanophore expression system to study ligand-GPCR interactions (LERNER 1994; JAYAWICKREME and KOST 1997). To date, more than 100 different GPCRs (published and unpublished data) which couple through either

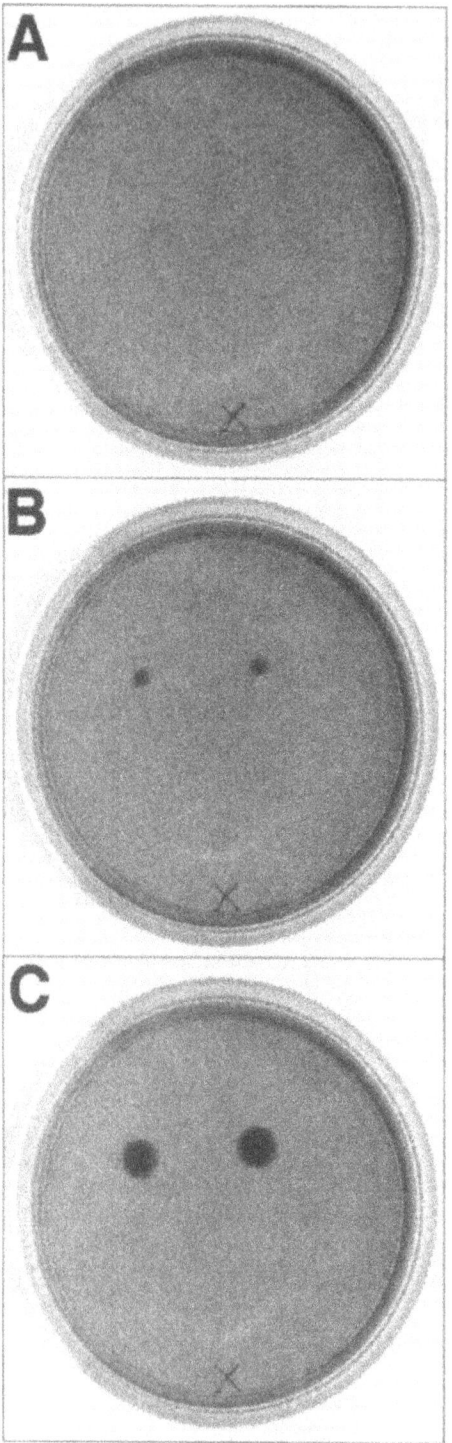

Fig. 7A–C. Response from two α-MSH peptide beads on a lawn of melanophore cells. α-MSH peptides were constructed on MBHA resin and partially cleaved as described for MUPL construction (JAYAWICKREME et al. 1998a,b). Beads were then placed on a lawn of melanophore cells overlaid with a thin layer of agarose in a petri dish. Responses at: **A** 0 min; **B** 5 min; **C** 10 min are shown

G_s, G_q, or G_i/G_o pathways have been successfully expressed and studied in melanophores.

The expression of G_s-coupled human β_2-adrenoreceptor in melanophores (POTENZA et al. 1992) was one of the first studies conducted to demonstrate its utility as a reporter assay. This study confirmed G_s-coupling and proper pharmacological validation of the receptor in melanophore system. Soon, the studies were extended to demonstrate the capability of G_q-coupled murine bombesin (gastrin releasing peptide) and rat NK-1 receptors to cause pigment translocation in melanophores (GRAMINSKI et al. 1993). In both cases, the proper pharmacological profile of the recombinant receptor and its ability to generate second messengers DAG and IP_3 via coupling to G_q protein upon stimulation with the agonists was obtained. Subsequent demonstration of the ability of G_i-coupled recombinant human D_2 receptor to couple in melanophore pigment aggregation pathway with proper pharmacological profile to known ligands (POTENZA et al. 1994) indicated the melanophore system's strength and potential to study any GPCR.

I. Characterization of Novel GPCRs

Since its development, a large number of GPCRs have been studied using melanophore. Expression studies on human D_3 receptor in melanophore system (POTENZA et al. 1994) provided the first evidence of demonstrating its coupling to G_i signaling pathway and provided pharmacological profiles for a range of ligands.

During the past two years melanophore technology has been a valuable tool for identifying the signaling pathways for newly discovered receptors and to establish their pharmacological profiles. These studies were among the first to identify the coupling of human CXC-chemokine receptor-4 (CXCR-4) to a G_i protein for initiating cellular signaling (CHEN et al. 1998). The pharmacological characterization of the interaction of SDF-1α and the receptor specific antibody, 12G5, with the CXCR4 receptor was also described. Two other recent studies reported the identification of signaling pathways for the human GALR2 and GALR3 receptor subtypes (KOLAKOWSKI et al. 1998). Using both melanophore and aequorin luminescence assays it was demonstrated that human GALR2 was coupled to the phospholipase C/protein kinase C pathway. In contrast, GALR3, when expressed in melanophore cells, caused pigment aggregation in responding to agonist peptides, suggesting communication of its intracellular signal by inhibition of adenylyl cyclase through G_i pathway (KOLAKOWSKI et al. 1998). Similar studies have been described in identifying signaling pathway for the recently cloned mouse leukotriene B4 receptor (m-BLTR) (MARTIN et. al. 1999). In melanophores transiently expressing the m-BLTR, LTB4 induced the aggregation of pigment granules, confirming the inhibition of cAMP production.

Another elegant example is the demonstration of $GABA_B$ receptor function in melanophores. GPCRs are commonly thought to bind their cognate ligands and elicit functional responses as monomeric receptors. This study (NG et al. 1999) was among the first few to describe the importance of a co-receptor, gb2, for the functional activity of the $GABA_B$ receptor (gb1a), suggesting a new mechanism for GPCR function. Either receptor, gb1a or gb2, when expressed alone in melanophores, did not show functional activity to the GABA ligand. These findings indicate that melanophores are a suitable system for studying receptor ligand interactions for GPCRs which function through heterodimerization.

With hundreds of novel receptors being identified as a result of the human genome project, melanophore technology will continue to serve as a valuable tool for identify signaling pathways and ligands to numerous novel GPCRs.

II. Lawn Format Screen System

Another feature of the recombinant melanophore system is its ability to be used in an open lawn format to screen synthetic combinatorial libraries (JAYAWICKREME et al. 1998a,b). The ability to detect functional cellular responses arising from individual beads in a synthetic peptide combinatorial library (SPCL) was first demonstrated using a lawn of melanophore cells expressing recombinant murine bombesin receptor (JAYAWICKREME et al. 1994a).

To obtain the functional cellular responses from individual beads on a lawn of cells, peptides were liberated from their solid supports so the molecules are free to interact with cellular targets. Peptide cleavage was done in a dry state so the problem of signal interference due to mixing of peptide molecules, particularly agonists and antagonists, was avoided. In addition, peptides were released from their supports in a controlled manner so that fractions were available for multiple independent tests, thus eliminating the need for iterative library analysis and resynthesis. The screening system has been successfully used to identify novel 7-mer peptide agonists for the bombesin receptor using a SPCL designed based on the 14 amino acid bombesin peptide sequence (JAYAWICKREME et al. 1994a). A similar study has been conducted to identify novel 8-mer and 9-mer peptide antagonists to α-MSH receptor using a focused SPCL designed based on the 13 amino acid α-MSH sequence (JAYAWICKREME et al. 1994b). In addition, due to the large number of analogs screened, this study generated information on structure activity relationships governing the ligand-receptor interactions.

In another example, a truly random tri-peptide library designed using 48 different amino acids (223,488 analogs) was successfully used to identify antagonists to human MCR1 receptor (QUILLAN et al. 1995). These studies clearly demonstrate that, at least for SPCLs, the lawn format screens provide a tremendous advantage over the solution phase screens with respect to the throughput, cost effectiveness, and infrastructure necessary to organize millions of individual molecules. The application of the lawn format screen technology is likely to extend towards synthetic small molecule libraries (SSMLs) as well. Even though the technology is well described for the SPCLs, its application to SSMLs is still limited due to the necessity of a tag for decoding the structure and the lack of availability of a universal bead chemistry to synthesize various types of small molecules. While it is beyond the scope of this chapter, it is worth mentioning a successful application of SSMLs to identify novel antimicrobial agents. This study (SILEN et al. 1998) describes the construction and screening of a 46,656 member triazine based SSML encoded with secondary amine tagging, allowing for rapid structural analysis of the compounds of interest, to identify novel antimicrobials on a lawn format assay. With the continuous ongoing developments in tag technology and bead-based chemistry, it is likely that in the future more lawn format compatible SSM libraries may become available for screens.

Melanophore lawn format technology, using either SPCLs or SSMLs, may serve as a valuable technology to identify ligands to thousands of receptors that are being identified through human genome project.

III. Single Transmembrane Receptors

In addition to the successful studies on GPCRs, the demonstration of the ability of platelet-derived growth factor (PDGF) β-receptor to couple to pigment translocation pathway expanded the scope of the melanophore technology towards studying tyrosine kinase type receptors (GRAMINSKI and LERNER 1994).

Unlike G protein-coupled receptors, these receptors have a single transmembrane domain and signal transduction is induced by receptor dimerization following ligand binding. Although PDGF has been shown to initiate more than one signal transduction pathway (VALIUS and KAZLAUSKAS 1993), activation of PLC-γ1 is crucial since this leads to the activation of protein kinase C (PKC) through the production of the second messengers diacylglycerol (DAG) and inositol trisphosphate via phosphatidylinositol-4,5-bisphosphate hydrolysis. Thus it is reasonable to hypothesize that melanophore cells express an endogenous PLC-γ1, and the activation of the transiently expressed PDGF β-receptor would presumably lead to the PKC activation and pigment dispersion. The demonstration of successful expression of an epidermal growth factor receptor (EGFR) (CARRITHERS 1999) which belongs to the same family strengthens this view. These studies suggest that the melanophore pigment translocation assay may provide a generic means to functionally study ligand-receptor interactions for receptor tyrosine kinases.

Recently the extension of melanophore system's capability to study single transmembrane receptors beyond receptor tyrosin kinases has been described. This study (CARRITHERS et al. 1999) describes the development of a rapid, functional assay for the erythropoietin receptor (EPOR), a member of the cytokine receptor family. EPOR itself does not couple to melanophore pigment translocation pathway. However, a receptor chimera composed of the extracellular portion of the EPOR and the transmembrane and intracellular region of the human epidermal growth factor receptor can mediate pigment dispersion in melanophores when treated with erythropoietin. The work demonstrates a successful step towards developing a generic melanophore-based assay to study transmembrane receptors.

There are an increasing number of examples that demonstrate the involvement of single transmembrane receptors in controlling a great number of biological and physiological events and, with the completion of the human genome project, more and more members of this family will be identified. Thus further research to develop this type of assay as a generic means to study ligand-receptor interactions for single transmembrane receptors would be a worthwhile investment towards drug discovery research.

IV. Receptor-Ligand Interaction Studies

With the examples shown throughout the previous sections of this chapter, the melanophore system is a value tool for studying receptor-ligand interactions. Among other applications in receptor-ligand interactions, chimeric receptors, receptor mutation, effector and cofactor molecule interaction, and receptor desensitization studies may greatly benefit from melanophore cells due to their versatile nature for use as a transient expression system. For example, a study conducted on chimeric olfactory receptors using melanophores provided insight to the evaluation of features important for the function of olfactory receptors and hence for their signaling. In this study (McCLINTOCK et al. 1997; McCLINTOCK and LERNER 1997), the G protein-coupling domains of the β_2-adrenoreceptor were replaced with homologous domains of putative olfactory receptors to produce chimeric receptors which were able to stimulate melanosome dispersion in melanophores, a G protein-mediated pathway.

Another recent application is its successful use to study the interaction of agouti and agouti-related protein with melanocortin receptors. The study insight concerned the antagonism of melanocortin receptors by agouti related protein (OLLMANN et al. 1997). More recently, the competitive antagonism of agonist binding by the carboxyl-terminal portion of Agouti protein (OLLMANN and BARSH 1999), and the down-regulation of melanocortin receptor signaling by an unknown mechanism that requires residues in the amino terminus of the Agouti protein (OLLMANN and BARSH 1999) was shown. Further, these studies resolved questions regarding signaling by Agouti protein. Evidence was presented for a novel signaling mechanism whereby α-MSH and Agouti protein or Agouti-related protein function as independent ligands that inhibit each other's binding and transduce opposite signals through a single receptor (OLLMANN et al. 1998).

In another work the melanophore assay system has been used as a screening tool to study functional interaction of bivalent peptide ligands targeted to GPCRs (CARRITHERS and LERNER 1996). It is thought that multivalent ligands targeted to one or more receptors may lead to a powerful way of targeting drugs to specific cell types. This work describes the development of bivalent peptide ligands to GPCRs receptors that interact more potently with the receptor than the respective monovalent peptide ligands.

E. Summary

Melanophore technology has the ability to serve as a reporter assay to study numerous 7TM receptors regardless of their coupling pathways. The beauty of the system is its use of a single cell type and a simple readout for many G protein signaling pathways. Its fast, robust, and steady signal extends its application beyond the traditional microtiter well format and into lawn format screens. Further studies to explore the melanophore system's ability to be applied to other gene families in addition to GPCRs, and single transmembrane receptors, may further extend its usefulness. As new gene sequences (MARCHESE et al. 1999) are identified through human genome sequencing projects, melanophores are positioned to assist in their characterization.

References

Abe K, Butcher RW, Nicholson WE, Baird CE, Liddle RA, Liddle GW (1969) Adenosine 3',5'-monophosphate (cyclic AMP) as the mediator of the actions of melanocyte stimulating hormone (MSH) and norepinephrine on the frog skin. Endocrinology 84:362–368

Abe K, Robison GA, Liddle GW, Butcher RW, Nicholson WE, Baird CE (1969) Role of cyclic AMP in mediating the effects of MSH, norepinephrine, and melatonin on frog skin color. Endocrinology 85:674–682

Akira E, Ide H (1987) Differentiation of neural crest cells of Xenopus laevis in clonal culture. Pigment Cell Research 1:28–36

Carrithers MD, Lerner MR (1996) Synthesis and characterization of bivalent peptide ligands targeted to G-protein-coupled receptors. Chemistry and Biology 3:537–542

Carrithers MD, Marotti LA, Yoshimura A, Lerner MR (1999) A melanophore-based screening assay for erythropoietin receptors. Journal of Biomolecular Screening 4:9–14

Chen WJ, Jayawickreme C, Watson C, Wolfe L, Holmes W, Ferris R, Armour S, Dallas W, Chen G, Boone L, Luther M, Kenakin T (1998) Recombinant human CXC-chemokine receptor-4 in melanophores are linked to G_i protein: seven transmembrane coreceptors for human immunodeficiency virus entry into cells. Molecular Pharmacology 53:177–181

Cozzi B, Rollag MD (1992) The protein-phosphatase inhibitor okadaic acid mimics MSH-induced and melatonin-reversible melanosome dispersion in Xenopus laevis melanophores. Pigment Cell Research 5:148–154

Daniolos A, Lerner AB, Lerner MR (1990) Action of light on frog pigment cells in culture. Pigment Cell Res 3:38–43

de Graan PNE, Gispen WH, van de Veerdonk FCG (1985) α-Melanotropin-induced changes in protein phosphorylation in melanophores. Mol Cell Endocrinology 42:119–125

de Graan PNE, Oestreicher AB, Zwiers H, Gispen WH, van de Veerdonk FCG (1985) Characterization of α-MSH induced changes in the phosphorylation of a 53 kDa protein in Xenopus melanophores. Mol Cell Endocrinology 42:127–133

Ebisawa T, Karne S, Lerner MR, Reppert SM (1994) Expression cloning of a high-affinity melatonin receptor from Xenopus dermal melanophores. Proc Natl Acad Sci 91:6133–6137

Fukuzawa T, Ide H (1983) Proliferation in vitro of melanophores from Xenopus laevis Journal of Experimental Zoology 226:239–244

Graminski GF, Jayawickreme CK, Potenza MN, Lerner MR (1993) Pigment dispersion in frog melanophores can be induced by a phorbol ester or stimulation of a recombinant receptor that activates phospholipase C. J Biol Chem 268:5957–5964

Graminski GF, Lerner MR (1994) A Rapid bioassay for platelet-derived growth factor β-receptor tyrosine kinase function. BIO/TECHNOLOGY 12:1008–1011

Halaban R, Ghosh S, Duray P, Kirkwood JM, Lerner AB (1986) Human melanocytes cultured from nevi and melanomas. Journal of Investigative Dermatology 87:95–101

Harris JI, Lerner AB (1957) Amino acid sequence of α-melanocyte stimulating hormone. Nature 179:1346–1347

Ide H (1974) Proliferation of amphibian melanophores in vitro. Developmental Biology 41:380–4

Jayawickreme CK, Graminski GF, Quillan JM, Lerner MR (1994) Creation and functional screening of a multi-use peptide library. Proc Natl Acad Sci 91:1614–1618

Jayawickreme CK, Quillan JM, Graminski GF, Lerner MR (1994) : Discovery and structure-function analysis of α-melanocyte-stimulating hormone antagonists. J Biol Chem 47:29846–29854

Jayawickreme CK, Kost TA (1997) Gene expression systems in the development of high-throughput screens. Current Opinion in Biotechnology 8:629–634

Jayawickreme CK, Jayawickreme SP, Lerner MR (1998) Functional screening of multi-use peptide libraries using melanophore bioassay. Methods in Molecular Biology; Combinatorial peptide library protocols, Ed. Cabilly S. 87:107–118

Jayawickreme CK, Jayawickreme SP, Lerner MR (1998) Generation of Multi-use peptide libraries for functional screenings. Methods in Molecular Biology; Combinatorial peptide library protocols, Ed. Cabilly S. 87:119–128.

Jockers R, Petit L, Lacroix I, de Coppet P, Barrett P, Morgan PJ, Guardiola B, Delagrange P, Marullo S, Strosberg AD (1997) Novel isoforms of Mel1c melatonin receptors modulating intracellular cyclic guanosine 3',5'-monophosphate levels. Molecular Endocrinology 11:1070–1081

Karne S, Jayawickreme CK, Nguyen TP, Anderson ML, Lerner MR (1991) Characterization of G-proteins in Xenopus melanocytes. Neuroscience Abstracts 17:607

Karne S, Jayawickreme CK, Lerner MR (1993) Cloning and characterization of an endothelin-3 specific receptor (ETc receptor) from Xenopus laevis dermal melanophores. J Biol Chem 268:19126–19133

Kolakowski LF Jr, O'Neill GP, Howard AD, Broussard SR, Sullivan KA, Feighner SD, Sawzdargo M, Nguyen T, Kargman S, Shiao LL, Hreniuk DL, Tan CP, Evans J, Abramovitz M, Chateauneuf A, Coulombe N, Ng G, Johnson MP, Tharian A, Khoshbouei H, George SR, Smith RG, O'Dowd BF (1998) Journal of Neurochemistry 71:2239–2251

Kondo H, Ide H (1983) Long-term cultivation of amphibian melanophores. In vitro ageing and spontaneous transformation to a continuous cell line. Experimental Cell Research 149:247–256

Lee TH, Lerner AB (1956) Isolation of melanocyte stimulating hormone from hog pituitary gland. J Biol Chem 221:943–959

Lerner AB (1961) Sci Am 205:99–108

Lerner MR, Reagan J, Gyorgyi T, Roby-Schemkovitz A (1988) Olfaction by melanophores: what does it mean? Proc Natl Acad Sci USA 85:261–264

Lerner MR, Potenza MN, Graminski GF, McClintock T, Jayawickreme CK, Karne S (1993) A new tool for investigating G-protein coupled receptors. Ciba Foundation Symposium; The molecular basis of smell and taste transduction 179:76–87

Lerner MR (1994) Tools for investigating functional interactions between ligands and G-protein-coupled receptors. TINS 17:142–146

Marchese A, George SR, Kolakowski LF, Lynch KR, O'Dowd BF (1999) Novel GPCRs and their endogenous ligands; expanding the boundaries of physiology and pharmacology. TIPS 20:370–375

Martin V, Ronde P, Unett D, Wong A, Hoffman TL, Edinger AL, Doms RW, Funk CD (1999) Leukotriene binding, signaling, and analysis of HIV coreceptor function in mouse and human leukotriene B_4 receptor-transfected cells. Journal of Biological Chemistry. 274:8597–8603

Marotti LA, Jayawickreme CK, Lerner MR (1999) Functional characterization of a receptor for vasoactive-intestinal-peptide-related peptides in cultured dermal melanophores from Xenopus laevis. Pigment Cell Research 12:89–97

McClintock TS, Graminski GF, Potenza MN, Jayawickreme CK, Roby-Shemkovitz A, Lerner MR (1993) Functional expression of recombinant G-protein-coupled receptors monitored by video imaging of pigment movement in melanophores. Anal Biochem 209:298–305

McClintock TS, Rising JP, Lerner MR (1996) Melanophore pigment dispersion responses to agonists show two patterns of sensitivity to inhibitors of cAMP-dependent protein kinase and Protein kinase C. J Cellular Physiology, 167:1–7

McClintock TS, Landers TM, Gimelbrant AA, Fuller LZ, Jackson BA, Jayawickreme CK, Lerner MR (1997) Functional expression of olfactory-adrenergic receptor chimeras and intracellular retention of heterologously expressed olfactory receptors. Mol Brain Res 48:270–278

McClintock TS, Lerner MR (1997) Functional analysis by imaging of melanophore pigment dispersion of chimeric receptors constructed by recombinant polymerase chain reaction. Brain Research Protocols 2:59–68

Miyashita Y, Moriya T, Yokosawa N, Hatta S, Arai J, Kusunoki S, Toratani S, Yokosawa H, Fujii N, Asami K (1996) Light-sensitive response in melanophores of Xenopus laevis. II. Rho is involved in light-induced melanin aggregation. Journal of Experimental Zoology 276:125–131

Moriya T, Miyashita Y, Arai J, Kusunoki S, Abe M, Asami K (1996) Light-sensitive response in melanophores of Xenopus laevis. I. Spectral characteristics of melanophore response in isolated tail fin of Xenopus tadpole. Journal of Experimental Zoology 276:11–18

Mullins UL, Fernandes PB, Eison AS (1997) Melatonin agonists induce phosphoinositide hydrolysis in Xenopus laevis melanophores. Cellular Signaling 9:169–173

Nilsson H, Wallin M (1997) Evidence for several roles of dynein in pigment transport in melanophores. Cell Motility and the Cytoskeleton 38:397–409

Ng GY, Clark J, Coulombe N, Ethier N, Hebert TE, Sullivan R, Kargman S, Chateauneuf A, Tsukamoto N, McDonald T, Whiting P, Mezey E, Johnson MP, Liu Q, Kolakowski LF Jr, Evans JF, Bonner TI, O'Neill GP (1999) Identification of a $GABA_B$ receptor subunit, gb2, required for functional $GABA_B$ receptor activity. Journal of Biological Chemistry 274:7607–7610

Ollmann MM, Wilson BD, Yang YK, Kerns JA, Chen Y, Gantz I, Barsh GS (1997) Antagonism of central melanocotin receptors in vitro and in vivo by agouti-related protein. Science 278:135–138

Ollmann MM, Lamoreux ML, Wilson BD, Barsh GS (1998) Interaction of Agouti protein with the melanocortin 1 receptor in vitro and in vivo. Genes and Development. 12:316–330

Ollmann MM, Barsh GS (1999) Down-regulation of melanocortin receptor signaling mediated by the amino terminus of Agouti protein in Xenopus melanophores. J Biol Chem 274:15837–15846

Pickering H, Sword S, Vonhoff S, Jones R, Sugden D (1996) Analogues of diverse structure are unable to differentiate native melatonin receptors in the chicken retina, sheep pars tuberalis and Xenopus melanophores. British Journal of Pharmacology 119:379–387

Potenza MN, Lerner MR (1991) A recombinant vaccinia virus infects *Xenopus* melanophores. Pigment Cell Res 4:186–192

Potenza MN, Lerner MR (1992) A rapid quantitative bio-assay for evaluating the effect of ligands upon receptors that modulate cAMP levels in a melanophore cell line. Pigment Cell Res 5:372–378

Potenza MN, Graminski GF, Lerner MR (1992) A method for evaluating the effects of ligands upon G_s protein-coupled receptors using a recombinant melanophore-based bio-assay. Anal Biochem 206:315–322

Potenza MN, Graminski GF, Schmauss C, Lerner MR (1994) Functional characterization of Human D_2 and D_3 Dopamine Receptors. J Neurosci 14:1463–1476

Potenza MN, Lerner MR (1994) Characterization of a serotonin receptor endogenous to frog melanophores. Naunyn-Schmiedeberg's Arch Pharmacol 349:11–19

Provencio I, Jiang G, De Grip WJ, Hayes WP, Rollag MD (1998) Melanopsin: An opsin in melanophores, brain, and eye. Proc Natl Acad Sci 95:340–345

Quillan JM, Jayawickreme CK, Lerner MR (1995) Combinatorial diffusion assay used to identify topically active melanocyte-stimulating hormone receptor antagonists. Proc Natl Acad Sci 92:2894–2898

Reilein AR, Tint IS, Peunova NI, Enikolopov GN, Gelfand VI (1998) Regulation of organelle movement in melanophores by protein kinase A (PKA), protein kinase C (PKC), and protein phosphatase 2A (PP2A). Journal of Cell Biology. 142:803–813

Rodionov VI, Gyoeva FK, Gelfand VI (1991) Kinesin is responsible for centrifugal movement of pigment granules in melanophores. Proc Natl Acad Sci 88:4956–4960

Rogers SL, Tint IS, Fanapour PC, Gelfand VI (1997) Regulated bi-directional motility of melanophore pigment granules along microtubules in vitro. Proc Natl Acad Sci 94:3720–3725

Rogers SL, Gelfand VI (1998) Myosin cooperates with microtubule motors during organelle transport in melanophores. Current Biology 8:161–164

Rogers SL, Tint IS, Gelfand VI (1998) In vitro motility assay for melanophore pigment organelles. Methods in Enzymology 298:361–372

Rollag MD, Adelman MR (1993) Actin and tubulin arrays in cultured Xenopus melanophores responding to melatonin. Pigment Cell Research 6:365–371

Rozdzai MN, Haimo LT (1986) Bi-directional pigment granule movements of melanophores are regulated by protein phosphorylation and dephosphorylation. Cell 47:1061–1070

Rubina KA, Starodubov SM, Nikeryasova EN, Onishchenko GE (1999) Microtubule-organizing centers in the mitotic melanophores of Xenopus laevis larvae in vivo: ultrastructural study. Pigment Cell Research 12:98–106

Sammak PJ, Adams SR, Harootunian AT, Schliwa M, Tsien RY (1992) Intracellular cyclic AMP, not calcium, determines the direction of vesicle movement in melanophores: Direct measurement by fluorescent ratio imaging. J Cell Biol 117:57–72

Scheenen WJ, de Koning HP, Jenks BG, Vaudry H, Roubos EW (1994) The secretion of alpha-MSH from xenopus melanotropes involves calcium influx through omega-conotoxin-sensitive voltage-operated calcium channels. Journal of Neuroendocrinology 6:457–464

Scheenen WJ, Jenks BG, Willems PH, Roubos EW (1994) Action of stimulatory and inhibitory alpha-MSH secretagogues on spontaneous calcium oscillations in melanotrope cells of Xenopus laevis. Pflugers Archiv – European Journal of Physiology. 427(3–4):244–251

Seldenrijk R, Hup DR, de Graan PN, van de Veerdonk FC (1979) Morphological and physiological aspects of melanophores in primary culture from tadpoles of Xenopus laevis. Cell and Tissue Research 198:397–409

Sieber-Blum M, Chokshi HR (1985) In vitro proliferation and terminal differentiation of quail neural crest cells in a defined culture medium. Experimental Cell Research 158:267–272

Silen JL, Lu AT, Solas DW, Gore MA, MacLean D, Shah NH, Coffin JM, Bhinderwala NS, Wang Y, Tsutsui KT, Look GC, Campbell DA, Hale RL, Navre M, DeLuca-Flaherty CR (1998) Screening for novel antimicrobials from encoded combinatorial libraries by using a two-dimensional agar format. Antimicrobial Agents and Chemotherapy 42:1447–1453

Sugden D (1991) Aggregation of pigment granules in single cultured Xenopus laevis melanophores by melatonin analogues. Br J Pharmacol 104:922–927

Sugden D (1992) Effect of putative melatonin receptor antagonists on melatonin-induced pigment aggregation in isolated Xenopus laevis melanophores. European Journal of Pharmacology 213:405–408

Sugden D, Rowe SJ (1992) Protein Kinase C activation antagonizes melatonin-induced pigment aggregation in Xenopus laevis melanophores. J Cell Biol 119:1515–1521

Teh MT, Sugden D (1999) The putative melatonin receptor antagonist GR128107 is a partial agonist on Xenopus laevis melanophores. British Journal of Pharmacology 126:1237–1245

Thaler CD, Haimo LT (1992) Control of organelle transport in melanophores: Regulation of Ca^{2+} and cAMP levels. Cell Motil. Cytoskel 22:175–184

Valius M, Kazlauskas A (1993) Phospholipase C-gamma 1 and phosphatidylinositol 3 kinase are the downstream mediators of the PDGF receptor's mitogenic signal Cell 73:321–334

Subject Index

accessory proteins, G protein coupled receptor system   324–325(tables)
acetylcholine   2–3
– arterial muscle, effect on   9
– blood vessels, effect on   8
– large coronary arteries, effect on   18(fig.)
– paradox, human microcoronary arteries   21–28
"acetylcholine test"   17–18
adenosine   63
adenosine A2 receptors   66–67
adenylate cyclase
– activity, receptor density   168–169
– $\alpha 2$-adrenoceptors   66–68
adenylyl cyclase   73
– effector enzymes   380–382
aequorin(s)   397–398, 402–403
– reporter gene cell line, construction   406(fig.)
affinity   4
– ligands   120–121
agonism
– inverse   167–178
– – activity, spontaneous   168–172
– – background   167–178
– – characterization   349–350
– – models   177–178
– – overexpression   168
– – receptor function modulation by agonists and inverse agonists   174–177
– – spontaneous receptor activity, mutations and diseases   173–174
– therapeutic vs. secondary   208–210
agonists   2
– affinity/efficacy driven   205
– efficacy   6–7
– – methods of estimating   201(table)
– full   224
– inverse   167–168
– negative   167
– partial   224
– physiological   5
– potency   6–7
– – ratios, limitations   201–204
– receptor interaction   5
– recombinant signaling system, analysis   321(table)
– weak   224
Agouti protein   434
allosteric equilibrium   227–231
– ion-channel example illustration   229(fig.)
– microscopic interpretation   235–238
allosteric modulation   220
$\alpha 1$-adrenoceptors   52–54
$\alpha 2$-adrenoceptors   55–56
– adenylate cyclase   66–68
α-adrenoceptors   16
α-melanocyte stimulating hormone (MSH) receptor   421–422
alprenolol   83, 169, 171
amines, endogenous   73
amlodipine   45–46(figs)
analytical dilution assays   1–2
angina
– variant   17
– – coronary spastic angina, with   28
– – role of receptors mediating coronary artery contraction   16–28
angiogram   16
angiotensin-I converting enzyme inhibitor (ACEI)   28–29, 35
angiotensin-receptor activation   298–302
animal disease, models of human disease   15

"animal units" 1–3
antagonism
– competitive *see* competitive antagonism
– functional 64–66
– non-competitive *see* non-competitive antagonism
antagonists 3–6, 219
– chemical 261
– competitive 261
– negative 167
– neutral 286
– non-competitive 261, 263–265(figs)
AR activation, small organics as ligands 291–298
– endogenous ligands, binding contacts 291–293
– epinephrine activation 293–298
arachidonic acid metabolites 62–63
Ari G complex, evidence for 158–163
arrhythmias 103
– β2-adrenoceptor mediated 103
asthma, allergic 10
atenolol 29, 104
atheromatous plaque 17
atopic allergy 10
atrial fibrillation 105, 107
atrial histamine receptors 100–101
atrial hyperresponsiveness 98
atrial myocytes, human 105
atropine 3
Aug-II receptor 298–299
– activation 300–302
– peptides and non-peptides binding 299–300
autacoids 5
autocrine activation 62–63
Autographa californica nuclear polyhedrosis virus (AcNPV) 338

baculoviruses 338
β1-adrenoceptors
– cardiostimulant effect 73
– function 74–75
– localization 74
– physiological, pathophysiological and therapeutic relevance 101–104
– polymorphism 77
β1-receptors 7
β2-adrenoceptors
– adipose tissue 372
– cardiostimulant effect 73
– function 74–75
– heart 372

– inotropic hyperresponsiveness to salbutamol 97(fig.)
– localization 74
– physiological, pathophysiological and therapeutic relevance 101–104
– selective coupling 75–80
β2-receptors 7
β3-adrenergic receptor agonists, algorithm application 212–214
β3-adrenoceptors
– adipocytes 80, 82
– cardiac, functional role 80–82
– cardiodepressant effect 73
– cardiodepression, evidence against and for 81–82
– cardiostimulation, evidence against and for 80–81
β4-adrenoceptors 73
– physiological, pathophysiological and therapeutic relevance 104–105
– putative, cardiostimulant effects through 82–90
– – β1-adrenoceptor, special state 89–90
– – β3-adrenoceptor, resemblance with 84–87
– – endogenous agonist, unknown identity 87–88
– – partial agonists, non-conventional 82–84
β-adrenoceptor antagonists 104, 169
β-adrenoceptor kinase-1 (βARK-1) 102
β-adrenoceptor kinase-1 inhibitor (βARKct) 102
β-adrenoceptor melanophores 422
β-amyloid precursor protein 327
Bezold-Jarisch reflex 90
binding 6
bioassays, past uses and future potential 1–11
– analytical 2
– quantitative 2, 4–5
biological response 218–219
biosystem 3
BR activation and rhodopsin, light as the ligand 287–291
bradykinin 18–19
BRL-37344 81–82
bupranolol 85
burimamide 175
butylcarazolol 83

Subject Index

calcitonin gene-related peptide (CGRP) 35
– concentration-relaxation curves to acetylcholine 37(fig.)
calcium
– calmodulin complex 52
– vascular smooth muscle contraction 52
calcium channel antagonists 43
– inotropic responses, right atrial trabecule 45(fig.)
calcium channels, store-operated 52–53
CAM 286
carcinoid heart disease 105
cardiac oxygen consumption 103
caveolins 327
CGP-12177 83
– agonist, non-conventional 84(fig.)
– β_2 adrenoceptor, agonist properties 90
– hydrophilicity 88
– inotropic/chronotropic/lusitropic effect 86(fig.)
– pharmacological properties, cardiac tissues 89
chest pain, acute, at rest 16
chromatography 2
chronic heart failure 34–37
– treatment, β_1-adrenoceptor blockers 103–104
cilazapril 28–29
cisapride 92, 94–95
– supraventricular arrhythmia 105
Clark's model 283
co-agonism 267–271
– competitive antagonism, effect on 271–273
– evidence for 267–268
– experimental data 274–278
– implications of 278–279
– non-competitive antagonist, effect on response of 273–274
– theory 268–270
combichem 8
competitive antagonism 4, 6
– co-agonism, effect on 271–273
congestive heart failure, vascular rectivity 28–34
contraction, receptors mediating 16–28
convulsion, hypoglycaemia induced 1
coronary arteries
– small
– – acetylcholine receptors location 26(fig.)
– – receptors mediating 21–28
coronary-bypass-graft surgery
– internal mammary artery 38–41
– saphenous vein 41
– vascular conduits, pharmacology 37–41
COS cells 316–317
coupling cofactor 326
CRE-luciferase reporter genes 402, 403(fig.)
cubic ternary complex (CTC) model 148, 152–156, 159(fig.)
– response as a fraction of total receptor 156(table)
– transitions within 155(fig.)
CXC-chemokine receptor-4 431
cyanopindolol 83

diet, high-cholesterol 17, 18(fig.)
diffusion 3
digoxin 35
dihydropyridine derivatives 43
diltiazem 38
– internal mammary artery 40(fig.)
diuretics 35
downregulation, receptors 175
drug-receptor kinetics 147
drug-receptor models 147–163
dual reporter assays 405–409

efficacy
– affinity, distinction between 218–219
– biological definitions 221–226
– – agonism and antagonism, scale 223–225
– – signal strength, nature 221–222
– – signal transduction, steps 225–226
– – stimulus-response relationship 222–223, 226–226
– functional proteins, concept generality 219
– molecular definitions 227–238
– – affinity and efficacy, molecular link between 227
– – allosteric equilibrium 227–231
– – allosteric equilibrium, microscopic interpretation 235–238
– – free-energy coupling 227–231
– – macroscopic perturbations, receptor 231–233
– – proteins, functional and physical states 233–235
– – thermodynamic definitions 227–231

- molecular, stochastic model 238–254
- - $\beta 2$ adrenoceptors, efficacy and fluorescence changes, relationship 224–228
- - conformational space, protein motion and fluctuations 238
- - ligand efficacy, probabilistic interpretation 241–244
- - macroscopic changes, constitutively active adrenoceptors 248–254
- - macroscopic constance, derivation 238–240
- - microscopic states, probability distribution 238–240
- - physical states and biological function, relationship between 244
efficacy, ligands 120–121
- parameters contributing 126–144
- - endosomal pH, dissociation rate constants 134–139
- - individual receptor-ligand complex, life time 126–131
- - receptor-desensitization rate constant Kx 131–134
- - ternary complex model, rate constants 139–143
efficacy, receptor molecular nature 183–186
- agonist side effects, prediction algorithm 206–215
- agonists, relative efficacies, measurement 204–206
- negative 186–188
- positive 186–188
- relative, operational measurement 188–201
- - agonist potency ratios, limitations 201–204
- - binding studies 188–192
- - function 192–201
endocytosis 134
endosomes
- receptor and ligand sorting 135(fig.)
- receptors and ligands inside, model for sorting 136(fig.)
endothelial dysfunction, hypertension, cause of 31–33
endothelin-1 17, 31, 45
endothelin-C receptor 422
endothelium-derived relaxing factor (EDRF)
- coronary artery (large), effect on 16–21

- hypertension, endothelial dysfunction 31–33
epidermal growth factor (EGF)
- endosomal sorting outcomes 138(fig.)
- receptor 433
- system 135
epinephrine, cardiostimulation, $\beta 1/2$ adrenoceptors 74
ergometrine 16
ergonovine 16
erythropoietin 2
erythropoietin receptor 433–434
extended ternary complex model 151–152, 159(fig.)

felodipine 45–46(figs)
force 3
forskolin 55
free-energy coupling 227–231

G protein-coupled receptor (GPCR)
- cardiac cross-talk 95–101
- signal transfer 218
- stoichiometry, relative 370–375
- structure/function, common 283–286
G protein-coupled receptor (GPCR)-G protein fusion proteins 367–370
G protein-coupled receptor (GPCR)-G protein-effector stoichiometries, systems to modulate 364–365
G protein-coupled signaling cascades, cellular distribution 365–367
Gaddum equation 4
GALR 213, 431
GCRE-luciferase reporter assay 411
gene therapy, heart failure 103
glucagon-like peptide (GLP-1) receptor 410
glyceryl trinitrate 38
- internal mammary artery 40(fig.)
- relaxation response, saphenous vein 43–44(figs)
Gs α 376–380
Gs protein 7, 335
- activity:
- - direct measures 337–338
- - effector modulation as a measure 336–337
- coupled receptors, human heart 73–107
- four families, α subunits classification 336(fig.)

Subject Index

- mammalian receptors, reconstruction 345–350
- – G protein activation 350, 355
- ternary complex 147–148
guanosine triphosphate γS shift 188–191

heart (human), Gs protein-coupled receptors 73–107
HEK-293 cells 316–317
helodermin 423
High Five cell line 338
high performance liquid chromatography 169
high-affinity selection binding 191–192, 193(fig.)
His-256 300
histamine 2
- H2 receptors 100
- heart rate, effect on 73
hormone 5
5-HT4 receptors 90–95
- physiological, pathophysiological and therapeutic relevance 105–107
5-HT4-like receptors 91–95
human calcitonin receptor type-2 (hCTR2) 192, 194(fig.)
human epicardial coronary artery
- acetylcholine responses 20(fig.)
- bradykinin, effect on 20(fig.)
- substance P, effect on 20(fig.)
- sumatriptan effect 19(fig.)
human vascular receptors 15–46
human vascular-to-cardiac tissue selectivity, VOCC antagonists 41–46
5-hydroxytryptamine (5-HT)
- heart beat, effect on 73
- human atrium, antagonism of the positive inotropic effects 93(fig.)
- see also serotonin
hypercholesterolaemia 31
hypoglycaemia 1

ICI-118,551 171, 286
inducible reporter genes 393–397
inositol triphosphate (IP3) 52–53
insect cell expression systems 338–345
- coupling using radioligand binding, quantitation 343–345
- endogenous G proteins, interactions of receptors 342–343
- insect cell lines, receptors expression 338–342

insulin, discovery 1
interferon 2
internal mammary artery 38–41
iodocyanopindolol 83
isolated tissue, pharmacodynamic analysis 15–46
isoprenaline 45

Jansen-type metaphyseal chondrodysplasia 173

ketanserin 16

law of mass action 4
Lawn format, melanophore cells 428
Lawn format screen system 432
leukotrienes 10
ligand-induced perturbation 218
ligand-receptor binding 119
LuFLIPRase assay 407–408

m2 muscarinic acetylcholine receptor 375
macroscopic linkage 231
melanin 415
melanophore recombinant receptor systems 415–435
- α-melanocyte stimulating hormone receptor 421–422
- β-adrenoceptor 422
- endogenous receptor signaling 419–421
- endothelin-c receptor 422
- melanophore assay technology 423–429
- melatonin-1c receptor 421
- receptor studies and applications 429–435
- serotonin receptor 422
- signaling pathways 417–419
- VIP receptor 423
melatonin-1c receptor 421
membrane receptors, signal transfer and conformational change 217–218
messenger molecules 5
"messenger-receptor" system 5
mesulergine 169
methane thiosulfonate ethylammonium 297
method of Furchgott 195–196, 210
method of Stephenson 196–197
methoxamine 45
metoprolol 103
mianserin 169

mibefrodil 43, 45–46(figs)
microtiter well format, melanophore cells 427–428
mobile receptors 147
Monte Carlo models 125
multi-use peptide libraries (MUPLs), preparation 429(fig.)
Mulvany-Halpern myograph 21
myocardial contractility, VOCC antagonists 42–46
myosin 52
myosin light-chain kinase 52

nadolol 82
neuromodulin (GAP-43) 327
New York Heart Association classification 34–35
nifedipine 17, 38, 45–46(figs), 65–66
– internal mammary artery 40(fig.)
nisoldipine 61
nitrates 35
nitric oxide 8–9, 81–82
– coronary arteries, large 16–21
nociceptin 405
– extracellular signal-regulated kinase, activation 407(fig.)
non-competitive antagonism
– definition 261–266
– response of co-agonists, effect of 273–274
norepinephrine 2
– cardiostimulation, β1/2 adrenoceptors 74
– composition 291–292

occupancy theory 147
"organ bath" experiment 3
orphan receptor 224
oxprenolol 83
oxymetazoline 204

P2Y receptors 63
PACAP receptor 408
papaverine 39–41
paracrine activation 62–63
parathyroid hormone receptor 410
PE 41
perindopril 29
pertussis toxin 58, 66, 81
pharmacology
– analytical 3
– quantitative 1
phenoxybenzamine 65–66, 204
phenylephrine 65–66

– intracoronary, α-adrenoceptor stimulation 28
phospholamban 74, 87
phosphorylation, atrial relaxation 91
pilocarpine 3–4
pinacidyl 62
pindolol 83, 104
– tachycardia, beneficial 104
plasmid vectors 315
platelet-derived growth factor (PDGF) β receptor 433
plethysmography, venous distensibility 33–34
– chronic heart failure 34–35
potency ratious 201–204
– agonists 201–204
– – differences, different receptor densities and/or coupling 205(fig.)
prazocin 67
prenalterol, properties as a β-adenergic receptor agonist 206(table)
primary hypertension, vascular reactivity 28–34
– endothelial dysfunction 31–33
– forearm veins 33–34
– remodelling 29–31
Prinzmetal's angina 16
propranolol 85, 104
prostacyclin 31, 63
protein C 74, 87
protein kinase C 52–53
– effects 54(fig.)
purines 62–63
putative β-4 adrenoceptor 85

radioimmunoassay 2
receptor models 147–148
– derived from biological and molecular definitions of efficacy 232(table)
– parameters 149(table)
– pharmacologic, interrelationship between 157(fig.)
receptor-G protein interactions, selectivity 335–336
receptor-ligand binding
– diffusion-versus reaction-controlled events 123–124
– equilibrium 122–123
– kinetic models 122–123
– model structures and dose-response curves 124–126
– modelling 121–122
– steady state 122–123
receptor(s) 3

Subject Index

- coupling constant 207
- efficacy 373–374
- reverse 224
- spare 224
recombinant-signaling systems 313–328
- assembly 314–318
- - stable transfection 314–316
- - transient expression systems 316–318
- drug-receptor interactions 318–328
- - accessory proteins, influence 323–328
- - cell-type specific signaling events 318–323
- perspective 328
relative β50 values, algorithm for calculation 210–212
Renilla luciferase 405–407
renzapride 92, 94–95
reporter gene assays
- drug discovery 401
- receptor pharmacology 401–411
reporter gene(s)
- stable expression 400–401
- transient expression 400
reporter proteins 397–400
reporter system
- detection methods
- - enzymatic 392–393
- - non-enzymatic 393
- intracellular signaling, measurement 393–400
- - inducible reporter genes 393–397
- - reporter proteins 397–400
- reporter gene system 391
- - pharmacology, use in 401–411
resistance 3
rhodopsin 284–285
- activation mechanism 288–291
- BR activation, light as the ligand 287–291
- structural information 287–288
rolipram 55

saphenous vein 38–41
- injectin/spraying of GTN and verapamil solution 44(fig.)
Schild plot 4, 41–46
secretin 5
selectivity 5–6
septic shock 10
serotonin (5-HT)
- atheromatous vessel, effect on 18(fig.)

- coronary arteries, large 16–21
- - effect on receptors 16
- *see also* 5-hydroxytryptamine
signal transduction 223
- steps 225–226
signal transfer, membrane receptor 217–218
single transmembrane receptors, melanophore 433
sodium nitroprusside 31, 35–36, 65
SR-58611 81–82
stroke 105, 107
substance P 18–19
sumatriptan, human epicardial coronary artery, effect on 19(fig.)
superoxide anion 31
syncope, neurocardiogenic 104

tachycardia
- beneficial, orthostatic hypotension 104
- 5-HT evoked 90
ternary complex model
- G protein-coupled receptors (GPCR) 147–163
- - cubic ternary complex (CTC) model, application 152–163
- - extended 151–152
- - receptor function 148–149
- - two-state theory 149–151
- rate constants 139–143
tert-butylpindolol 83
thromboxane A2/prostaglandin H2 31
thyroid-stimulating hormone receptor 336
thyrotropin-releasing hormone (TRH) receptor 409
timolol 104
TN-5 cells 338
transforming growth factor α (TGF-α) 137–138
triethylamine 295
troponin I 74, 87
tulbulin 327
two-state receptor theory 150–151

U46619 17, 26
- coronary arteries, large, effect on 19(fig.)
UK14304 368

vascular smooth muscle 52
- myogenic activation 61–62

vasoconstriction, α-adrenoceptor-
  mediated 51–68
– adenylate cyclase and α2-
  adrenoceptors 66–68
– antagonists, functional 64–66
– autocrine and paracrine activation
  62–64
– coupling mechanisms
– – intact animal 56–59
– – molecular and cellular levels 52–56
– myogenic activation 61–62
– roconciliation 59–61

ventricular fibrillation 103
verapamil 38, 43, 45–46(figs)
– internal mammary artery 40(fig.)
– relaxation response, saphenous vein
  43–44(figs)
VIP-related peptides, receptor 423
"virtual" constant Ko 235
VOCC antagonists 41–46

wasp venom mastaparan 327

Xenopus laevis melanophores 415–416

GPSR Compliance

The European Union's (EU) General Product Safety Regulation (GPSR) is a set of rules that requires consumer products to be safe and our obligations to ensure this.

If you have any concerns about our products, you can contact us on

ProductSafety@springernature.com

In case Publisher is established outside the EU, the EU authorized representative is:

Springer Nature Customer Service Center GmbH
Europaplatz 3
69115 Heidelberg, Germany

www.ingramcontent.com/pod-product-compliance
Ingram Content Group UK Ltd.
Pitfield, Milton Keynes, MK11 3LW, UK
UKHW021255180426
11947UKWH00010B/791